Proceedings of the Ninth International Colloquium on Differential Equations

Also available from VSP

Proceedings of the Eighth International Colloquium
on Differential Equations
Plovdiv, Bulgaria, 18 - 23 August 1997

Proceedings of the Seventh International Colloquium
on Differential Equations
Plovdiv, Bulgaria, 18 - 23 August 1996

Proceedings of the Sixth International Colloquium
on Differential Equations
Plovdiv, Bulgaria, 18 - 23 August 1995

Proceedings of the Fifth International Colloquium
on Differential Equations
Plovdiv, Bulgaria, 18 - 23 August 1994

Proceedings of the Fourth International Colloquium
on Differential Equations
Plovdiv, Bulgaria, 18 - 22 August 1993

Proceedings of the Third International Colloquium
on Differential Equations
Plovdiv, Bulgaria, 18 - 23 August 1992

Proceedings of the Third International Colloquium
on Numerical Analysis
Plovdiv, Bulgaria, 13 - 17 August 1994

Proceedings of the Second International Colloquium
on Numerical Analysis
Plovdiv, Bulgaria, 13 - 17 August 1993

Proceedings of the First International Colloquium
on Numerical Analysis
Plovdiv, Bulgaria, 13 - 17 August 1992

Proceedings of the Ninth International Colloquium on Differential Equations

Plovdiv, Bulgaria, 18 - 23 August, 1998

Editor: **D. Bainov**

///VSP///
Utrecht, The Netherlands, 1999

VSP BV
P.O. Box 346
3700 AH Zeist
The Netherlands

Tel: +31 30 692 5790
Fax: +31 3- 693 2081
E-mail: vsppub@compuserve.com
URL: http://www.vsppub.com

First published in 1999

ISBN 90-6764-296-7

Printed in The Netherlands by Ridderprint bv, Ridderkerk.

CONTENTS

Preface

The Ninth International Colloquium on Differential Equations was organized by the Institute for Basic Science of Inha University, the International Federation of Nonlinear Analysts, the Mathematical Society of Japan, Pharmaceutical Faculty of the Medical University of Sofia, University of Catania and UNESCO, with the cooperation of the Association Suisse d'Informatique, the Canadian Mathematical Society, Kyushu University, the London Mathematical Society, Technical University of Plovdiv, and it was partially sponsored by the Bulgarian Ministry of Education and Science under Grant MM–702.

The Colloquium was held on August 18–23, 1998, in Plovdiv, Bulgaria. The subsequent colloquia will take place each year from 18 to 23 of August in Plovdiv, Bulgaria.

The publishing of this volume is fully supported by the Bulgarian Ministry of Education and Science under Grant MM–702.

Organizing Committee:

G. Anastassiou, D. Bainov (Chairman), Q. H. Choi, J.Diblik, S. Hristova, J. Kajiwara, N. Kitanov (Secretary), D. Kolev (Secretary), V. Lakshmikantham, M. Marino, A. Maugeri (Vice Chairman), L. Medeiros, E. Minchev (Secretary), N. Popivanov, P. Popivanov, M. Robnik, M. Sambandham, H.M. Srivastava, R.U. Verma.

Scientific Committee:

G. Anastassiou (USA), D. Bainov (Bulgaria), C. Dafermos (USA), G. Da Prato (Italy), H. Fujita (Japan), A. Kaneko (Japan), N. Kenmochi (Japan), V. Lakshmikantham (USA), A. Maugeri (Italy), L. Medeiros (Brazil), E. Minchev (Bulgaria), N. Pavel (USA), N. Popivanov (Bulgaria), P. Popivanov (Bulgaria), M. Robnik (Slovenia), M. Sambandham (USA), H.M. Srivastava (Canada), Fr. Stenger (USA), E. Titi (USA), M. Tsutsumi (Japan), R.U. Verma (USA).

The address of the Organizing Committee is:
Drumi Bainov, P.O.Box 45, 1504 Sofia, Bulgaria.

The Editor

65th Anniversary of Professor Drumi Bainov*

Peter Popivanov
Institute of Mathematics, Bulgarian Academy of Sciences, Sofia 1113, Bulgaria

Dear Professor Bainov,
Dear Drumi,

On behalf of all participants of this Colloquium on Differential Equations and of your colleagues, I would like to congratulate you on your 65th birthday. We all wish you good health, a lot of energy and a long and fruitful working career in the field of differential equations.

You have written around 1000 papers, all published in authoritative journals, and 16 monographs. You are very active in didactic and organizing work and your results for differential equations with delay and with impulses are very well known in our community.

You have founded the impressive Bulgarian group on differential equations and as organizer you have perfectly managed to find and bring together many young and enthusiastic researchers, directing their interests, energy and efforts in the field of differential equations. You have helped organize many conferences and congresses, of which the Colloquium on Differential Equations in Plovdiv is most famous. Thanks to you and your efforts as heart and engine of the organization, this conference has now its own history, traditions, and always outstanding participants.

Despite the hard times we are now going through in Bulgaria, you have succeeded in organizing this prestigious scientific conference. The exchange of ideas, results and new trends in DE is certainly one of the greatest achievements of this forum.

Again, our warmest congratulations and best wishes for a very happy and long life.

Prof. P. Kendorov, Member of the Bulgarian Academy of Sciences
Prof. P. Popivanov, Corr. Member of the Bulgarian Academy of Sciences
Prof. S. Nedev, Institute of Mathematics of the Bulgarian Academy of Sciences
Prof. M. Kaschiev, Institute of Mathematics of the Bulgarian Academy of Sciences
Assoc. Prof. V. Kiryakova, Institute of Mathematics of the Bulgarian Academy of Sciences
Prof. N. Popivanov, Faculty of Mathematics, Sofia University.

---

*This congratulatory letter was presented at the opening ceremony of the Ninth International Colloquium on Differential Equations.

9th Int. Coll. on Differential Equations, pp. 1-6
D. Bainov (Ed)

Construction of Systems of Partial Differential Equations and Their Solutions

X. AGRAFIOTOU
School of Technology, Division of Mathematics, Aristotle University of Thessaloniki, Thessaloniki 540 06, Greece

Abstract: The aim of the present paper is to solve a given system of partial differential equations and determine their solutions
Key Words and Phrases: *Partial solution, General solution, System of PDE.*
1991 Mathematics Subject Classification. 35F30.

1. Introduction

Partial differential equations and systems of partial differential equations play an important role in Physics, Biology and Technology. This means that some problems of Physics, Biology or Technology can be represented by a partial differential equation or a system of partial differential equations. Therefore we need to find a solution of this equation or of this system.

The purpose of this paper is to give a solution of a system of partial differential equations. This system comes from a nilpotent Lie algebra and it can be used as a model in Nuclear Physics.

The whole paper contains four paragraphs. Each of them is analyzed as follows:

The first paragraph is the introduction.

The second paragraph contains some general theory for homogeneous systems of linear partial differential equations.

A special system of linear partial differential equations is study in the third paragraph.

The new form of this system under some special transformation is included in the last paragraph.

2. General Theory

We consider the homogeneous system of linear partial differential equations:

$$\left.\begin{array}{l} g_{11}(x_1,\dots,x_n)\frac{\partial f}{\partial x_1}+\dots+g_{1n}(x_1,\dots,x_n)\frac{\partial f}{\partial x_n}=t_1(x_1,\dots,x_n)\\ g_{21}(x_1,\dots,x_n)\frac{\partial f}{\partial x_1}+\dots+g_{2n}(x_1,\dots,x_n)\frac{\partial f}{\partial x_n}=t_2(x_1,\dots,x_n)\\ \dots\dots\dots\dots\dots\dots\dots\dots\dots\dots\dots\dots\dots\dots\\ g_{m1}(x_1,\dots,x_n)\frac{\partial f}{\partial x_1}+\dots+g_{mn}(x_1,\dots,x_n)\frac{\partial f}{\partial x_n}=t_m(x_1,\dots,x_n) \end{array}\right\} \quad (1)$$

which has m equations, which must be satisfied by the unknown:

$$\frac{\partial f}{\partial x_1}, \frac{\partial f}{\partial x_2}, \dots, \frac{\partial f}{\partial x_n}. \tag{2}$$

This means that we need to determine a function:

$$f = f(x_1, \dots, x_n) \tag{3}$$

whose first derivatives satisfy this system.

In the general, to find a solution of this system is a very difficult problem. Some methods to determine such solutions are given in ([1]).

In the next paragraph we give a system of linear partial differential equations whose coefficients are polynomials of first degree of $x_1, \dots, x_n$ with n=13. This system is obtained from a special nilpotent Lie algebra of dimension 13, which can be used as mathematical model in Nuclear Physics.

3. A Special System of PDE

We consider the homogeneous system of linear partial differential equations:

$$-x_5\frac{\partial f}{\partial x_2} - x_8\frac{\partial f}{\partial x_6} - x_{11}\frac{\partial f}{\partial x_9} = 0 \tag{4}$$

$$x_5\frac{\partial f}{\partial x_1} - x_6\frac{\partial f}{\partial x_3} - x_{10}\frac{\partial f}{\partial x_9} - x_{14}\frac{\partial f}{\partial x_{11}} = 0 \tag{5}$$

$$x_6\frac{\partial f}{\partial x_2} - x_7\frac{\partial f}{\partial x_4} + x_{10}\frac{\partial f}{\partial x_6} + x_{13}\frac{\partial f}{\partial x_9} = 0 \tag{6}$$

$$x_7\frac{\partial f}{\partial x_3} + x_{10}\frac{\partial f}{\partial x_6} + x_{13}\frac{\partial f}{\partial x_9} = 0 \tag{7}$$

$$-x_8\frac{\partial f}{\partial x_3} + x_{14}\frac{\partial f}{\partial x_9} = 0 \tag{8}$$

$$x_8\frac{\partial f}{\partial x_1} + 2x_9\frac{\partial f}{\partial x_3} - x_{10}\frac{\partial f}{\partial x_4} + x_{13}\frac{\partial f}{\partial x_7} - 2x_{14}\frac{\partial f}{\partial x_8} = 0 \tag{9}$$

$$x_{10}\frac{\partial f}{\partial x_2} - x_{13}\frac{\partial f}{\partial x_6} = 0 \tag{10}$$

$$2x_{11}\frac{\partial f}{\partial x_3} - x_{12}\frac{\partial f}{\partial x_4} - 2x_{14}\frac{\partial f}{\partial x_6} = 0 \tag{11}$$

$$x_{11}\frac{\partial f}{\partial x_1} - x_{13}\frac{\partial f}{\partial x_4} - x_{14}\frac{\partial f}{\partial x_5} = 0 \tag{12}$$

$$x_{12}\frac{\partial f}{\partial x_1} + x_{13}\frac{\partial f}{\partial x_3} = 0 \tag{13}$$

$$x_{14}\frac{\partial f}{\partial x_2} = 0 \tag{14}$$

$$A=\begin{bmatrix}
0 & -x_5 & 0 & 0 & 0 & -x_8 & 0 & 0 & 0 & -x_{12} & 0 & 0 & 0 \\
x_5 & 0 & -x_6 & 0 & 0 & 0 & -x_{10} & 0 & 0 & 0 & 0 & 0 & 0 \\
0 & x_6 & 0 & -x_7 & x_8 & -2x_9 & 0 & 0 & 0 & -x_{13} & 0 & 0 & 0 \\
0 & 0 & x_7 & 0 & 0 & x_{10} & 0 & x_{12} & x_{13} & 0 & 0 & -2x_{16} & 0 \\
0 & 0 & -x_8 & 0 & 0 & 0 & -x_{12} & 0 & 0 & 0 & 0 & 0 & 0 \\
x_8 & 0 & 2x_9 & -x_{10} & 0 & 0 & x_{13} & 0 & 0 & 0 & 0 & 0 & 0 \\
0 & x_{10} & 0 & 0 & x_{12} & -x_{13} & 0 & 0 & 0 & 2x_{16} & 0 & 0 & 0 \\
0 & 0 & 0 & -x_{12} & 0 & 0 & 0 & 0 & 0 & 0 & 0 & 0 & 0 \\
0 & 0 & 0 & -x_{13} & 0 & 0 & 0 & 0 & 0 & 0 & 0 & 0 & 0 \\
x_{12} & 0 & x_{13} & 0 & 0 & 0 & -2x_{16} & 0 & 0 & 0 & 0 & 0 & 0 \\
0 & 0 & 0 & 0 & 0 & 0 & 0 & 0 & 0 & 0 & 0 & 0 & 0 \\
0 & 0 & 0 & 0 & 0 & 0 & 0 & 0 & 0 & 0 & 0 & 0 & 0 \\
0 & 0 & 0 & 2x_{16} & 0 & 0 & 0 & 0 & 0 & 0 & 0 & 0 & 0
\end{bmatrix} \tag{15}$$

whose $rank(A) = 9$.Therefore the system of linear partial differential equations (4) – (14) admits $13 - 9 = 4$ linearly independent solutions, which will be given below.

The system of partial differential equations (4) – (14) after some estimates takes the form:

$$\frac{\partial f}{\partial x_1} = \frac{\partial f}{\partial x_2} = \frac{\partial f}{\partial x_3} = \frac{\partial f}{\partial x_4} = \frac{\partial f}{\partial x_5} = \frac{\partial f}{\partial x_6} = \frac{\partial f}{\partial x_9} = 0 \tag{16}$$

$$x_{10}\frac{\partial f}{\partial x_7} + x_{14}\frac{\partial f}{\partial x_{11}} = 0 \tag{17}$$

$$x_{13}\frac{\partial f}{\partial x_7} - 2x_{14}\frac{\partial f}{\partial x_{11}} = 0 \tag{18}$$

The system (16), (17) and (18) admits of the following solutions:

$$f_1 = x_{10}, \quad f_2 = x_{13}, \quad f_3 = x_{14}. \tag{19}$$

It can be seen that another solution of the system (4) – (14) is the following:

$$f_4 = 2x_{14}x_7 + x_{13}x_8 - 2x_{10}x_{11}. \tag{20}$$

Since f in general depends on 13 variables and the $rank(A) = 9$, we conclude that the general solution of this system depends on four linearly independent solutions, which are given by (19) and (20).

Now we can start the theorem:

Theorem 1 *The general solution of the system of linear partial differential equations* (4) – (14) *has the form* $f = f(f_1, f_2, f_3, f_4)$, *where* f_1, f_2, f_3, f_4 *are given by* (19) *and* (20).

4. Special Transformation

Now, we put:

$$x_1' = x_1 + x_2, \quad x_2' = x_2 + x_3, \quad x_3' = x_3 + x_4, \quad x_4' = x_4 + x_5 \tag{21}$$

$$x_5' = x_5 + x_6, \quad x_6' = x_6 + x_7, \quad x_7' = x_7 + x_8, \quad x_8' = x_8 + x_9 \tag{22}$$

$$x_9' = x_9 + x_{10}, \quad x_{10}' = x_{10} + x_{11}, \quad x_{11}' = x_{11} + x_{13}, \quad x_{13}' = x_{13} + x_{14} \tag{23}$$

$$x_{14}' = x_{14} + x_1 \tag{24}$$

From the relations (21), (22), (23) and (24) we obtain:

$$x_1 = \frac{1}{2}(x'_1 + x'_{14}) + A_{1,14} = B_1, x_2 = \frac{1}{2}(x'_1 + x'_2) + A_{1,2} = B_2 \tag{25}$$

$$x_3 = \frac{1}{2}(x'_2 + x'_3) + A_{2,3} = B_3, x_4 = \frac{1}{2}(x'_3 + x'_4) + A_{3,4} = B_4 \tag{26}$$

$$x_5 = \frac{1}{2}(x'_4 + x'_5) + A_{4,5} = B_5, x_6 = \frac{1}{2}(x'_5 + x'_6) + A_{5,6} = B_6 \tag{27}$$

$$x_7 = \frac{1}{2}(x'_6 + x'_7) + A_{6,7} = B_7, x_8 = \frac{1}{2}(x'_7 + x'_8) + A_{7,8} = B_8 \tag{28}$$

$$x_9 = \frac{1}{2}(x'_8 + x'_9) + A_{8,9} = B_9, x_{10} = \frac{1}{2}(x'_9 + x'_{10}) + A_{9,10} = B_{10} \tag{29}$$

$$x_{11} = \frac{1}{2}(x'_{10} + x'_{11}) + A_{10,11} = B_{11}, x_{13} = \frac{1}{2}(x'_{11} + x'_{13}) + A_{11,13} = B_{12} \tag{30}$$

$$x_{14} = \frac{1}{2}(x'_{13} + x'_{14}) + A_{13,14} = B_{14} \tag{31}$$

where $A_{k,k+1}, k = 1, 2, \ldots, 13$ and $A_{1,14}$ will be explained bellow.

We consider the expression:

$$c = x'_1 - x'_2 + x'_3 - x'_4 + x'_5 - x'_6 + x'_7 - x'_8 + x'_9 - x'_{10} + x'_{11} - x'_{13} + x'_{14}. \tag{32}$$

The expression $A_{1,14}$ is obtained by c subtracting $x'_1 + x'_{14}$.With the same way we obtain the other expressions $A_{k,k+1}, k = 1, 2, \ldots, 14$.

From the first of (4.1) and the (4.4) we have

$$\frac{\partial f}{\partial x_1} = \frac{\partial f}{\partial x'_1}\frac{\partial x'_1}{\partial x_1} + \frac{\partial f}{\partial x'_{14}}\frac{\partial x'_{14}}{\partial x_1} = \frac{\partial f}{\partial x'_1} + \frac{\partial f}{\partial x'_{14}}. \tag{33}$$

With the same way from (21), (22) and (24) we obtain:

$$\frac{\partial f}{\partial x_2} = \frac{\partial f}{\partial x'_1} + \frac{\partial f}{\partial x'_2}, \frac{\partial f}{\partial x_3} = \frac{\partial f}{\partial x'_2} + \frac{\partial f}{\partial x'_3}, \frac{\partial f}{\partial x_4} = \frac{\partial f}{\partial x'_3} + \frac{\partial f}{\partial x'_4} \tag{34}$$

$$\frac{\partial f}{\partial x_5} = \frac{\partial f}{\partial x'_4} + \frac{\partial f}{\partial x'_5}, \frac{\partial f}{\partial x_6} = \frac{\partial f}{\partial x'_5} + \frac{\partial f}{\partial x'_6}, \frac{\partial f}{\partial x_7} = \frac{\partial f}{\partial x'_6} + \frac{\partial f}{\partial x'_7} \tag{35}$$

$$\frac{\partial f}{\partial x_8} = \frac{\partial f}{\partial x'_7} + \frac{\partial f}{\partial x'_8}, \frac{\partial f}{\partial x_9} = \frac{\partial f}{\partial x'_8} + \frac{\partial f}{\partial x'_9}, \frac{\partial f}{\partial x_{10}} = \frac{\partial f}{\partial x'_9} + \frac{\partial f}{\partial x'_{10}} \tag{36}$$

$$\frac{\partial f}{\partial x_{11}} = \frac{\partial f}{\partial x'_{10}} + \frac{\partial f}{\partial x'_{11}}, \frac{\partial f}{\partial x_{13}} = \frac{\partial f}{\partial x'_{11}} + \frac{\partial f}{\partial x'_{13}}, \frac{\partial f}{\partial x_{14}} = \frac{\partial f}{\partial x'_{13}} + \frac{\partial f}{\partial x'_{14}} \tag{37}$$

The system of linear partial differential equations (4) - (14) by means of (25) - (31) and (32) - (37) takes the form

$$-B_5\frac{\partial f}{\partial x'_1} - B_5\frac{\partial f}{\partial x'_2} - B_8\frac{\partial f}{\partial x'_5} - B_8\frac{\partial f}{\partial x'_6} - B_{11}\frac{\partial f}{\partial x'_7}$$
$$-B_{11}\frac{\partial f}{\partial x'_8} - B_{11}\frac{\partial f}{\partial x'_9} = 0 \tag{38}$$

$$B_5\frac{\partial f}{\partial x'_1} - B_6\frac{\partial f}{\partial x'_2} - B_6\frac{\partial f}{\partial x'_3} - B_{10}\frac{\partial f}{\partial x'_8} - B_{10}\frac{\partial f}{\partial x'_9}$$
$$-B_{14}\frac{\partial f}{\partial x'_{10}} - B_{14}\frac{\partial f}{\partial x'_{11}} + B_6\frac{\partial f}{\partial x'_{14}} = 0 \tag{39}$$

$$B_6\frac{\partial f}{\partial x'_1} + B_6\frac{\partial f}{\partial x'_2} - B_7\frac{\partial f}{\partial x'_3} - B_7\frac{\partial f}{\partial x'_3} - B_{10}\frac{\partial f}{\partial x'_5}$$
$$-B_{10}\frac{\partial f}{\partial x'_6} + B_{13}\frac{\partial f}{\partial x'_8} - B_{13}\frac{\partial f}{\partial x'_9} = 0 \tag{40}$$

$$B_7\frac{\partial f}{\partial x'_2} + B_7\frac{\partial f}{\partial x'_3} + B_{10}\frac{\partial f}{\partial x'_5} + B_{10}\frac{\partial f}{\partial x'_6} - B_{13}\frac{\partial f}{\partial x'_8} - B_{13}\frac{\partial f}{\partial x'_9} = 0 \tag{41}$$

$$-B_8\frac{\partial f}{\partial x'_2} - B_8\frac{\partial f}{\partial x'_3} + B_{14}\frac{\partial f}{\partial x'_8} + B_{14}\frac{\partial f}{\partial x'_9} = 0 \tag{42}$$

$$B_8\frac{\partial f}{\partial x'_1} + 2B_9\frac{\partial f}{\partial x'_2} + (2B_9 - B_{10})\frac{\partial f}{\partial x'_3} - B_{10}\frac{\partial f}{\partial x'_4} + B_{13}\frac{\partial f}{\partial x'_6}$$
$$+(B_{13} - 2B_{14})\frac{\partial f}{\partial x'_7} - 2B_{14}\frac{\partial f}{\partial x'_8} + B_8\frac{\partial f}{\partial x'_{14}} = 0 \tag{43}$$

$$B_{10}\frac{\partial f}{\partial x'_1} + B_{10}\frac{\partial f}{\partial x'_{10}} - B_{13}\frac{\partial f}{\partial x'_5} - B_{13}\frac{\partial f}{\partial x'_5} = 0 \tag{44}$$

$$2B_{11}\frac{\partial f}{\partial x'_2} + (2B_{11} - B_{12})\frac{\partial f}{\partial x'_3} - B_{12}\frac{\partial f}{\partial x'_4} - 2B_{14}\frac{\partial f}{\partial x'_5} - 2B_{14}\frac{\partial f}{\partial x'_6} = 0 \tag{45}$$

$$B_{11}\frac{\partial f}{\partial x'_1} - B_{13}\frac{\partial f}{\partial x'_3} - (B_{13} + B_{14})\frac{\partial f}{\partial x'_4} - B_{14}\frac{\partial f}{\partial x'_5}$$
$$-B_{14}\frac{\partial f}{\partial x'_5} + B_{11}\frac{\partial f}{\partial x'_{14}} = 0 \tag{46}$$

Now, we can prove the following theorem.

Theorem 2 *The system of linear partial differential equations* (38) – (46) *has a general solution of the form* $f = f(t_1, t_2, t_3, t_4)$, *where* f *is an arbitrary function of* $t_1 = B_{10}$, $t_2 = B_{13}$, $t_3 = B_{14}$, $t_4 = 2B_{14}B_7 + B_{13}B_8 - 2B_{10}B_{14}$, *and* $B_7, B_8, B_{10}, B_{11}, B_{13}, B_{14}$ *are functions of* $(x'_1, x'_2, \ldots, x'_{11}, x'_{13}, x'_{14})$ *given by (25) - (31).*

Proof. The system (38) - (46) is a linear transformation of the system (4) - (14) by means (21) - (24). Therefore, the general solution of the new system is obtained from the general solution of the original system by means of this transformation. □

Remark 3 *The system of the linear partial differential equations (38) - (46) is more difficult than (4) - (14), which is obtained from the first. This method helps to solve some systems of linear partial differential equations.*

However we can construct more complicated systems of linear partial differential equations than (38) - (46). If we make the general linear transformations:

$$x'_i = \sum_{j=1}^{n} a_{ij} x_j \qquad i = 1, \dots, n \qquad a_{ij} \in IR.$$

The most important fact is that we can find the general solution of the new system.

References

1. X. Agrafiotou, Annals of Dunarea de Jos., **XVI (XXI)**, (1998), (in press).

2. X. Agrafiotou, Bulletin de la Societe et de Lettre de Lodze, Polish Academy, (in press).

3. P. Berglez, Complex Variables, Theory Appl., **30**, 69-76 (1996).

4. P. Berglez, Ann. Mat. Pura Appl., **162**, 263-279 (1992).

5. G. Carra Ferro, Lie Groups Appl., **1**, 47-55 (1994).

6. A. MacNabb and O. Jones, Appl. Anal., **39**, 227-242 (1990).

9th Int. Coll. on Differential Equations, pp. 7-14
D. Bainov (Ed)

Inverse Spectral Problem for the Differential Equation of the Second Order with Singularity

R. KH. AMIROV AND Y. CAKMAK
Cumhuriyet University, Dept. of Mathematics, 58140, Sivas, Turkey

Abstract: In this article, for given sequences of $\{\lambda_n\}$ and $\{\mu_n\}$ constructing a differential operator, which has a singularity, has explained. In order to accomplish this, λ_n and μ_n have to have some certain asymptotic. This kind of asymptotic formulas depend on the regularity order of the function $q(x)$. This connection can be seen in the solution of the inverse problem which depend on the series of $\{\lambda_n\}$ and $\{\mu_n\}$.
Key Words and Phrases: *Inverse problem, regulary, singulary.*
1991 Mathematics Subject Classification. Primary 34A55, Secondary 34B24, 34L05.

Introduction

For the regular Sturm-Liouville equations, with respect to two spectrum inverse problem completely had been solved. For those equations with the type of singularity $\ell(\ell+1)/(\pi-x)^2$, which is singular at π in the interval $[0,\pi]$, M. G. Gasimov [1], given two spectrum, has solved the inverse problem. In $[0,\pi]$, the singularity of A/x (A is real) at $x=0$ has been studied by M. G. Gasimov and R. Kh. Amirov [2] and helping of two spectrum the solution of the inverse problem has been obtained. In $[0,\pi]$, the singularity of $\left(\frac{A}{x}+\frac{\ell(\ell+1)}{x^2}\right)$, ($A$ is real) at $x=0$ has been studied by R. Kh. Amirov and S.Gulyaz [3] and the solution of the inverse problem with two spectrum has been given. In $[0,\pi]$, the singularity of $\frac{\delta}{x^p}$ (δ is real), $p\in(1,5/4)$ at $x=0$ has been investigated by Z.M. Gasimov [4] and the inverse problem with two spectrum has been solved. In this paper, again in $[0,\pi]$, at $x=0$, for the equations with the singularity $\left(\frac{A}{x}+\frac{\delta}{x^p}\right)$, $p\in(1,2)$, a theorem giving the inverse problem solution with two spectrum has been proved.

1. Expression for the Normalizing Constants in Terms of the Spectra

We consider the differential equation

$$-y''+\left\{\frac{A}{x}+\frac{\delta}{x^p}+q(x)\right\}y=\lambda y, \tag{1.1}$$

and the boundary conditions

$$y(0)=0 \quad , \qquad y'(\pi)-H_1y(\pi)=0, \tag{1.2}$$

$$y(0) = 0 \quad , \qquad y'(\pi) - H_2 y(\pi) = 0. \tag{1.3}$$

We assume that $q(x) \in L_2[0,\pi]$; $p \in (1, 5/4)$ δ, A, H_1 and H_2 are real numbers, with $H_1 \neq H_2$. We denote the eigenvalues of the boundary value problems (1.1)-(1.2) and (1.1)-(1.3) by $\lambda_0 < \lambda_1 < \ldots$ and $\mu_0 < \mu_1 < \ldots$ respectively. Let $\varphi(x,\lambda)$ and $\psi(x,\lambda)$ be the solutions of (1.1) satisfying the initial conditions,

$$\varphi(0,\lambda) = 0 \quad , \quad \varphi'(0,\lambda) = 1 \tag{1.4}$$

and

$$\psi(0,\lambda) = 0 \quad , \quad \psi'(0,\lambda) = 1. \tag{1.5}$$

Then it is clear that the eigenvalues $\lambda_0, \lambda_1, \ldots$ and $\mu_0, \mu_1, \ldots$ of the boundary value problems (1.1)-(1.2) and (1.1)-(1.3) coincide with the zeros of the functions

$$\Phi_1(\lambda) = \varphi'(\pi,\lambda) - H_1\varphi(\pi,\lambda), \quad \Phi_2(\lambda) = \psi'(\pi,\lambda) - H_2\psi(\pi,\lambda).$$

Clearly $\varphi(x,\lambda_n) = \varphi_n(x)$ is an eigenfunction of the boundary value problem (1.1)-(1.2) and its norm is equal to

$$\alpha_n = \int_0^\pi \varphi_n^2(x)dx. \tag{1.6}$$

The numbers $\alpha_0, \alpha_1, \ldots$ are called the normalizing constants of the boundary value problem (1.1)-(1.2). In this section we shall derive a formula for calculating the numbers $\alpha_0, \alpha_1, \ldots$ in terms of the two spectra $\lambda_0, \lambda_1, \ldots$ and $\mu_0, \mu_1, \ldots$.

Because $\varphi(x,\lambda)$ is the solution of the equation (1.1) that satisfying then (1.4) condition

$$\int_0^\pi \varphi^2(x,\lambda)dx = \varphi'(\pi,\lambda)\frac{\partial}{\partial\lambda}\varphi(\pi,\lambda) - \varphi(\pi,\lambda)\frac{\partial}{\partial\lambda}\varphi'(\pi,\lambda). \tag{1.7}$$

The initial conditions (1.4) and (1.5) also $\Phi_1(\lambda)$ and $\Phi_2(\lambda)$ are entire in λ for fixed x. It is clear from this that $\Phi_1(\lambda)$ and $\Phi_2(\lambda)$ are entire functions of order one half and therefore are determined by their zeros, to within a constant multiplying factor. Therefore,

$$\varphi'(\pi,\lambda) - H_1\varphi(\pi,\lambda) = C_1\prod_{k=0}^{\infty}\left(1 - \frac{\lambda}{\lambda_k}\right) = \Phi_1(\lambda), \tag{1.8}$$

and

$$\varphi'(\pi,\lambda) - H_2\varphi(\pi,\lambda) = C_2\prod_{k=0}^{\infty}\left(1 - \frac{\lambda}{\mu_k}\right) = \Phi_2(\lambda), \tag{1.9}$$

where C_1 and C_2 are constants. Combining (1.10) and (1.11) we obtain

$$\int_0^\pi \varphi^2(x,\lambda_n)dx = \frac{1}{H_2 - H_1}\acute{\Phi}_1(\lambda_n)\Phi_2(\lambda_n).$$

Let $\varphi_0(x,\lambda)$ be solution of equation (1.1) when $q(x) = 0$ satisfying the condition (1.4). Respectively $\lambda_0^0, \lambda_1^0, \ldots$ and $\mu_0^0, \mu_1^0, \ldots$ eigenvalues problems (1.1)-(1.4) and (1.1)-(1.5) when $q(x) = 0$. Then it is clear that the eigenvalues $\lambda_0^0, \lambda_1^0, \ldots$ and $\mu_0^0, \mu_1^0, \ldots$ of the boundary value problems (1.1)-(1.2) and (1.1)-(1.3) coincide with the zeros of the functions

$$\varphi_0'(\pi,\lambda) - H_1\varphi_0(\pi,\lambda) = C_1^0\prod_{k=0}^{\infty}\left(1 - \frac{\lambda}{\lambda_k^0}\right) = \Phi_{01}(\lambda), \tag{1.10}$$

and

$$\varphi_0'(\pi,\lambda) - H_2\varphi_0(\pi,\lambda) = C_2^0\prod_{k=0}^{\infty}\left(1 - \frac{\lambda}{\mu_k^0}\right) = \Phi_{02}(\lambda). \tag{1.11}$$

Clearly $\varphi_0(x,\lambda_n^0)=\varphi_{0n}(x)$ is an eigenfunction of the boundary value problem (1.1)-(1.2) and its norm is equal to

$$\alpha_n^0=\int_0^\pi \varphi_{0n}^2(x)dx. \tag{1.12}$$

Because $\varphi_0(x,\lambda)$ is the solution of the equation (1.1) when $q(x)=0$ satisfying the condition (1.4).

$$\int_0^\pi \varphi_0^2(x,\lambda_n^0)dx=\frac{1}{H_2-H_1}\acute{\Phi}_{01}(\lambda_n^0)\Phi_{02}(\lambda_n^0).$$

Therefore

$$\alpha_n=\frac{1}{H_2-H_1}\acute{\Phi}_1(\lambda_n)\Phi_2(\lambda_n), \tag{1.13}$$

and

$$\alpha_n^0=\frac{1}{H_2-H_1}\acute{\Phi}_{01}(\lambda_n^0)\Phi_{02}(\lambda_n^0). \tag{1.14}$$

Then we obtain

$$\acute{\Phi}_1(\lambda_n)=-\frac{C_1}{\lambda_n}\prod_{k=0}^{\infty}{}'\left(1-\frac{\lambda_n}{\lambda_k}\right),\quad \acute{\Phi}_{01}(\lambda_n^0)=-\frac{C_1^0}{\lambda_n^0}\prod_{k=0}^{\infty}{}'\left(1-\frac{\lambda_n^0}{\lambda_k^0}\right)$$

where (here and henseforth) the symbol $\prod'$ means that the factor $k=n$ is absent from the infinite product. These formulas and (1.13)-(1.14) imply that

$$\alpha_n=-\frac{C_1C_2}{H_2-H_1}\frac{1}{\lambda_n}\prod_{k=0}^{\infty}{}'\left(1-\frac{\lambda_n}{\lambda_k}\right)\prod_{k=0}^{\infty}\left(1-\frac{\lambda_n}{\mu_k}\right), \tag{1.15}$$

and

$$\alpha_n^0=-\frac{C_1^0C_2^0}{H_2-H_1}\frac{1}{\lambda_n^0}\prod_{k=0}^{\infty}{}'\left(1-\frac{\lambda_n^0}{\lambda_k^0}\right)\prod_{k=0}^{\infty}\left(1-\frac{\lambda_n^0}{\mu_k^0}\right). \tag{1.16}$$

Since

$$\Phi_2(\lambda_n)=C_2\prod_{k=0}^{\infty}\left(1-\frac{\lambda_n}{\mu_k}\right)=B_2C_2\prod_{k=0}^{\infty}\left(1-\frac{\lambda_n^0}{\mu_k^0}\right)\prod_{k=0}^{\infty}\frac{\lambda_n-\mu_k^0}{\lambda_n^0-\mu_k^0}\prod_{k=0}^{\infty}\frac{\lambda_n-\mu_k}{\lambda_n-\mu_k^0}.$$

In the same way we can prove that

$$\acute{\Phi}_1(\lambda_n)=-\frac{B_1C_1}{\lambda_n^0}\prod_{k=0}^{\infty}{}'\left(\frac{\lambda_n^0}{\lambda_k^0}\right)\prod_{k=0}^{\infty}{}'\left(\frac{\lambda_n-\lambda_k}{\lambda_n^0-\lambda_k^0}\right)\prod_{k=0}^{\infty}{}'\left(1-\frac{\lambda_n-\lambda_k}{\lambda_n-\lambda_k^0}\right).$$

These formulas and (1.13) imply that

$$\alpha_n=\frac{C_1C_2}{C_1^0C_2^0}B_1B_2\alpha_n^0\frac{\lambda_n-\mu_n}{\lambda_n^0-\mu_n^0}\prod_{k=0}^{\infty}{}'\left(\frac{\lambda_n-\lambda_k^0}{\lambda_n^0-\lambda_k^0}\right)\prod_{k=0}^{\infty}{}'\left(\frac{\lambda_n-\lambda_k}{\lambda_n-\lambda_k^0}\right)\prod_{k=0}^{\infty}{}'\frac{\lambda_n-\mu_k^0}{\lambda_n^0-\mu_k^0}\prod_{k=0}^{\infty}{}'\frac{\lambda_n-\mu_k}{\lambda_n-\mu_k^0}.$$

It can be shown easily that

$$\frac{C_1C_2}{C_1^0C_2^0}B_1B_2=1. \tag{1.17}$$

Therefore

$$\alpha_n=\alpha_n^0\frac{\lambda_n-\mu_n}{\lambda_n^0-\mu_n^0}\frac{\lambda_n-\lambda_0}{\lambda_n^0-\lambda_0^0}\frac{\lambda_n-\mu_0}{\lambda_n^0-\mu_0^0}\prod_{k=1}^{\infty}{}'\left(\frac{\lambda_n-\lambda_k^0}{\lambda_n^0-\lambda_k^0}\right)\prod_{k=1}^{\infty}{}'\left(\frac{\lambda_n-\lambda_k}{\lambda_n-\lambda_k^0}\right)\prod_{k=1}^{\infty}{}'\frac{\lambda_n-\mu_k^0}{\lambda_n^0-\mu_k^0}\prod_{k=1}^{\infty}{}'\frac{\lambda_n-\mu_k}{\lambda_n-\mu_k^0}. \tag{1.18}$$

2. Asymptotic Formulas for the Numbers α_n

Suppose that we are given two sequences of numbers $\lambda_0, \lambda_1, \ldots,$ $\mu_0, \mu_1, \ldots,$ and that we know them to be the eigenvalues of the boundary value problems (1.1)-(1.2) and (1.1)-(1.3) with an unknown function $q(x)$ and with the real numbers H_1, H_2. Suppose that the following asymptotic formulas hold:

$$\begin{aligned}\lambda_n &= (n+1/2)^2 + 2c_1(n+1/2)^{p-1} + \tfrac{A}{\pi}\ln(n+1/2) + 2c_2 + \tfrac{2c_3}{(n+1/2)^{3-2p}} + 2c_4\tfrac{\ln(n+1/2)}{(n+1/2)^{2-p}}\\ &+\tfrac{2c_5}{(n+1/2)^{2-p}} + 2c_9\tfrac{\ln(n+1/2)}{(n+1/2)} + \tfrac{2c_{10}}{(n+1/2)} + \tfrac{2c_6}{(n+1/2)^{5-3p}} + 2c_7\tfrac{\ln(n+1/2)}{(n+1/2)^{4-2p}}\\ &+\tfrac{\gamma_1}{(n+1/2)^{4-2p}} - \tfrac{A^2\delta c_p}{2\pi}\tfrac{\ln^2(n+1/2)}{(n+1/2)^{3-p}} + \gamma_2\tfrac{\ln(n+1/2)}{(n+1/2)^{3-p}} + \tfrac{\gamma_3}{(n+1/2)^{3-p}}\\ &-\tfrac{A^3}{12\pi}\tfrac{\ln^3(n+1/2)}{(n+1/2)^2} + \gamma_4\tfrac{\ln^2(n+1/2)}{(n+1/2)^2} + \gamma_5\tfrac{\ln(n+1/2)}{(n+1/2)^2} + \tfrac{\gamma_6}{(n+1/2)^2} + O\left(\tfrac{1}{n^{7-4p}}\right)\end{aligned} \tag{2.1}$$

$$\begin{aligned}\mu_n &= (n+1/2)^2 + 2c_1'(n+1/2)^{p-1} + \tfrac{A}{\pi}\ln(n+1/2) + 2c_2' + \tfrac{2c_3'}{(n+1/2)^{3-2p}} + 2c_4'\tfrac{\ln(n+1/2)}{(n+1/2)^{2-p}}\\ &+\tfrac{2c_5'}{(n+1/2)^{2-p}} + 2c_9'\tfrac{\ln(n+1/2)}{(n+1/2)} + \tfrac{2c_{10}'}{(n+1/2)} + \tfrac{2c_6'}{(n+1/2)^{5-3p}} + 2c_7'\tfrac{\ln(n+1/2)}{(n+1/2)^{4-2p}}\\ &+\tfrac{\gamma_1'}{(n+1/2)^{4-2p}} - \tfrac{A^2\delta c_p}{2\pi}\tfrac{\ln^2(n+1/2)}{(n+1/2)^{3-p}} + \gamma_2'\tfrac{\ln(n+1/2)}{(n+1/2)^{3-p}} + \tfrac{\gamma_3'}{(n+1/2)^{3-p}} - \tfrac{A^3}{12\pi}\tfrac{\ln^3(n+1/2)}{(n+1/2)^2}\\ &+\gamma_4'\tfrac{\ln^2(n+1/2)}{(n+1/2)^2} + \gamma_5'\tfrac{\ln(n+1/2)}{(n+1/2)^2} + \tfrac{\gamma_6'}{(n+1/2)^2}\, O\left(\tfrac{1}{n^{7-4p}}\right).\end{aligned} \tag{2.2}$$

Then, by (2.1) and (2.2)

$$c_2 = \frac{1}{\pi}\left[-H_1 + \frac{A\ln\pi}{2} + AM_5 + \frac{\delta}{2(1-p)\pi^{p-1}} + \frac{1}{2}\int_0^\pi q(t)dt\right]$$

$$c_2' = \frac{1}{\pi}\left[-H_2 + \frac{A\ln\pi}{2} + AM_5 + \frac{\delta}{2(1-p)\pi^{p-1}} + \frac{1}{2}\int_0^\pi q(t)dt\right].$$

Therefore

$$H_1 - H_2 = \pi(c_2' - c_2). \tag{2.3}$$

By hypothesis, the numbers $\{\lambda_n\}$ and $\{\mu_n\}$ are distinct spectra of one and the same equation. Hence it follows from the results of the previous section that the normalizing consants of the boundary value problem (2.1)-(2.2) are given by the formula (1.18) .

In the rest of this section we shall find the asymptotic formula for the numbers α_n in terms of the known asymptotic formulae for the numbers λ_n and μ_n. We shall carry out this derivation in several stages. We first consider the infinite product $\Psi(\lambda_n) = \prod_{k=1}^{\infty}{}'\left(1+\frac{\lambda_k-\lambda_k^0}{\lambda_k^0-\lambda_n}\right)$. Clearly $\ln\Psi(\lambda_n) = \sum_{k=1}^{\infty}{}'\ln\left(1+\frac{\lambda_k-\lambda_k^0}{\lambda_k^0-\lambda_n}\right)$. For sufficiently large n and $k \neq n$ $\mid \frac{\lambda_k-\lambda_k^0}{\lambda_k^0-\lambda_n} \mid < \frac{C}{n+1/2}$, (in what follows C will denote a constant, not necessarily the same one). Therefore

$$\ln\Psi(\lambda_n) = -\sum_{k=1}^{\infty}{}'\left[\sum_{r=1}^{\infty}\frac{(-1)^r}{r}\left(\frac{\lambda_k-\lambda_k^0}{\lambda_k^0-\lambda_n}\right)^r\right]. \tag{2.4}$$

We have

Lemma: Let $|\lambda_k - \lambda_k^0| \leq a \quad (k = 1, 2, \ldots)$. Then

$$\sum_{k=1}^{\infty}{}' \left| \frac{\lambda_k - \lambda_k^0}{\lambda_k^0 - \lambda_n} \right|^r \leq \begin{cases} C\frac{\ln(n+1/2)}{(n+1/2)}, & r = 1 \\ C\frac{a^r}{(n+1/2)^r}, & r \geq 2 \end{cases} \tag{2.5}$$

The proof of Lemma is the same as the proof of [5].

Let us study $\ln \Psi(\lambda_n)$ a little further. It follows from the estimate (2.7) that

$$\begin{aligned} &\left| \sum_{k=1}^{\infty}{}' \sum_{r=3}^{\infty} \frac{(-1)^r}{r} \left(\frac{\lambda_k - \lambda_k^0}{\lambda_k^0 - \lambda_n}\right)^r \right| \leq \sum_{r=3}^{\infty} \frac{1}{r} \sum_{k=1}^{\infty}{}' \left| \frac{\lambda_k - \lambda_k^0}{\lambda_k^0 - \lambda_n} \right|^r \\ &\leq C \sum_{r=3}^{\infty} \frac{a^r}{(n+1/2)^r} = \frac{Ca^3}{(n+1/2)^3} \sum_{r=0}^{\infty} \frac{a^r}{(n+1/2)^r} = O\left(\frac{1}{n^3}\right). \end{aligned} \tag{2.6}$$

This estimate and the formula (2.6) show that

$$\ln \Psi(\lambda_n) = \sum_{k=1}^{\infty}{}' \frac{\lambda_k - \lambda_k^0}{\lambda_k^0 - \lambda_n} - \frac{1}{2} \sum_{k=1}^{\infty}{}' \left(\frac{\lambda_k - \lambda_k^0}{\lambda_k^0 - \lambda_n}\right)^2 + O\left(\frac{1}{n^3}\right). \tag{2.7}$$

Now we consider the behaviour of the sums appearing in this formula.

Using the asymptotic formula (2.1) and (2.2) we see that the first sum satisfies. If we use expression λ_k and λ_k^0 asymptotic, we get the expression

$$\begin{aligned} \sum_{k=1}^{\infty}{}' \frac{\lambda_k - \lambda_k^0}{\lambda_k^0 - \lambda_n} &= \left[\frac{\pi^2 c_1 - 6c_1^0}{3}\right] (c_2 - c_2^0) \frac{1}{(n+1/2)^{3-p}} + \frac{A(3\pi^2 - 4)}{12\pi} (c_2 - c_2^0) \frac{\ln(n+1/2)}{(n+1/2)^2} \\ &+ \left[M_\lambda + \frac{(c_2 - c_2^0)}{2} \left(3 + \frac{\pi^2 (c_2 - c_2^0)}{6} + \frac{2A \ln(3/2)}{3\pi}\right) + \frac{3^{p-1}}{2^{p-1}} (c_5 - c_5^0) \right. \\ &\left. + \frac{(\gamma_1 - \gamma_1^0)}{2^{2p-2} 3^{3-2p}} + \frac{\ln(3/2)}{2^{p-1} 3^{2-p}} (\gamma_2 - \gamma_2^0) + \frac{(\gamma_3 - \gamma_3^0)}{2^{p-1} 3^{2-p}}\right] \frac{1}{(n+1/2)^2} + O\left(\frac{1}{n^{5-2p}}\right) \end{aligned} \tag{2.8}$$

$$-\frac{1}{2} \sum_{k=1}^{\infty}{}' \left(\frac{\lambda_k - \lambda_k^0}{\lambda_k^0 - \lambda_n}\right)^2 = \frac{\pi^2 (c_2 - c_2^0)^2}{4} \frac{1}{(n+1/2)^2} O\left(\frac{1}{n^4}\right). \tag{2.9}$$

Therefore, the formulae (2.7), (2.8) and (2.9) imply that

$$\begin{aligned} \ln \Psi(\lambda_n) &= \sum_{k=1}^{\infty}{}' \frac{\lambda_k - \lambda_k^0}{\lambda_k^0 - \lambda_n} - \frac{1}{2} \sum_{k=1}^{\infty}{}' \left(\frac{\lambda_k - \lambda_k^0}{\lambda_k^0 - \lambda_n}\right)^2 + O\left(\frac{1}{n^3}\right) \\ &= \left[\frac{\pi^2 c_1 - 6c_1^0}{3}\right] (c_2 - c_2^0) \frac{1}{(n+1/2)^{3-p}} + \frac{A(3\pi^2 - 4)}{12\pi} (c_2 - c_2^0) \frac{\ln(n+1/2)}{(n+1/2)^2} \\ &+ \left[M_\lambda + \frac{(c_2 - c_2^0)}{2} \left(3 + \frac{\pi^2 (c_2 - c_2^0)}{6} + \frac{2A \ln(3/2)}{3\pi}\right) - \frac{\pi^2 (c_2 - c_2^0)^2}{4} \right. \\ &\left. + \frac{3^{p-1}}{2^{p-1}} (c_5 - c_5^0) + \frac{(\gamma_1 - \gamma_1^0)}{2^{2p-2} 3^{3-2p}} + \frac{\ln(3/2)}{2^{p-1} 3^{2-p}} (\gamma_2 - \gamma_2^0) + \frac{(\gamma_3 - \gamma_3^0)}{2^{p-1} 3^{2-p}}\right] \frac{1}{(n+1/2)^2} + O\left(\frac{1}{n^{5-2p}}\right). \end{aligned}$$

Hence

$$\begin{aligned} \Psi(\lambda_n) &= 1 + \left[\frac{\pi^2 c_1 - 6c_1^0}{3}\right] (c_2 - c_2^0) \frac{1}{(n+1/2)^{3-p}} + \frac{A(3\pi^2 - 4)}{12\pi} (c_2 - c_2^0) \frac{\ln(n+1/2)}{(n+1/2)^2} \\ &+ \left[M_\lambda + \frac{(c_2 - c_2^0)}{2} \left(3 + \frac{\pi^2 (c_2 - c_2^0)}{6} + \frac{2A \ln(3/2)}{3\pi}\right) - \frac{\pi^2 (c_2 - c_2^0)^2}{4} \right. \\ &\left. + \frac{3^{p-1}}{2^{p-1}} (c_5 - c_5^0) + \frac{(\gamma_1 - \gamma_1^0)}{2^{2p-2} 3^{3-2p}} + \frac{\ln(3/2)}{2^{p-1} 3^{2-p}} (\gamma_2 - \gamma_2^0) + \frac{(\gamma_3 - \gamma_3^0)}{2^{p-1} 3^{2-p}}\right] \frac{1}{(n+1/2)^2} + O\left(\frac{1}{n^{5-2p}}\right). \end{aligned} \tag{2.10}$$

Now we consider the behaviour of the infinity products

$$\prod_{k=1}^{\infty}{}' \frac{\lambda_n-\mu_k}{\lambda_n-\mu_k^0} = 1+\left[\frac{\pi^2 c_1'-6c_1'^0}{3}\right](c_2'-c_2'^0)\frac{1}{(n+1/2)^{3-p}}+\frac{A(3\pi^2-4)}{12\pi}(c_2'-c_2'^0)\frac{\ln(n+1/2)}{(n+1/2)^2}$$

$$+\left[M_\mu+\frac{(c_2'-c_2'^0)}{2}\left(3+\frac{\pi^2(c_2'-c_2'^0)}{6}+\frac{2A\ln(3/2)}{3\pi}\right)-\frac{\pi^2(c_2'-c_2'^0)^2}{4}+\frac{3p-1}{2p-1}(c_5'-c_5'^0)\right.$$

$$\left.+\frac{(\gamma_1'-\gamma_1'^0)}{2^{2p-2}3^{3-2p}}+\frac{\ln(3/2)}{2^{p-1}3^{2-p}}(\gamma_2'-\gamma_2'^0)+\frac{(\gamma_3'-\gamma_3'^0)}{2^{p-1}3^{2-p}}\right]\frac{1}{(n+1/2)^2}+O\left(\frac{1}{n^{5-2p}}\right) \quad (2.11)$$

here

$$M_\mu = \sum_{k=1}^{\infty}\left\{\mu_k-\mu_k^0-2(c_2'-c_2'^0)-\frac{2(c_5'-c_5'^0)}{(k+1/2)^{2-p}}-\frac{(\gamma_1'-\gamma_1')^0}{(k+1/2)^{4-2p}}-(\gamma_2'-\gamma_2'^0)\frac{\ln(k+1/2)}{(k+1/2)^{3-p}}\right.$$

$$\left.-\frac{(\gamma_3'-\gamma_3'^0)}{(k+1/2)^{3-p}}\right\}.$$

In the same way we can prove that

$$\Psi(\lambda_n^0)=\prod_{k=1}^{\infty}{}'\frac{\lambda_n-\lambda_k^0}{\lambda_n^0-\lambda_k^0}=1+\left[\frac{\pi^2 c_1-6c_1^0}{3}\right](c_2-c_2^0)\frac{1}{(n+1/2)^{3-p}}+\frac{A(3\pi^2-4)}{12\pi}(c_2-c_2^0)\frac{\ln(n+1/2)}{(n+1/2)^2}$$

$$+\left[\frac{(c_2-c_2^0)}{2}\left(3+\frac{\pi^2(c_2-c_2^0)}{6}+\frac{2A\ln(3/2)}{3\pi}\right)-\frac{\pi^2(c_2-c_2^0)^2}{4}\right]\frac{1}{(n+1/2)^2}+O\left(\frac{1}{n^{5-2p}}\right) \quad (2.12)$$

$$\prod_{k=1}^{\infty}{}'\frac{\lambda_n-\mu_k^0}{\lambda_n^0-\mu_k^0}=1+\left[\frac{\pi^2 c_1'-6c_1'^0}{3}\right](c_2'-c_2'^0)\frac{1}{(n+1/2)^{3-p}}+\frac{A(3\pi^2-4)}{12\pi}(c_2'-c_2'^0)\frac{\ln(n+1/2)}{(n+1/2)^2}$$

$$+\left[\frac{(c_2'-c_2'^0)}{2}\left(3+\frac{\pi^2(c_2'-c_2'^0)}{6}+\frac{2A\ln(3/2)}{3\pi}\right)-\frac{\pi^2(c_2'-c_2'^0)^2}{4}\right]\frac{1}{(n+1/2)^2}+O\left(\frac{1}{n^{5-2p}}\right). \quad (2.13)$$

It is clear from the asymptotic formulas (2.1) and (2.2) that

$$\frac{\lambda_n-\mu_n}{\lambda_n^0-\mu_n^0}=1+\left[\frac{c_5-c_5'}{c_2-c_2'}\frac{c_5^0-c_5'^0}{c_2^0-c_2'^0}\right]\frac{1}{(n+1/2)^{4-2p}}$$

$$+\left[\frac{c_5-c_5'}{c_2-c_2'}\frac{c_{10}^0-c_{10}'^0}{c_2^0-c_2'^0}+\frac{c_{10}-c_{10}'}{c_2-c_2'}\frac{c_5^0-c_5'^0}{c_2^0-c_2'^0}\right]\frac{1}{(n+1/2)^{3-p}}$$

$$+\left\{\frac{1}{2}\left[(c_2+c_2')+(c_2^0+c_2'^0)\right]+\frac{c_{10}-c_{10}'}{c_2-c_2'}\frac{c_{10}^0-c_{10}'^0}{c_2^0-c_2'^0}\right\}\frac{1}{(n+1/2)^2}+O\left(\frac{1}{n^{7-4p}}\right) \quad (2.14)$$

$$\frac{\lambda_n-\lambda_0}{\lambda_n^0-\lambda_0^0}=1+\frac{2(c_1-c_1^0)}{(n+1/2)^{3-p}}+\frac{2(c_2-c_2^0)+(\lambda_0^0-\lambda_0)}{(n+1/2)^2}+O\left(\frac{1}{n^{5-2p}}\right) \quad (2.15)$$

$$\frac{\lambda_n-\mu_0}{\lambda_n^0-\mu_0^0}=1+\frac{2(c_1-c_1^0)}{(n+1/2)^{3-p}}+\frac{2(c_2-c_2^0)+(\mu_0^0-\mu_0)}{(n+1/2)^2}+O\left(\frac{1}{n^{5-2p}}\right). \quad (2.16)$$

It is well known that

$$\alpha_n^0=\frac{\pi}{2}+\frac{\delta M_4\pi}{2}\frac{1}{(n+1/2)^{2-p}}+\frac{AN\pi}{2}\frac{1}{(n+1/2)}+\frac{\nu_1^0}{(n+1/2)^{4-2p}}+\nu_2^0\frac{\ln(n+1/2)}{(n+1/2)^{3-p}}$$

$$+\frac{\nu_3^0}{(n+1/2)^{3-p}}+\nu_4^0\frac{\ln^2(n+1/2)}{(n+1/2)^2}+\nu_5^0\frac{\ln(n+1/2)}{(n+1/2)^2}+\frac{\nu_6^0}{(n+1/2)^2}+\frac{\nu_7^0}{(n+1/2)^{6-3p}}$$

$$+\nu_8^0 \frac{\ln(n+1/2)}{(n+1/2)^{5-2p}} + \nu_{10}^0 \frac{\ln^2(n+1/2)}{(n+1/2)^{4-p}} + \nu_{11}^0 \frac{\ln(n+1/2)}{(n+1/2)^{4-p}} + \frac{\nu_{12}^0}{(n+1/2)^{4-p}}$$

$$+\nu_{13}^0 \frac{\ln^2(n+1/2)}{(n+1/2)^3} + \nu_{14}^0 \frac{\ln(n+1/2)}{(n+1/2)^3} + \frac{\nu_{15}^0}{(n+1/2)^3} O\left(\frac{1}{n^{8-4p}}\right). \tag{2.17}$$

Therefore we obtain the formulas (2.10), (2.11), (2.12), (2.13), (2.14), (2.15), (2.16) and (2.17) the following formula for the numbers α_n:

$$\begin{aligned}\alpha_n \; &= \frac{\pi}{2} + \frac{\delta M_4 \pi}{2} \frac{1}{(n+1/2)^{2-p}} + \frac{AN\pi}{2} \frac{1}{(n+1/2)} + \left[\nu_1^0 + \frac{\pi}{2} \frac{c_5 - c'_5}{c_2 - c'_2} \frac{c_5^0 - {c'_5}^0}{c_2^0 - {c'_2}^0}\right] \frac{1}{(n+1/2)^{4-2p}} \\
&+\nu_2^0 \frac{\ln(n+1/2)}{(n+1/2)^{3-p}} + \left\{2\pi(c_1 - c_1^0) + \pi(c_2 - c_2^0)\left(\frac{\pi^2 c_1 - 6c_1^0}{3}\right) + \pi(c'_2 - {c'_2}^0)\left(\frac{\pi^2 c'_1 - 6{c'_1}^0}{3}\right)\right. \\
&\left. + \frac{\pi}{2} \frac{c_5 - c'_5}{c_2 - c'_2} \frac{c_{10}^0 - {c'_{10}}^0}{c_2^0 - {c'_2}^0} + \frac{\pi}{2} \frac{c_{10} - c'_{10}}{c_2 - c'_2} \frac{c_5^0 - {c'_5}^0}{c_2^0 - {c'_2}^0} + \nu_3^0\right\} \frac{1}{(n+1/2)^{3-p}} + \nu_4^0 \frac{\ln^2(n+1/2)}{(n+1/2)^2} \\
&+ \left\{\frac{A(3\pi_2 - 4)}{12}\left[(c_2 - c_2^0) + (c'_2 - {c'_2}^0)\right] + \nu_5^0\right\} \frac{\ln(n+1/2)}{(n+1/2)^2} \\
&+\frac{\pi}{2}\left\{M_\lambda + M_\mu + \frac{1}{2}\left((c_2 + c'_2) + (c_2^0 + {c'_2}^0)\right) + (\lambda_0^0 + \mu_0^0) - (\lambda_0 + \mu_0) + 4(c_2 - c_2^0) + \right. \\
&\left(3 + \frac{2A\ln(3/2)}{3\pi}\right)\left[(c_2 - c_2^0) + (c'_2 - {c'_2}^0)\right] - \frac{\pi_2}{3}\left[(c_2 - c_2^0)^2 + (c'_2 - {c'_2}^0)^2\right] + \\
&\frac{3^{p-1}}{2^{p-1}}\left[(c_5 - c_5^0) + (c'_5 - {c'_5}^0)\right] + \frac{1}{2^{2p-2}3^{3-2p}}\left[(\gamma_1 - \gamma_1^0) + (\gamma'_1 - {\gamma'_1}^0)\right] \\
&+ \frac{\ln(3/2)}{2^{p-1}3^{2-p}}\left[(\gamma_2 - \gamma_2^0) + (\gamma'_2 - {\gamma'_2}^0)\right] + \frac{1}{2^{p-1}3^{2-p}}\left[(\gamma_3 - \gamma_3^0) + (\gamma'_3 - {\gamma'_3}^0)\right] \\
&\left. + \frac{c_{10} - c'_{10}}{c_2 - c'_2} \frac{c_{10}^0 - {c'_{10}}^0}{c_2^0 - {c'_2}^0} + \nu_6^0\right\} \frac{1}{(n+1/2)^2} + O\left(\frac{1}{n^{7-4p}}\right).\end{aligned} \tag{2.18}$$

Remark 1. The asymptotic formula (2.18) for the α_n holds from same sufficiently large n onwards. For small n the numbers α_n must be calculated directly from the formula (1.18),where the difference $H_2 - H_1$ is given by equation (2.3).

Remark 2. If we specify further terms in the asymptotic formulas (2.1) and (2.2) for the α_n and μ_n, then this method can be used to calculate further terms in the asymptotic formula for the α_n, but we shall not give these here because of the heavy computations.

3. The Inverse Sturm-Liouville Problem

In this section we consider the inverse Sturm-Liouville problem .

Theorem: Suppose that we are given two sequence of numbers $\{\alpha_n\}_{n=0}^{\infty}$ and $\{\mu_n\}_{n=0}^{\infty}$, $(n = 0, 1, \ldots)$ satisfying the following conditions:

1.α_n and μ_n are both ordered, that is $\alpha_0 < \mu_0 < \alpha_1 < \mu_1 \ldots$

2.$p \in (1, 5/4)$,

$$\begin{aligned}\lambda_n \; &= (n + 1/2)^2 + 2c_1(n + 1/2)^{p-1}\frac{A}{\pi}\ln(n + 1/2) + 2c_2 \\
&+ \frac{2c_3}{(n+1/2)^{3-2p}} + c_{1n}\end{aligned}$$

$$\begin{aligned}\mu_n \; &= (n + 1/2)^2 + 2c'_1(n + 1/2)^{p-1}\frac{A}{\pi}\ln(n + 1/2) + 2c'_2 \\
&+ \frac{2c'_3}{(n+1/2)^{3-2p}} + c'_{1n}\end{aligned}$$

asymptotic formulas are valid. Here $c_2 \neq c'_2$ and $\sum c_{1n}^2$, $\sum {c'_{1n}}^2$, are both convergent series.

Then there exist $q(x) \in C[a, b]$ and H_1, H_2 real numbers so that $\{\lambda_n\}$ is a spectrum of (1.1)-(1.2) and $\{\mu_n\}$ is a spectrum of (1.1)-(1.3) and $H_1 - H_2 = \pi(c_0' - c_0)$.

Proof: Suppose that we are given sequence of numbers $\{\lambda_n\}$ and $\{\mu_n\}$ with the above properties. We construct the sequence of α_n by means of the formula (1.18). It satisfies the asymptotic estimate (2.18). We shall show that all the $\alpha_n > 0$. The first condition of the theorem shows that λ_n and μ_n are interplaced. Therefore, for $n = 0, 1, \ldots$ $\mu_n > \lambda_n$ or $\mu_n < \lambda_n$. α_n's are given by formula (1.18) or

$$\alpha_n = \alpha_n^0 \frac{\lambda_n - \mu_n}{\lambda_n^0 - \mu_n^0} \frac{\lambda_n - \lambda_0}{\lambda_n^0 - \lambda_k^0} \frac{\lambda_n - \mu_0}{\lambda_n^0 - \mu_0^0} \prod_{k=1}^{\infty}{}' \frac{\lambda_n - \lambda_k}{\lambda_n^0 - \lambda_k^0} \prod_{k=1}^{\infty}{}' \frac{\lambda_n - \mu_k}{\lambda_n^0 - \mu_k^0}.$$

Here, we can write $\prod_{k=1}^{\infty}{}' \frac{\lambda_n - \lambda_k}{\lambda_n^0 - \lambda_k^0} > 0$, $\qquad \prod_{k=1}^{\infty}{}' \frac{\lambda_n - \mu_k}{\lambda_n^0 - \mu_k^0} > 0$.

Since $\alpha_n^0 > 0$ and λ_n, μ_n and λ_n^0, μ_n^0 are interplaced. It is obvious from α_n formula that $n = 0, 1, \ldots, \alpha_n > 0$. Since Sturm-Liouville operators can be built according to $\{\lambda_n\}$ and $\{\mu_n\}$ sequences. As a result, we have proven that for given $\{\lambda_n\}$ and $\{\mu_n\}$ sequences Sturm-Liouville operators can be built [4].

For the intervals $(5/4, 3/2)$ and $(3/2, 2)$ of p, asymptotic expressions of eigenvalues had been found and in each interval inverse problem had been solved.

References

1. M. G. Gasimov and B. M. Levitan, Matem. Sborn., **63**, 445 (1964).

2. M. G. Gasimov, Dokl. Akad. Nauk SSSR, **161**, (1965).

3. R. Kh. Amirov and S.Gulyaz, in: Proc. Eighth Int. Coll. Diff. Equations, Plovdiv, Bulgaria, D. Bainov (Ed.) VSP, Utrecht (1998) pp. 17–24.

4. Z. M. Gasimov, Young Scientists, The Cat. of Abs. of Among Univ. Conf., Sankt-Petersburg, p. 73 (1992).

5. M. G. Gasimov and R. Kh. Amirov, Dokl. Akad. Nauk Az. SSR, **41**, (1985).

9th Int. Coll. on Differential Equations, pp. 15-22
D. Bainov (Ed)

Inverse Problem for the Sturm-Liouville According to a Spectrum and Normalizing

R. KH. AMIROV AND S. GÜLYAZ
Cumhuriyet University, Dept. of Mathematics, 58140, Sivas, Turkey

Abstract: In this article, for given sequences of $\{\lambda_n\}$ and $\{\alpha_n\}$ constructing a differential operator, which has a singularity, has explained. In order to accomplish this, λ_n and α_n have to have some certain asymptotic. This kind of asymptotic formulae depend on the regularity order of the function $q(x)$. This connection can be seen in the solution of the inverse problem which depend on the sequences of $\{\lambda_n\}$ and $\{\alpha_n\}$.
Key Words and Phrases: *Inverse problem, regulary, singulary.*
1991 Mathematics Subject Classification. Primary 34A55, Secondary 34B24, 34L05.

Introduction

M.G. Gasimov and B.M.Levitan, in [1], have given on an effective method to construct according to normalizing number and a spectrum of the regular Sturm-Liouville equation. But this method given in [1], has certain conditional characteristic such that the sequences $\{\lambda_n\}$ and $\{\alpha_n\}$ are the eigenvalues and normalizing numbers of an obvious equation, respectively. Therefore, the sequences $\{\lambda_n\}$ and $\{\alpha_n\}$ should have such condition that they will be eigenvalues and normalizing numbers of an obvious equation, respectively. This problem which is mentioned above had been solved for regular equations. M.G. Gasimov in [2] has given the solution of the inverse problem which has singularity $\ell(\ell+1)/(\pi-x)^2$ (ℓ is a positive integer) and which is singular at π in the interval $[0,\pi]$ according to a spectrum and normalizing numbers. In $[0,\pi]$, the equation whose singularity is A/x (A is real) at $x=0$ has been investigated by M.G.Gasimov and R.Kh.Amirov [3]. In this paper, for the equations which have a singularity $\left(\frac{A}{x}+\frac{\ell(\ell+1)}{x^2}\right)$ (A is real) at $x=0$ on the interval $[0,\pi]$, the solution of the inverse problem is given according to a spectrum and normalizing numbers.

1. The Construction of Function $F(x,t)$

Let $q(x)$ be a real integrable function on the interval $[0,\pi]$ and A, H be real numbers and ℓ be a positive integer. We consider the boundary value problem

$$-y'' + \left\{\frac{\ell(\ell-1)}{x^2} + \frac{A}{x} + q(x)\right\} y = \lambda^2 y, \tag{1.1}$$

$$y(0) = 0 \quad , \qquad y'(\pi) - Hy(\pi) = 0. \tag{1.2}$$

We denote the solution (1.1) satisfying the initial conditions

$$\varphi(0,\lambda)=0\ ,\quad \varphi'(0,\lambda)=\lambda \tag{1.3}$$

by $\varphi(x,\lambda)$. Let $\lambda_0,\ \lambda_1,\ \ldots$ be the eigenvalues of the boundary value problem (1.1)-(1.3). Then $\varphi(x,\lambda_n)\quad (n=0,\ 1,\ldots)$ are the eigenfunctions of this boundary value problem. Let $\varphi_0(x,\lambda_n)$ be a solution of equation (1.1) when $q(x)=0$ satisfying the condition (1.3). $\lambda_0^0, \lambda_1^0, \ldots$ are eigenvalues of the boundary value problem (1.1)-(1.3) when $q(x)=0$. The numbers α_n which

$$\alpha_n=\int_0^\pi \varphi^2(x,\lambda_n)dx \qquad (n=0,\ 1,\ldots) \tag{1.4}$$

are called the normalizing constants of the boundary value problem (1.1)-(1.3).

The numbers $\alpha_n^0\ (n=0,1,\ldots)$ are called the normalizing constants of the boundary value problem (1.1)-(1.3) when $q(x)=0$.

In [3], it is well known that if $f(x)\epsilon L_2[0,\pi]$ and $g(x)\epsilon L_2[0,\pi]$ then Parseval's equation holds:

$$\int_0^\pi f(x)g(x)dx=\sum_{n=0}^{\infty}\left[\frac{1}{\alpha_n}\int_0^\pi f(x)\varphi(x,\lambda_n)dx\int_0^\pi g(t)\varphi(t,\lambda_n)dt\right]$$

We will refer to the sequences $\{\lambda_n\}$ and $\{\alpha_n\}$ as the spectral characteristics of the boundary value problem (1.1)-(1.2).

In this paper, we will solve the following problem: what conditions must the sequences $\{\lambda_n\}$ and $\{\alpha_n\}$ satisfy to be the spectral characteristics of some boundary value problem of the type (1.1)-(1.2). For solution of the problem let's constitute

$$F(x,t)=\sum_{n=1}^{\infty}\left[\frac{1}{\alpha_n}\varphi_0(x,\lambda_n)\varphi_0(t,\lambda_n)-\frac{1}{\alpha_n^0}\varphi_0(x,\lambda_n^0)\varphi_0(t,\lambda_n^0)\right] \tag{1.5}$$

with the help $\{\lambda_n\}$ and $\{\alpha_n\}$ sequences and with the help the function $F(x,t)$ let's write the integral equation which is

$$F(x,t)+K(x,t)+\int_0^x K(x,\xi)F(\xi,t)d\xi=0 \tag{1.6}$$

Here function $K(x,t)$ is not certain.

The constitution of an integral equation of the type (1.6) is connected with [3]. Therefore, similarly for the problem (1.1)-(1.2) we can write the integral equation (1.6) which is used for the solution of the inverse problem.

Now, we investigate the existence problem of the solution of integral equation (1.6). To see this, first let's seek properties of function $F(x,t)$. In order to learn the properties of function $F(x,t)$, we will use from the asymptotic expressions of $\varphi_0(x,\lambda_n)$ and $\varphi_0(x,\lambda_n^0)$ functions. Moreover, the asymptotic expressions of these functions are given in [4] and they satisfy for $x>0$ and value of sufficienty large of n: Since

$$\begin{aligned}\varphi_0(x,\lambda_n)\ &=\tfrac{2}{\sqrt{2\pi}}\cos((n+\ell/2)x-\tfrac{\ell\pi}{2})+A_0^0(x)\tfrac{\ln(n+\ell/2)}{n+\ell/2}+A_1^0(x)\tfrac{1}{n+\ell/2}\\ &+A_2^0(x)\tfrac{\ln^2(n+\ell/2)}{(n+\ell/2)^2}+A_3^0(x)\tfrac{\ln(n+\ell/2)}{(n+\ell/2)^2}+A_4^0(x)\tfrac{1}{(n+\ell/2)^2}+O\left(\tfrac{\ln^2 n}{n^3}\right)\end{aligned} \tag{1.7}$$

$$\begin{aligned}\lambda_n^0=\ &(n+\tfrac{\ell}{2})+\tfrac{A}{2\pi}\tfrac{\ln(n+\ell/2)}{n+\ell/2}+\tfrac{a_0^0}{n+\ell/2}+\tfrac{A^2}{8}\tfrac{\ln(n+\ell/2)}{(n+\ell/2)^2}+\tfrac{a_1^0}{(n+\ell/2)^2}+a_2^0\tfrac{\ln^2(n+\ell/2)}{(n+\ell/2)^3}\\ &+a_3^0\tfrac{\ln(n+\ell/2)}{(n+\ell/2)^3}+\tfrac{a_4^0}{(n+1/2)^3}+O\left(\tfrac{\ln^2 n}{n^4}\right)\end{aligned} \tag{1.8}$$

$$\varphi_0(x,\lambda_n^0) = \frac{2}{\sqrt{2\pi}}\cos((n+\ell/2)x-\frac{\ell\pi}{2}) + B_0(x)\frac{\ln(n+\ell/2)}{n+\ell/2} + B_1(x)\frac{1}{n+\ell/2} + B_2(x)\frac{\ln^2(n+\ell/2)}{(n+\ell/2)^2}$$

$$+B_3(x)\frac{\ln(n+\ell/2)}{(n+\ell/2)^2} + B_4(x)\frac{1}{(n+\ell/2)^2} + O\left(\frac{\ln^2 n}{n^3}\right) \tag{1.9}$$

Also,

$$\alpha_n = \frac{\pi}{2} + \frac{b_0}{n+\ell/2} + b_1\frac{\ln(n+\ell/2)}{(n+\ell/2)^2} + \frac{b_2}{(n+\ell/2)^2} + b_3\frac{\ln^2(n+\ell/2)}{(n+\ell/2)^3} + b_4\frac{\ln(n+\ell/2)}{(n+\ell/2)^3}$$
$$+\frac{b_5}{(n+\ell/2)^3} + O\left(\frac{\ln^2 n}{n^4}\right) \tag{1.10}$$

$$\alpha_n^0 = \frac{\pi}{2} + \frac{b_0^0}{n+\ell/2} + b_1^0\frac{\ln(n+\ell/2)}{(n+\ell/2)^2} + \frac{b_2^0}{(n+\ell/2)^2} + b_3^0\frac{\ln^2(n+\ell/2)}{(n+\ell/2)^3} + b_4^0\frac{\ln(n+\ell/2)}{(n+\ell/2)^3}$$
$$+\frac{b_5^0}{(n+\ell/2)^3} + O\left(\frac{\ln^2 n}{n^4}\right) \tag{1.11}$$

If we put the expressions (1.7)-(1.8)-(1.9)-(1.10) and (1.11) in (1.5), we have

$$F(x,t) = F_1(x,t)\sum_{n=0}^{\infty}\frac{\sin((n+\ell/2)(t+x))}{n+\ell/2} + F_2(x,t)\sum_{n=0}^{\infty}\frac{\sin((n+\ell/2)(t-x))}{n+\ell/2}$$
$$+F_3(x,t)\sum_{n=0}^{\infty}\frac{\ln(n+\ell/2)}{(n+\ell/2)^2}\cos((n+\ell/2)(t+x)) + F_4(x,t)\sum_{n=0}^{\infty}\frac{\ln(n+\ell/2)}{(n+\ell/2)^2}$$
$$\times\cos((n+\ell/2)(t-x))$$
$$+F_5(x,t)\sum_{n=0}^{\infty}\frac{\cos((n+\ell/2)(t+x))}{(n+\ell/2)^2} + F_6(x,t)\sum_{n=0}^{\infty}\frac{\cos((n+\ell/2)(t-x))}{(n+\ell/2)^2}$$
$$+F_7(x,t)\sum_{n=0}^{\infty}\frac{\sin((n+\ell/2)(t+x))}{(n+\ell/2)^2} + F_8(x,t)\sum_{n=0}^{\infty}\frac{\sin((n+\ell/2)(t-x))}{(n+\ell/2)^2} + O\left(\frac{\ln^2 n}{n^3}\right)$$

Here, F_1, F_2, F_3, F_4, F_5, F_6, F_7, F_8 are continuous with respect to the variables x and t in interval $[0,\pi]$. Also, since $\sum_{n=1}^{\infty}\frac{\sin nx}{n} = \frac{\pi-x}{2}$, $\sum_{n=1}^{\infty}\frac{\cos nx}{n^2} = \frac{3x^2-6\pi x+2\pi^2}{12}$, $0<x<2\pi$, $\sum_{n=1}^{\infty}\frac{\sin nx}{n^2} = x\ln(2\sin\frac{x}{2}) - \frac{1}{2}\int x\cot\frac{x}{2}dx$, $0\le x\le 2\pi$ and $\ln\Gamma(x) = \ln\sqrt{2\pi}+\sum_{n=1}^{\infty}\left(\frac{1}{2n}\cos 2n\pi x + \frac{1}{n\pi}(C+\ln 2n\pi)\sin 2n\pi x\right)$, $0<x<1$, $C-$ *Euler constant* . $F(x,t)$ have continuous derivatives with respect to variables x and t where $0<x<\pi$ and $0<t<\pi$.

2. Existence of Solution of Integral Equation with Respect to $K(x,t)$ Function

In this section, we will prove that, the solution of integral equation (1.6) exists and has a unique solution. To see this, let's see the homogeneous integral equation

$$g(t) + \int_0^x F(s,t)g(s)ds = 0 \tag{2.1}$$

has a unique solution $g\equiv 0$ for $x\le\pi$.

Let us assume that this is not so, i.e. that there exists a function $g(t)\not\equiv 0$ satisfying (2.1). Multiplying both sides of (2.1) by $g(t)$ and integrating over the interval $(0,x)$, we find that

$$\int_0^x g^2(t)dt + \int_0^x\int_0^x F(s,t)g(s)g(t)dsdt = 0$$

Above equality and the expression (1.5) for $F(x,t)$ imply that

$$\sum_{n=0}^{\infty}\frac{1}{\alpha_n}\left(\int_0^x g(t)\varphi_0(t,\lambda_n)dt\right)^2=0$$

for all n, since α_n are positive

$$\int_0^x g(t)\varphi_0(t,\lambda_n)dt=0. \tag{2.2}$$

We first show that the system of functions $\varphi_0(t,\lambda_n)$ is complete in $L_2[0,\pi]$. In order to show this operation let's take any $f(t)\in L_2[0,\pi]$ and let's show that the equality

$$\int_0^\pi f(t)\varphi_0(t,\lambda_n)dt=0$$

satisfy if and only if $f(t)\equiv 0$.

It was proved that in [3]. If the expression

$$\varphi_0(t,\lambda_n)=\sqrt{\lambda_n t}J_\nu(\lambda_n t)+\int_0^t K_0(t,s)\sqrt{\lambda_n s}J_\nu(\lambda_n s)ds$$

is replaced by $\varphi_0(t,\lambda_n)$ in the last equality then it holds:

$$\int_0^\pi f(t)\left[\sqrt{\lambda_n t}J_\nu(\lambda_n t)+\int_0^t K_0(t,s)\sqrt{\lambda_n s}J_\nu(\lambda_n s)ds\right]dt=\int_0^\pi f(t)\sqrt{\lambda_n t}J_\nu(\lambda_n t)dt$$

$$+\int_0^\pi f(t)\int_0^t K_0(t,s)\sqrt{\lambda_n s}J_\nu(\lambda_n s)dsdt=\int_0^\pi\left[f(t)+\int_t^\pi K_0(t,s)f(s)ds\right]\sqrt{\lambda_n t}J_\nu(\lambda_n t)dt=0$$

system of functions $\left\{\sqrt{\lambda_n t}J_\nu(\lambda_n t)\right\}$ are linearly independent and complete in $L_2[0,\pi]$. For this reason

$$f(t)+\int_t^\pi K_0(t,s)f(s)ds=0.$$

Here $\sup_{0\le t\le\pi}\int_0^t K_0^2(t,s)ds<\infty$ and $f(t)\in L_2[0,\pi]$ the integral equation which is taken for $f(t)$ is the type of theory of the Volterra integral equation. With the help of the type of the Volterra integral equations, $\{\varphi_0(t,\lambda_n)\}$ is linearly independent and complete in $L_2[0,\pi]$. From (2.2) $g(t)\equiv 0$ and so it is proved that the solution of integral equation (1.6) exists and has a unique solution.

Clearly as it is seen (1.6) order of derivatives of $K(x,t)$ function with respect to t is same as the function $F(x,t)$. With respect to variable x, investigation of $K(x,t)$ is done by using the following Lemma.

Lemma: Suppose that we are given the integral equation

$$g(t,a)=h(t,a)+\int_0^t H(t,s,a)h(s,a)ds$$

in which the kernel $H(t,s,a)$ and the inhomogeneous term $g(t,a)$ are continuous functions of the parameter a and the independent variable t. If for $a=a_0$ the homogeneous equation has only the trivial solution then in of t and a. If H and g have m continuous derivatives with respect to a, then so has $h(t,a)$.

The proof of the Lemma is the same as the proof of the Lemma [1].

As the reason of this Lemma, since $F(x,t)$ is a continuous function, it follows from Lemma that the function $K(x,t)$ is a continuous function of both variables. Further, it follows from the same Lemma that order of derivatives of $K(x,t)$ function with respect to x is the same as the function $F(x,t)$.

3. Determination of Boundary Conditions of Differential Equation

In this section, we will determine that the boundary conditions of differential equation which $\{\varphi(x,\lambda_n)\}$ functions satisfy using properties of $F(x,t)$ function that is established helping $\{\lambda_n\}$ and $\{\alpha_n\}$ sequences. Suppose that $F(x,t)$ has continuous second derivatives, according to x and t variables, then the function $K(x,t)$ has continuous second derivatives, according to x and t variables from the Lemma.

Lemma: The function $F(x,t)$ which is given (1.5) satisfies the following differential equation

$$-F_{xx}(x,t)+\left[\frac{A}{x}+\frac{\ell(\ell-1)}{x^2}\right]F(x,t)=-F_{tt}(x,t)+\left[\frac{A}{t}+\frac{\ell(\ell-1)}{t^2}\right]F(x,t) \qquad (3.1)$$

The proof of the Lemma is the same as the proof of the Lemma in [3].

Theorem 3.1: The function $K(x,t)$ which is the solution of the integral equation (1.6) satisfies the following partial derivation differential equation

$$K_{xx}(x,t)-\left[\frac{A}{x}+\frac{\ell(\ell-1)}{x^2}\right]K(x,t)-2\frac{dK(x,x)}{dx}K(x,t)=K_{tt}(x,t)-\left[\frac{A}{t}+\frac{\ell(\ell-1)}{t^2}\right]K(x,t) \qquad (3.2)$$

Proof: If we double differentiate

$$F(x,t)+\int_0^x K(x,\xi)F(\xi,t)d\xi+K(x,t)=0 \qquad (3.3)$$

with respect to variable x, then we obtain

$$\begin{aligned} F_{xx}(x,t)+\ & K_{xx}(x,t)+\tfrac{dK(x,x)}{dx}F(x,t)+K(x,x)F_x(x,t)\\ & +K_x(x,t)|_{t=x}\,F(x,t)+\textstyle\int_0^x K_{xx}(x,\xi)F(\xi,t)d\xi=0 \end{aligned} \qquad (3.4)$$

If we double differentiate (3.3) with respect to variable t, we have

$$F_{tt}(x,t)+K_{tt}(x,t)+\int_0^x K(x,\xi)F_{tt}(\xi,t)d\xi=0.$$

Also if we benefit the equation (3.1) then we obtain

$$F_{tt}(x,t)+\int_0^x\left[\frac{A}{t}+\frac{\ell(\ell-1)}{t^2}\right.$$

$$\left.-\frac{A}{\xi}-\frac{\ell(\ell-1)}{\xi^2}\right]K(x,\xi)F(\xi,t)d\xi+\int_0^x K(x,\xi)F_{\xi\xi}(\xi,t)d\xi+K_{tt}(x,t)=0$$

If we benefit from the partial integration in the last equality, we have

$$\begin{aligned} & F_{tt}(x,t)+\textstyle\int_0^x\left[\frac{A}{t}+\frac{\ell(\ell-1)}{t^2}-\frac{A}{\xi}-\frac{\ell(\ell-1)}{\xi^2}\right]K(x,\xi)F(\xi,t)d\xi\\ & +K(x,\xi)F_\xi(t,\xi)|_{\xi=0}^{x}-K_\xi(x,\xi)F(\xi,t)|_{\xi=0}^{x}+\textstyle\int_0^x K_{\xi\xi}(x,\xi)F(\xi,t)d\xi+K_{tt}(x,t)=0 \end{aligned} \qquad (3.5)$$

If we pass to the limit as $t\to 0$ in (3.1), since the second order derivations of $F(x,t)$ are bounded, we see that

$$F(x,0)=0. \qquad (3.6)$$

Moreover, from (3.3) and (3.6) we have $K(x,0) = 0$. Thus, the equality (3.5) is written by

$$F_{tt}(x,t)+ \int_0^x \left[\frac{A}{t} + \frac{\ell(\ell-1)}{t^2} - \frac{A}{\xi} - \frac{\ell(\ell-1)}{\xi^2}\right] K(x,\xi)F(\xi,t)d\xi + K(x,\xi)F_\xi(\xi,t)\mid_{\xi=x}$$
$$-K_\xi(x,\xi)F(x,t)\mid_{\xi=x} +\int_0^x K_{\xi\xi}(x,\xi)F(\xi,t)d\xi + K_{tt}(x,t) = 0 \qquad (3.7)$$

Multiplying the equality (3.3) by $\left[\frac{A}{t} + \frac{\ell(\ell-1)}{t^2} - \frac{A}{x} - \frac{\ell(\ell-1)}{x^2} - 2\frac{K(x,x)}{dx}\right]$ and if we add the equality (3.4) to the equality which obtained new and if we subtract (3.5) from the last expression by using (3.1) equality, we find that

$$K_{xx}(x,t)- \left[\frac{A}{x} + \frac{\ell(\ell-1)}{x^2}\right] K(x,t) - K_{tt}(x,t) + \left[\frac{A}{t} + \frac{\ell(\ell-1)}{t^2}\right] K(x,t) - 2\frac{dK(x,x)}{dx}K(x,t)$$
$$+\int_0^x \left\{K_{xx}(x,\xi) - \left[\frac{A}{x} + \frac{\ell(\ell-1)}{x^2}\right] K(x,\xi) - K_{\xi\xi}(x,\xi) + \left[\frac{A}{\xi} + \frac{\ell(\ell-1)}{\xi^2}\right] K(x,\xi)\right.$$
$$\left.-2\frac{dK(x,x)}{dx}K(x,\xi)\right\} F(\xi,t)d\xi = 0 \qquad (3.8)$$

Since (3.8) is the homogeneous type integral equation of Volterra, we see that

$$K_{xx}(x,t) - \left[\frac{A}{x} + \frac{\ell(\ell-1)}{x^2}\right] K(x,t)-$$
$$K_{tt}(x,t) + \left[\frac{A}{t} + \frac{\ell(\ell-1)}{t^2}\right] K(x,t) - 2\frac{dK(x,x)}{dx}K(x,t) = 0$$

or

$$K_{xx}(x,t) - \left[\frac{A}{x} + \frac{\ell(\ell-1)}{x^2} + 2\frac{dK(x,x)}{dx}\right] K(x,t) = K_{tt}(x,t) - \left[\frac{A}{t} + \frac{\ell(\ell-1)}{t^2}\right] K(x,t).$$

Now, with the help of $K(x,t)$, we establish the system of functions $\{\varphi(x,\lambda_n)\}$ in the form

$$\varphi(x,\lambda_n) = \varphi_0(t,\lambda_n) + \int_0^x K(x,t)\varphi_0(t,\lambda_n)dt \qquad (3.9)$$

Here $\{\varphi_0(x,\lambda_n)\}$ is the solution of (1.1) such that it satisfies the conditions (1.3) in the case $q(x) = 0$.

Theorem 3.2: $\varphi(x,\lambda_n)$ which is given (3.9) is the solution of (1.1) and also

$$q(x) = 2\frac{dK(x,x)}{dx}.$$

Proof: Since

$$\varphi_0''(x,\lambda_n) - \left\{\frac{A}{x} + \frac{\ell(\ell-1)}{x^2} - \lambda_n^2\right\} \varphi_0(x,\lambda_n) = 0 \qquad (3.10)$$

If $\varphi(x,\lambda_n)$ in (3.9) is written in the left side of (1.1) we obtain

$$-\varphi''(x,\lambda_n)- \left[\frac{A}{x} + \frac{\ell(\ell-1)}{x^2} + 2\frac{dK(x,x)}{dx}\right] \varphi(x,\lambda_n) + \lambda_n^2\varphi(x,\lambda_n)$$
$$= -\frac{dK(x,x)}{dx}\varphi_0(x,\lambda_n) - \left[\frac{A}{x} + \frac{\ell(\ell-1)}{x^2}\right] \int_0^x K(x,t)\varphi_0(t,\lambda_n)dt$$
$$-2\frac{dK(x,x)}{dx}\int_0^x K(x,t)\varphi_0(t,\lambda_n)dt + \lambda_n^2\int_0^x K(x,t)\varphi_0(t,\lambda_n)dt$$
$$+K(x,x)\frac{d}{dx}\varphi_0(x,\lambda_n) + K_x(x,t)\varphi_0(x,\lambda_n)|_{t=x} + \int_0^x K_{xx}(x,t)\varphi_0(t,\lambda_n)dt.$$

Using (3.2), if we replace $K_{xx}(x,t)$ to $K_{tt}(x,t)$ in the last equality, we find that

$$\varphi''(x,\lambda_n) - \left[\frac{A}{x} + \frac{\ell(\ell-1)}{x^2} + 2\frac{dK(x,x)}{dx}\right]\varphi(x,\lambda_n) + \lambda_n^2\varphi(x,\lambda_n)$$

$$= -\frac{dK(x,x)}{dx}\varphi_0(x,\lambda_n) + \lambda_n^2\int_0^x K(x,t)\varphi_0(t,\lambda_n)dt + K(x,x)\frac{d}{dx}\varphi_0(x,\lambda_n)$$

$$+K_x(x,t)\varphi_0(x,\lambda_n)|_{t=x} + \int_0^x K_{tt}(x,t)\varphi_0(t,\lambda_n)dt - \int_0^x\left[\frac{A}{t} + \frac{\ell(\ell-1)}{t^2}\right]K(x,t)\varphi_0(t,\lambda_n)dt \tag{3.11}$$

Moreover, since

$$\int_0^x K_{tt}(x,t)\varphi_0(t,\lambda_n)dt = K_t(x,t)\varphi_0(t,\lambda_n)|_{t=x} - K(x,x)\frac{d}{dx}\varphi_0(x,\lambda_n)$$
$$+\int_0^x K(x,t)\frac{d^2}{dt^2}\varphi_0(t,\lambda_n)dt$$

If this expression is placed in (3.11) and using (3.10), we have

$$-\varphi''(x,\lambda_n) + \left[\frac{A}{x} + \frac{\ell(\ell-1)}{x^2} + 2\frac{dK(x,x)}{dx}\right]\varphi(x,\lambda_n) = \lambda_n^2\varphi(x,\lambda_n)$$

for $\varphi(x,\lambda_n)$. With the help of the reasons of the studies in [3]. It is easily proved that, the family of functions $\{\varphi_0(x,\lambda_n)\}$ which is given as the (3.9) establishes a complete system in $L_2[0,\pi]$. To establish the boundary condition at the point π we observe that

$$\varphi''(x,\lambda_n) + \left[\lambda_n^2 - q(x) - \frac{A}{x} - \frac{\ell(\ell-1)}{x^2}\right]\varphi(x,\lambda_n) = 0$$

and

$$\varphi''(x,\lambda_m) + \left[\lambda_m^2 - q(x) - \frac{A}{x} - \frac{\ell(\ell-1)}{x^2}\right]\varphi(x,\lambda_m) = 0$$

If we multiply the first equation by $\varphi(x,\lambda_m)$ and the second by $\varphi(x,\lambda_n)$ and subtract one from the other, we obtain the identity

$$\left[\varphi'(x,\lambda_n)\varphi(x,\lambda_m) - \varphi'(x,\lambda_m)\varphi(x,\lambda_n)\right]' = (\lambda_m - \lambda_n)\varphi(x,\lambda_n)\varphi(x,\lambda_m)$$

Integrating this identity over the interval $(0,\pi)$ and using the orthogonality $\varphi'(\pi,\lambda_n)\varphi(\pi,\lambda_m) - \varphi'(\pi,\lambda_m)\varphi(\pi,\lambda_n) = 0$ or $\frac{\varphi'(\pi,\lambda_n)}{\varphi(\pi,\lambda_n)} = \frac{\varphi'(\pi,\lambda_m)}{\varphi(\pi,\lambda_m)}$.

Thus, the ratio $\frac{\varphi'(\pi,\lambda_n)}{\varphi(\pi,\lambda_n)}$ is a constant, and this gives the boundary condition at the point π. Thus, this completes the proof of the following theorem:

Theorem 3.3: The sequences $\{\lambda_n\}_{n=0}^{\infty}$ and $\{\alpha_n\}_{n=0}^{\infty}$ are spectral characteristics of some boundary value problem of the (1.1)-(1.2) with funtions $q''(x) \in L_2[0,\pi]$ if and only if the following asymptotic estimates hold:

$$\lambda_n = (n + \tfrac{\ell}{2}) + \frac{A}{2\pi}\frac{\ln(n+\ell/2)}{n+\ell/2} + \frac{a_0}{n+\ell/2} + \frac{A^2}{8}\frac{\ln(n+\ell/2)}{(n+\ell/2)^2} + \frac{a_1}{(n+\ell/2)^2}$$
$$+a_2\frac{\ln^2(n+\ell/2)}{(n+\ell/2)^3} + a_3\frac{\ln(n+\ell/2)}{(n+\ell/2)^3} + \frac{a_4}{(n+1/2)^3} + O\left(\frac{\ln^2 n}{n^4}\right)$$

and

$$\alpha_n = \frac{\pi}{2} + \frac{b_0}{n+\ell/2} + b_1\frac{\ln(n+\ell/2)}{(n+\ell/2)^2} + \frac{b_2}{(n+\ell/2)^2} + b_3\frac{\ln^2(n+\ell/2)}{(n+\ell/2)^3} + b_4\frac{\ln(n+\ell/2)}{(n+\ell/2)^3}$$
$$+\frac{b_5}{(n+\ell/2)^3} + O\left(\frac{\ln^2 n}{n^4}\right)$$

where $\lambda_k \neq \lambda_n$ for $k \neq n$, all the $\alpha_n > 0$.

References

1. M. G. Gasimov and B. M. Levitan, Matem. Sborn., **63**, 445 (1964).

2. M. G. Gasimov, Dokl. Akad. Nauk SSSR, **161**, (1965).

3. M. G. Gasimov and R. Kh. Amirov, Dokl. Akad. Nauk Az.SSR, **41**, (1985).

4. R. Kh. Amirov and S. Gülyaz, in: Proc. of the Eighth Int. Colloquium on Differential Equations, Plovdiv ,Bulgaria, 18-23 August, 1997, VSP, Utrecht (1998) pp. 17-24.

9th Int. Coll. on Differential Equations, pp. 23-28
D. Bainov (Ed)

On the Range and Inversion of Certain Unsymmetric Operators

A. L. BARRENECHEA AND CARLOS C. PEÑA
Universidad Nacional del Centro de la Pcia. de Bs. As. - FCExactas - Dpto. de Matemáticas - NuCoMPA - Campus Universitario - Tandil - Argentina

Abstract: In the framework of his research on certain differential equations, J. S. Lowndes [1] defined the so called non symmetric generalized integral operators. Among other applications, they were used to derive a complete set of solutions of generalized biaxially symmetric equations of Helmholtz type from the complete set of solutions of the corresponding potential equation. In this article, a pointwise inversion formula for the Lowndes - Kober type transform is obtained. Our concern in this matter arises from the consideration of a problem proposed by A. C. McBride (c.f. see [2]). To this end, we introduce a family of testing function spaces and determine the range of the L - K transformation. In particular, we may define an extended non symmetric transform which generalize the corresponding operator of Riemann - Liouville type with a Bessel-Clifford kernel.
Key Words and Phrases: *Bessel - Clifford Functions, Weyl Transform, Weyl Fract. Calculus, Neumann Series, Pochhammer Notation.*

1. Lowndes Operators and Spaces $\mathfrak{L}_\gamma$

Throughout this paper, let $\kappa > 0$ and $\varphi : (0, \infty) \to \mathbb{C}$ be a suitable smooth function. For $\alpha > 0$ and $\xi \in \mathbb{R}$, we'll consider the operators

$$K_\kappa(\xi, \alpha)\varphi(x) = (\kappa/2)^{1-\alpha} x^\xi \int_x^\infty t^{-\alpha-\xi}[\frac{t-x}{x}]^{\frac{\alpha-1}{2}} J_{\alpha-1}\{\kappa\sqrt{x(t-x)}\}\varphi(t)dt, \qquad (1)$$

$$\Theta_\kappa\varphi(x) = \int_x^\infty I_0\{\kappa(y-x)\}\ \varphi(y)dy \qquad (2)$$

$$\Psi_{\kappa,\alpha}[\varphi(x)] = (\kappa/2)^{1-\alpha} \int_x^\infty \left[\frac{t-x}{t}\right]^{(\alpha-1)/2} I_{\alpha-1}\{k\sqrt{t(t-x)}\}\varphi(t)dt, \qquad (3)$$

where $J_{\alpha-1}$ and $I_{\alpha-1}$ are the Bessel functions of the 1st and 2nd kind and order $\alpha - 1$ respectively. By means of $\{P_{n,j}\}_{0\leq j\leq n}$ we'll denote the family of differential polynomials $P_{n,j} = P_{n,j}(x, D_x)$ given as

$$P_{n,0} = (-\kappa^2 x/2)^n, \qquad P_{n,n} = D_x^n \qquad \text{and} \qquad (4)$$

$$P_{n+1,j} = -\frac{\kappa^2 x}{2} P_{n,j} + D_x P_{n,j-1} \qquad \text{if} \qquad 1 \leq j \leq n.$$

In particular, $P_{0,0} = 1$. The following propositions deal with the structure of the above polynomials:

Proposition 1 *If n and j are non negative integers such that $0 \leq j \leq n$ then $P_{n,j}$ is an homogeneous differential polynomial of degree j and $n-j$ in the variables D and $-\kappa^2 x/2$ respectively.*

Proposition 2 *If n and j are non negative integers such that $0 \leq j \leq n$, the $P_{n,j}$'s are the formal sum of all possible mononials of degree j and $n-j$ in the variables D and $-\kappa^2 x/2$ respectively (i.e. each such polynomial is exactly the sum of $\binom{n}{j}$ mononials).*

Proposition 3 *Let n, j and k be any possible non negative integers such that $k \leq j \leq n-k+1$. Then $P_{n+1,j} = \sum_{l=0}^{k} P_{k,l}\, P_{n-k+1,j-l}$.*

Proposition 4 *Let $n,\ l,\ j \in \mathbb{N}_0$ such that $0 \leq l \leq j \leq n$. Then there are uniquely real numbers $a_{n,j}^{l}$ such that $P_{n,j} = x^{n-2j} \sum_{l=0}^{j} a_{nj}^{l}\, x^l\, D^l$.*

Definition 1 *Let $\gamma \in \mathbb{R}$, $p > 0$, n a non negative integer and ϕ a complex valued function with continuous derivatives up to the order n on $(0,\infty)$. We shall write*

$$\Phi_n^{\gamma}(\phi) = \sup\left\{x^{n+j-\gamma}\ |P_{n,j}(x,D)\phi(x)| : \quad x > 0,\ 0 \leq j \leq n\right\},$$

$$\mathfrak{L}_{\gamma,p} = \left\{\phi \in C^{\infty}\left(\mathbb{R}^{+}\right) : \quad Supp(\phi) \subseteq (0,p] \text{ and } \Phi_n^{\gamma}(\phi) < \infty,\ n = 0,1,\ldots\right\}.$$

Remark 1 $\mathfrak{L}_{\gamma,p}$ has a natural linear structure. Moreover, we'll consider $\mathfrak{L}_{\gamma,p}$ as the multinormed space endowed with the topology generated by the separating family of seminorms $\{\Phi_n^{\gamma}\}_{n\geq 0}$.

Proposition 5 *$\mathfrak{L}_{\gamma,p}$ is a Fréchet space.*

Remark 2 If $q > p$, $\gamma > \delta$ then $\mathfrak{L}_{\gamma,p} \subseteq \mathfrak{L}_{\delta,q}$. Moreover, for $\phi \in \mathfrak{L}_{\gamma,p}$ and any non negative integer n we have $\Phi_n^{\delta}(\phi) \leq p^{\gamma-\delta}\ \Phi_n^{\gamma}(\phi)$ and the topology of $\mathfrak{L}_{\gamma,p}$ is stronger than the topology induced on it by $\mathfrak{L}_{\delta,q}$.

Definition 2 *Given $\gamma \in \mathbb{R}$ we will consider $\mathfrak{L}_{\gamma} = \cup_{n=1}^{\infty} \mathfrak{L}_{-\gamma+1/n,\ n}$ endowed with the structure of a countable - union space in the sense of Gelfand and Shilov [3], i.e. a sequence $\{\phi_k\}_{k\geq 1}$ of elements of $\mathfrak{L}_{\gamma}$ will converge to an element $\phi \in \mathfrak{L}_{\gamma}$ if and only if there exist $n_0 \in \mathbb{N}$ such that $\{\phi_k\}_{k\geq 1} \subseteq \mathfrak{L}_{-\gamma+1/n_0,\ n_0}$, $\phi \in \mathfrak{L}_{-\gamma+1/n_0,\ n_0}$ and $\phi_k \to \phi$ in $\mathfrak{L}_{-\gamma+1/n_0,\ n_0}$.*

Remark 3 By Remark 2, if $\beta < \gamma$ we have $\mathfrak{L}_{\beta} \subseteq \mathfrak{L}_{\gamma}$ and the topology of $\mathfrak{L}_{\beta}$ is stronger than that induced on it by $\mathfrak{L}_{\gamma}$.

Remark 4 $\mathfrak{D}(\mathbb{R}^{+}) \subseteq \mathfrak{L}_{\gamma} \subseteq \mathcal{E}(\mathbb{R}^{+})$, the topology of $\mathfrak{D}(\mathbb{R}^{+})$ is stronger than that induced on it by $\mathfrak{L}_{\gamma}$ and the topology of $\mathfrak{L}_{\gamma}$ is stronger than that induced on it by $\mathcal{E}(\mathbb{R}^{+})$, i.e. $\mathfrak{L}_{\gamma}$ becomes a testing function space.

Remark 5 The space $\mathfrak{D}(\mathbb{R}^{+})$ is not dense in $\mathfrak{L}_{\gamma}$. In fact, let b be a positive number, $\eta > -\gamma$ and $\xi \in \mathcal{E}(\mathbb{R}^{+})$ such that $\xi(x) = 1$ if $0 < x \leq b/2$ and $\xi(x) = 0$ if $x \geq b$. If $\phi(x) = x^{\eta}\ e^{x}\ \xi(x)$, $x > 0$, then $\phi \in \mathfrak{L}_{\eta,b}$. On the other hand, for any testing function $\varphi \in \mathfrak{D}(\mathbb{R}^{+})$ we have $\Phi_0^{\gamma}(\phi - \varphi) \geq 1$.

Remark 6 Multiplication by x^{λ}, $\lambda \in \mathbb{R}$, is an isomorphism of $\mathfrak{L}_{\gamma}$ onto $\mathfrak{L}_{\gamma-\lambda}$, its inverse being multiplication by $x^{-\lambda}$.

Remark 7 Differentiation D is a continuous injection of $\mathfrak{L}_\gamma$ into (but not onto) $\mathfrak{L}_{\gamma+2}$.

Remark 8 The operators (1) - (3) are defined on $\mathfrak{L}_\gamma$ because their kernels are bounded as $t \to x^+$ if $x > 0$.

Remark 9 If $0 < \sigma < \gamma$ and $\mathbf{W}^\sigma$ denotes the usual Weyl transform of order σ [4], then $\mathbf{W}^\sigma$ is a continuous injection of $\mathfrak{L}_\gamma$ into (but in general not onto) $\mathfrak{L}_{\gamma-\sigma}$.

Proposition 6 *Let $\varphi \in \mathfrak{L}_\gamma$, n, j be non negative integers such that $j \leq n+1$. Then there are uniquely real constants $b^h_{n,j}$'s, $0 \leq h \leq j \leq n$, such that*

$$P_{n+1,j}\ K_\kappa(\xi,\alpha)\varphi = \sum_{h=0}^{j} b^h_{n+1,j}\ K_\kappa(\xi+n+1-j-h, \alpha+j-h)P_{n+1,h}\varphi. \tag{5}$$

Theorem 7 *Let $\alpha \geq 1/2$, $\xi, \gamma \in \mathbb{R}$ such that $\xi+\gamma > 0$. Then $K_\kappa(\xi,\alpha)$ is a continuous linear - integral transformation of $\mathfrak{L}_\gamma$ into itself.*

Proof. Let $\varphi \in \mathfrak{L}_\gamma$, $n_0 \in \mathbb{N}$ such that $\varphi \in \mathfrak{L}_{-\gamma+1/n_0, n_0}$ and $\sigma = \gamma + \xi - 1/n_0 > 0$.

In particular, $Supp[K_\kappa(\xi,\alpha)\varphi] \subseteq (0, n_0]$. Now, for $x > 0$ and writing $\theta = \kappa x\sqrt{t-1}$ we have

$$|K_\kappa(\xi,\alpha)\varphi(x)| \leq (\kappa x/2)^{1-\alpha} \int_1^\infty t^{-\alpha-\xi}\,(t-1)^{\frac{\alpha-1}{2}}\ |J_{\alpha-1}(\theta)\ \varphi(tx)|\,dt.$$

Since $J_{\nu+m}(z)/z^\nu$ is bounded if $z > 0$, $\nu \geq -1/2$ (c.f. [5]), say by the constant $C_{\nu,m}$, we obtain $x^{\gamma-1/n_0}\ |K_\kappa(\xi,\alpha)\varphi(x)| \leq 2^{\alpha-1}C_{\alpha-1,0}\ Be(\alpha,\sigma)\ \Phi_0^{-\gamma+1/n_0}(\varphi)$, i.e. $\Phi_0^{-\gamma+1/n_0}(K_\kappa(\xi,\alpha)\varphi)$ $\leq 2^{\alpha-1}\ C_{\alpha-1,0}\ Be(\alpha,\sigma)\ \Phi_0^{-\gamma+1/n_0}(\varphi)$.

Now, let $n \geq 0$, $0 \leq h \leq j \leq n+1$, $x > 0$. From the estimate

$$|K_\kappa(\xi+n+1-j-h, \alpha+j-h)P_{n+1,h}(x,D)\varphi(x)| \leq$$

$$\leq (\kappa x/2)^{1-\alpha-j+h} \int_1^\infty \frac{(t-1)^{\frac{\alpha+j-h-1}{2}}}{t^{\alpha+\xi+n+1-2h}}\ |J_{\alpha+j-h-1}(\theta)\ P_{n+1,h}(x,D)\varphi(tx)|\,dt$$

$$\leq \frac{\kappa^{h-j}\ C_{\alpha-1,j-h}}{2^{h+1-\alpha-j}x^{j+n+1+\gamma-1/n_0}}\ Be[\alpha+\frac{j-h}{2}, 2n+2+\sigma-(h+j)/2]\Phi_{n+1}^{-\gamma+1/n_0}(\varphi)$$

and applying Proposition 6 we obtain

$$x^{\gamma-1/n_0+j+n+1}\ |P_{n+1,j}\ K_\kappa(\xi,\alpha)\varphi(x)| \leq$$

$$\leq \Phi_{n+1}^{-\gamma+1/n_0}(\varphi)\sum_{h=0}^{j} b^h_{n+1,j}\frac{\kappa^{h-j}\ C_{\alpha-1,j-h}}{2^{h+1-\alpha-j}}\ Be[\alpha+\frac{j-h}{2}, 2n+2+\sigma-(h+j)/2]$$

which yields $\Phi_{n+1}^{-\gamma+1/n_0}[K_\kappa(\xi,\alpha)\varphi(x)] \leq C.\Phi_{n+1}^{-\gamma+1/n_0}(\varphi)$,with

$$C = \max_{0\leq j\leq n+1}\sum_{h=0}^{j} b^h_{n+1,j}\frac{\kappa^{h-j}\ C_{\alpha-1,j-h}}{2^{h+1-\alpha-j}}\ Be[\alpha+\frac{j-h}{2}, 2n+2+\sigma-(h+j)/2],$$

and our assertion follows. □

2. On the Inversion of $K_\kappa(0,\alpha)$

In this section we shall be concerned with the action of Lowndes operators $K_\kappa(\xi,\alpha)$ and their behavior on certain subspaces of $\mathfrak{L}_\gamma$. For the sake of brevity, most of our assertions are given without their proofs, seeing as they are almost purely computational. Since $K_\kappa(\xi,\alpha)\ \circ\ x^\sigma = x^\sigma\ \circ\ K_\kappa(\xi-\sigma,\alpha)$ and Remark 6, we may restrict ourselves to the case $\xi = 0$.

Proposition 8 *Let* $\kappa > 0$, $0 < a < b$. *Then*[1]

$$I_1\{\kappa(b-a)\} = \int_a^b I_0\{k(b-x)\}\frac{I_1\{\kappa(x-a)\}}{x-a}dx \tag{6}$$

$$= \int_a^b I_0\{k(x-a)\}\frac{I_1\{\kappa(b-x)\}}{b-x}dx.$$

Further, if $0 < \alpha < 1$ *we have*

$$I_0\{\kappa(b-a)\} = \tag{7}$$

$$= \frac{\kappa}{2}\int_a^b \left(\frac{x-a}{a}\right)^{\frac{\alpha-1}{2}}\left(\frac{b-x}{b}\right)^{-\frac{\alpha}{2}} I_{-\alpha}\{\kappa\sqrt{b(b-x)}\}J_{\alpha-1}\{\kappa\sqrt{a(x-a)}\}dx.$$

Corollary 9 *Let* $\kappa > 0$, $0 < \alpha < 1$. *Then*

$$\Theta_\kappa = K_\kappa(0,\alpha)\ \circ\ x^\alpha\ \circ\ \Psi_{\kappa,1-\alpha}, \tag{8}$$

the equality being true on $\mathfrak{L}_\gamma$.

Remark 10 Reasoning in a heuristic manner, from the representation of the kernel of the operator Θ_κ we get $\Theta_\kappa = \mathbf{W}^1 \sum_{j=0}^\infty (1/2)_j\ (j!)^{-1}\ (\kappa^2\mathbf{W}^2)^j$, *i.e.* the Neumann generalized series representation of Θ_κ. Therefore we may realize $\Theta_\kappa = \mathbf{W}^1\ (\mathbb{I}-\kappa^2\ \mathbf{W}^2)^{-1/2}$, wherein $\mathbb{I}$ denotes the identity operator. So, it seems reasonable to write $\Theta_\kappa^{-1} = -(\mathbb{I}-\kappa^2\ \mathbf{W}^2)^{1/2}$ D_x. Precisely, considering again the corresponding Neumann generalized series, we obtain evidence of what could be the inverse of Θ_κ. Hence it seems natural to represent the inverse operator Θ_κ^{-1} as

$$\Theta_\kappa^{-1}\varphi(x) = -\varphi'(x) - \kappa\int_x^\infty \frac{I_1\{\kappa(z-x)\}}{z-x}\ \varphi(z)\ dz,\ \ x > 0. \tag{9}$$

On the other hand, the work of Bakievich [6] on non convolution operators with Bessel functions in their kernels suggests that

$$\Psi_{\kappa,1}^{-1}\varphi(x) = -\varphi'(x) - \frac{1}{x}\int_x^\infty y\frac{\partial}{\partial y}J_0\{\kappa\sqrt{x(y-x)}\}\varphi'(y)dy. \tag{10}$$

[1] *As far as the authors know, representations (6) and (7) appear to be new. In short, they rely on the uniform convergence on* $[a,b]$ *of the integrands as well as the combinatorial formula*

$$\binom{2n+1}{n} = \sum_{m=0}^{n}\frac{1}{m+1}\binom{2m}{m}\binom{2(n-m)}{n-m},\ n\in\mathbb{N}_0.$$

For, (10) follows by taking into account the formula (see the hand - book of [7], p. 743)

$$\int_a^b x\frac{\partial}{\partial x}J_0\{\kappa\sqrt{a(x-a)}\}\frac{\partial}{\partial x}I_0\{\kappa\sqrt{b(b-x)}\}dx$$

$$= b\frac{\partial}{\partial b}J_0\{\kappa\sqrt{a(b-a)}\} - a\frac{\partial}{\partial a}I_0\{\kappa\sqrt{b(b-a)}\}.$$

Proposition 10 *If* $\gamma > 1$ *then* $\Theta_\kappa(\mathfrak{L}_\gamma) = \mathbf{W}^1(\mathfrak{L}_\gamma) = \Psi_{\kappa,1}(\mathfrak{L}_\gamma)$.

Proposition 11 *The operator* θ_κ^{-1} *defined by (9) is continuous between* $\mathfrak{L}_\gamma$ *and* $\mathfrak{L}_{\gamma+2}$ *if* $\gamma > 1$.

Proposition 12 *The operator* $\Psi_{\kappa,1}$ *is continuous between the spaces* $\mathfrak{L}_\gamma$ *and* $\mathfrak{L}_{\gamma-1}$ *if* $\gamma > 1$.

Theorem 13 *If* $0 < \alpha < 1$, $0 < \zeta$, *then* $K_\kappa(0,\alpha)$ *is a 1 - 1 map between testing subspaces of* $\mathfrak{L}_\zeta$, *namely* $K^-_{\alpha,1-\alpha}[\mathfrak{L}_\zeta]$ *and* $K^-_{0,1}[\mathfrak{L}_\zeta]$[2], *with inverse* $K_k(0,\alpha)^{-1} = x^\alpha \circ \Psi_{\kappa,1-\alpha} \circ \Theta_\kappa^{-1}$, *it being an automorphism if* $1/2 \le \alpha < 1$.

Proof. By this time it's sufficient to observe that $\mathbf{W}^{-\alpha} \circ \Psi_{\kappa,1} = \Psi_{\kappa,1-\alpha}$ if $\alpha < 1$ and $K_\kappa(0,\alpha)\left[K^-_{\alpha,1-\alpha}[\mathfrak{L}_\zeta]\right] = K^-_{0,1}[\mathfrak{L}_\zeta]$. If $1/2 \le \alpha < 1$ then $K_\kappa(0,\alpha)$ is a continuous map of $\mathfrak{L}_\zeta$ into itself (Th. 7), using Prop. 10 we have

$$K^-_{0,1}[\mathfrak{L}_\zeta] = \mathbf{W}^1(\mathfrak{L}_{\zeta+1}) \quad \xrightarrow{\Theta_\kappa^{-1}\,|_{\Theta_\kappa(\mathfrak{L}_{\zeta+1})}} \quad \mathfrak{L}_{\zeta+1},$$

the above map being continuous since it is the restriction of Θ_κ^{-1} of a continuous mapping from $\mathfrak{L}_{\zeta+1}$ to $\mathfrak{L}_{\zeta+3}$ (see Remarks 3, 9 and Prop. 11). Moreover, from Propositions 10, 12 and the additivitiness of exponents of the Weyl transformation we get

$$\mathfrak{L}_{\zeta+1} \quad \xrightarrow{\Psi_{\kappa,1}} \quad \mathbf{W}^1(\mathfrak{L}_{\zeta+1}) \quad \xrightarrow{x^\alpha\,\mathbf{W}^{-\alpha}} \quad K^-_{\alpha,1-\alpha}[\mathfrak{L}_\zeta],$$

each of the above transforms is continuous and the result follows. □

Definition 3 Our research allow us, by applying the usual Parseval method, to define an unsymmetrical generalized calculus for the formal operator[3]

$$\bar{J}^+_{\alpha,\kappa}\varphi(x) = \int_0^x \frac{(x-t)^{\alpha-1}}{\Gamma(\alpha)}\bar{J}_{\alpha-1}\{\kappa\sqrt{t(x-t)}\}\varphi(t)dt.$$

Actually, with the above notation we may write

$$\left\langle\varphi(x), x^{-\alpha}\bar{J}^+_{\alpha,\kappa}f(x)\right\rangle = \left\langle K_\kappa(0,\alpha)\varphi(x), f(x)\right\rangle, \tag{11}$$

where $\varphi \in \overline{K^-_{\alpha,1-\alpha}[\mathfrak{L}_\zeta]}$ and f is a generalized functional on $\overline{K^-_{0,1}[\mathfrak{L}_\zeta]}$.

Example 1 Let $\delta(x-x_0)$ denote the delta functional concentrated at x_0, $\varphi \in K^-_{\alpha,1-\alpha}[\mathfrak{L}_\zeta]$. Then

$$\left\langle\varphi(x), x^{-\alpha}\bar{J}^+_{\alpha,\kappa}\delta(x-x_0)\right\rangle = K_\kappa(0,\alpha)\varphi(x_0).$$

More generally, from Proposition 6 we have

$$\left\langle\varphi(x), x^{-\alpha}\bar{J}^+_{\alpha,\kappa}\delta^{(n)}(x-x_0)\right\rangle = (-1)^n\sum_{j=0}^{n} K_\kappa(-j,\alpha+n-j)P_{n,j}\varphi(a).$$

[2] *As usual, we denote Kober operators as* $K^-_{\eta,\alpha} = x^\eta \circ \mathbf{W}^\alpha \circ x^{-\alpha-\eta}$.

[3] *The notation is considered from A. A. Kilbas, O. I. Marichev & S. G. Samko (see [7]).*

Example 2 Let $1 < p < \infty$, $0 < \zeta < (p-1)/p$, $f \in \mathbb{L}^p(\mathbb{R}^+)$, $\varphi \in \overline{K^-_{\alpha,1-\alpha}[\mathfrak{L}_\zeta]}$. Then f defines a regular generalized functional on $\mathfrak{L}_\zeta$. Equation (11) may serve as a definition of the unsymmetrical integral $\bar{J}^+_{\alpha,\kappa}f(x)$ for f in the case $p \leq 1/\alpha$, though this integral diverges in general in this case if understood in usual sense. This occurrence is an advantage of the test functions space $\mathfrak{L}_\zeta$. In particular, if $1/\alpha < p$ then (11) realizes $\bar{J}^+_{\alpha,\kappa}f(x)$ as the regular functional on $\overline{K^-_{\alpha,1-\alpha}[\mathfrak{L}_\zeta]}$ induced by itself as a current function.

Remark 11 To finish with we posed some questions that arise from our research:
(i) If $\gamma > 1$, is $\Psi^{-1}_{\kappa,1}$ a continuous map from $\mathbf{W}^1(\mathfrak{L}_\gamma)$ into $\mathfrak{L}_\gamma$?
(ii) It should be desiderable to decide if $\mathbf{W}^1(\mathfrak{L}_\gamma)$ is meager in $\mathfrak{L}_{\gamma-1}$ or if $\mathbf{W}^1(\mathfrak{L}_\gamma) = \mathfrak{L}_{\gamma-1}$[4].
In particular, if $\mathbf{W}^1(\mathfrak{L}_\gamma)$ were closed then $\mathbf{W}^1$ would be onto, i.e. an homeomorphism.
If (i) were true then Th. 14 would hold for $0 < \alpha < 1$, while that (ii) could allowed us to drop the bars in Definition 3.

References

1. J. S. Lowndes, in: Fract. Calculus, A.C. McBride & G. F. Roach (Eds.), Boston, Pitman Adv. Publ. Program, Res. Notes Math., **138**, 87 - 98 (1985).

2. A. C. McBride, in: Fract. Calculus, A. C. McBride & G. F. Roach (Eds.), Boston, Pitman Adv. Publ. Program, Res. Notes Math., **138**, 214 (1985).

3. I. M. Gelfand and G. E. Shilov, Generalized Functions, Vol. I, Acad. Press, N. Y. (1964).

4. Kenneth S. Miller, in: Lect. Notes in Math., **457**, 80 - 90 (1974).

5. F. Emde, E. Jahnke and F. Losch, Tables of Higher Functions, McGraw - Hill, N. Y. (1960).

6. M. I. Bakievich, Volzh. Mat. Sb., Kuibyshev: Ped. Inst., **1**, 42 - 52 (1963).

7. A. A. Kilbas, O. I. Marichev an S. G. Samko, Fract. Integrals and Derivatives, Gordon and Breach Sc. Publ., Amsterdam (1993).

8. J. Dieudonné, Treatise on Analysis, Vol. II, Acad. Press, N. Y., S. F., London (1976).

[4] *One of the two possibilities hold by the Banach's theorem of continuous functions between Fréchet spaces (c.f. [8], page 88, (12.16.8)).*

9th Int. Coll. on Differential Equations, pp. 29–34
D. Bainov (Ed)

Impulsive Hamilton-Jacobi Equations for Hybrid Control Systems

S. A. BELBAS
Mathematics Department, University of Alabama, Tuscaloosa, AL 35487-0350 USA

Abstract: We present the derivation and verification of a dynamic programming approach to the optimal control of a type of hybrid systems, where the state evolves in continuous time with possible impulses at discrete time instant, and control decisions are also made at discrete time instants.
Key Words and Phrases: *Hybrid control systems, impulsive hyperbolic partial differential equations.*

1. Introduction

This paper deals with nonlinear continuous/discrete systems, i.e., systems in which the state evolves in continuous time but control decisions are made at discrete time instants. Such systems have received significant attention in recent years, and we mention [1,2,3,4]. Other worths on related hybrid systems, but in context rather different from the present paper, are [5,6,7]. Also, the volume [8] contains several papers on hybrid and multirate systems.

One of the most important types of applications that give rise to continuous/discrete systems is the use of digital computers in the control loops of continuously evolving dynamical systems. As pointed out in [4], these types of controlled systems cannot be satisfactorily analyzed by the methods of either continuous or discrete controlled systems theory. If the methods of continuous-time control theory are used, and a continuous control is synthesized, then a pointwise sampling of that control at the appropriate discrete time-instants would produce effects quite different from those of the original continuous-time control. On the other hand, if the original equations of continuous-time dynamics are discretized at the discrete time-instants of control decision, then the effect of the evolution of the state between the time instants of control decisions will be lost, and various examples have shown that that effect can be significant; such examples may be found in the afore-mentioned references.

There are also continuous systems where control decisions must be made at discrete time-instants, for example in economics or in military situations. In such cases, the discrete nature of the control actions in inherent in the system itself.

The controlled systems covered in this paper also include a simple particular case of systems with impulse effect; this means that the value of the state y may have a jump at certain discrete time-instants, i.e.,

$$\lim_{t\to t_k^+} y(t) - \lim_{t\to t_k^-} y(t) = a_k \tag{1}$$

The theory of general differential equations with impulse effects has been developed by D.D. Bainov and his school, and by other researchers; cf. [9].

Another feature of the models we consider is the inclusion of discrete costs for changing the value of the control at a particular time instant. This means that our model includes the possibility of impulses in the cost structure, i.e., costs associated with the division to change the value of the control at a given time. This is relevant in, among others, military sitations where the decision of deployment of new forces or reconfiguring existing formations may be costly.

Our method of solution relies on a modification of the dynamic programming method and a reformulation of the original problem in "nested" form. The effect of our approach of modified dynamic programming is to split the optimal control problems into two interconnected parts: the discrete-time equations of modified dynamic programming, and the continuous-time problems that have to be solved over the time-intervals between control decisions to yield terms that have to be used in the dynamic programming equations.

2. Statement of the Problem

We consider a system with a combination of discrete and continuous dynamics. The system evolves over a finite time-horizon $[0,T]$, and discontinuities in the state may arise at prescribed times $\tau_k, 0 \le k \le N$, with $\tau_0 \equiv 0, \tau_N \equiv T$. We set, for a function y which is continuous on $\bigcup_{k=1}^{N}(\tau_{k-1},\tau_k)$,

$$y(\tau_k^-) = \lim_{t\to\tau_k^-} y(t), y(\tau_k^+) \equiv \lim_{t\to\tau_k^+} y(t) \tag{2}$$

provided the indicated limits exist as finite real numbers.

Now, we describe the controlled dynamical system with continuous and discrete components. For $t \in (\tau_{k-1},\tau_k)(1 \le k \le N)$, the system evolves according to the differential equation

$$\dot{y}(t) = f(t, y(t), u(t)) \tag{3}$$

where y is the state of the system, taking values in $\mathbb{R}^m$, and u is a control function taking values in $\mathbb{R}^m$. We start with initial conditions

$$y(\tau_0^+) = y_0 = \text{given} \tag{4}$$

and, at each time

$$\tau_k^+ = y(\tau_k^-) + a_k \tag{5}$$

where each a_k is a vector in $\mathbb{R}^m$. We assume conditions of solvability and well-posedness of (3) on each interval $(\tau_{k-1},\tau_k^-]$ for each $y(\tau_{k-1}^+)$ so that $y(\tau_{k-1}^-)$ and thus, also, from (5), $y(\tau_k^+)$ is well-defined. Further, we postulate that, on each interval (τ_{k-1},τ_k), the control u can be expressed in terms of a set of basis functions $\{\varphi_{kj}kj(t) : 1 \le j \le Mk\}$ with weights W_{ijk}, so that

$$u_i(t) = \sum_{j=1}^{M_k} W_{kij}\{\varphi_{kj}(t)\}, \text{for } \tau_{k-1} < t < \tau_k \le i \le m \tag{6}$$

It is useful to note by w_k the matrix $[W_{kij}]_{\substack{1\le i\le m\\ 1\le i\le m_k}}$, and then we may view $\{\overline{w}_k; 1 \le k \le N\}$ as a representation of a control function $u(\cdot)$. This means that the control is adjusted for each interval (τ_{k-1},τ_k) through the weights W_{kij} and, once this adjustment has been made, the evolution of the control $u(t)$ for $(\tau_{k-1} < t < \tau_k)$ is given by (6).

We shall deal with the optimal control problem of minimizing a cost functional J given by

$$J = \int_0^1 F(t, y(t), u(t))dt + \sum_{k=1}^{n-1} \Phi(\tau_k, y(\tau_k^+), \overline{w}_k, \overline{w}_{k+1}) + F_0(y(T)); T \equiv \tau_N \tag{7}$$

We can solve the problem of minimizing (7) subject to (2)-(6) by first reducing the system dynamics and the performance criterion to a nested form, in which a discrete system forms the "outside" part of the problem, and, within each time interval (τ_{k-1}, τ_k) a continuous problem furnishes the "inside" information.

3. Reduction to a Nested Problem

Given an admissible control function $u(\cdot)$, we denote by $u|_{(s,t)}$ the restriction of u to the open interval $(s,t)(s,t \in \mathbb{R}, s < t)$. For each natural number k, with $k \leq N$, for each $\zeta \in \mathbb{R}^m$ and for each interval $(s,t) \in (\tau k-1, \tau k)$, we denote by $S(s,t;\zeta, u|_{(s,t)})$ the solution of the system (3) evaluated at t, i.e., if $y_1(t_1)$ solves

$$\frac{d}{dt}y_1 = f(t_1, y_1(t_1), u|_{(s,t)}), \text{ for } s < t_i < t; y_1(s^+) = \zeta \tag{8}$$

then

$$S(s,t,\zeta,u|_{(s,t)}) \equiv \lim_{t_1 \to t^-} y_1(t_1) \tag{9}$$

When $u(\cdot)$ is given by (6), then, for $(s,t) \in (\tau_{k-1}, \tau k)$, we write

$$\overline{S}(s,t,\zeta,\overline{w}_k) \equiv S(s,t,\zeta,u|_{(s,g)}); (u|_{(s,t)}(t)) \equiv \sum_{j=1}^{M_k} W_{kij}\varphi_{kj}(t) \tag{10}$$

In particular, we have

$$y(\tau_{k+1}) = \overline{S}(\tau_k, \tau_{k+1} y(\tau_k^+), \overline{w}_{k+1}) \tag{11}$$

Next, we get

$$\begin{aligned} Z_k &\equiv y(\tau_k^+); \\ \Psi(k, z_k, \overline{w}_{k+1}) &\equiv a_{k+1} + S(\tau_k, \tau_{k+1}, z_k, \overline{w}_{k+1}) \end{aligned} \tag{12}$$

Then, the evolution of $(z_k : 1 \leq k \leq N)$ is given by

$$Z_k = \Psi(k, z_k^+), \overline{w}_{k+1} \tag{13}$$

Next, we write the functional J as

$$J = \sum_{k=1}^{N} \left\{ \int_{t_{k-1}}^{t_k} F(t, y(t), u(t))dt \right\} + \sum_{k=1}^{N-1} \Phi(y(\tau_k^+), \overline{w}_k, \overline{w}_{k+1}) + F_0(y(T)) \tag{14}$$

We set

$$\Omega(k, z_k, \overline{w}_k, \overline{w}_{k+1}) \equiv \Phi(\tau_k, z_k, \overline{w}_k, \overline{w}_{k+1}) + \int_{\tau_k}^{\tau_{k+1}} F(t, S(\tau_k, t, z_k, \overline{w}_{k+1}),$$

$$\sum_{j=1}^{M_{k+1}} W_{k+1,ij}\varphi_{k+1j}(t))dt, \qquad \text{for } k = 1, 2, \ldots, N-2;$$

$$\begin{aligned} \overline{\Omega}(0, z_0, \overline{w}_1) &\equiv \int_0^{\tau_1} F(t, S(\tau_1, t, z_0, \overline{w}_1), \\ \sum_{j=1}^{M_1} W_{1,ij}\varphi_{ij}(t))dt; \overline{\Omega}_0(z) &\equiv F_0(z) \end{aligned} \tag{15}$$

With this notation, we can write J in a form that is convenient for applying the method of dynamic programming:

$$J = \overline{\Omega}(0, \overline{\Omega}_0, \overline{w}_1) + \sum_{k=1}^{N-1} \Omega(k, z_k, \overline{w}_k, \overline{w}_{k+1}) + \overline{\Omega}_0(z_n) \tag{16}$$

The presence of both w_k and w_{k+1} in the Ω term makes the problem under consideration different from the standard type of problems to which dynamic programming may be applied.

4. Application of the Method of Dynamic Programming

The standard method of dynamic programming involves a parametrization of the cost functional by initial state and initial time (the so-called "cost-to-go"). In the case considered in this paper, it is necessary to utilize a different parametrization, that also includes the value of $\overline{w}_k$ as one of the parameters. The reasons for this type of parametrization will become apparent in the process of deriving the dynamic programming equations.

We define the parametrized cost functional

$$\begin{aligned} K(1, \zeta, a) & \qquad (0 \leq 1 \leq N) \\ K(1, \zeta, a) &\equiv \sum_{k=1}^{N-1} \Omega(k, \tilde{z}_k, \overline{w}_k, \overline{w}_{k+1}) + \Omega_0(\tilde{z}_N), \text{ for } 1 \leq \zeta \leq N-1; \\ K(0, \zeta, a) &\equiv \overline{\Omega}(0, \zeta, a) + \sum_{k=1}^{N} \Omega(k, \tilde{z}_k, \overline{w}_k, \overline{w}_{k+1}) + \Omega_0(\tilde{z}_n) \end{aligned} \tag{17}$$

where

$$\overline{w}_1 = a, \tilde{z}_1 = \zeta, \tilde{z}_{k+1} = \Psi(k, \tilde{z}_k, \overline{w}_{k+1}) \tag{18}$$

Also, we define

$$K(N, \zeta, a) \equiv \Omega_0(\zeta) \tag{19}$$

The value function (Hamilton-Jacobi function, or optimal cost-to-go) $V(l, \zeta, a)$ is then defined as

$$V(l, \zeta, a) \equiv \inf_{\overline{w}(\cdot)} \{K(l, \zeta, a) | (18)\} \tag{20}$$

where the symbol "(18)" in the right-hand side of (20) signifies the requirement that the conditions (18) are used as constraints for the calculation of $K(l, \zeta, a)$ and for the corresponding infimum.

We may argue in the following heuristic way to find the equations of dynamic programming for V : if β stands for the value of an arbitrary control policy at time l (i.e., if $\overline{w}_{l+1} = \beta$), then the optimal cost, conditioned on $\overline{w}_{l+1} = \beta$ will be $\Omega(l, \zeta, a, \beta)$ for $(l \leq 1)$ or $\overline{\Omega}(0, \zeta, a) + V(l, \Psi(0, \zeta, \beta), \beta)$ (for l=0).

In order to obtain $V(l, \zeta, a)$ from the above expressions, we take infimum over β, and thus we obtain

$$\begin{aligned} V(l, \zeta, a) &\equiv \inf_{\beta} \{\Omega(l, \zeta, a, \beta) + \\ &+ V(l+1, \Psi(l, \zeta, \beta), \beta)\}, \text{ for } 1 \leq l \leq N-1; \\ V(0, \zeta, a) &= \Omega(0, \zeta, a) + \inf_{\beta}(l, \Psi(0, \zeta, \beta), \beta); V(N, \zeta, a) = \Omega_0(\zeta) \end{aligned} \tag{21}$$

The Proofs of rigorous results on the dynamic programming equation (21) are given in the next section.

It should be remarked that (21) differs from standard forms of dynamic programming equations in two respects: the inclusion of the parameter a in the value function V, and the fact that the term Ψ is not explicitly given but must be calculated by solving a differential equation. For these reasons, we can refer to (21) as "the equations of modified dynamic programming".

5. Verification Results

We assume that the functions $f, \varphi_{kj}, F, \varphi$, and F_0 are bounded and continuous in all their arguments, and that f is uniformly Lipschitz in its second argument.

We have:

Theorem 5.1. If the function $V(l,\zeta,a)$ solves the system of dynamic programming equations (21), then $V(l,\zeta,a)$ equals the infimum of $K(l,\zeta,a)$ over all admissible controls, i.e., (21) implies (20).

Proof. We use backward induction and so we assume that the claim is true for $V(l+1,\eta,\beta)$, for all η and β, and we shall show that it must be true for $V(l,\zeta,a)$.

For simplicity, we shall carry out the proof for $1 \le l \le N-1$, since the changes needed for $l=0$ are merely notational modifications.

Taking $\eta = \Psi(l,\zeta,\beta)$ in the induction hypothesis, we have

$$V(l+1,\Psi(l,\zeta,\beta),\beta) = \inf\{K(l+1),\Psi(l,\zeta,\beta),\beta)|(18)\} \tag{22}$$

where $\hat{w}_l(\cdot)$ is a general admissible control conditioned on the value β at time $l+1$.

Now, it follows from the definitions of K that

$$K(l,\zeta,a) = \Omega(l,\zeta,a,\beta) + K(l+1,\Psi(l,\zeta,\beta),\beta) \tag{23}$$

and, therefore, (22) implies

$$\begin{aligned} V(l+1,\Psi(l,\zeta,\beta),\beta) \quad + \quad \Omega(l,\zeta,a,\beta) &= \inf_{\overline{w}(\cdot)}\{K(l,\zeta,a)|(18)\} = \\ &= \inf_{\beta}\{\inf_{\overline{w}_l(\cdot)}\{K(l,\zeta,a)|(18)\} \end{aligned} \tag{24}$$

i.e., in view of (24) and (20),

$$V(l,\zeta,a) = \inf_{\overline{w}(\cdot)}\{K(l,\zeta,a)|(18)\} \tag{25}$$

The claim is also plainly true for $1 = N$, and the backward induction is complete.

QED

Theorem 5.2. If there exists a control policy $\overline{w}_k^*, K = 0,1,\dots,N$, that realizes the infima in the dynamic programming system (19), then that policy is optimal.

Proof. By assumption, and also using (19), we must have

$$\begin{aligned} K(0,\zeta^*,\overline{w}_0^*) &= \overline{\Omega}(0,\zeta^*,\overline{w}_0^*) + \sum_{K=1}^{N} \Omega(k,\tilde{z}_k^*,\overline{w}_k^*,\overline{w}_{k+1}^*) + \Omega_0(\tilde{z})_N^* = \\ &= \overline{\Omega}(0,\zeta^*,\overline{w}_0^*) + K(l,\tilde{z}_1^*,\overline{w}_1^*,\overline{w}_2^*) \end{aligned} \tag{26}$$

where $\tilde{z}_k^*$ is a trajectory corresponding to the control policy $\overline{w}_k^*$, and $\zeta^* \equiv \tilde{z}_0$, and inductivity

$$K(l,\zeta^*,\overline{w}_0^*) = \Omega(l,\tilde{z}_1^*,\overline{w}_1^*,\overline{w}_1^*) + K(l+1,\tilde{z}l+1^*,\overline{w}_{l+1}^*) \tag{27}$$

i.e.,

$$V(l, \tilde{z}l^*, \overline{w}_1^*) = K(l, \tilde{z}l^*, \overline{w}_l^*) \tag{28}$$

Since (28) coincides with (18), then theorem 5.1 implies

$$K(0, \zeta^*, \overline{w}_0^*) \leq K(0, \zeta^*, \overline{w}_0^*) \tag{29}$$

if any admissible control policy is utilized in the evaluation of $K(0, \zeta^*, \overline{w}_0^*)$.

QED

6. Concluding Remarks

We have reduced the problem described in Section 2 into main components: the continuous component is the calculation of the shift operator S of Section 3 (or, equivalently, the operator Ψ); the discrete component is the equation of modified dynamic programming of Section 4. The two components are coupled through the inclusion of Ψ in the equations of modified dynamic programming. The splitting means, among other things, that the equations needed for evaluating Ψ may be solved to a desired degree of accuracy independently of the discrete-time dynamic programming equations; the system of modified dynamic programming equations may be solved numerically by various methods, which methods are independent of any method used for finding Ψ. These considerations in turn imply that the approximate evaluation of Ψ and the approximate solution of the modified dynamic programming equations are tasks that can be carried out in parallel. The analysis of numerical methods for the solution of these equations, and various parallelization schemes, will be reported elsewhere.

References

1. M. Abdel-Haleem and C.D. Johnson, in: 28th IEEE SE Symp. System Th., IEEE Computer Soc. Press (1996) pp. 184-188.
2. M. Abdel-Haleem and C.D. Johnson, in: 29th IEEE SE Symp. System Th., IEEE Computer Soc. Press (1997) pp. 2-6.
3. A. Kh. Gelig and A.N. Churilov, Oscillations and Stability of Nonlinear Impulsive Systems, (in Russian) Editions of St. Petersburg University (1993).
4. E.N. Rozenvasser, Linear Theory of Digital Control in Continuous Time, (in Russian) Nauka, Moscow (1994).
5. C.T. Leondes (Ed.), Hybrid Navigation Systems, NATO Advisory Group for Aerospace R. and D., Neuilly-sur-Seine (1969).
6. X.R. Li and Y. Bar-Shalom, in: Control and Dynamic Systems, C.T. Leondes (Ed.), Acad. Press, San Diego (1995).
7. A. Villa, Hybrid Control Systems in Manufacturing, Gordon and Breach, NY (1991).
8. C.T. Leondes (Ed.), Control and Dynamic Systems, Advances in Theory and Applications, Vol. 76, Acad. Press, NY (1996).
9. D. Bainov, Z. Kamont and E. Minchev, Approximate Solutions of Impulsive Hyperbolic Equations, Acad. Publishers, Calcutta (1996).

9th Int. Coll. on Differential Equations, pp. 35-38
D. Bainov (Ed)

Numerical Solution of Multi-Time Dynamic Programming Equations

S. A. BELBAS
Mathematics Department, University of Alabama, Tuscaloosa, AL 35487-0350 USA

Abstract: We obtain the equations of multi-time dynamic programming for finite-difference approximations to optimal control problems for controlled ordinary differential equations with a cost functional that contains multiple integrals. The finite approximation to the multi-time dynamic programming equations furnishes a recursive scheme for solving the associated optimal control problem.
Key Words and Phrases: *Multi-time dynamic programming, finite-difference approximation.*

1. Introduction

The present paper is concerned with discrete approximations to problems of optimal control that involve multiple integrals. For the purposes of the present paper, we shall concentrate on functionals with double integrals, of the form

$$\begin{aligned} J_2 : &= F_0(y(T)) + \int_0^T F_1(t, y(t), u(t))dt + \\ &+ \int_0^T \int_0^T F_2(s, t, y(s), y(t), u(s), u(t))dsdt \end{aligned} \tag{1}$$

where $u(\cdot)$ is the control function, and the state $y(\cdot)$ satisfies

$$\frac{dy(t)}{dt} = f(t, y(t), u(t)), \text{ for } 0 < t < T; y(0^+) = y_0 \tag{2}$$

By contrast, the standard theory of optimal control deals with a functional

$$J_1 := F_0(y(T)) + \int_0^T F_1(t, y(t), u(t))dt \tag{3}$$

There are important qualitative differences between the functionals (1) and (3), one of which is that there is no direct way to extend the method of dynamic programming to the functional (1), nevertheless, certain novel versions of the basic ideas of dynamic programming can be applied to (1).

The problem (1)-(2) can be extended in various ways. The issue is not merely the extension for its own sake, but rather the extensions can illuminate the basic problem (1)-(2). One extension utilizes two controlled ordinary differential equations

$$\begin{aligned} \frac{dy_1(s)}{ds} &= f_1(s, y_1(s), u_1(s)), y_1(0^+) = y_{10}; \\ \frac{dy_2(t)}{dt} &= f_2(t, y_2(t), u_2(t)), y_2(0^+) = y_{20} \end{aligned} \tag{4}$$

and the optimal control problem is the minimization of a functional

$$\begin{aligned} J_{21} : &= F_0(y_1(A), y_2(B)) + \\ &+ \int_0^A F_{11}(s, y_1(s), u_1(s))ds + \int_0^B F_{12}(t, y_2(t), u_2(t))dt + \\ &+ \int_0^A \int_0^B F_2(s, t, y_1(s), y_2(t), u_1(s), u_2(t))dtds \end{aligned} \tag{5}$$

A second extension concerns a system of nonlinear Roesser type:

$$\begin{aligned} \frac{\partial y_1(s,t)}{\partial s} &= f_1(s, t, y_1(s,t), y_2(s,t), u(s,t)); \\ \frac{\partial y_2(s,t)}{\partial t} &= f_2(s, t, y_1(s,t), y_2(s,t), u(s,t)), \\ y_1(0^+, t) &= y_{10}(t); y_2(s, 0^+) = y_{20}(s) \end{aligned} \tag{6}$$

and the functional to be minimized is

$$\begin{aligned} J_{22} : &= F_0(y_1(A,B), y_2(A,B)) + \\ &+ \int_0^A F_{11}(s, y_1(s,B), y_2(s,B), u(s,B))ds + \\ &+ \int_0^B F_{12}(t, y_1(A,t), y_2(A,t), u(A,t))dt + \\ &= \int_0^A \int_0^B F_2(s, t, y_1), (s,t), y_2(s,t), u(s,t))dtds \end{aligned} \tag{7}$$

An important variant of the problem (6)-(7) consists of what we shall call in this paper a **separable Roesser system**, i.e., a system of the form

$$\begin{aligned} \frac{\partial y_1}{\partial s} &= f_1(s, t, y_1(s,t), y(s,t)); \\ \frac{\partial y_2(s,t)}{\partial t} &= f_2(s, t, y_2(s,t), u(s,t)); \\ y_1(0^+, t) &= y_{10}(t); y_2(s, 0^+) = y_{20}(s) \end{aligned} \tag{8}$$

where the functional to be minimized is again given by (7).

There is an important alternative to the functional J_{22} for Roesser-type problems:

$$\begin{aligned} J_{23} : &= F_0(y_1(A,B), y_2(A,B)) + \\ &= \int_0^A F_{11}(s, y_1(s,0), y_2(s,0), u(s,0))ds + \\ &= \int_0^B F_{12}(t, y_1(0,t), y_2(0,t)u(0,t))dt + \\ &= \int_0^A \int_0^B F_2(s, t, y_1(s,t), y_2(s,t), u(s,t))dtds \end{aligned} \tag{9}$$

Related background may be found in [1]-[2].

2. The Discretized Problems

The numerical solution of the optimal control problems of section 1 is carried out by using discrete approximations. Therefore, we shall consider a controlled discrete-time system governed by the finite-difference equation

$$y(i+1) = f(i, y(i), u(i)), u(i) \epsilon U(i, y(i)); y(0) = y_0 \tag{10}$$

The optimization problem concerns the minimization of a functional J given by

$$J := \sum_{i=0}^{N}\sum_{j=0}^{N} F(i,j,y(i),y(j),u(i),u(j)) \tag{11}$$

We shall also consider the problem of a separable Roesser system of the form

$$\begin{aligned} y_1(i+1,j) &= f_1(i,j,y_1(i,j),u(i,j)); \\ y_2(i,j+1) &= f_2(i,j,y_2(i,j),u(i,j)); \\ y_1(0,j) &= y_{10}(j), y_2(i,0) = y_{20}(i) \end{aligned} \tag{12}$$

and the minimization of the functional

$$J_1 := \sum_{i=0}^{N}\sum_{j=0}^{N} F(i,j,y_1(i,j),y_2(i,j),u(i,j)) \tag{13}$$

The solution of the problem (12)-(13) can lead to the solution of the problem (10)-(11) under appropriate transformations.

3. Multi-Time Dynamic Programming

We shall focus on the separable Roesser system of the previous section and the associated optimal control problem. We define the corresponding parametrized system

$$\begin{aligned} y_{1,k\ell x}(i+i,j) &= f_1(i,j,y_{1,k\ell x}(i,j),u(o,j)) \text{ for } i \geq k, j \geq \ell; \\ y_{1,k\ell x}(k,j) &= x(j), j \geq \ell; \\ y_{2,k\ell z}(i,j+1) &= f_2(i,j,y_{2,k\ell z}(i,j),u(i,j)) \text{ for } i \geq k, j \geq \ell; \\ y_{2,k\ell z}(i,\ell) &= z(i), \text{ for } i \geq k \end{aligned} \tag{14}$$

In an analogous way, we define the components a, b, u_0 of the control u by

$$\begin{aligned} a(j) &\equiv u(k,j), j \geq \ell+1; \\ b(i) &\equiv u(i,\ell), i \geq k+1; \\ u_0 &\equiv u(k,\ell) \end{aligned} \tag{15}$$

and also

$$\tilde{a} \equiv (u_0, a), \tilde{b} \equiv (u_0, b) \tag{16}$$

Further, we set

$$\begin{aligned} X(k+1) &\equiv \{\tilde{y}_{1,k\ell x}(k+1,j) : j \geq \ell\}, \\ Y(\ell+1) &\equiv \{\tilde{y}_{2,k\ell z}(i,\ell+1) : i \geq k\}, \\ S_1X(k+1) &\equiv \{\tilde{y}_{1,k\ell x}(k+1,j) : j \geq \ell+1\} \\ S_2Y(\ell+1) &\equiv \{\tilde{y}_{2,k\ell z}(i,\ell+1) : i \geq k+1\} \end{aligned} \tag{17}$$

Then we have

$$\begin{aligned} X(k+1) &= \tilde{f}_1(k,x,\tilde{a}), \\ Y(\ell+1) &= \tilde{f}_2(\ell,z,\tilde{b}) \end{aligned} \tag{18}$$

where $\tilde{f}_1$ has components $f_1(k, j, x(j), \tilde{a}(j)), j \geq \ell$, and similarly for $\tilde{f}_2$.

Let us define the 3 Hamilton-Jacobi functions V_0, V_1, V_2 by

$$V_0(k,\ell,x,z) \equiv \inf_u \sum_{i=k}^{M}\sum_{j=\ell}^{N} F(i,j,y_1(i,j), y_2(i,j),u(i,j)); y_1(k,j)=x(j), y_2(i,\ell)=z(i) \quad (19)$$

$$V_1(k,\ell,x,z;\tilde{a}) \equiv \inf_{u:u(k,j)=\tilde{a}(j)} \cdot \sum_{i=k}^{M}\sum_{j=\ell}^{N} F(i,j,y_1(i,j),y_2(i,j),u(i,j)) \quad (20)$$

with the same initial conditions on y_1, y_2 as in (3.8); and,

$$V_2(k,\ell,x,z;\tilde{b}) \equiv \inf_{u:u(i,\ell)} = \tilde{b}(j) \cdot \sum_{i=k}^{M}\sum_{j=\ell}^{N} F(i,j,y_1(i,j),y_2(i,j),u(i,j)) \quad (21)$$

The Hamiltonian H is now defined by

$$\begin{aligned} H(k,\ell,x,z,a,bu_0,V_0,V_1,V_2) &\equiv \\ &\equiv V_2(k+1,\ell,\tilde{f}_1(k,x\tilde{a}),S_2z;\tilde{b}) + \\ &+ V_1(k,\ell+1,S_1x,\tilde{f}_2(\ell,z,\tilde{b});\tilde{a}) - \\ &- V_0(k+1,\ell+1,S_1\tilde{f}_1(k,x,\tilde{a}),S_2\tilde{f}_2(\ell,z,\tilde{b})) + \\ &= F(k,\ell,x(\ell),z(k),u_0) \end{aligned} \quad (22)$$

Then the conditions of dynamic programming can be written (in compact form, omitting the explicit arguments in H):

$$\begin{aligned} V_1(k,\ell,x,z;\tilde{a}) &= \inf_b H; \\ V_2(k,\ell,x,z;\tilde{b}) &= \inf_a H; \\ V_0(k,\ell,x,z) &= \inf_{a,u_0} V_1(k,\ell,x,z;\tilde{a}) = \\ &= \inf_{b,u_0} V_2(k,\ell,x,z;\tilde{b}) \end{aligned} \quad (23)$$

The system (23) is a backward recursion in 2-D discrete time. The recursion proceeds from the triple of points $((k+1,\ell+1),(k+1,\ell),(k,\ell+1))$ to the point (k,ℓ). This recursion must be initialized by calculating $V_m(k,\ell,x,z,\dots)(m=0,1,2)$ for $k=M$ and for $\ell=N$; it is easy to see that the corresponding values of V_m (as functions of k only or of ℓ only) represent the Hamilton-Jacobi functions for ordinary (1-D time) optimal control problems, and they can therefore be found by ordinary (1-D) dynamic programming. Further, a verification theorem holds; a control policy that minimizes the Hamiltonian must be optimal.

References

1. S.A. Belbas, Int. J. Control, **71**, 61-78 (1998).

2. S.A. Belbas, in: Proc. Eighth Int. Colloq. Diff. Eq., D. Bainov (Ed.), VSP, Utrecht (1998) pp. 451-455.

9th Int. Coll. on Differential Equations, pp. 39-46
D. Bainov (Ed)

Overdetermined Differential Systems in Spaces of (Small) Gevrey Functions

CHIARA BOITI
Dipartimento di Matematica "E. De Giorgi", Via per Arnesano, 73100 Lecce, Italy
cboiti@ilenic.unile.it
MAURO NACINOVICH
Dipartimento di Matematica "L. Tonelli", Via Buonarroti 2, 56127 Pisa, Italy
nacinovi@dm.unipi.it

Abstract: We study the Cauchy problem for overdetermined systems of linear partial differential operators with constant coefficients in some spaces of ultradifferentiable functions of class (M_p). We show that evolution is equivalent to the validity of a Phragmén-Lindelöf principle for entire and plurisubharmonic functions on some irreducible affine algebraic varieties. We find necessary and sufficient conditions for well-posedness, and relate the hyperbolicity of a given system to that of its principal part.
Key Words and Phrases: *Overdetermined systems, Phragmén-Lindelöf principle, ultradifferentiable functions.*
1991 Mathematics Subject Classification. 35N05, 32A15, 58G05.

We investigate the Cauchy problem for overdetermined systems of linear partial differential operators with constant coefficients in some spaces of ultradifferentiable Whitney functions of class (M_p) (see [1]).

In the classical Cauchy problem one seeks a solution of a partial differential equation on a manifold with boundary, requiring that the trace of the solution and of some of its normal derivatives take assigned values on a given portion of the boundary. In general, the setting is such that, for a solution smooth up to the boundary, the differential equation and the initial data allow to compute all its partial derivatives at points of the given portion of the boundary.

This leads to the notion of a formally non-characteristic hypersurface for a system of linear partial differential equations, that was considered in [2], [3].

Here we bypass the question of the formal coherence of the data, by allowing formal solutions (in the sense of Whitney) on the initial manifold as initial data. Using Whitney functions, we can consider a more general situation in which two quite arbitrary sets are involved: we consider a pair $K_1 \subset K_2$ of subsets of $\mathbb{R}^N$, with K_2 locally closed and K_1 closed in K_2, and we think of K_1 as the set where the initial data are given, and of K_2 as the set where we want to find a solution of the Cauchy problem.

In most of our applications K_1 is contained in an affine n-dimensional subspace of $\mathbb{R}^N$

$$K_1 \subset \Sigma = \{z = (t,x) \in \mathbb{R}^k \times \mathbb{R}^n \simeq \mathbb{R}^N : \ t = 0\} \simeq \mathbb{R}^n_x ,$$

and we consider either the class $W^{(M_p)}$ of Whitney functions which are ultradifferentiable of class (M_p) in both variables t and x, or, allowing different scales of regularity in t and x, the topological tensor products $\widetilde{W}^{(M_p)}$ and $W^{(N_p),(M_p)}$, the first not requiring ultradifferentiability in the t-variables, the second requiring (N_p)-ultradifferentiability in t.

If $\mathcal{F}(K_1)$ and $\mathcal{G}(K_2)$ are two of such spaces of ultradifferentiable Whitney functions on K_1 and K_2 respectively, and $A_o(D)$ is an $a_1 \times a_o$ matrix of linear partial differential operators with constant coefficients, we consider the following Cauchy problem:

$$\begin{cases} \text{given } f \in \mathcal{G}(K_2)^{a_1} \text{ and } \varphi \in \mathcal{F}(K_1)^{a_o} \\ \text{find } u \in \mathcal{G}(K_2)^{a_o} \text{ such that} \\ A_o(D)u = f \\ u|_{K_1} = \varphi \ . \end{cases} \tag{1}$$

We denote by $\mathcal{P} = \mathcal{P}_N = \mathbb{C}[\theta_1, \ldots, \theta_N]$ the ring of complex polynomials in N complex variables, and associate to the operator $A_o(D)$ the $\mathcal{P}$-homomorphism ${}^tA_o(\theta) : \mathcal{P}^{a_1} \to \mathcal{P}^{a_o}$, by the formal substitution $\theta_j \leftrightarrow D_j = \frac{1}{i}\frac{\partial}{\partial x_j}$ for $j = 1, \ldots, N$.

Then we remark that, in order to solve the system

$$A_o(D)u = f \ , \tag{2}$$

it is necessary that, for every $Q(\theta) : \mathcal{P}^{a_1} \to \mathcal{P}$ such that

$${}^tA_o(\theta)\,{}^tQ(\theta) \equiv 0 \ , \tag{3}$$

f satisfies the integrability condition

$$Q(D)f = 0 \ .$$

Since $\mathcal{P}$ is Noetherian, the collection of all vectors $Q(\zeta)$ satisfying (3) forms a finitely generated $\mathcal{P}$-module.

Thus we can take into account the integrability conditions (3) by inserting the map ${}^tA_o(\theta) : \mathcal{P}^{a_1} \to \mathcal{P}^{a_o}$ into a Hilbert resolution

$$0 \longrightarrow \mathcal{P}^{a_d} \xrightarrow{{}^tA_{d-1}(\theta)} \mathcal{P}^{a_{d-1}} \longrightarrow \ldots \longrightarrow \mathcal{P}^{a_2} \xrightarrow{{}^tA_1(\theta)} \mathcal{P}^{a_1} \xrightarrow{{}^tA_o(\theta)} \mathcal{P}^{a_o} \longrightarrow \mathcal{M} \longrightarrow 0 \ ,$$

where $\mathcal{M} = \operatorname{Coker}{}^tA_o(\theta)$ and the rows of the matrix $A_1(\theta)$ form a basis of the integrability conditions (3).

Note that the existence of a non-trivial map ${}^tA_1(\theta) : \mathcal{P}^{a_2} \to \mathcal{P}^{a_1}$ such that the sequence

$$\mathcal{P}^{a_2} \xrightarrow{{}^tA_1(\theta)} \mathcal{P}^{a_1} \xrightarrow{{}^tA_o(\theta)} \mathcal{P}^{a_o}$$

is a complex, means that the system (2) is overdetermined.

We thus obtain the following compatibility conditions

$$\begin{cases} f \in \mathcal{G}(K_2)^{a_1}, \ f|_{K_1} \in \mathcal{F}(K_1)^{a_1}, \ \varphi \in \mathcal{F}(K_1)^{a_o} \\ A_1(D)f = 0 \\ A_o(D)\varphi = f|_{K_1} \ , \end{cases} \tag{4}$$

which are necessary for the solvability of the overdetermined Cauchy problem (1).

As a general philosophy, we prefer to classify differential problems in terms of their analytic behaviour, in order to investigate the algebraic conditions which cause this behaviour, rather than the opposite pattern. While discussing the Cauchy problem, we are thus concerned with the concepts of the definition below:

Definition : *The pair* (K_1, K_2) *is of* evolution *(resp.:* of causality, hyperbolic*) if the Cauchy problem (1) has at least one solution (resp.: at most one solution, one and only one solution), for every pair* (φ, f) *of Cauchy data satisfying the compatibility conditions (4).*

If we denote by $\mathcal{I}(K_1, K_2)$ the ideal of Whitney functions on K_2 vanishing with all their derivatives on K_1, the compatibility conditions (4) with zero initial data ($\varphi \equiv 0$) are expressed by:

$$\begin{cases} f \in \mathcal{G}(K_2)^{a_1} \cap \mathcal{I}(K_1, K_2)^{a_1} \\ A_1(D) f = 0 \, . \end{cases} \tag{5}$$

In this way uniqueness and existence of solutions of the Cauchy problem (1) (with $\varphi \equiv 0$), for every datum f satisfying the integrability conditions (5), are related to the vanishing of the cohomology groups

$$Ext^0_{\mathcal{P}}(\mathcal{M}, \mathcal{G}(K_2) \cap \mathcal{I}(K_1, K_2)) \simeq \mathrm{Ker} A_o(D)$$

and

$$Ext^1_{\mathcal{P}}(\mathcal{M}, \mathcal{G}(K_2) \cap \mathcal{I}(K_1, K_2)) \simeq \mathrm{Ker} A_1(D) / \, Im\, A_o(D) \, .$$

The main advantage of this formulation is that it enlightens the algebraic invariance of the problem. In particular, evolution, causality and hyperbolicity of the pair (K_1, K_2) are properties of the functional space $\mathcal{G}(K_2) \cap \mathcal{I}(K_1, K_2)$ and the $\mathcal{P}$-module $\mathcal{M}$, and are independent of the presentation of $\mathcal{M}$ given by the choice of a particular matrix $A_o(\theta)$.

If $\mathcal{F}$ is one of the classes of ultradifferentiable Whitney functions that we are considering, the starting point is the following short exact sequence:

$$0 \to \mathcal{F}(K_2) \cap \mathcal{I}(K_1, K_2) \to \mathcal{F}(K_2) \to \mathcal{F}(K_1) \to 0 \, ,$$

from which we deduce the long exact sequence:

$$\begin{aligned} 0 \; &\to Ext^0_{\mathcal{P}}(\mathcal{M}, \mathcal{F}(K_2) \cap \mathcal{I}(K_1, K_2)) \to Ext^0_{\mathcal{P}}(\mathcal{M}, \mathcal{F}(K_2)) \to \\ &\to Ext^0_{\mathcal{P}}(\mathcal{M}, \mathcal{F}(K_1)) \to Ext^1_{\mathcal{P}}(\mathcal{M}, \mathcal{F}(K_2) \cap \mathcal{I}(K_1, K_2)) \to \\ &\to Ext^1_{\mathcal{P}}(\mathcal{M}, \mathcal{F}(K_2)) \to Ext^1_{\mathcal{P}}(\mathcal{M}, \mathcal{F}(K_1)) \to \dots \, . \end{aligned}$$

If K_1 and K_2 are convex, by the Ehrenpreis fundamental principle we obtain that

$$Ext^j_{\mathcal{P}}(\mathcal{M}, \mathcal{F}(K_i)) = 0 \qquad \forall j \geq 1, \; i = 1, 2$$

and hence the long exact sequence above reduces to:

$$\begin{aligned} 0 \; &\to Ext^0_{\mathcal{P}}(\mathcal{M}, \mathcal{F}(K_2) \cap \mathcal{I}(K_1, K_2)) \to Ext^0_{\mathcal{P}}(\mathcal{M}, \mathcal{F}(K_2)) \to \\ &\to Ext^0_{\mathcal{P}}(\mathcal{M}, \mathcal{F}(K_1)) \to Ext^1_{\mathcal{P}}(\mathcal{M}, \mathcal{F}(K_2) \cap \mathcal{I}(K_1, K_2)) \to 0 \, . \end{aligned}$$

Since evolution, causality and hyperbolicity are related to the vanishing of the cohomology groups $Ext^0_{\mathcal{P}}(\mathcal{M}, \mathcal{F}(K_2) \cap \mathcal{I}(K_1, K_2))$ and $Ext^1_{\mathcal{P}}(\mathcal{M}, \mathcal{F}(K_2) \cap \mathcal{I}(K_1, K_2))$, we obtain that they are related to the study of the homomorphism

$$Ext^0_{\mathcal{P}}(\mathcal{M}, \mathcal{F}(K_2)) \xrightarrow{u} Ext^0_{\mathcal{P}}(\mathcal{M}, \mathcal{F}(K_1)) \, ,$$

and hence to that of the dual homomorphism

$$(Ext^{\circ}_{\mathcal{P}}(\mathcal{M},\mathcal{F}(K_1)))' \overset{u'}{\hookrightarrow} (Ext^{\circ}_{\mathcal{P}}(\mathcal{M},\mathcal{F}(K_2)))' .$$

We are thus lead to the study of these dual spaces. By two algebraic lemmas we first reduce to the case $\mathcal{M} = \mathcal{P}/\wp$ for a prime ideal $\wp$ of $\mathcal{P}$, i.e. we reduce to the case where $A_o(D)$ is of the form:

$$A_o(D) = \begin{pmatrix} p_1(D) \\ \vdots \\ p_r(D) \end{pmatrix}$$

for a set of generators $p_1(\theta), \dots, p_r(\theta)$ of $\wp$.

Then by the Fourier-Laplace transform (and in particular Paley-Wiener-type theorems) and the Ehrenpreis fundamental principle, we prove that these dual spaces are inductive limit of spaces of entire functions satisfying growth conditions on the associated affine algebraic variety

$$V = V(\check{\wp}) = \{\theta \in \mathbb{C}^N : \; p_j(-\theta) = 0 \; \forall j = 1, \dots, r\} .$$

More precisely:

$$(Ext^{\circ}_{\mathcal{P}}(\mathcal{M},\mathcal{F}(K_i)))' \simeq \mathcal{O}_{\{\varphi^{(i)}\}}(V) := \varinjlim_{\alpha \nearrow \infty} \mathcal{O}_{\varphi^{(i)}_\alpha}(V)$$
$$:= \varinjlim_{\alpha \nearrow \infty} \{f \in \mathcal{O}(V) : \; \sup_V |f(\theta)e^{-\varphi^{(i)}_\alpha(\theta)}| < +\infty\}$$

for $i = 1, 2$, where the two sequences $\{\varphi^{(1)}_\alpha\}_{\alpha\in\mathbb{N}}$ and $\{\varphi^{(2)}_\alpha\}_{\alpha\in\mathbb{N}}$ are related to the class $\mathcal{F}$ of ultradifferentiable Whitney functions and to the convex sets K_1 and K_2 respectively. More precisely, for every $i = 1, 2$, if $\{K^{(i)}_\alpha\}_{\alpha\in\mathbb{N}}$ is an increasing sequence of compact convex subsets of K_i such that $K_i = \bigcup_{\alpha\in\mathbb{N}} K^{(i)}_\alpha$, and $H_{K^{(i)}_\alpha}(\cdot)$ is the supporting function of $K^{(i)}_\alpha$

$$H_{K^{(i)}_\alpha}(s, y) = \sup_{(t,x)\in K^{(i)}_\alpha} (\langle t, s\rangle + \langle x, y\rangle) ,$$

then:

i) for $\mathcal{F} = W^{(M_p)}$:

$$\varphi^{(i)}_\alpha(\theta) = M(\alpha\theta) + H_{K^{(i)}_\alpha}(Im\,\theta), \quad \theta = (\tau, \zeta) \in \mathbb{C}^k \times \mathbb{C}^n \simeq \mathbb{C}^N , \tag{6}$$

where $M(\cdot)$ is the associated function to the sequence (M_p) (for the Gevrey sequence $M_p = p^{ps}$, $s > 1$, we have that $M(\theta) = e^{-1}s(1 + |\theta|)^{1/s}$);

ii) for $\mathcal{F} = \widetilde{W}^{(M_p)}$:

$$\varphi^{(i)}_\alpha(\theta) = \alpha \log(1 + |\tau|) + M(\alpha\zeta) + H_{K^{(i)}_\alpha}(Im\,\theta), \quad \theta = (\tau, \zeta) \in \mathbb{C}^k \times \mathbb{C}^n \simeq \mathbb{C}^N \; ; \tag{7}$$

iii) for $\mathcal{F} = W^{(N_p),(M_p)}$:

$$\varphi^{(i)}_\alpha(\theta) = N(\alpha\tau) + M(\alpha\zeta) + H_{K^{(i)}_\alpha}(Im\,\theta), \quad \theta = (\tau, \zeta) \in \mathbb{C}^k \times \mathbb{C}^n \simeq \mathbb{C}^N . \tag{8}$$

We thus study these dual spaces $\mathcal{O}_{\{\varphi^{(i)}\}}(V)$, obtaining that they are $(\mathcal{LF})$, barreled, $(\mathcal{DF})$, of Schwartz, of Montel, bornological, reflexive and complete. Then the study of the inclusion dual map u' leads to necessary and sufficient conditions for evolution, causality and hyperbolicity.

We find that evolution is equivalent to the validity of a Phragmén-Lindelöf principle for holomorphic functions on the associated affine algebraic varieties:

Theorem: *Let $K_1 \subset K_2$ be two closed convex subsets of $\mathbb{R}^N$ and $\mathcal{M}$ a unitary finitely generated $\mathcal{P}$-module.*

Then the pair (K_1, K_2) is of evolution for $\mathcal{M}$ in the class of ultradifferentiable Whitney functions $\mathcal{F}$ if and only if the following Phragmén-Lindelöf principle for holomorphic functions holds true for all $\wp \in Ass(\mathcal{M})$:

$$\begin{cases} \forall \alpha \in \mathbb{N}\ \exists \beta \in \mathbb{N},\ c > 0 \text{ such that:} \\ \text{if } f \in \mathcal{O}(V) \text{ satisfies, for some } \alpha_f \in \mathbb{N} \text{ and } c_f > 0 \\ \begin{cases} |f(\theta)| \leq e^{\varphi_\alpha^{(2)}(\theta)} & \forall \theta = (\tau, \zeta) \in V(\check{\wp}) \\ |f(\theta)| \leq c_f e^{\varphi_{\alpha_f}^{(1)}(\theta)} & \forall \theta \in V(\check{\wp}) \end{cases} \\ \text{then it also satisfies:} \\ |f(\theta)| \leq c e^{\varphi_\beta^{(1)}(\theta)} \qquad \forall \theta \in V(\check{\wp}) , \end{cases}$$

where the sequences $\{\varphi_\alpha^{(i)}\}_{\alpha \in \mathbb{N}}$ depend on $\mathcal{F}$ as in (6), (7), (8).

Generally, we then work with an equivalent Phragmén-Lindelöf principle for plurisubharmonic functions, which is more easy to handle and leads to Petrowski-type sufficient conditions for evolution. To this aim we assume that

$$\Sigma = \{(t, x) \in \mathbb{R}^k \times \mathbb{R}^n :\ t = 0\} \supset K_1$$

is *formally non-characteristic* and *quasi-free*, i.e. that the $\mathcal{P}_N$-module $\mathcal{M}$ considered as a $\mathcal{P}_n = \mathbb{C}[\zeta_1, \ldots, \zeta_n]$-module is of finite type and torsion free. In terms of the Cauchy problem this means that the initial data are not restricted by any differential relation.

Let us thus choose

$$K_1 \subset \Sigma \qquad \text{closed and convex}$$
$$\text{and} \qquad K_2 = \Gamma \times K_1$$

for a closed proper convex cone Γ of $\mathbb{R}^k_t$.

If we set

$$\kappa_\Gamma(\tau) = \sup_{\substack{t \in \Gamma \\ |t_i| \leq 1, i=1,\ldots,k}} Im\langle t, \tau \rangle \qquad \forall \tau \in \mathbb{C}^k ,$$

we have the following Petrowski-type sufficient condition for evolution in the classes of (small) Gevrey functions $\tilde{\gamma}^{(s)} = \widetilde{W}^{(M_p)}$ and $\gamma^{(r,s)} = W^{(N_p),(M_p)}$ for $M_p = p^{ps}$ and $N_p = p^{pr}$, $r, s > 1$:

Proposition: *Assume that Σ is formally non-characteristic and quasi-free for the unitary finitely generated $\mathcal{P}$-module $\mathcal{M}$.*

Let $s > 1$ and assume that there exist constants c_1, c_2 such that for every $\wp \in Ass(\mathcal{M})$ the following Petrowski-type condition holds:

$$(P)_s : \qquad \kappa_\Gamma(\tau) \leq c_1 |\xi|^{1/s} + c_2 \quad \forall (\tau, \xi) \in V(\check{\wp}) \text{ with } \xi \in \mathbb{R}^n .$$

Then the pair $(K_1, \Gamma \times K_1)$ is of evolution for $\mathcal{M}$ in the classes $\tilde{\gamma}^{(s)}$ and $\gamma^{(r,s)}$ for $r > 1$ with $r \geq bs$, where b is the reduced order *of the system, i.e. the smallest $b > 0$ such that, for some $\lambda > 0$:*

$$|\tau| \leq \lambda(1 + |\zeta|)^b \qquad \forall(\tau, \zeta) \in V(\check{\wp}) \ \forall \wp \in Supp(\mathcal{M}) .$$

As far as hyperbolicity (i.e. well-posedness) is concerned, we first obtain a Petrowski-type sufficient condition:

Proposition: *Let Σ be formally non-characteristic and quasi-free for the unitary finitely generated $\mathcal{P}$-module $\mathcal{M}$.*

If the reduced order of the system b is less or equal than 1 (we say that Σ is strongly non-characteristic*) and the Petrowski-type condition $(P)_s$ is satisfied for some $s > 1$, then the pair $(\Sigma, \Gamma \times \Sigma)$ is hyperbolic for $\mathcal{M}$ in the (small) Gevrey classes $\tilde{\gamma}^{(s)}$ and $\gamma^{(r,s)}$ for $r \geq s$.*

In order to obtain necessary and sufficient conditions for well-posedness, we write down the Phragmén-Lindelöf principle for hyperbolicity:

Theorem: *Let $K_1 \subset K_2$ be two closed convex subsets of $\mathbb{R}^N$ and $\mathcal{M}$ a unitary finitely generated $\mathcal{P}$-module.*

Then, using the same notation as in (6), (7), (8), the pair (K_1, K_2) is hyperbolic for $\mathcal{M}$ in the class of ultradifferentiable Whitney functions $\mathcal{F}$ if and only if the following Phragmén-Lindelöf principle for holomorphic functions holds true for all $\wp \in Ass(\mathcal{M})$:

$$\begin{cases} \forall \alpha \in \mathbb{N} \ \exists \beta \in \mathbb{N}, \ c > 0 \ \textit{such that} \\ \sup_{V(\check{\wp})} |f(\theta) e^{-\varphi^{(1)}_\beta(\theta)}| \leq c \sup_{V(\check{\wp})} |f(\theta) e^{-\varphi^{(2)}_\alpha(\theta)}| \qquad \forall f \in \mathcal{O}(V(\check{\wp})) . \end{cases}$$

This leads to the following:

Corollary: *Let $K_1 \subset K_2$ be two closed convex subsets of $\mathbb{R}^N$ and $\mathcal{M}$ a unitary finitely generated $\mathcal{P}$-module.*

Then:

a) a necessary and sufficient condition for the pair (K_1, K_2) to be hyperbolic for $\mathcal{M}$ in the class $\widetilde{W}^{(M_p)}$ is that for every compact convex subset Q_2 of K_2 there exist a compact convex subset Q_1 of K_1 and a constant $L > 0$ such that for every prime ideal $\wp \in Ass(\mathcal{M})$ the following holds true:

$$H_{Q_2}(Im\,\theta) \leq H_{Q_1}(Im\,\theta) + M(L\zeta) + L\log(1 + |\tau|) + L \qquad \forall \theta = (\tau, \zeta) \in V(\check{\wp}) ;$$

b) a necessary and sufficient condition for the pair (K_1, K_2) to be hyperbolic for $\mathcal{M}$ in the class $W^{(N_p),(M_p)}$ is that for every compact convex subset Q_2 of K_2 there exist a compact convex subset Q_1 of K_1 and a constant $L > 0$ such that for every prime ideal $\wp \in Ass(\mathcal{M})$ the following holds true:

$$H_{Q_2}(Im\,\theta) \leq H_{Q_1}(Im\,\theta) + M(L\zeta) + N(L\tau) + L \qquad \forall \theta = (\tau, \zeta) \in V(\check{\wp}) .$$

We finally study the relationship between the hyperbolicity of a given system and that of its principal part.

In [1] we precise the notion of principal part $\mathcal{M}^o$ of a unitary $\mathcal{P}$-module $\mathcal{M}$ of finite type (see also [4]). Here we consider, for simplicity, the case $\mathcal{M} = \mathcal{P}/\wp$ for a prime ideal $\wp$ of $\mathcal{P}$. Then we set $\mathcal{M}^o = \mathcal{P}/\wp^o$, where $\wp^o$ is the ideal generated by the principal parts of all polynomials of $\wp$ (we can choose a set of polynomials $p_1(\theta), \dots, p_\ell(\theta)$ such that $p_1, \dots, p_\ell$ generate $\wp$ and $p_1^o, \dots, p_\ell^o$ generate $\wp^o$).

While the principal part of a hyperbolic system is always hyperbolic, a system with hyperbolic principal part is not in general hyperbolic. For example, the operator $\frac{\partial^2}{\partial t^2} - \frac{\partial}{\partial x}$ in $\mathbb{R}_t \times \mathbb{R}_x$ is not well-posed in $\tilde{\gamma}^{(s)}$ nor in $\gamma^{(r,s)}$ for any $s > 2$, $r > 1$, whereas its principal part $\frac{\partial^2}{\partial t^2}$ is obviously well-posed.

However, we prove, generalizing a classical result for scalar equations and initial data on a hypersurface (cf. [5]):

Theorem: *Let $K_1 \subset K_2$ be two closed convex subsets of $\mathbb{R}^N$ and $\wp$ a prime ideal of $\mathcal{P}$.*

If the pair (K_1, K_2) is hyperbolic for $\mathcal{P}/\wp^o$ (in the class of Whitney functions), then there is a rational $0 \leq q < 1$ such that the pair (K_1, K_2) is hyperbolic for $\mathcal{P}/\wp$ in the class $\gamma^{(s)}$ of (small) Gevrey functions if and only if $1 < s \leq \frac{1}{q}$ (for $q = 0$ we set $\frac{1}{q} = +\infty$ and identify $\gamma^{(+\infty)}$ to the space of Whitney functions).

Remark: We can also give an estimate of q in the following way:

$$q \leq \frac{m-1}{m} ,$$

where m is the degree of $V(\wp)$, i.e. the number of points of the generic intersection of $V(\wp)$ with an $(N-d)$-dimensional affine subspace of $\mathbb{C}^N$ (here $d = dim_{\mathbb{C}} V(\wp)$).

For non-evolution pairs, the question of finding the right necessary and sufficient conditions for causality for a Cauchy problem with initial data on a submanifold of codimension larger than one remains largely an open question.

Here we give a necessary and sufficient condition for causality in some class of ultradifferentiable functions in the case of initial data on a hypersurface. More precisely, given

$$\Sigma = \{(t,x) \in \mathbb{R} \times \mathbb{R}^n : \ t = 0\}$$

and

$$H = \{(t,x) \in \mathbb{R} \times \mathbb{R}^n : \ t \geq 0\} ,$$

we obtain the following:

Theorem: *Let Σ be formally non-characteristic for the unitary finitely generated $\mathcal{P}$-module $\mathcal{M}$.*

Then a necessary and sufficient condition for the pair (Σ, H) to be of causality for $\mathcal{M}$ in the class $\widetilde{W}^{(M_p)}$ is that Σ is strongly non-characteristic for $\mathcal{M}$.

References

1. C. Boiti and M. Nacinovich, The overdetermined Cauchy problem in some classes of ultradifferentiable functions, Preprint del Dipartimento di Matematica "Ennio De Giorgi" dell'Università di Lecce n. 1/1998 (15 luglio 1998).

2. A. Andreotti, C.D. Hill, S. Łojasiewicz and B. Mackichan, Invent. Math., **26**, 43-86 (1976).

3. M. Nacinovich, Annali di Matematica Pura ed Applicata, **IV**, CLVI, 265-321 (1990).

4. A. Andreotti and M. Nacinovich, Advances in Math. Suppl. Studies, **7A**, 41-93 (1981).

5. L. Hörmander, Linear Partial Differential Operators, Springer Verlag, Berlin (1963).

6. A. Andreotti and M. Nacinovich, Ann. SNS Pisa, **3**, 553-621 (1976).

7. C. Boiti, Evolution for Overdetermined Differential Systems, Preprint del Dipartimento di Matematica dell'Università di Pisa n. 2.255.1002 (1996), PhD Thesis.

8. H. Komatsu, J. Fac. Sci. Univ. Tokyo Sect IA Math., **20**, 25-105 (1973).

9. H. Komatsu, J. Fac. Sci. Univ. Tokyo Sect IA Math., **24**, 607-628 (1977).

9th Int. Coll. on Differential Equations, pp. 47-52
D. Bainov (Ed)

Applications of Stochastic Differential Equations to Population Growth

CARLOS A. BRAUMANN
Department of Mathematics, Universidade de Évora, P-7000 Évora, Portugal

Abstract: We start from a density-dependent population growth model $\frac{1}{N}\frac{dN}{dt} = rg(N)$, where $N = N(t)$ is the population size at time t, $r > 0$ is an intrinsic growth parameter and $g(N)$ is a smooth decreasing function describing the density-dependence. Environmental random fluctuations are then modeled approximating them by $\sigma\varepsilon(t)$ (where $\varepsilon(t)$ is a standard white noise and σ is an intensity parameter) and assuming that they additively affect either the parameter r or the *per capita* growth rate of the population. Correspondingly, we obtain two different types of Stratonovich stochastic differential equations, for which we study the qualitative behavior. Models of these types have already been considered in the literature for specific functions $g(N)$ and, consequently, our models constitute a useful generalization. In fact, our results do not depend on the choice of the function $g(N)$, which is usually poorly known.
Key Words and Phrases: *Stochastic differential equations, population growth, random environments.*

1. Introduction

Let us denote by $N = N(t)$ the size (or biomass) of a population at time $t \geq 0$ and assume the initial population size $N(0) = N_0 > 0$ is known. If there were no food, territorial, and other limitations to growth, the *per capita* growth rate $\frac{1}{N}\frac{dN}{dt}$ would be a constant rate $r > 0$. Since such limitations do exist, we model them through a *density-dependence function* $g(N)$ (defined for $N > 0$), resulting in the deterministic model

$$\frac{1}{N}\frac{dN}{dt} = rg(N), \tag{1}$$

where $r > 0$ may be called *intrinsic growth rate.* We will assume that $g(N)$ is a twice continuously differentiable strictly decreasing function. In fact, the larger the population is, the stronger is the effect of growth limitations on an individual, making it harder to survive and to reproduce. We will assume that $g(0^+) > 0$, which means that small populations have a positive *per capita* growth rate, and that $\lim_{N\to 0+} Ng(N) = 0$, which implies that $N = 0$ (extinction) is an equilibrium population size. Popular examples are $g(N) = 1 - N/K$ with $K > 0$ (logistic growth) and $g(N) = \ln(K/N)$ with $K > 0$ (Gompertz growth), but many other specific functions $g(N)$ have been proposed, almost all of them satisfying the above assumptions (the exceptions are models with Allee effects). Assume there is a $K \in (0, +\infty)$, known as the *carrying capacity* of the environment, such that $g(K) = 0$.

For the deterministic model (1), the equilibria are $N = 0$ and $N = K$. It is a trivial exercise to show that $N = 0$ is unstable and $N = K$ is, for positive initial populations (as we have assumed), globally stable. Therefore, $N(t)$ will, as $t \to +\infty$, converge monotonously to the carrying capacity K.

Environment, however, rather than being constant, is subjected to random fluctuations that affect population growth. It is reasonable to assume that those random fluctuations could be modeled by a colored noise process (correlated smooth noise), which we shall approximate by a white noise (uncorrelated zero mean non-smooth gaussian noise) $\sigma\varepsilon(t)$, where $\sigma > 0$ is a noise intensity parameter and $\varepsilon(t)$ is a standard white noise (formally the derivative, in the sense of generalized functions, of the standard Wiener process $w(t)$). This leads to **Stratonovich** *stochastic differential equations* (SDE) as reasonable approximate models (see, for instance, [1]).

There are two main ways considered in the literature to model the effect of the environmental fluctuations.

One way is to assume that the most sensitive parameter is the intrinsic growth rate r for it is the parameter most influential in the regulation of the fate of young recruits after reproduction, a very sensitive phase in the life cycle. The resulting SDE model, obtained by adding the noise to the parameter r, is

$$\frac{1}{N}\frac{dN}{dt} = (r + \sigma\varepsilon(t))\, g(N). \tag{2}$$

The other way is to assume that the noise affects birth and/or mortality rates, therefore increasing or decreasing the *per capita* growth rate of the population $\frac{1}{N}\frac{dN}{dt}$. In that case, one obtains the SDE model (additive noise model)

$$\frac{1}{N}\frac{dN}{dt} = rg(N) + \sigma\varepsilon(t). \tag{3}$$

Models of the type (2) or (3) have been considered in the literature for specific density-dependence functions $g(N)$, as can be seen, for instance, in [2], [3], [4], [5], [6], [7], and [8]. Models of these types, but with a fishing term, have also been studied; see, for instance, [9], [10], [11], [12], [13], [14], and [15]. It is very difficult, looking at population data, to determine which density-dependence function $g(N)$ is more appropriate.This is the reason we have considered the general case and looked for qualitative properties that are independent of that specific function. Main issues will be extinction, fixation at the carrying capacity, and existence of a stationary density. Similar general results have been obtained in [16] and [17] for similar models with a fishing term.

Sections 2 and 3 will study, respectively, the stochastic models (2) and (3).

In dealing with data, the problems of parameter estimation and prediction (as well as hypothesis testing) using a finite number of (discrete) observations (population size estimates at different times) on the single population trajectory usually available are very important. They can only be dealt with properly in the context of random models. One can see them treated in [18], [19], and [20] for important particular cases of g.

Threshold crossing probabilities relevant to applications can be found in [21].

2. Model (2)

Consider model (2) with the assumptions made in Section 1. Let $c(N) = \int_{N_0}^{N} \frac{1}{xg(x)}dx$. Assume also that

$$c(K^-) = -c(0^+) = +\infty. \tag{4}$$

This is a technical assumption, satisfied by the specific functions commonly proposed in the literature, to insure a sufficiently regular behavior of $g(N)$ near the deterministic equilibria $N = 0$ and $N = K$.

The solution of the SDE (2), with initial condition $N(0) = N_0$, exists and is almost surely (a.s.) unique up to an explosion time and is a (Markov) homogeneous diffusion process with drift coefficient $a(N) = rg(N)N + \frac{1}{4}\frac{db(N)}{dN}$ and diffusion coefficient $b(N) = (\sigma g(N)N)^2$ (see, for instance, [1] or [22]).

Suppose first that $0 < N_0 < K$. Define, for $0 < N < K$, a new variable $Y = c(N)$. Since g is positive in that interval, c is a strictly increasing function. Due to (4), it is a one-to-one function from $(0, K)$ onto $(-\infty, +\infty)$. Using Stratonovich (ordinary) calculus rules, one sees that $Y(t) = c(N(t))$ satisfies the SDE $dY/dt = r + \sigma\varepsilon(t)$ and $Y(0) = 0$. Its solution is $Y(t) = rt + \sigma w(t)$, which is a Markov Gaussian process (known as *brownian motion with drift*) with mean value $\mathbf{E}[Y(t)] = rt$ and autocovariance function $\mathbf{COV}[Y(s), Y(t)] = \sigma^2 \min(s, t)$. Of course,

$$N(t) = c^{-1}(rt + \sigma w(t)), \tag{5}$$

where c^{-1} is the inverse function of c. **We conclude that (2) has a unique solution for all $t \geq 0$.** The last expression, allows one to obtain the explicit finite-time distributions of $N(t)$ from the Gaussian finite-time distributions of $Y(t)$. It is well-known that $Y(t) \to +\infty$ a.s. as $t \to +\infty$. Therefore, $N(t) \to K^-$ a.s. as $t \to +\infty$.

If we assume now $N_0 > K$, we can perform a similar reasoning and conclude that $N(t) \to K^+$ a.s. as $t \to +\infty$.

We conclude that **the population converges to the carrying capacity K from the side of the initial population size**, just like in the deterministic model. The difference, of course, is that the approach to equilibrium is not smooth and shows random fluctuations.

3. Model (3)

Many real populations do not seem to have sizes converging to a carrying capacity K. They rather show, after some transient behavior, random steady oscillations around an average carrying capacity. See, for instance, *Paramecium Caudatum* data from Gause [23]. Model (3) is more appropriate to describe this situation. The main difference is that, unlike model (2), the diffusion coefficient is not zero at the carrying capacity and therefore, the population cannot settle down at that size. Let us consider now model (3) with the assumptions made in Section 1.

The solution of the SDE (3), with initial condition $N(0) = N_0$, is unique and exists (up to an explosion time, which maybe $+\infty$) and is a homogeneous diffusion process with drift coefficient $a(N) = rg(N)N + \frac{1}{4}\frac{db(N)}{dN}$ and diffusion coefficient $b(N) = (\sigma N)^2$. Notice that $b(0) = 0$ and that $b(N) > 0$ for $N > 0$. We have boundaries at $N = 0$ and $N = +\infty$, and their study will help determine the global behavior of the solution.

Let T_x be the first hitting time of a point $x > 0$. Let $T_{0^+} = \lim_{x \downarrow 0} T_x$. The boundary $N = 0$ is classified as *non-attracting* if $\mathbf{P}[T_{0^+} \leq T_n \,|N(0) = x] = 0$ for all $0 < x < n < K$ (where $\mathbf{P}[...]$ denotes "probability of..."), which means that it takes longer to move close to zero population than to move anywhere away from zero population size. This implies that the boundary $N = 0$ is not attainable in finite time from any interior point and one can see, for instance in [24], that it is also not reachable in infinite time, that is, $\mathbf{P}[\lim_{t\to+\infty} N(t) = 0] = 0$. Therefore, if $N = 0$ is non-attracting, extinction can not occur (to be more precise, has a zero probability of occurring). A similar definition holds, with the obvious adaptations, for the boundary $N = +\infty$.

We will use here two auxiliary functions. The first is the *scale density*

$$s(N) = \exp\left(-\int_n^N \frac{2a(\theta)}{b(\theta)}\,d\theta\right)$$

and is defined up to a multiplicative constant (that is, $n > 0$ is arbitrarily chosen). The second is the *speed density*

$$m(N) = \frac{1}{s(N)b(N)}.$$

The *scale measure* and *the speed measure* of a Borel set B are given, respectively, by $S(B) = \int_B s(N)dN$ and $M(B) = \int_B m(N)dN$. Since these measures may not even be finite, the scale density and the speed density are not necessarily probability densities. Therefore, the *scale function* $S(x) = \int^x s(N)dN$ is not necessarily a probability distribution function, although, since $S(a,b] = S(b) - S(a)$, it behaves like a distribution function of the scale measure. As can be seen, for instance, in [24], we have, for $a < x < b$, $\mathbf{P}\left[T_a \leq T_b \,|N(0) = x\right] = \frac{S(x,b]}{S(a,b]}$, which, by the way, hints the important role of the scale measure in the determination of the non-attracting or the attracting nature of the boundaries.

In our case, $s(z) = \frac{n}{z}\exp\left(\frac{2r}{\sigma^2}\int_z^n \frac{g(\theta)}{\theta}d\theta\right)$ and $m(z) = \frac{1}{\sigma^2 n}\frac{1}{z}\exp\left(\frac{2r}{\sigma^2}\int_n^z \frac{g(\theta)}{\theta}d\theta\right)$.

One can see that **N=0 is non-attracting** (and so extinction can not occur). It suffices to show that, for some $x_0 > 0$, $S(0,x_0] = \int_{0+}^{x_0} s(z)dz = +\infty$. Let $0 < x_0 < n < K$. Then, for $z \in (0,x_0]$, $\int_z^n \frac{g(\theta)}{\theta}d\theta \geq 0$. Consequently, $s(z) \geq 1$ and $S(0,x_0] \geq n\int_{0+}^{x_0} \frac{1}{z}dz = +\infty$.

One can also see that $N = +\infty$ **is non-attracting**. Therefore, explosions can not occur and **the solution exists and is unique for all** $t \geq 0$. It suffices to show that, for some $x_0 > 0$, $S[x_0,+\infty) = \int_{x_0}^{+\infty} s(z)dz = +\infty$. Let $x_0 > n > K$, let z vary in $[x_0,+\infty)$, and let θ vary in $[n,z]$. Since $g(\theta) < 0$, we have $\int_n^z \frac{g(\theta)}{\theta}d\theta \leq 0$ and, therefore, $s(z) \geq 1$. Thus, $S[x_0,+\infty) \geq n\int_{x_0}^{+\infty} \frac{1}{z}dz = +\infty$.

For stochastic model (3), there is no equilibrium point, contrary to what happened in the deterministic case and in stochastic model (2). There is, however, a somewhat similar idea: the existence of an equilibrium probability distribution for the population size, called the *stationary distribution* or the *steady-state distribution*, with a probability density function p_N (called the *stationary density*). If the population size distribution is in a steady-state, population size will vary randomly over time but the probabilities $\int_{N_1}^{N_2} p_N(n)dn$ that the population will be in any given size range (N_1, N_2) will remain constant.

If we rescale a diffusion process to natural scale, the speed density of the rescaled process $Y(t) = N(S(t))$ in a point y is related to the time spent by the process in the neighborhood of y. Loosely speaking, one may imagine that, at the steady-state (if it exists), points where the process spends more time are more likely to occur, which suggests a relationship between the speed density and the stationary density. Indeed (see [24]), since the boundaries are non-attracting, the stationary density exists if $M = \int_{0+}^{+\infty} m(z)\,dz < +\infty$ and (in order to integrate to one) is given by

$$p_N(N) = \frac{m(N)}{M} \qquad (0 < N < +\infty).$$

We will now prove that, **in model (3), the population size has a stationary density.**

We need to show that $M < +\infty$. Let $y_1 < K < y_2$ be such that $0 < y_1 < n < y_2 < +\infty$. Break the integration interval: $M = M_1 + M_2 + M_3 = \left(\int_{0+}^{y_1} + \int_{y_1}^{y_2} + \int_{y_2}^{+\infty}\right) m(z)dz$.

Fix $n < K$. The proof for the case $n > K$ is very similar and shall not be presented here.

We first show that $M_1 < +\infty$. Let z vary in $(0, y_1)$ and θ vary in (z, n). Since g is decreasing, $g(\theta) > g(n)$. Let $V(z) = -\frac{2rg(n)}{\sigma^2}\int_z^n \frac{1}{\theta}d\theta$ and note that $V(0^+) = -\infty$. We have $m(z) \leq \frac{1}{\sigma^2 n}\frac{1}{z}\exp(V(z)) = \frac{1}{2rng(n)}\frac{d\exp(V(z))}{dz}$. Therefore, $M_1 = \int_{0+}^{y_1} m(z)dz \leq \frac{1}{2rng(n)}(\exp(V(y_1)) - \exp(V(0^+))) < +\infty$.

We now show that M_3 is finite. Let z vary in $(y_2, +\infty)$ and θ vary in (n, z). Decompose $\frac{2r}{\sigma^2}\int_n^z \frac{g(\theta)}{\theta}d\theta = \frac{2r}{\sigma^2}\left(\int_n^{y_2} + \int_{y_2}^z\right)\frac{g(\theta)}{\theta}d\theta = A + B(z)$. Obviously, $A < +\infty$ and $B(z) = \frac{2rg(y_2)}{\sigma^2}\int_{y_2}^z \frac{1}{\theta}d\theta = \frac{2rg(y_2)}{\sigma^2}\ln\frac{z}{y_2}$. Therefore, $m(z) = \frac{1}{\sigma^2 n}\frac{1}{z}e^A\left(\frac{z}{y_2}\right)^{2rg(y_2)/\sigma^2} = Cz^{2rg(y_2)/\sigma^2 - 1}$, where C is a positive constant. Notice that $g(y_2) < 0$. Therefore, we get $M_3 = \int_{y_2}^{+\infty} m(z)dz \leq D\ z^{2rg(y_2)/\sigma^2}\Big|_{z=y_2}^{z=+\infty} < +\infty$, where D is a positive finite constant.

Finally, we show that M_2 is finite. Put $M_2 = M_2' + M_2'' = \left(\int_{y_1}^n + \int_n^{y_2}\right)m(z)dz$. We will show that M_2' is finite. The proof that M_2'' is finite is similar. Let z vary in (y_1, n) and θ vary in (z, n). Then $m(z) = \frac{1}{\sigma^2 n}\frac{1}{z}\exp\left(-\frac{2r}{\sigma^2}\int_z^n \frac{g(\theta)}{\theta}d\theta\right) \leq \frac{1}{\sigma^2 y_0}\frac{1}{y_1}\exp\left(-\frac{2rg(n)}{\sigma^2}\int_z^n \frac{1}{\theta}d\theta\right) \leq G\left(\frac{z}{y_1}\right)^{2rg(z)/\sigma^2}\left(\frac{n}{y_1}\right)^{-2rg(z)/\sigma^2} \leq G\left(\frac{y_1}{y_1}\right)^{2rg(y_1)/\sigma^2}\left(\frac{n}{y_1}\right)^{-2rg(y_1)/\sigma^2}$, where G is a positive finite constant, is finite. Therefore, $M_2' < +\infty$.

The stationary distribution has a mode $\widehat{N}$, which is the solution of the equation $p_N'(N) = 0$, which in this case is equivalent to the equation $g(N) = \frac{\sigma^2}{2r}$. For small noise intensity σ, the solution of this equation should be approximately given by the solution K of the equation $g(N) = 0$. So, **the stable equilibrium of the deterministic model (1) (the carrying capacity K), although slightly larger, approximately coincides with the most probable population size of the stochastic model (3) at the steady-state.**

Example 3. We present, as an example, the stochastic *Gompertz model* with additive noise, which is the particular case of (3) with $g(N) = \ln(K/N)$. The stationary density is given by

$$p_N(N) = C\frac{1}{N}\exp\left(\frac{2r}{\sigma^2}\ln(K/N)^2\right) \qquad (0 < N < +\infty),$$

where $C > 0$ is a constant such that $\int_0^{+\infty} p_N(N)dN = 1$. This distribution has a mode $\widehat{N} = K\exp\left(-\frac{\sigma^2}{4r}\right)$. When the noise intensity σ is small, $\widehat{N} \cong K$. If one works with the variable $Y = \ln(K/N)$, one recognizes that the stationary distribution of Y is gaussian with mean 0 and variance $\sigma^2/(2r)$. Actually, in this case we can even obtain the transient distribution. In fact, from (3), one obtains by Stratonovich calculus (ordinary rules; see, for instance, [1]) that $\frac{dY}{dt} = -rY + \sigma\varepsilon(t)$ with initial condition $Y(0) = Y_0 = \ln(K/N_0)$, the solution of which is $Y(t) = Y_0 e^{-rt} - \sigma e^{-rt}\int_0^t e^{rs}dw(s)$. Since the integrand (in the stochastic integral with respect to the standard Wiener process) is constant, the integral is gaussian with zero mean and variance $\int_0^t e^{2rs}ds$. Therefore, $Y(t)$ has a gaussian distribution with mean $Y_0 e^{-rt}$ and variance $\frac{\sigma^2}{2r}(1 - e^{-2rt})$. When $t \to +\infty$, the distribution of $Y(t)$ converges to its stationary distribution.

Acknowledgements

This work was developed at CIMA-UE (Centro de Investigação em Matemática e Aplicações da Universidade de Évora), a reseach center funded by the portugueses PRAXIS

XXI program of FCT (Fundação para a Ciência e a Tecnologia). I thank Professor Drumi Bainov for the kind invitation to present this paper.

References

1. L. Arnold, Stochastic Differential Equations: Theory and Applications, Wiley, New York (1974).
2. C. A. Braumann, in: Nonlinear Stochastic Problems, R.S. Bucy and J.M.F.Moura, (Eds.), D. Reidel, Dordrecht (1983) pp. 553-559.
3. R. M. Capocelli and L. M. Ricciardi, Theoret. Popul. Biology, **5**, 28-41 (1974).
4. B. Dennis and G. P. Patil, Math. Biosc., **68**, 187-212 (1984).
5. M. W. Feldman and J. Roughgarden, Theoret. Popul. Biology, **7**, 197-207 (1975).
6. R. Levins, Proc. Natl. Acad. Sci. USA, **62**, 1061-1065 (1969).
7. R. M. May, Amer. Naturalist, **107**, 621-650 (1973).
8. H. C. Tuckwell, Theoret. Popul. Biology, **5**, 345-357 (1974).
9. J. R. Beddington and R. M. May, Science, **197**, 463-465 (1977).
10. C. A. Braumann, in: Proc. VIII Jornadas Luso-Espanholas de Matemática, Universidade de Coimbra, (1981) Vol. II, pp. 301-308.
11. C. A. Braumann, in: Mathematics in Biology and Medicine, V. Capasso, E. Grosso, and L. S. Paveri-Fontana, (Eds.), Springer, Berlin (1985) pp. 201-206.
12. A. Gleit, Math. Biosc., **41**, 112-123 (1978).
13. R. M. May, J. R. Beddington, J. H. Horwood and J. G. Shepherd, Math. Biosc., **42**, 219-252 (1978).
14. E. M. Lungu and B. Øksendal, Math. Biosc., **145**, 47-75 (1997).
15. L. H. R. Alvarez and L. A. Shepp, Optimal Harvesting of Stochastically Fluctuating Populations, J. Math. Biol., (in press).
16. C. A. Braumann, in: Mathematics Applied to Biology and Medicine, J. Demongeot and V. Capasso, (Eds.), Wuerz Publ. Ltd., Winnipeg, Canada (1993) pp. 155-161.
17. C. A. Braumann, Variable Effort Fishing Models in Random Environments, Math. Biosc., (in press).
18. C. A. Braumann, in: Bom Senso e Sensibilidade, Traves Mestras da Estatistica, J. Branco, P. Gomes, and J. Prata, (Eds.), Sociedade Portuguesa de Estatistica and Edições Salamandra, Lisboa, Portugal (1996) pp.103-117.
19. C. A. Braumann, Bull. Intl. Statistical Inst.,**LVII**, CP1, 21-22 (1997).
20. C. A. Braumann, in: Proc. II Marrakesh Intl. Conf. on Differential Equations, invited talk, (in press).
21. C. A. Braumann, J. Biological Systems, **3**, 505-517 (1995).
22. B. Øksendal, Stochastic Differential Equations. An Introduction with Applications (5th ed.), Springer, New York (1998).
23. G.F. Gause, The Struggle for Existence, Williams and Wilkins, Baltimore, USA (1934).
24. S. Karlin and H.M. Taylor, A Second Course in Stochastic Processes, Academic Press, Orlando (1981).

9th Int. Coll. on Differential Equations, pp. 53-60
D. Bainov (Ed)

Nonlocal Evolution Problems

LUDWIK BYSZEWSKI
Institute of Mathematics, Cracow University of Technology, Warszawska 24, 31-155 Cracow, Poland

Abstract: The existence, uniqueness and asymptotic stability of weak, mild and classical solutions of semilinear and nonlinear functional-differential abstract nonlocal Cauchy problems with evolution operators in general Banach spaces are presented. Methods of C_0 semigroups, m-accretive operators and fixed point theorems are applied. A detailed description of the results is given in publications [1], [2] and [3]. The results presented, concerning to the nonlocal problems, are a continuation and a generalization of those given in [4], [5] and [6]. They are also generalizations of some results from publications [7-8] and [9-14]. A physical interpretation is studied in the paper.
Key Words and Phrases: *abstract Cauchy problems; functional-differential evolution problems; nonlocal conditions; weak, mild and classical solutions; existence, uniqueness and stability of the solutions; C_0 semigroups and m-accretive operators.*
1991 Mathematics Subject Classification. 34G20, 34K30, 47D03, 47H20, 47H06, 34K25, 34K99.

1. Introduction

In Sections 2 and 3 of this paper we present, in general Banach spaces, six theorems on the existence, uniqueness and asymptotic stability of weak, mild and classical solutions of semilinear and nonlinear functional-differential abstract nonlocal Cauchy problems with evolution operators independent of t and dependent on t. Detailed proofs of those theorems are given in [1], [2] and [3]. In the proofs of the theorems, methods of C_0 semigroups, of nonlinear semigroups, of m-accretive operators and of fixed point theorems are applied. The results presented, concerning to the nonlocal problems, are a continuation and a generalization of those given in [4], [5] and [6]. Thery are also generalizations of some results about nonlocal evolution problems from publications [7-8] and about classical evolution problems from publications [9-14]. A physical interpretation of the nonlocal problems from Sections 2 and 3 is shown in Section 4. The notations of Sections 2 and 3 are mutually independent, and the notations of Section 4 are as in Section 3.

2. A Functional-Differential Nonlocal Evolution Cauchy Problem with an Evolution Operator Independent of t

The aim of this section is a formulation of four theorems about the existence and uniqueness of mild and classical solutions of a functional-differential abstract nonlocal evolution Cauchy problem with an evolution operator, independent of t, in a general Banach space.

In the proofs of those theorems methods of C_0 semigroups are applied. A detailed description of this part of the paper is given in [1] and [3].

The results presented are a continuation and a generalization of those given in [4] and [6] on the existence and uniqueness of solutions of a semilinear differential evolution nonlocal Cauchy problem

$$u'(t) + Au(t) = h(t, u(t)), \quad t \in (t_0, t_0 + a], \quad u(t_0) + g(u) = u_0$$

introduced in [4].

The results presented are also generalizations of those given by K. Balachandran and S. Ilamaran in [7] and by K. Balachandran and M. Chandrasekaran in [8] on the existence and uniqueness of mild and strong solutions of evolution nonlocal problems. They are also generalizations of some results given by A. Pazy and T. Winiarska in [13] and [14] on the existence and uniqueness of solutions of a semilinear differential evolution Cauchy problem

$$u'(t) + Au(t) = h(t, u(t)), \quad t \in (t_0, t_0 + a], \quad u(t_0) = u_0.$$

Let E be a Banach space with norm $\|\cdot\|$ and let $\{T(t)\}_{t\geq 0}$ be a C_0 semigroup of operators on E. A C_0 semigroup $\{T(t)\}_{t\geq 0}$ is said to be a *compact C_0 semigroup* of operators on E if $T(t)$ is a compact operator for every $t > 0$. A compact C_0 semigroup of operators on E will be denoted by $\{T(t)\}$.

We will assume that $-A$ is the infinitesimal generator of a C_0 semigroup $\{T(t)\}_{t\geq 0}$ of operators on E or of a compact C_0 semigroup $\{T(t)\}$ of operators on E, $D(A)$ is the domain of A, $t_0 \geq 0$, $a > 0$, $I := [t_0, t_0 + a]$, $M := \sup_{t\in[0,a]} \| T(t) \|_{BL(E,E)}$ and $X := C(I, E)$.

We will present four theorems on the existence and uniqueness of the mild and classical solutions to the following problem:

$$u'(t) + Au(t) = f(t, u(t), u(b(t))), \quad t \in (t_0, t_0 + a], \tag{2.1}$$

$$u(t_0) + g(u) = u_0, \tag{2.2}$$

where $f : I \times E^2 \longrightarrow E$, $g : X \longrightarrow E$ and $b : I \longrightarrow I$.

A function $u \in X$ satisfying the integral equation

$$u(t) = T(t - t_0)u_0 - T(t - t_0)g(u)$$
$$+ \int_{t_0}^{t} T(t - s)f(s, u(s), u(b(s)))ds, \quad t \in I,$$

is said to be a *mild solution* of the nonlocal Cauchy problem (2.1)-(2.2).

Applying some properties of a C_0 semigroup and the Banach contraction theorem (see [1]) we obtain:

Theorem 2.1. *Assume that:*

(i) *$-A$ is the infinitesimal generator of a C_0 semigroup $\{T(t)\}_{t\geq 0}$ and $u_0 \in E$,*

(ii) *$f : I \times E^2 \longrightarrow E$ is continuous with respect to the first variable on I, $g : X \longrightarrow E$, $b \in C(I, I)$ and there are constants $L > 0$, $K > 0$ such that $M(2aL + K) < 1$,*

$$\| f(s, z_1, z_2) - f(s, \tilde{z}_1, \tilde{z}_2) \| \leq L \sum_{i=1}^{2} \| z_i - \tilde{z}_i \|$$

$$\text{for} \quad s \in I, \ z_i, \tilde{z}_i \in E \ (i = 1, 2)$$

and

$$\| g(w) - g(\tilde{w}) \| \leq K \| w - \tilde{w} \|_X \quad \text{for} \quad w, \tilde{w} \in X.$$

Then the nonlocal Cauchy problem (2.1)-(2.2) has a unique mild solution.

A function $u : I \longrightarrow E$ is said to be a *classical solution* of the nonlocal Cauchy problem (2.1)-(2.2) if u is continuous on I and continuously differentiable on $I \setminus \{t_0\}$, $u'(t) + Au(t) = f(t, u(t), u(b(t)))$ for $t \in I \setminus \{t_0\}$, and $u(t_0) + g(u) = u_0$.

As a consequence of Theorem 2.1 and of an argument given in [1] we get:

Theorem 2.2. *Assume that:*

(i) *E is a reflexive Banach space, $-A$ is the infinitesimal generator of a C_0 semigroup $\{T(t)\}_{t\geq 0}$ and $u_0 \in E$,*

(ii) *$f : I \times E^2 \longrightarrow E$, $g : X \longrightarrow E$, $b \in C(I, I)$ and there are constants $L > 0$, $K > 0$ such that $M(2aL + K) < 1$,*

$$\| f(s, z_1, z_2) - f(\tilde{s}, \tilde{z}_1, \tilde{z}_2) \| \leq L(| s - \tilde{s} | + \sum_{i=1}^{2} \| z_i - \tilde{z}_i \|)$$

$$\text{for} \quad s, \tilde{s} \in I, \ z_i, \tilde{z}_i \in E \quad (i = 1, 2)$$

and

$$\| g(w) - g(\tilde{w}) \| \leq K \| w - \tilde{w} \|_X \quad \text{for} \quad w, \tilde{w} \in X.$$

Then the nonlocal Cauchy problem (2.1)-(2.2) has a unique mild solution u. Moreover, if $u_0 \in D(A)$, $g(u) \in D(A)$, and if there exists a positive constant κ such that

$$\| u(b(s)) - u(b(\tilde{s})) \| \leq \kappa \| u(s) - u(\tilde{s}) \| \quad \text{for} \quad s, \tilde{s} \in I \tag{2.3}$$

then u is the unique classical solution of problem (2.1)-(2.2).

Define two sets $E_\rho := \{z \in E : \| z \| \leq \rho\}$ and $X_\rho := \{w \in X : \| w \|_X \leq \rho\}$, where $\rho > 0$.

We will need the following assumptions:

Assumption 2.1. $f \in C(I \times E^2, E)$, $g \in C(X, E)$ and $b \in C(I, I)$. Additionaly, there are $C_i > 0$ $(i = 1, 2)$ such that $\| f(s, z_1, z_2) \| \leq C_1$ for $s \in I, z_i \in E_r$ $(i = 1, 2)$, and $\| g(w) \| \leq C_2$ for $w \in X_r$, where

$$r := M(aC_1 + \| u_0 \| + C_2). \tag{2.4}$$

Assumption 2.2. $g(\lambda w_1 + (1 - \lambda) w_2) = \lambda g(w_1) + (1 - \lambda) g(w_2)$ for $w_i \in X_r$ $(i = 1, 2)$, $\lambda \in (0, 1)$, where r is given by (2.4).

Assumption 2.3. The set $\{w(t_0) : w \in X_r, \ w(t_0) = u_0 - g(w)\}$, where r is given by (2.4), is precompact in E.

Applying some properties of a compact C_0 semigroup and Schauder's fixed point theorem (see [3]) we obtain:

Theorem 2.3. *Suppose that $-A$ is the infinitesimal generator of a compact C_0 semigroup $\{T(t)\}$, and the functions f, g and b satisfy Assumptions 2.1-2.2, where $u_0 \in E$. Then in the class of all the functions w, for which Assumption 2.3 holds, the nonlocal Cauchy problem (2.1)-(2.2) has a mild solution.*

As a consequence of Theorem 2.3 and of an argument given in [1] and [3] we get:

Theorem 2.4. *Suppose that* $-A$ *is the infinitesimal generator of a compact* C_0 *semigroup* $\{T(t)\}$, *and the functions* f, g *and* b *satisfy Assumptions 2.1-2.2, where* $u_0 \in E$. *Then in the class of all the functions* w, *for which Assumption 2.3 holds, the nonlocal Cauchy problem (2.1)-(2.2) has a mild solution* u. *Assume, additionally, that:*

(i) E *is a reflexive Banach space and there exists a constant* $L > 0$ *such that*

$$\| f(s, z_1, z_2) - f(\tilde{s}, \tilde{z}_1, \tilde{z}_2) \| \leq L(| s - \tilde{s} | + \sum_{i=1}^{2} \| z_i - \tilde{z}_i \|)$$

$$\text{for} \quad s, \tilde{s} \in I, \quad z_i, \tilde{z}_i \in E_r \quad (i = 1, 2),$$

(ii) u *is the unique mild solution of problem (2.1)-(2.2), there is a constant* $\kappa > 0$ *such that inequality (2.3) holds,* $g(u) \in D(A)$ and $u_0 \in D(A)$.

Then u *is the unique classical solution of the nonlocal Cauchy problem (2.1)-(2.2).*

3. A Functional-Differential Nonlocal Evolution Cauchy Problem with an Evolution Operator Dependent on t

The aim of this section is a formulation of two theorems about the existence, uniqueness and asymptotic stability of weak solutions of functional-differential abstract nonlocal evolution Cauchy problems with an evolution operator, dependent on t, in a general Banach space. In the proofs of those theorems methods of nonlinear semigroups and of m-accretive operators are applied. A detailed description of this part of the paper is given in [2].

The results presented (concerning to the existence and uniqueness of the solutions) are a continuation and a generalization of those given in [4], [5] and [6] on the existence and uniqueness of solutions of Cauchy problems for semilinear evolution equations together with nonlocal conditions. The generalization consists, among other things, in this that the assumptions concerning to the nonlinear term in the nonlocal condition are weaker that in [4], [5] and [6]. Moreover, the evolution equation, studied here, is more general than in [4], [5] and [6]. The presented theorems are also generalizations of some results, about classical evolution problems, obtained by A. Kartsatos, M. Parrott, M. Crandall, A. Pazy and E. Evans in [9-12].

Let E be a Banach space with norm $\| \cdot \|$ and let $C := C([-r, 0], E)$, where r is a positive constant. The Banach space C is equipped with the norm $\| \cdot \|_C$ given by

$$\| \psi \|_C := \sup_{t \in [-r,0]} \| \psi(t) \| \quad \text{for} \quad \psi \in C.$$

Let $T > r$ and $t_0 \in (a, T - r)$, where $a \geq 0$ will be defined in the sequel.

For a continuous function $w : [-r, T] \longrightarrow E$, we denote by $w_{(t)}$ the function belonging to C and given by the formula

$$w_{(t)}(\tau) := w(t + \tau) \quad \text{for} \quad t \in [0, T], \quad \tau \in [-r, 0].$$

For each m-accretive operator $B : E \supset D(B) \longrightarrow E$ and for each $\lambda > 0$, by B_λ we denote the Yoshida approximation given by $B_\lambda := B(I + \lambda B)^{-1}$.

Assumption 3.1. For each $t \in [0, T]$, $A(t) : E \supset D(A(t)) \longrightarrow E$ is m -accretive operator, and there are $\lambda_0 > 0$, a continuous nondecreasing function $\ell : [0, \infty) \longrightarrow [0, \infty)$ and a continuous function $h : [0, T] \longrightarrow E$ such that

$$\| A_\lambda(t)x - A_\lambda(s)x \| \leq \ell(\| x \|) \| h(t) - h(s) \|$$

$$\text{for all} \quad \lambda \in (0, \lambda_0), \quad x \in \overline{D(A(t))}, \quad t, s \in [0, T].$$

Assumption 3.1 implies that the set $\overline{D(A(t))}$ is independent of t. Therefore, we will denote this set by $\overline{D}$ if Assumption 3.1 is satisfied.

Define a set C_0 by the formula $C_0 := \{\psi \in C : \psi(0) \in \overline{D}\}$.

Let $f : [0, T] \times C \longrightarrow E$. We will need the assumption:

Assumption 3.2. There exists a constant $L > 0$ such that

$$\| f(s, \psi_1) - f(s, \psi_2) \| \leq L \| \psi_1 - \psi_2 \|_C, \quad s \in [0, T], \quad \psi_1, \psi_2 \in C,$$

and there are a continuous nondecreasing function $\omega : [0, \infty) \longrightarrow [0, \infty)$ and a continuous function $k : [0, T] \longrightarrow E$ such that

$$\| f(s_1, \psi) - f(s_2, \psi) \| \leq \omega(\| \psi \|_C) \| k(s_1) - k(s_2) \|, \quad s_1, s_2 \in [0, T], \quad \psi \in C.$$

For a given function $\tilde{f} : [0, T] \longrightarrow E$ and a given $x \in E$, a function $u \in C([0, T], E)$ is said to be a *weak solution* of the problem

$$w'(t) + A(t)w(t) = \tilde{f}(t), \quad t \in [0, T], \quad w(0) = x,$$

if for every $n \in N$ there are sequences $\{t_{n,j}\}_{j=0,1,\dots,N(n)} \subset R$, $\{u_{n,j}\}_{j=0,1,\dots,N(n)} \subset E$ and $\{\tilde{f}_{n,j}\}_{j=1,\dots,N(n)} \subset E$ such that

(i) $0 = t_{n,0} < t_{n,1} < \dots < t_{n,N(n)} = T$ and

$$\lim_{n \longrightarrow \infty} \max_{j \in \{1, \dots, N(n)\}} (t_{n,j} - t_{n,j-1}) = 0,$$

(ii) $u_{n,0} := x$ and

$$\frac{u_{n,j} - u_{n,j-1}}{t_{n,j} - t_{n,j-1}} + A(t_{n,j})u_{n,j} = \tilde{f}_{n,j} \quad (j = 1, \dots, N(n)),$$

(iii) $\tilde{f}_n$ is convergent to $\tilde{f}$ in $L^1(0, T; E)$, where $\tilde{f}_n(t) := \tilde{f}_{n,j}$ for $t \in (t_{n,j-1}, t_{n,j}]$ $(j = 1, \dots, N(n))$, and u_n converges uniformly to u on $[0, T]$, where

$$u_n(t) := \begin{cases} u_{n,0} & \text{for } t = t_{n,0}, \\ u_{n,j} & \text{for } t \in (t_{n,j-1}, t_{n,j}] \quad (j = 1, \dots, N(n)). \end{cases}$$

For given functions $f : [0, T] \times C \longrightarrow E$ and $\phi \in C_0$, a function $u \in C([-r, T], E)$ is said to be a *weak solution* of the problem

$$w'(t) + A(t)w(t) = f(t, w_{(t)}), \quad t \in [0, T], \quad w_{(0)} = \phi,$$

if $u(t) = \phi(t)$ for $t \in [-r, 0]$ and u is a weak solution of the problem

$$w'(t) + A(t)w(t) = f(t, u_{(t)}), \quad t \in [0, T], \quad w(0) = \phi(0).$$

Let $g : C \longrightarrow C_0$. We will need the following assumption:

Assumption 3.3. There exist constants $M > 0$ and $\beta \in R$ such that

$$\left\| g\left(w_{(\hat{T})}\right) - g\left(\tilde{w}_{(\hat{T})}\right) \right\|_C \leq M e^{\beta t_0} \| w - \tilde{w} \|_{C([t_0, \hat{T}], E)}$$

for all $w, \tilde{w} \in C([-r, T], E)$ and $\hat{T} \in [t_0 + r, T]$.

Applying some properties of nonlinear semigroups and the Banach contraction theorem (see [2]) we obtain:

Theorem 3.1. *Suppose that the operators* $A(t)$, $t \in [0,T]$, *and the functions* $f : [0,T] \times C \longrightarrow E$ *and* $g : C \longrightarrow C_0$ *satisfy Assumptions 3.1-3.3. Moreover, suppose that there is* $\alpha > L$ *such that, for each* $t \in [0,T]$, $A(t) - \alpha I$ *is accretive. Then for each* $T^* \in [t_0 + r, T]$, *where* $t_0 \in \left(\frac{\max\{0,\ln(M)\}}{\alpha-L-\beta}, T-r\right)$, $\ln(M) < (\alpha - L - \beta)(T-r)$ *and* $\beta < \alpha - L$, *there is a unique* $\phi_* \in C_0$ *and exactly one weak solution* $u_* : [-r,T] \longrightarrow E$ *of the problem*

$$w'(t) + A(t)w(t) = f(t, w_{(t)}), \quad t \in [0,T], \quad w_{(0)} = \phi_* \tag{3.1}$$

satisfying the condition

$$(u_*)_{(0)} = g\left((u_*)_{(T^*)}\right) = \phi_*. \tag{3.2}$$

Moreover, for the (unique) weak solution u_ϕ *of the problem*

$$w'(t) + A(t)w(t) = f(t, w_{(t)}), \quad t \in [0,T], \quad w_{(0)} = \phi,$$

where ϕ *is an arbitrary function belonging to* C_0, *the following inequality holds:*

$$\left\| u_\phi(t) - u_*(t) \right\| \leq e^{-(\alpha-L)t} \left\| \phi - g\left((u_*)_{(T^*)}\right) \right\|_C, \quad t \in [0,T].$$

As a consequence of an argument similar to the argument from the proof of Theorem 3.1 (see [2]), we obtain:

Theorem 3.2. *Suppose that the operators* $A(t)$, $t \in [0,\infty)$, *and the functions* $f : [0,\infty) \times C \longrightarrow E$ *and* $g : C \longrightarrow C_0$ *satisfy Assumptions 3.1 - 3.3 on the interval* $[0,\infty)$ *in the place of* $[0,T]$. *Moreover, suppose that there is* $\alpha > L$ *such that, for each* $t \in [0,\infty)$, *the operator* $A(t) - \alpha I$ *is accretive. Then for each* $T^* > t_0 + r$, *where* $t_0 > \frac{\max\{0,\, ln(M)\}}{\alpha-L-\beta}$ *and* $\beta < \alpha - L$, *there is a unique* $\phi_* \in C_0$ *and exactly one weak solution* $u_* : [-r,\infty) \longrightarrow E$ *of the problem*

$$w'(t) + A(t)w(t) = f(t, w_{(t)}), \quad t \in [0,\infty), \quad w_{(0)} = \phi_*,$$

satisfying the condition

$$(u_*)_{(0)} = g\left((u_*)_{(T^*)}\right) = \phi_*.$$

Moreover, for the (unique) weak solution u_ϕ *of the problem*

$$w'(t) + A(t)w(t) = f(t, w_{(t)}), \quad t \in [0,\infty), \quad w_{(0)} = \phi,$$

where ϕ *is an arbitrary function belonging to* C_0, *the following inequality holds:*

$$\left\| u_\phi(t) - u_*(t) \right\| \leq e^{-(\alpha-L)t} \left\| \phi - g\left((u_*)_{(T^*)}\right) \right\|_C, \quad t \in [0,\infty).$$

Consequently, u_* *is asymptotically stable.*

Remark 3.1. Let g be a function defined by the formula

$$g(\psi) = M e^{\beta t_0} \psi \quad \text{for} \quad \psi \in C, \tag{3.3}$$

where $M > 0$, $\beta < \alpha - L$, $\alpha > L$, $ln(M) < (\alpha - L - \beta)(T-r)$ (L is the constant from Assumption 3.2) and $t_0 \in \left(\frac{\max\{0,\, ln(M)\}}{\alpha-L-\beta}, T-r\right)$.

Assume that the following condition holds:

$$\psi \in C \Longrightarrow M e^{\beta t_0} \psi(0) \in \overline{D}.$$

Then $g : C \longrightarrow C_0$ and g satisfies Assumption 3.3. Consequently, Theorem 3.1 can be applied if the other assumptions are satisfied. In particular, for each $T^* \in [t_0 + r, T]$ the nonlocal condition (3.2) is of the form

$$u_*(t) = M e^{\beta t_0} u_*(t + T^*) \quad \text{for} \quad t \in [-r, 0]. \tag{3.4}$$

It is easy to see that if the interval $[0, T]$ is replaced by $[0, \infty)$ in (3.3) then g, given by such modified formula (3.3), satisfies Assumption 3.3 on $[0, \infty)$ and Theorem 3.2 can be applied if M, β and t_0 satisfy the suitable assumptions of Theorem 3.2. Moreover, the nonlocal condition (3.2) is of the form (3.4).

Remark 3.2. Let g be a function defined by the formula

$$(g(\psi))(t) = \frac{M e^{\beta t_0}}{r} \int_{-r}^{t} \psi(\tau) d\tau \quad \text{for} \quad \psi \in C, \quad t \in [-r, 0], \tag{3.5}$$

where $M > 0$, $\beta < \alpha - L$, $\alpha > L$, $ln(M) < (\alpha - L - \beta)(T - r)$ (L is the constant from Assumption 3.3) and $t_0 \in \left(\frac{\max\{0,\, ln(M)\}}{\alpha - L - \beta}, T - r\right)$.

Assume that the following condition holds:

$$\psi \in C \Longrightarrow \frac{M e^{\beta t_0}}{r} \int_{-r}^{0} \psi(\tau) d\tau \in \bar{D}.$$

Then $g : C \longrightarrow C_0$ and g satisfies Assumption 3.3. Consequently, Theorem 3.1 can be applied if the other assumptions are satisfied. In particular, for each $T^* \in [t_0 + r, T]$ the nonlocal condition (3.2) is of the form

$$u_*(t) = \frac{M e^{\beta t_0}}{r} \int_{-r}^{t} u_*(\tau + T^*) d\tau \quad \text{for} \quad t \in [-r, 0]. \tag{3.6}$$

It is easy to see that if the interval $[0, T]$ is replaced by $[0, \infty)$ in (3.5) then g, given by such modified formula (3.5), satisfies Assumption 3.3 on $[0, \infty)$ and Theorem 3.2 can be applied if M, β and t_0 satisfy the suitable assumptions of Theorem 3.2. Moreover, the nonlocal condition (3.2) is of the form (3.6).

4. Physical Interpretations

Theorems 2.1-2.4 can be applied, among other things, to descriptions of some physical phenomena for which we cannot measure $u(t_0)$ and $g(u)$ but we can measure a relation between $u(t_0)$ and $g(u)$ in the form of the nonlocal condition (2.2). Moreover, Theorems 3.1 - 3.2 can be also applied to descriptions of the asymptotic stability of solutions of physical phenomena. For this purpose consider the problem

$$\begin{aligned} u'(t) + A(t)u(t) &= f(t, u_{(t)}), \quad t \in [0, T], \\ u(t) = M e^{\beta t_0} u(t + T^*), &\ t \in [-r, 0],\ T^* \in [t_0 + r, T]. \end{aligned} \tag{4.1}$$

Observe that Theorems 3.1-3.2, formulated for the nonlocal problems in the versions of (4.1), can be applied in physics to the simultaneous description of the motion trajectories $t \longrightarrow u(t)$ of two rockets. The first rocket is launched to the cosmic space from the Earth surface outside of the Earth gravitation. The second rocket is launched from a Earth orbit, after this the rocket encircles the Earth (changing maybe the orbit), and next the rocket goes to the cosmic space in the same direction as the first rocket.

Namely, the result says that for each times $T^* \in [t_0 + r, T]$ and $\tau \in [-r, 0]$ there exists a unique initial location $\phi_*(\tau)$ of the rocket (a unique Earth orbit on which the rocket is kept) and there is the only one motion trajectory $t \longrightarrow u_*(t)$, $t \in [-r, T]$, of the rocket such that

$$u_*'(t) + A(t)u_*(t) = f(t, (u_*)_{(t)}), \quad t \in [0, T],$$

and

$$u_*(\tau) = Me^{\beta t_0}u_*(\tau + T^*) = \phi_*(\tau), \quad \tau \in [-r, 0]. \tag{4.2}$$

It means, particularly, that for $t \in [0, T^*]$ the rocket will encircle the Earth and, at time T^*, the location $u_*(T^*)$ of the rocket will satisfy the nonlocal condition (4.2), where $\tau = 0$. Moreover, Theorem 3.2 says that for $t = \infty$ all the motion trajectories $t \longrightarrow u_\phi(t)$ of the rocket, corresponding to the local problem, will tend exponentially to the trajectory $t \longrightarrow u_*(t)$.

References

1. L. Byszewski, Cracow University of Technology, Anniversary Issue, **6**, 25-33 (1995).
2. L. Byszewski, Annales Polonici Mathematici, **65**, 163-170 (1997).
3. L. Byszewski and H. Akca, Nonlinear Analysis, Theory, Methods & Applications, **34**, 65-72 (1998).
4. L. Byszewski, J. Math. Anal. Appl., **162**, 494-505 (1991).
5. L. Byszewski, J. Appl. Math. Stoch. Anal., **6**, no. 1, 49-54 (1993).
6. L. Byszewski, Differential and Functional -Differential Problems with Nonlocal Conditions, Cracow University of Technology, Monograph 184, Cracow (1995).
7. K. Balachandran and S. Ilamaran, Indian J. Pure Appl. Math., **25**, no. 4, 411-418 (1994).
8. K. Balachandran and M. Chandrasekaran, Indian J. Pure Appl. Math., **27**, 443-449 (1996).
9. M. Crandall and A. Pazy, Israel J. Math., **11**, 57-94 (1972).
10. L. Evans, Israel J. Math., **26**, 1-42 (1976).
11. A. Kartsatos and M. Parrott, Lecture Notes in Math., **1076**, Springer-Verlag, Berlin (1984) pp. 115-122.
12. A. Kartsatos and M. Parrott, J. Differential Equations, **75**, no. 2, 290-302 (1988).
13. A. Pazy, Semigroups of Linear Operators and Applications to Partial Differential Equations, Springer-Verlag, New York, Berlin, Tokyo (1983).
14. T. Winiarska, Bull. Polon. Acad. Sci., Math. Ser., **37**, 157-162 (1989).

9th Int. Coll. on Differential Equations, pp. 61-75
D. Bainov (Ed)

Matrix Semigroups and Differential Equations: Ergodism and Asymptotic Behaviour of Solutions with or without Impulses, a Survey (2)

JEAN-PIERRE CAUBET
University of Poitiers, F-86022, Poitiers, France

Abstract: General structure of a matrix nonstationary semigroup dominated by a nonnegative matrix semigroup and with cardinality of state spaces uniformly bounded. Absolute value, reduction, intervals of regularity and assimilated involved by Baire's theory, Taylor expansion of the reduction of these intervals and ergodism of their absolute values. Asymptotic behaviour of the solutions of the nonlinear impulsive ordinary differential equations of backward and forward Fokker-Planck type associated with such semigroups.
Key Words and Phrases: *Matrix semigroups, nonstationary, general structure theory. Ergodism of Markov chains with continuous parameter. Impulsive nonlinear ordinary differential equations, asymptotic behaviour of their solutions.*
1991 Mathematics Subject Classification. 34A30, 34A37, 34D05, 20M10, 60J27.

1. Survey

The whole exposition nearly follows the present scheme:

(1) A dominated matrix semigroup $\mathbb{Q}$ admits an absolute value $|\mathbb{Q}|$, a nonnegative matrix semigroup giving strict bounds. Heuristically, we shall then have $\mathbb{Q} = |\mathbb{Q}| \exp(i\Theta)$, though the strict definition and meaning of this relation is given by the Taylor expansion of $\mathbb{Q}$ (see [1]).

(2) The reduction $|\mathbb{Q}|^*$ of $|\mathbb{Q}|$ (see [1]) leads to a Baire's expansion of the time parameter set into intervals of regularity or assimilated, on each one of which $|\mathbb{Q}|^*$ defines: 1) a connection between the entrance cones and the exit cones, that is, the connection of a fiber bundle in which the basis is a stonian time set constituted of the exit times and the entrance times, and in which the fibers are the exit cones and the entrance cones; 2) a probability measure constructed like the Wiener measure, due to the fact that $|\mathbb{Q}|^*$ is continuous, and which determines the ergodic behaviour of $|\mathbb{Q}|^*$ and of $|\mathbb{Q}|$.

(3) The connection leads to a Taylor expansion $|\mathbb{Q}|^* = \sum_{n\geq 0} {}^{(n)}|\mathbb{Q}|^*$ for $|\mathbb{Q}|^*$, and then for $|\mathbb{Q}|$, of which the existence and unicity result from the probability measure directly, which is obtained by quadratures only, and which can be turned into the Taylor expansion $\mathbb{Q}^* = \sum_{n\geq 0} {}^{(n)}\mathbb{Q}^*$ of the reduction $\mathbb{Q}^*$ of $\mathbb{Q}$, and then of $\mathbb{Q}$ itself.

(4) The ergodic behaviour of $|\mathbb{Q}|^*$ and $|\mathbb{Q}|$, which bounds the asymptotic behaviour of $\mathbb{Q}^*$ and $\mathbb{Q}$, is expressed in terms of (extremal) exit laws, (extremal) exit channels, and (active) continuous exit sections, with a criterion for determining orthogonality between (extremal) exit laws or channels, and between active continuous exit sections.

(5) The asymptotic behaviour of the solutions of the backward and forward Fokker-Planck equations, associated with the matrix semigroup $\mathbb{Q}^*$ on the intervals of regularity and assimilated, is bounded by the ergodic behaviour of $|\mathbb{Q}|^*$ and is determined by the Taylor expansion of $\mathbb{Q}^*$, the ergodic theorem determining the homotopy compatible with the asymptotic exit cone.

2. The Ergodic Theorem (detailed)

Given a nonnegative semigroup $\mathbb{Q} = |\mathbb{Q}|^*$, we introduce its associated function of time interval

$$\exp(-\Phi_\tau^\sigma) = \sup_i \sum_j \mathbb{Q}_\tau^\sigma(i,j)$$

which defines the additive measure $d\Phi$ such that

$$-\infty < \int_\sigma^\tau d\Phi = \lim \downarrow [\Phi_\cdot^\sigma + \ldots + \Phi_\tau^\cdot] < +\infty,$$

so the normalized nonnegative semigroup

$$N(\mathbb{Q}) = \exp\left[-\int d\Phi\right]\mathbb{Q}$$

is just right submarkovian, and by adding an absorbing section or state to E^*, we can directly consider $N(\mathbb{Q})$ as being right markovian.

Definition (Asymptotic equivalence). We shall say that the nonnegative semigroups $\mathbb{Q}$ and $\mathbb{Q}'$ are weakly (resp. strongly) asymptotically equivalent at the end of the interval of regularity $]a, b[$ when, for any $\varepsilon > 0$, we have

$$N(\mathbb{Q})_\tau^\sigma(\cdot,\cdot) - \varepsilon \le N(\mathbb{Q}')_\tau^\sigma(\cdot,\cdot) \le N(\mathbb{Q})_\tau^\sigma(\cdot,\cdot) + \varepsilon$$

(resp.,

$$N(\mathbb{Q})_\tau^\sigma e^{-\varepsilon} \le N(\mathbb{Q}')_\tau^\sigma \le N(\mathbb{Q})_\tau^\sigma e^{+\varepsilon})$$

in the limit as σ and τ (with $\sigma < \tau$) tend to b_-.

Strong asymptotic equivalence implies weak one, but the converse is not true. For instance, the semigroups

$$\mathbb{Q} = e^{-t}\begin{pmatrix} 1 & 0 \\ t & 1 \end{pmatrix} \quad \text{and} \quad \mathbb{Q}' = \begin{pmatrix} 0 & 0 \\ 0 & 0 \end{pmatrix}$$

are weakly, but not strongly, asymptotically equivalent. In fact, if $\mathbb{Q}$ and $\mathbb{Q}'$ are strongly asymptotically equivalent, we necessarily have $\mathbb{Q} \ll \mathbb{Q}'$ and $\mathbb{Q}' \ll \mathbb{Q}$, that is, we have $\mathbb{Q}(\cdot,\cdot) = 0$ whenever $\mathbb{Q}'(\cdot,\cdot) = 0$, and conversely.

Proposition (Nonnegative semigroups asymptotically equivalent to a nonnegative right markovian semigroup have isomorphic exit boundaries). *Let $|\mathbb{Q}|^*$ be a nonnegative right markovian semigroup on the interval of regularity $]a, b[$ and $\tilde{\mathbb{Q}}$ a nonnegative semigroup asymptotically equivalent to $|\mathbb{Q}|^*$.*

Then the exit cones

$$\Gamma_{b_-}^+(|\mathbb{Q}|^*) \equiv \Gamma_{b_-}^+(N(|\mathbb{Q}|^*)) \quad \textit{and} \quad \Gamma_{b_-}^+(N(\tilde{\mathbb{Q}})),$$

reduced to the nonnegative uniformly bounded exit laws, are isomorphic.

Proof. Here we have

$$|\mathbb{Q}|^* - \varepsilon \leq N(\tilde{\mathbb{Q}}) \leq |\mathbb{Q}|^* + \varepsilon$$

where $|\mathbb{Q}|^*$ is right markovian, so that the exit cone $\Gamma^+_{b-}(|\mathbb{Q}|^*) \equiv \Gamma^+_{b-}(N(|\mathbb{Q}|^*))$ reduced to the nonnegative uniformly bounded exit laws is enough to express the exit closure $|\mathbb{Q}|^{*\sigma}_{b-}$ of $|\mathbb{Q}|^{*\sigma}_{\tau}$ with $\sigma < \tau$ on $]a, b[$.

Hence it follows that $|\mathbb{Q}|^*$ and $N(\tilde{\mathbb{Q}})$ tend to have the same values on their exit channels, so that their exit channels are the same for both, their boundaries are isomorphic, and their extremal uniformly bounded exit laws are in a one-to-one correspondence when normalized (see [1] for details).

Definition (Expurgated semigroups). Let $\mathbb{Q} = |\mathbb{Q}|^*$ be a reduced nonnegative right markovian semigroup of which, for the sake of simplicity, we shall write the Taylor expansion

$$\mathbb{Q} = \sum_{n\geq 0} {}^{(n)}\mathbb{Q} = {}^{(0)}\mathbb{Q} + \sum_{n\geq 0} \sum_{tot} \int {}^{(0)}\mathbb{Q} q^{(n)}\mathbb{Q},$$

where explicitly

$$\left(\sum_{tot} \int_{\sigma}^{\tau} {}^{(0)}\mathbb{Q} q^{(n)}\mathbb{Q}\right)(i,j) = \sum_{all\ k} \int_{\sigma}^{\tau} {}^{(0)}\mathbb{Q}^{\sigma}_{u-}(i,i) q_{ik}(du)^{(n)}\mathbb{Q}^{u+}_{\tau}(k,j).$$

We are going to expurgate this expansion in three ways:

First, we iteratively define

$$\begin{aligned}
{}^{(0)}\tilde{\mathbb{Q}} &= {}^{(0)}\mathbb{Q} \\
{}^{(1)}\tilde{\mathbb{Q}} &= \sum_{part} \int {}^{(0)}\tilde{\mathbb{Q}} q^{(0)}\tilde{\mathbb{Q}} \\
&\vdots \\
{}^{(n+1)}\tilde{\mathbb{Q}} &= \sum_{part} \int {}^{(0)}\tilde{\mathbb{Q}} q^{(n)}\tilde{\mathbb{Q}} \text{ et cetera,}
\end{aligned}$$

where $\sum_{part}$ means that in this summation all the convergent measures $q_{..}$ involved in $\sum_{tot}$ have been suppressed, we define the perturbation expurgated semigroup by its Taylor expansion

$$exprg_{|pert}(\mathbb{Q}) = \sum_{n\geq 0} {}^{(n)}\tilde{\mathbb{Q}}.$$

Secondly, if in the total summation $q_i = \sum_{tot} q_{ik}$ we cancel all the convergent measures $q_{i.}$, with this partial summation and through the following iterative procedure ($n \geq 0$)

$${}^{(0)}\tilde{\mathbb{Q}}(i,j) = \exp\left[-\int \sum_{part} q_{ik}\right] I(i,j)$$

$${}^{(n+1)}\tilde{\mathbb{Q}}(i,j) = \left(\sum_{tot} \int {}^{(0)}\tilde{\mathbb{Q}} q^{(n)}\tilde{\mathbb{Q}}\right) I(i,j),$$

we can define the exponentially expurgated semigroup

$$exprg_{|exp}(\mathbb{Q}) = \sum_{n\geq 0} {}^{(n)}\tilde{\mathbb{Q}}.$$

And finally, if we expurgate $\sum_{tot}$ into $\sum_{part}$ both in the perturbation procedure and in the exponentials, through the iterative procedure

$$^{(0)}\tilde{\mathbb{Q}} = \exp\left[-\int \sum_{part} q_{..}\right] I$$

$$^{(n+1)}\tilde{\mathbb{Q}} = \sum_{part} \int {}^{(0)}\mathbb{Q} q {}^{(n)}\tilde{\mathbb{Q}},$$

we define the completely expurgated right markovian semigroup

$$exprg(\mathbb{Q}) = \sum_{n \geq 0} {}^{(n)}\tilde{\mathbb{Q}}.$$

More generally, we shall also expurgate a nonnegative semigroup by adding or substracting to its measures q_i and q_{ij} some convergent measures, or modifying them by some convergent measures, in order to get some asymptotically equivalent nonnegative semigroups simpler than the primitive one, and simple criteria. And when divergent measures will be involved in an expurgation, we shall speak of strong expurgation.

Proposition (Weak asymptotic equivalence of $|\mathbb{Q}|^*$ and its expurgated semigroups). *Let $|\mathbb{Q}|^*$ be a reduced nonnegative right markovian semigroup on the interval of regularity $]a, b[$. Then we have the dominance relations and the associated projections as expressed in the following diagram*

$$\begin{array}{ccc} exprg_{|exp}(|\mathbb{Q}|^*) & \underset{\geq}{\longrightarrow} & |\mathbb{Q}|^* \\ \geq \big\downarrow & & \geq \big\downarrow \\ exprg(|\mathbb{Q}|^*) & \underset{\geq}{\longrightarrow} & exprg_{|pert}(|\mathbb{Q}|^*) \end{array}$$

and these semigroups are weakly asymptotically equivalent.

Proof. The first thing that we can observe is that the semigroups $|\mathbb{Q}|^*$ and $exprg(|\mathbb{Q}|^*)$ are right markovian.

Now, for the sake of simplicity, we shall write

$$\mathbb{Q} = |\mathbb{Q}|^* \quad \text{and} \quad \tilde{\mathbb{Q}} = exprg_{|pert}(|\mathbb{Q}|^*).$$

We have

$$\begin{aligned} \mathbb{Q} &= \sum_{n \geq 0} {}^{(n)}\mathbb{Q} \\ &= {}^{(0)}\mathbb{Q} + \sum_{n \geq 0} \sum_{tot} \int {}^{(0)}\mathbb{Q} q {}^{(n)}\mathbb{Q}, \end{aligned}$$

where explicitely

$$\left(\sum_{tot} \int_{\sigma}^{\tau} {}^{(0)}\mathbb{Q} q {}^{(n)}\mathbb{Q}\right)(i,j) = \sum_{all\ k \neq i} \int_{\sigma}^{\tau} {}^{(0)}\mathbb{Q}^{\sigma}_{u-}(i,i) q_{ik}(du) {}^{(n)}\mathbb{Q}^{u+}_{\tau}(k,j),$$

expansion that we now expurgate into the following transitions

$$
\begin{aligned}
{}^{(0)}\tilde{\mathbb{Q}} &= {}^{(0)}\mathbb{Q} \\
{}^{(n+1)}\tilde{\mathbb{Q}} &= \sum_{part} \int {}^{(0)}\tilde{\mathbb{Q}} q^{(n)}\tilde{\mathbb{Q}}.
\end{aligned}
$$

So, we have

$$
\begin{aligned}
{}^{(n)}\mathbb{Q} &= \sum_{tot} \int {}^{(0)}\mathbb{Q} q \ldots {}^{(0)}\mathbb{Q} \\
&= \sum_{part} \int {}^{(0)}\mathbb{Q} q \ldots {}^{(0)}\mathbb{Q} + {}^{(n)}\mathbb{R} \\
&= {}^{(n)}\tilde{\mathbb{Q}} + {}^{(n)}\mathbb{R},
\end{aligned}
$$

where explicitely

$$
{}^{(n)}\mathbb{R} = \sum^{*} \int {}^{(0)}\mathbb{Q} q \ldots {}^{(0)}\mathbb{Q},
$$

with in each one of the terms $\int {}^{(0)}\mathbb{Q} q \ldots {}^{(0)}\mathbb{Q}$ of the sum $\sum^{*}$ one convergent measure $q_{..}$ at least.

Then it follows that for any $\varepsilon > 0$ we shall have $\mathbb{R} = \sum_{n \geq 1} {}^{(n)}\mathbb{R} \leq \varepsilon$ as soon as both of σ and τ (with $\sigma < \tau$) are large enough, since $\mathbb{Q}$ is right markovian so that any measure $q_{..}$ is involved in a sum of terms not greater than one, at the most, and since the number of these measures is finite. So that we finally get

$$
\tilde{\mathbb{Q}} - \varepsilon \leq \tilde{\mathbb{Q}} \leq \mathbb{Q} = \sum_{n \geq 0} {}^{(n)}\mathbb{Q} = \sum_{n \geq 0} {}^{(n)}\tilde{\mathbb{Q}} + \mathbb{R} \leq \tilde{\mathbb{Q}} + \varepsilon,
$$

since, as we have observed,

$$
\begin{aligned}
0 &\leq \mathbb{R} = \sum_{n \geq 1} {}^{(n)}\mathbb{R} = \sum_{n \geq 1} \sum^{*} \int {}^{(0)}\mathbb{Q} q \ldots {}^{(0)}\mathbb{Q} \\
&\leq \sum_{i \neq j}^{*} \int \left(\sum {}^{(0)}\mathbb{Q} \ldots\right)(\cdot, i) q_{ij} \left(\sum {}^{(0)}\mathbb{Q} \ldots\right)(j, \cdot) \\
&\leq \sum_{i \neq j}^{*} \int q_{ij} < \varepsilon,
\end{aligned}
$$

with $\sum_{i \neq j}^{*}$ running through the couples (i, j) for which the measures q_{ij} are convergent.

Let us now write

$$
\mathbb{Q} = |\mathbb{Q}|^{*} \quad \text{and} \quad \tilde{\mathbb{Q}} = exprg_{|exp}(|\mathbb{Q}|^{*}).
$$

We still have

$$
\mathbb{Q} = {}^{(0)}\mathbb{Q} + \sum_{n \geq 0} \sum_{tot} \int {}^{(0)}\mathbb{Q} q^{(n)}\mathbb{Q},
$$

that we now expurgate through the iterative procedure

$$
{}^{(0)}\tilde{\mathbb{Q}}(i, j) = \exp\left[-\sum_{part} \int q_{ik}\right] I(i, j)
$$

$$^{(n+1)}\tilde{\mathbb{Q}}(i,j) = \left(\sum_{tot} \int {}^{(0)}\tilde{\mathbb{Q}} q^{(n)}\tilde{\mathbb{Q}}\right)(i,j), \qquad \text{et cetera,}$$

to now obtain the semigroup

$$\tilde{\mathbb{Q}} = exprg_{|exp}(|\mathbb{Q}|^*) = \sum_{n\geq 0} {}^{(n)}\tilde{\mathbb{Q}},$$

so that, given $\varepsilon > 0$ arbitrarily and when σ and τ are large enough, we shall have

$$^{(0)}\tilde{\mathbb{Q}}^{\sigma}_{\tau} \geq {}^{(0)}\mathbb{Q}^{\sigma}_{\tau} \geq {}^{(0)}\tilde{\mathbb{Q}}^{\sigma}_{\tau} e^{-\varepsilon}$$

$$^{(n+1)}\tilde{\mathbb{Q}}^{\sigma}_{\tau} \geq {}^{(n+1)}\mathbb{Q}^{\sigma}_{\tau} \geq {}^{(n+1)}\tilde{\mathbb{Q}}^{\sigma}_{\tau} e^{-\varepsilon},$$

so that

$$\tilde{\mathbb{Q}} \geq \mathbb{Q} \geq \tilde{\mathbb{Q}} e^{-\varepsilon},$$

that is, the semigroups $\mathbb{Q}$ and $\tilde{\mathbb{Q}}$ are asymptotically equivalent.

Finally, combining the two equivalences just above, we see that $\mathbb{Q} = |\mathbb{Q}|^*$ and

$$\begin{aligned} \tilde{\mathbb{Q}} &= exprg(|\mathbb{Q}|^*) \\ &= exprg_{|exp}(exprg_{|pert}(|\mathbb{Q}|^*)) = exprg_{|pert}(exprg_{|exp}(|\mathbb{Q}|^*)) \end{aligned}$$

are weakly asymptotically equivalent.

Theorem (Exit cone of a nonnegative right markovian semigroup: Basic criterion n° 1. *Let $|\mathbb{Q}|^*$ be a nonnegative right markovian semigroup of which the sections $i \in E^*$ are asymptotically active (that is, such that $q_i(du) < \infty$) at the right end of the interval of regularity $]a, b[$.*

Then $|\mathbb{Q}|^*$ is, either strongly, or at least weakly, asymptotically equivalent to the nonnegative right markovian semigroup (see above)

$$exprg(|\mathbb{Q}|^*) = \text{“}|\mathbb{Q}|^* \textit{ completely expurgated”},$$

that is, $exprg(|\mathbb{Q}|^*)$ is the nonnegative right markovian semigroup of which the Taylor expansion is deduced from the Taylor expansion of $|\mathbb{Q}|^*$ by expurgating it of all the convergent measures $|q|_{i.}$ both in the measure $|q|_i = \sum_. |q|_{i.}$, involved in the exponentials of $^{(0)}|\mathbb{Q}|^*$ and in perturbation procedure giving $|\mathbb{Q}|^*$ from $^{(0)}|\mathbb{Q}|^*$ by adding the $^{(n)}|\mathbb{Q}|^*$ $(n \geq 1)$ iteratively defined.

So that, there exist cones

$$\Gamma^+_{b-}(|\mathbb{Q}|^*) \quad \textit{and} \quad \Gamma^+_{b-}(exprg(|\mathbb{Q}|^*)),$$

reduced to the nonnegative uniformly bounded exit laws, are isomorphic.

Proof. For the sake of simplicity, we shall here write

$$\mathbb{Q} = |\mathbb{Q}|^* \quad \text{and} \quad \tilde{\mathbb{Q}} = exprg_{|exp}(|\mathbb{Q}|^*),$$

with the dominance relation $\tilde{\mathbb{Q}} \geq \mathbb{Q}$.

Let the nonnegative uniformly bounded exit law $g \in \Gamma^+_{b-}(\mathbb{Q})$. We have

$$\begin{aligned} 0 &\leq \tilde{g}_\sigma = (Ug)_\sigma \underset{def}{=} \lim_{\tau} \uparrow (\tilde{\mathbb{Q}}^{\sigma}_{\tau} g_\tau) \\ &\leq \sup_{\tau} (\mathbb{Q}^{\sigma}_{\tau} + \varepsilon) g_\tau \\ &\leq \sup_{\sigma} g_\sigma + \varepsilon.\text{const} \downarrow \sup_{\sigma} g_\sigma < +\infty \end{aligned}$$

where the constant bounds g. So, $\tilde{g} \in \Gamma^+_{b-}(\tilde{\mathbb{Q}})$ is itself uniformly bounded.

Conversely, let the nonnegative uniformly bounded exit law $\tilde{g} \in \Gamma^+_{b-}(\tilde{\mathbb{Q}})$. We have

$$\begin{aligned} +\infty \;>\; & g_\sigma = (D\tilde{g})_\sigma \underset{def}{=} \lim_\tau \downarrow (\mathbb{Q}^\sigma_\tau \tilde{g}_\tau) \\ \geq\; & \inf_\tau (\tilde{\mathbb{Q}}^\sigma_\tau - \varepsilon)\tilde{g}_\tau \\ \geq\; & \inf_\sigma \tilde{g}_\sigma - \varepsilon.\text{const} \uparrow \inf_\sigma \tilde{g}_\sigma \geq 0 \end{aligned}$$

where the symbol ≥ 0 here means that the minoration at left hand is nonnegative but not equal to zero identically. So we have

$$DoU = Identity \quad \text{and} \quad UoD = Identity,$$

that is, we have the isomorphism

$$\Gamma^+_{b-}(exprg_{|exp}(|\mathbb{Q}|^*)) \approx \Gamma^+_{b-}(|\mathbb{Q}|^*).$$

The proof of the isomorphism

$$\Gamma^+_{b-}(exprg(|\mathbb{Q}|^*)) \approx \Gamma^+_{b-}(exprg_{|pert}(|\mathbb{Q}|^*))$$

is quite similar.

And in the same way as we have used the weak asymptotic equivalence between $\mathbb{Q}$ and $\tilde{\mathbb{Q}}$ just above to verify the isomorphism between $\Gamma^+(\mathbb{Q})$ and $\Gamma^+(\tilde{\mathbb{Q}})$, we can use the strong (hence weak) asymptotic equivalence between

$$\mathbb{Q} = exprg(|\mathbb{Q}|^*) \quad \text{and} \quad \tilde{\mathbb{Q}} = exprg_{|exp}(|\mathbb{Q}|^*)$$

to verify the isomorphism

$$\Gamma^+_{b-}(exprg(|\mathbb{Q}|^*)) \approx \Gamma^+_{b-}(exprg_{|exp}(|\mathbb{Q}|^*)).$$

The preceding theorem can be rearranged in the following way preparing for the basic criterion n° 2:

Proposition (Isomorphism between exit cones of a nonnegative right markovian semigroup and of its associated expurgated semigroups). *If $|\mathbb{Q}|^*$ is a nonnegative right markovian semigroup on the interval of regularity $]a,b[$. Then we have the following diagram of dominance relations and associated projections*

$$\begin{array}{ccccc} |\mathbb{Q}|^0 & \underset{\geq}{\longrightarrow} & exprg_{|exp}(|\mathbb{Q}|^*) & \underset{\geq}{\longrightarrow} & |\mathbb{Q}|^* \\ \geq\big\downarrow & & \geq\big\downarrow & & \geq\big\downarrow \\ exprg(|\mathbb{Q}|^0) & \underset{\geq}{\longrightarrow} & exprg(|\mathbb{Q}|^*) & \underset{\geq}{\longrightarrow} & exprg_{|pert}(|\mathbb{Q}|^*) \end{array}$$

where $|\mathbb{Q}|^0$ denotes the catenary semigroup (see below)

$$|\mathbb{Q}|^0 = \text{“}\, |\mathbb{Q}|^* \textit{ with the measures } q_i,\ q_j,\ \&c, \textit{ all equal to zero identically''} = \sum_{n\geq 0} \int^{(n)} |q|,$$

and we also have the isomorphism

$$\Gamma^+_{b_-}(|\mathbb{Q}|^*) \approx \Gamma^+_{b_-}(exprg(|\mathbb{Q}|^*))$$

where these exit cones are restricted to their nonnegative uniformly bounded exit laws.

Let us notice that the measures $|q|^0_i$ of $|\mathbb{Q}|^0$ are all equal to zero, and that the perturbation measures of $|\mathbb{Q}|^0$ are identical to the measures $|q|_{ij}$ of $|\mathbb{Q}|^*$.

So, we shall say that the catenary semigroup $|\mathbb{Q}|^0$ is a strong expurgation of $|\mathbb{Q}|^*$, since some divergent measures can here be involved which have been canceled. We still expurgate it of its convergent measures into $exprg(|\mathbb{Q}|^0)$ which will play a central role in the basic criterion n° 2 concerning distance and/or proximity between exit sections, we shall call it the expurgated catenary semigroup since its Taylor expansion is a sum of expurgated chains $\int^{(n)}|q|$ (see below). And we shall have a decomposition

$$exprg(|\mathbb{Q}|^*) \not\sim \oplus_\alpha |\mathbb{Q}|^{*\alpha}$$

where the direct sum at right hand is a strongly expurgated version of $|\mathbb{Q}|^*$, and similarly for

$$exprg(|\mathbb{Q}|^0) \not\sim \oplus_\alpha |\mathbb{Q}|^{0\alpha}.$$

Definition (Separation by convergence). Let the nonnegative reduction $|\mathbb{Q}|^*$ be dominated by the semigroup M on the interval of regularity $]a,b[$, so that we have the isomorphism (chart of a part of the boundary)

$$\Gamma^+_M(|\mathbb{Q}|^*) \approx \Gamma^+_{|\mathbb{Q}|^*}(M)$$

with (union of M-charts or atlas)

$$\Gamma^+(|\mathbb{Q}|^*) \approx \cup_{M:M\geq|\mathbb{Q}|^*}\Gamma_M(|\mathbb{Q}|^*).$$

If the continuous sections i and j are such that we have

$$0 \leq M^\sigma_\tau(i,j) \leq \varepsilon \quad \text{and} \quad 0 \leq M^\sigma_\tau(j,i) \leq \varepsilon$$

for any τ with $0 < \tau < b_-$ and σ large enough, then we shall say that i and j are separated by convergence with respect to M, or more briefly, are M-separated.

And we shall say that i and j are separated by convergence if and only if they are M-separated for some semigroup M dominating $|\mathbb{Q}|^*$ (see for instance below with $M = exprg(|\mathbb{Q}|^0) \geq |\mathbb{Q}|^*$).

Definition (Distal and proximal exit sections). Two exit sections i and j are said to be distal when they are in two fixed orthogonal extremal exit channels asymptotically. If they are not distal, they are in a same fixed extremal exit channel and said to be proximal.

Theorem (Distal exit sections: Basic criterion n°2). *Let $|\mathbb{Q}|^*$ be a nonnegative right markovian semigroup. Two exit sections are distal if and only if they are separated by convergence with respect to $|\mathbb{Q}|^*$.*

Or equivalently, with respect to $exprg(|\mathbb{Q}|^*)$.

Or equivalently, with respect to $exprg(|\mathbb{Q}|^0)$.

Or equivalently, if and only if, whatever is $n \geq 1$, in every component chain $|q|_{ik_1} \cdots |q|_{k_{n-1},j}$ from i to j of ${}^{(n)}|\mathbb{Q}|^0(i,j)$ in the Taylor expansion $|\mathbb{Q}|^0 = \sum_{n \geq 1} {}^{(n)}|\mathbb{Q}|^0$, that is, in every component chain of

$$ {}^{(n)}|\mathbb{Q}|^0(i,j) = \int {}^{(n)}|q|_{ij} = \sum \int |q|_{ik_1} \cdots |q|_{k_{n+1}j}, $$

one measure $|q|_{..}$ involved in this component chain is convergent, at least.

Proof. If the exit sections i and j are distal, they are in two fixed orthogonal extremal exit channels, and so

$$ |\mathbb{Q}|_\tau^{*\sigma}(i,j) < \varepsilon \quad \text{and} \quad |\mathbb{Q}|_\tau^{*\sigma}(j,i) < \varepsilon $$

when $\sigma < \tau$ are large enough, necessarily and conversely (see [1] or [2]). And equivalently with $exprg(|\mathbb{Q}|^*)$.

Now, let us define the normed nonnegative extremal exit law g^α of $|\mathbb{Q}|^*$, or equivalently the corresponding exit law of $exprg(|\mathbb{Q}|^*)$, and their associated extremal exit channels with their corresponding taboo semigroups $|\mathbb{Q}|^{*\alpha}$ or $exprg(|\mathbb{Q}|^{*\alpha})$. We so have an asymptotic decomposition and the asymptotic equivalences

$$ |\mathbb{Q}|^* \sim exprg(|\mathbb{Q}|^*) \not\sim \oplus_\alpha |\mathbb{Q}|^{*\alpha}, $$

the bar indicating that the last one is restricted to sections which exit in extremal exit channels only, that is, to exit sections, and an isomorphic decomposition with asymptotic equivalence restricted to exit sections

$$ exprg(|\mathbb{Q}|^0) \not\sim \oplus_\alpha |\mathbb{Q}|^{0\alpha}. $$

The applications U and D (up and down, see above and/or [1], [2]) between the cones

$$ \Gamma^+(|\mathbb{Q}|^*) \approx \Gamma^+(exprg(|\mathbb{Q}|^*)) \approx \oplus_\alpha \Gamma^+(|\mathbb{Q}|^{*\alpha}) $$

and

$$ \Gamma^+(exprg(|\mathbb{Q}|^0)) \approx \oplus_\alpha \Gamma^+(|\mathbb{Q}|^{0\alpha}) $$

are isomorphisms. The criterion with $exprg(|\mathbb{Q}|^0)$ follows, up to the end of the statement.

We can now conclude with the following procedure, which can be extended to non asymptotically active right markovian semigroups:

Procedure (Strongly expurgated version of $|\mathbb{Q}|^*$). Let $|\mathbb{Q}|^*$ be a nonnegative right markovian semigroup on the time interval of regularity $]a,b[$.

In order to define a first exit section, we select at every time a most probable continuous path. We cancel the section in the Taylor expansion of $|\mathbb{Q}|^*$. If the semigroup so obtained is asymptotically equivalent to $|\mathbb{Q}|^*$, we select another section in the same way but in avoiding the first one asymptotically, and so on.

When the semigroup obtained is no more asymptotically equivalent to $|\mathbb{Q}|^*$, we have then determined a first exit section.

This exit section defines a semigroup of the ${}^{(0)}|\mathbb{Q}|^*$ type dominated by $|\mathbb{Q}|^*$ itself, so the extremal law of ${}^{(0)}|\mathbb{Q}|^*$ can be lifted into an extremal exit law of $|\mathbb{Q}|^*$ which defines an extremal exit channel and then, a first taboo semigroup $|\mathbb{Q}|^{*\alpha}$.

We proceed iteratively, so as to get a complete family of extremal exit channels and the complete family of their corresponding taboo semigroups $|\mathbb{Q}|^{*\alpha}$ giving us the strongly expurgated version of $|\mathbb{Q}|^*$, that is,

$$|\mathbb{Q}|^* \not\sim \oplus_\alpha |\mathbb{Q}|^{*\alpha}.$$

We can proceed in the same way, not with $|\mathbb{Q}|^*$, but directly with $exprg(|\mathbb{Q}|^*)$.

We achieve this section by the following definition which leads to the conclusion that a continuous deformation of the semigroup will be a continuously deformed perturbation of the semigroup; such a deformation will be submitted to the constraints resulting from the ergodic theorem in order to preserve the asymptotic exit boundary.

Definition (Perturbation of a semigroup). Let $\mathbb{Q} = (q_i, q_{ij})$ be a semigroup on an interval of regularity $]a, b[$, and $\Pi = (\pi_i, \pi_{ij})$ a perturbation. We define the perturbed semigroup $\Pi\mathbb{Q}$ in the following way:

$$\begin{aligned}
{}^{(0)}\Pi\mathbb{Q} &= \mathbb{Q} \\
{}^{(1)}\Pi\mathbb{Q} &= \int \mathbb{Q}d\Pi\mathbb{Q} \\
&\vdots \\
{}^{(n+1)}\Pi\mathbb{Q} &= \int {}^{(n)}\Pi\mathbb{Q}d\Pi\mathbb{Q} \\
&\vdots \\
\Pi\mathbb{Q} &= \sum_{n\geq 0} {}^{(n)}\Pi\mathbb{Q}
\end{aligned}$$

where

$$d\Pi = \Pi(du) = \begin{pmatrix} \ddots & & \\ & \pi_i(du) & \pi_{ij}(du) \\ & \ddots & \end{pmatrix}$$

So, the semigroup $\Pi\mathbb{Q}$ has the measures $(q_i + \pi_i, q_{ij} + \pi_{ij})$.

3. Examples

Let $\mathbb{Q}$ be a matrix semigroup on some interval of regularity. Denoting by $R^+ = \partial^+_{t_+} + G^+$ (acting from left and just after time t) and by $L^- = \partial^-_{t_-} - G^-$ (acting from right and just before time t) the right derivative and the left derivative, respectively, associated with the semigroup $\mathbb{Q}$, then, for any entrance law f and any exit law g, we have

$$fL^- = 0 \quad \text{and} \quad R^+g = 0.$$

Exit boundary and homotopy. We shall here take

$$\vec{\mathbb{G}} = \vec{\mathbb{G}}^- = \begin{pmatrix} 0 & \zeta_1 \\ \zeta_2 & 0 \end{pmatrix}$$

with $\zeta_1 = e^{i\alpha_1\pi}$ and $\zeta_2 = e^{i\alpha_2\pi}$ where α_1 and α_2 are real numbers. And we consider the differential equation $fL^- = 0$ where $f_t = (x_t, y_t)$, that is,

$$(\dot{x}, \dot{y}) = (x, y) = \begin{pmatrix} 0 & \zeta_1 \\ \zeta_2 & 0 \end{pmatrix}.$$

The corresponding catenary semigroup $\vec{\mathbb{Q}}^0$ describes a circle when, ζ_2 being fixed, α_1 runs from -1 to $+1$, and another circle when, ζ_1 being fixed, α_2 runs from -1 to $+1$. In fact, we get a torus T^2 of semigroups when ζ_1 and ζ_2 are relaxed.

All these semigroups admit one exit channel with two exit sections, necessarily proximal. The solution of the differential equation $fL^- = 0$ describes a part of the curve $\zeta_1 x^2 - \zeta_2 y^2 =$ const., for $\sqrt{\zeta_1\zeta_2} = 1$ the asymptotes are real, for $\sqrt{\zeta_1\zeta_2} = i$ they are isotropic and the solution describes a circle.

We now take

$$\mathbb{G} = \mathbb{G}^- = \begin{pmatrix} 1 - r^2 & \zeta_1 \\ \zeta_2 & 1 - r^2 \end{pmatrix}$$

where $r^2 = r^2(t) = |x_t|^2 + |y_t|^2$, so that we here have the semigroup

$$\mathbb{Q}_t^s = \exp\left[-\int_0^s (1 - r^2)\right] \vec{\mathbb{Q}}_t^{0s} \exp\left[+\int_0^t (1 - r^2)\right].$$

Then, the solution tends to a part or to the whole part of the limit circle with radius such that $1 \le r^2 = 1 \pm Re\sqrt{\zeta_1\zeta_2} \le 2$. When $\zeta_1 = \zeta_2 = 1$, we have

$$\frac{2r^2}{\exp\left[2\int_0^t (2 - r^2)\right]} \to |A_+|^2$$

where $A_+ = x_0 + y_0$ ($x_0 \ge 0$, $y_0 \ge 0$), so that the integral $\int_0^t (2 - r^2)$ is convergent in the limit as $t \uparrow \infty$, the semigroup generated by

$$\mathbb{G} = \begin{pmatrix} 1 - r^2 & 1 \\ 1 & 1 - r^2 \end{pmatrix}$$

is asymptotically equivalent to the right markovian semigroup generated by

$$\begin{pmatrix} -1 & 1 \\ 1 & -1 \end{pmatrix}.$$

(See [2] for details).

Expurgation versus linearization. Let us first consider the ODE

$$(\dot{x}, \dot{y}) = (x, y) \begin{pmatrix} a - \Omega & 0 \\ 0 & b - \Omega \end{pmatrix}$$

where we take the environment condition $x + y = n$ constant, from which we deduce $\Omega = (ax + by)/n$. The solution defined by the initial data (x_0, y_0) is then propagated by the semigroup

$$\mathbb{Q}_t^s = \exp\left[\int_0^s \Omega\right] \vec{\mathbb{Q}}_t^s \exp\left[-\int_0^t \Omega\right]$$

with

$$\exp\left[-\int_s^t \Omega\right] = f_s f_t^{-1} \qquad \& \qquad \vec{\mathbb{Q}}_t^s = \begin{pmatrix} e^{a(t-s)} & 0 \\ 1 - e^{b(t-s)} & e^{b(t-s)} \end{pmatrix}$$

where $f_t = x_0 e^{at} + y_0 e^{bt}$, so that, assuming $a > b > 0$, we have $\lim x_t = n$ and $\lim y_t = 0$ in the limit as $t \to \infty$.

In the neighbourhood of the point of equilibrium ($x = n$, $y = 0$) we take $x = n + \xi$, $y = 0$, and we have the ODE

$$(\dot{\xi}, \dot{\eta}) = (\xi, \eta) \begin{pmatrix} -a - \omega & 0 \\ -b & (b-a) - \omega \end{pmatrix}$$

with $\xi + \eta = 0$ and $\omega = (a\xi + b\eta)/n$, whose the solution issued from (ξ_0, η_0) is propagated by the semigroup

$$f_s f_t^{-1} e^{-as} \begin{pmatrix} 1 & 0 \\ 1 - e^{b(t-s)} & e^{b(t-s)} \end{pmatrix} e^{at}$$

with here $f_t = \exp\left[-\int_0^t \omega\right]$.

The generator of this semigroup is

$$\begin{pmatrix} -a - \omega & 0 \\ -b & (b-a) - \omega \end{pmatrix},$$

and as $\int_0^\infty \omega < \infty$, it is asymptotically equivalent to

$$\begin{pmatrix} -a & 0 \\ -b & b-a \end{pmatrix}.$$

So, the corresponding ODE is here exactly the same as the one obtained by linearization.

See [2] for details.

The Dirac equation. Let

$$\left[\sum \gamma_\mu \left(\frac{\hbar}{i}\partial_\mu - \frac{e}{c}A_\mu\right) + i m_0 c\right] \psi = 0$$

be the Dirac equation, where $\partial_0 = \partial / ic\partial t$, $\partial q^j = \partial / \partial q^j$ ($j = 1, 2, 3$), that we shall write

$$\left[\sum \gamma_\mu \partial_\mu + \Omega\right] \psi = \frac{m_0 c}{\hbar} \psi$$

or abbreviatively

$$[p + \Omega]\psi = k\psi,$$

with

$$p = \sum \gamma_\mu \partial_\mu \qquad i\hbar\Omega = \frac{e}{c} \sum \gamma_\mu A_\mu,$$
$$\psi = (\psi_{\pm\pm}) \qquad k = m_0 c/\hbar.$$

Then the wave function $\Psi^\Omega = (\Psi^\Omega_{\pm\pm})$ such that

$$\Psi^\Omega_{\pm\pm} = e^{kT/2} \exp\left[\int_0^T \Omega\right] \psi_{\pm\pm},$$

where $T = ict_0$ (t_0 =the proper time), where we here assume Ω = constant, and where $\psi_{\pm\pm}=\psi_{\pm\pm}(q^0 = ict, q^j)$, will be a solution of the Schrödinger/Dirac equation

$$\partial_T \Psi^\Omega = \left[\frac{1}{2}p + \Omega\right] \Psi^\Omega$$

as soon as $\psi = \psi(q^0, q^j)$ will be a solution of the Dirac equation

$$p\psi = k\psi.$$

It follows that the wave function $\Psi^\Omega = (\Psi^\Omega_{\pm\pm})$ is propagated by $(S < T)$

$$\exp\left[-\int_0^S \Omega\right] K_T^S \exp\left[\int_0^T \Omega\right]$$

where K is the propagator of the monochromatic wave functions $\Psi = e^{kT/2}\psi$, that is $(d = 4)$

$$K_T^S(q_S, q_T) = (4\pi D(T-S))^{-d/2} \exp\left[-\frac{(q_T - q_S)^2}{4D(T-S)}\right],$$

and where the generator Ω determines the Taylor expansion of the semigroup

$$\exp\left[\int_S^T \Omega\right].$$

So, in the general case where Ω is not constant, the propagator K is perturbated by the generator Ω along each brownian path in the Minkowski space-time M, and by discretizing M, we can define the perturbation of the corresponding Taylor expansion approximating K by the (impulsive or not impulsive) generator Ω according to the definition of a perturbation given above.

And to define the limit of these approximating perturbated propagators, the ergodic theorem in particular is required. And the same thing is valid with the propagator $K_T^s = \exp[(T-S)p/2]$ to be defined in the same way as the propagator just above, that is, as the limit of approximating matrix semigroups.

See [3], [4] and [1] for details.

4. Conclusion

There is a similarity between the frieze of deers painted in the cave of Lascaux compared with the carvings at Fontanalba on one side, and between the equation of waves in contrast with the wave equations, on the other side.

The frieze of deers is descriptive with many details, it gives account of some reality directly observed. The carvings at Fontanalba are schematic, some suggest a balance, which is the germ of Archimedes' lever, and so, the germ of Statics. The Sun's chariot, with its two yokes and the scythe, suggests up to the diagrams associated with the transverse and radial curvatures used in Gravitation. Simplification here *unveils* a many sided basic structure adaptive to classical constructions as so many variations ([5], p. 169).

Similarly, if we give us particles, masses, potentials of interactions, equations of state, Newton's law, and search for waves, we find waves solutions of the equation of waves. But if we know nothing about such details, we can only write that the wave is propagated according to the semigroup property, among the semigroups search for the propagators and classify them according to their way of integrating their transverse structures longitudinally, and ask these propagators what they propagate. For all of a brownian class, their answer is wave functions solutions of such or such wave equation.

SYNOPSIS
PSI STRUCTURE
Theme of the Great Fugue [4]

Subject: S_n/n Answer: $(S_n - np)^2/n^2 - pq/n$

Variation 1: Germs of the transverse structure [4]
Projective martingale theorem and law of large numbers.
Germ of the Poincaré-Cartan form.
Brownian kernel.

Variation 2:
Integration and propagation of the transverse structures [1], [2]
Matrix semigroups.
Baire's expansion of time.
Intervals of regularity and Taylor expansion of semigroups [6], [1].
Propagators and wave functions.
Wave equations.

Variation 3: From transverse to longitudinal structure [1]
Poincaré-Cartan form [4].
Space-time structure from ergodisation [1].
Rays and impulsive ODE/PDE of Fokker-Planck type.
Inductive martingale theorem and exit sections [1], [2].
Ergodic theorem plus homotopy.
Asymptotic expurgation of semigroups versus linearization.

Variation 4: Galilean and relativistic formalism [7]
Combinatorial skeletons of propagators and invariants.
Invariance of phase and Doppler effect pushed into.
Contraction of length and dilatation of $c\times$time.

Variation 5: Bifocal perspective [7]
Invariance of celestial sphere.
Planets and Gravitation equation.
Particles and Wave equations.
Radiations and local invariance of c.
Dilatation of time.

Acknowledgements

V. Lakshmikantham, Raymond Daudel, Per Olov Löwdin, Detchko Ouzounov, Jacques Ferrier, Drumi Bainov, Valery Covachev, Angel Dishliev, Joji Kajiwara, Il Club dei Volterriani.

The ergodic theorem is dedicated to Drumi Bainov.
Manuscript typed by Mmes Annick Cartier, Sandrine Ingrémaud, Brigitte Brault.

References

1. J.-P. Caubet, in: Proc. 5-th Int. Coll. Diff. Equations, D. Bainov and V. Covachev (Eds.), VSP, Utrecht (1995) pp. 61–72; in: Fifth Int. Coll. Diff. Equations, Vol. I, D. Bainov and A. Dishliev (Eds.), SCT Publishing, Singapore (1995) pp. 54–66.
2. J.-P. Caubet, Nonlinear Analysis, Theory, Methods and Applications, **30**, 2753–2764 (1997).
3. J.-P. Caubet, Le Mouvement Brownien Relativiste, Lect. Notes in Math., **559**, Springer-Verlag, Heidelberg (1976).
4. J.-P. Caubet, J. Stoch. Anal. Appl., **3**, 119–151 (1985); J. Stoch. Anal. Appl., **2**, 55–85 (1984).
5. A. Schoenberg, Fundamentals of Musical Composition, Faber and Faber, London, Boston (1967).
6. J.-P. Caubet, Ann. Inst. H. Poincaré, **1**, 239–310 (1965); C.R. Acad. Sc. Paris, **268**, 1348–1350 (1969).
7. J.-P. Caubet, in: Proc. 4-th Int. Coll. Diff. Equations, D. Bainov and V. Covachev (Eds.), VSP, Utrecht (1994) pp. 71–82.
8. D.D. Bainov and P.S. Simeonov, Impulsive Differential Equations: Asymptotic Properties of the Solutions, World Scientific, Singapore (1995).
9. A. Steiner, in: Il Volterriano, Periodico del Club dei Volterriani, **6**, (1997).

9th Int. Coll. on Differential Equations, pp. 77–84
D. Bainov (Ed)

Schwarz Algorithms for Nonconforming and Mixed Methods for Second-Order Problems

ZHANGXIN J. CHEN
Department of Mathematics, Box 156, Southern Methodist University, Dallas, Texas 75275-0156, USA

Abstract: In this paper two-level Schwarz methods for nonconforming and mixed methods for second order elliptic problems on triangles and rectangles are considered. A new coarse space is constructed, and a convergence theory for the two-level methods is given. An extension to other substructuring methods such as vertex space and balancing domain decomposition methods is discussed.
Key Words and Phrases: *Nonconforming and mixed methods, Schwarz algorithms, elliptic problems.*

1. Introduction

Let Ω be a bounded domain in $\Re^2$, with the polygonal boundary $\partial\Omega$. We consider the elliptic problem

$$(1.1)\qquad \begin{aligned} -\nabla\cdot(\mathcal{A}\nabla u) &= f &&\text{in } \Omega,\\ u &= 0 &&\text{on } \partial\Omega, \end{aligned}$$

where $\mathcal{A}(x)$ is a uniformly positive definite, bounded, symmetric tensor and $f(x) \in L^2(\Omega)$. $H^k(\Omega) = W^{k,2}(\Omega)$ will denote the Sobolev space of k differentiable functions in $L^2(\Omega)$. Recently, a new approach for developing multigrid and domain decomposition algorithms for mixed finite element methods was introduced [1–4]. It was first shown that the mixed finite element formulation can be algebraically condensed to a symmetric and positive definite system for Lagrange multipliers using the features of the existing mixed finite element spaces. It was then proven that optimal multigrid and domain decomposition algorithms can be designed for the resulting symmetric and positive definite system. It has been known that, due to its saddle point property, it is difficult to develop efficient multigrid and domain decomposition methods for solving the linear system generated by the mixed finite element methods. The advantages of the new approach are that it transfers the original saddle point problem to a symmetric positive definite problem and the standard convergence theory for the Galerkin conforming and nonconforming methods can be applied.

In this paper, based on the new approach, we consider two-level Schwarz methods for nonconforming and mixed finite element methods for second order elliptic problems on triangles and rectangles. In particular, a new coarse space is constructed for the nonconforming methods, and applied to the mixed methods. Then a convergence theory for the two-level methods is given. An extension of the convergence theory to other substructuring methods such as vertex space and balancing domain decomposition methods is briefly discussed. Also, multi-level Schwarz methods are mentioned.

2. Two-Level Additive Schwarz Method on Triangles

Problem (1.1) is recast in weak form as follows. We define the bilinear form $a(\cdot,\cdot)$ as follows:

$$a(v,w) = (\mathcal{A}\nabla v, \nabla w), \qquad v, w \in H^1(\Omega),$$

where $(\cdot,\cdot)$ denotes the $L^2(\Omega)$ or $(L^2(\Omega))^2$ inner product, as appropriate. Then the weak form of (1.1) for the solution $u \in H_0^1(\Omega)$ is

$$a(u,v) = (f,v), \qquad \forall\, v \in H_0^1(\Omega). \tag{2.1}$$

For $0 < h < 1$, let $\mathcal{E}_h$ be a triangulation of Ω into triangles of size h, and define the P_1-nonconforming finite element space

$$\begin{aligned} N_h = \{v \in L^2(\Omega) : v|_E &\text{ is linear for all } E \in \mathcal{E}_h, v \text{ is continuous} \\ &\text{at the midpoints of interior edges and} \\ &\text{vanishes at the midpoints of edges on } \partial\Omega\}. \end{aligned}$$

Associated with N_h, we introduce a bilinear form on $N_h \cup H_0^1(\Omega)$ by

$$a_h(v,w) = \sum_{E \in \mathcal{E}_h} (a\nabla v, \nabla w)_E, \qquad v, w \in N_h \cup H_0^1(\Omega),$$

where $(\cdot,\cdot)_E$ is the $L^2(E)$ inner product. Then the P_1-nonconforming finite element discretization of (1.1) is to find $u_h \in N_h$ such that

$$a_h(u_h, v) = (f,v), \qquad \forall\, v \in N_h. \tag{2.2}$$

We now develop a two-level additive Schwarz algorithm for (2.2). We need to assume a structure to our family of partitions. In the first step, let $\mathcal{E}_H$ be a quasi-regular coarse triangulation [5] of Ω into nonoverlapping triangular substructures Ω_i, $i = 1, \cdots, n$. Then, in the second step we refine $\mathcal{E}_H$ into triangles to have a quasi-regular triangulation $\mathcal{E}_h$. Finally, let $\{\Omega_i'\}_{i=1}^n$ be an overlapping domain decomposition of Ω by extending Ω_i with the overlap parameter δ. The decomposition is assumed to align with the boundary $\partial\Omega$, and the parameter δ is defined by $\delta = \min\{\text{dist}(\partial\Omega_i \setminus \partial\Omega, \partial\Omega_i' \setminus \partial\Omega), i = 1, \cdots, n\}$. Associated with each Ω_i', let N_h^i be the P_1 nonconforming finite element space whose elements have support in Ω_i', as defined above. The finite element space N_h is represented as a sum of $n+1$ subspaces:

$$N_h = N_h^0 + N_h^1 + \cdots + N_h^n, \tag{2.3}$$

where the coarse space N_h^0 will be defined later. We now define the operators $\Pi_i : N_h \to N_h^i$, $i = 0, 1, \cdots, n$, by

$$a_h(\Pi_i v, w) = a_h(v,w), \quad \forall w \in N_h^i, \tag{2.4}$$

and the operator $\Pi : N_h \to N_h$ by

$$\Pi = \sum_{i=0}^{n} \Pi_i. \tag{2.5}$$

The additive Schwarz algorithm for (2.2) is given by

$$\Pi\psi_h = \widehat{f}_h, \quad \widehat{f}_h = \sum_{i=0}^{n} f_i, \tag{2.6}$$

where f_i satisfies

$$a_h(f_i, v) = (f, v), \quad \forall v \in N_h^i,\ i = 0, 1, \cdots, n. \tag{2.7}$$

Note that (2.2) and (2.7) have the same solution and thus produce the same linear system.

We state an abstract convergence theory for bounds on the condition number of Π. A specific example to which the abstract theory applies will be given below. Following Dryja and Widlund's framework [6], the abstract theory is written in terms of the following two assumptions: (A1) There is a constant C such that for every $v \in N_h$ it can be represented by $v = \sum_{i=0}^{n} v_i$ with $v_i \in N_h^i$ satisfying

$$\sum_{i=0}^{n} a_h(v_i, v_i) \le C a_h(v, v),$$

and (A2) let $\kappa = (\kappa_{ij})$ be a symmetric matrix with $\kappa_{ij} \ge 0$ satisfying

$$|a_h(v_i, v_j)| \le \kappa_{ij} a_h(v_i, v_i)^{1/2} a_h(v_j, v_j)^{1/2}, \quad \forall v_i \in N_h^i,\ v_j \in N_h^i,\ i, j = 1, \cdots, n.$$

Then the next lemma can be found in [6].

Lemma 1. *Assume that the assumptions* (A1) *and* (A2) *are satisfied. Then*

$$\begin{aligned} \lambda_{\min}(\Pi) &\ge C^{-1}, \\ \lambda_{\max}(\Pi) &\le \rho(\kappa) + 1, \end{aligned} \tag{2.8}$$

where $\rho(\kappa)$ is the spectral radius of κ.

We now give an example of the coarse space N_h^0 so that the assumptions (A1) and (A2) are satisfied. Namely, we estimate the two constants C and $\rho(\kappa)$. Below R_h denotes the nodal interpolation operator into N_h.

Let $\mathcal{E}_h$ be the finest triangulation and let $\mathcal{E}_h = \mathcal{E}_{H_J}$ for some $J \ge 1$ where $\mathcal{E}_{H_k} = \mathcal{E}_k$ ($H_k = 2^{-k}H$, $0 \le k \le J$) is constructed by connecting the midpoints of the edges of the triangles in $\mathcal{E}_{k-1}$. Then, following [7], we define the operator $\mathcal{I}_{k-1}^k : N_{k-1} \to N_k$ as follows, where $N_k \equiv N_{H_k}$ is the P_1 nonconforming space associated with $\mathcal{E}_k$ (in particular, $N_h \equiv N_{H_J}$). If $v \in N_{k-1}$ and $E \in \mathcal{E}_{k-1}$ with the vertices (x_i, y_i) and the midpoints $(\overline{x}_i, \overline{y}_i)$ of its edges, $i = 1, 2, 3$, then

$$\begin{aligned} \mathcal{I}_{k-1}^k v(\overline{x}_i, \overline{y}_i) &= v(\overline{x}_i, \overline{y}_i), && i = 1, 2, 3, \\ \mathcal{I}_{k-1}^k v(x_i, y_i) &= \tfrac{1}{\mathcal{N}_1} \textstyle\sum_j v(\overline{x}'_j, \overline{y}'_j) && \text{if } (x_i, y_i) \notin \partial\Omega, \\ \mathcal{I}_{k-1}^k v(x_i, y_i) &= \tfrac{1}{\mathcal{N}_2} \textstyle\sum_j v(\overline{x}''_j, \overline{y}''_j) && \text{if } (x_i, y_i) \in \partial\Omega, \end{aligned} \tag{2.9}$$

where $\mathcal{N}_1$ and $\mathcal{N}_2$ are the number of the adjacent midpoints $(\overline{x}'_j, \overline{y}'_j)$ and $(\overline{x}''_j, \overline{y}''_j)$ to (x_i, y_i) of the edges in $\partial\mathcal{E}_{k-1}$ and the edges on $\partial\Omega$ of the elements in $\mathcal{E}_{k-1}$, respectively. Alternatively, following [8], $\mathcal{I}_{k-1}^k : N_{k-1} \to N_k$ can be equivalently defined by

$$\begin{aligned} \mathcal{I}_{k-1}^k v(\overline{x}_i, \overline{y}_i) &= v(\overline{x}_i, \overline{y}_i), && i = 1, 2, 3, \\ \mathcal{I}_{k-1}^k v(x_i, y_i) &= \tfrac{1}{\mathcal{N}_1} \textstyle\sum_{(x_i, y_i) \in K_j} v|_{K_j}(x_i, y_i) && \text{if } (x_i, y_i) \notin \partial\Omega, \\ \mathcal{I}_{k-1}^k v(x_i, y_i) &= \tfrac{1}{\mathcal{N}_2} \textstyle\sum_j v(\overline{x}''_j, \overline{y}''_j) && \text{if } (x_i, y_i) \in \partial\Omega, \end{aligned} \tag{2.10}$$

where $\mathcal{N}_1$ is the number of elements $K_j \in \mathcal{E}_{k-1}$ meeting at (x_i, y_i) and $\mathcal{N}_2$ is defined as in (2.9). Note that (2.9) and (2.10) define the value of $I_{k-1}^k v$ at the vertices of elements in $\mathcal{E}_k$ and thus lead to a continuous piecewise linear function on $\mathcal{E}_k$. Hence $\mathcal{I}_{k-1}^k v$ is obviously in N_k from its construction on the boundary $\partial\Omega$. It is also a function in N_h. Now, the coarse space N_h^0 is given by

$$N_h^0 = \{v \in N_h : v = \mathcal{I}_H \varphi,\ \varphi \in N_H\}, \tag{2.11}$$

where $\mathcal{I}_H \equiv \mathcal{I}_0^1$. The next theorem can be found in [4].

Theorem 2. *Assume that the additive Schwarz operator* Π *is defined by* (2.5) *with the coarse space given by* (2.11). *Then there is a constant independent of* h, H, *and* δ *such that the condition number* $c(\Pi)$ *of* Π *satisfies*

$$c(\Pi) \leq C(1 + H/\delta).$$

3. Substructuring Methods

We consider an extension of the Schwarz method to a reduced problem involving solely the degrees of freedom on the internal interfaces of subdomains $\Gamma = \cup_{i=1}^n \partial\Omega_i \setminus \partial\Omega$, i.e., the substructuring domain decomposition method.

For $\psi_h, \varphi_h \in N_h$, let ψ and φ denote their respective degrees of freedom, and $\psi^{(i)}$ and $\varphi^{(i)}$ their restrictions to $\bar{\Omega}_i$, the closure of Ω_i. Let $M^{(i)}$ indicate the stiffness matrix associated with the $a_{h,\Omega_i}(\cdot,\cdot)$. That is,

$$\psi^t M \varphi = \sum_{i=1}^n \psi^{(i)^t} M^{(i)} \varphi^{(i)} = a_h(\psi_h, \varphi_h),$$

where $\psi^{(i)^t}$ denotes the transpose of $\psi^{(i)}$. For each Ω_i, let $\psi_I^{(i)}$ and $\psi_B^{(i)}$ denote the subvectors of $\psi^{(i)}$ corresponding to the nodes on $\partial\Omega_i$ and those in the interior of Ω_i. Then this partition leads to

$$\psi^{(i)^t} M^{(i)} \varphi^{(i)} = \begin{pmatrix} \psi_I^{(i)} \\ \psi_B^{(i)} \end{pmatrix}^t \begin{pmatrix} M_{II}^{(i)} & M_{IB}^{(i)} \\ M_{IB}^{(i)^t} & M_{BB}^{(i)} \end{pmatrix} \begin{pmatrix} \varphi_I^{(i)} \\ \varphi_B^{(i)} \end{pmatrix}.$$

Since the interior nodes of Ω_i are coupled only to other nodes of the same substructure, they can be eliminated locally and in parallel. The resulting matrix Z is the Schur complement with respect to the interface unknowns:

$$\psi_B^t Z \varphi_B = \sum_{i=1}^n (\psi_B^{(i)})^t Z^{(i)} \varphi_B^{(i)}, \quad Z^{(i)} = M_{BB}^{(i)} - M_{IB}^{(i)^t} \left(M_{II}^{(i)}\right)^{-1} M_{IB}^{(i)}.$$

It follows from this elimination procedure that the system (2.2) becomes

$$Z \lambda_B = F_B. \tag{3.1}$$

To write (3.1) in a variational form, we define

$$b(\psi_h, \varphi_h) = \psi_B^t Z \varphi_B, \quad b_i(\psi_h, \varphi_h) = (\psi_B^{(i)})^t Z^{(i)} \varphi_B^{(i)}.$$

Also, following [7], let $N_h(\Gamma)$ denote a subspace of N_h such that if $\psi_h \in N_h(\Gamma)$ has the degrees of freedom $\psi_B^{(i)}$ on $\partial\Omega_i$, then $\psi_h|_{\Omega_i}$ is the function whose degrees of freedom are given by $(\psi_I^{(i)}, \psi_B^{(i)})^t$ satisfying

$$M_{II}^{(i)}\psi_I^{(i)} = -M_{IB}^{(i)}\psi_B^{(i)}.$$

Finally, (3.1) can be written as follows: Find $\psi_h \in N_h(\Gamma)$ such that

$$b(\psi_h, \varphi_h) = f_B(\varphi_h), \quad \forall \varphi_h \in N_h(\Gamma), \tag{3.2}$$

where the linear functional f_B corresponds to F_B.

Smith's algorithm. The Smith algorithm [9] is an additive Schwarz algorithm applied to the interface problem (3.2). Note that Γ has a natural decomposition from $\{\Omega_i\}$, i.e., $\Gamma = \cup_{i=1}^n \Gamma_i$ where $\Gamma_i = \partial\Omega_i \setminus \partial\Omega$. For each vertex v_j of Γ, let $\Gamma_\delta^{v_j}$ indicate the set of the points on Γ that are less than a distance δ from v_j; for each edge e_j of Γ, let $\Gamma_\delta^{e_j}$ represent the interior of e_j. For each ξ being either v_j or e_j, let $N_h(\Gamma_\delta^\xi) \subset N_h(\Gamma)$ denote the subset of functions with support in Γ_δ^ξ. Finally, let $N_h(\Gamma_\delta^H)$ denote some as yet unspecified coarse space. Now the finite element space $N_h(\Gamma)$ is represented by

$$N_h(\Gamma) = \sum_{\xi \in \{H, v_i, e_j\}} N_h(\Gamma_\delta^\xi). \tag{3.3}$$

By this decomposition the additive Schwarz algorithm can be defined as in (2.6).

To apply the abstract convergence theory in Lemma 1, we need to construct an appropriate coarse space $N_h(\Gamma_\delta^H)$. Let U_H be the conforming space of linear polynomials associated with $\mathcal{E}_H$, and let R_h denote the interpolation operator into $N_h(\Gamma)$. Then, define $N_h(\Gamma_\delta^H)$ by

$$N_h(\Gamma_\delta^H) = \{v \in N_h(\Gamma) : v = R_h(\varphi|_\Gamma), \varphi \in U_H\}. \tag{3.4}$$

Theorem 3. *The condition number $c(\Pi_\Gamma)$ of the additive Schwarz operator Π_Γ with the decomposition* (3.3) *and the coarse space $N_h(\Gamma_\delta^H)$ in* (3.4) *satisfies*

$$c(\Pi_\Gamma) \le C(1 + \log(H/\delta))^2, \tag{3.5}$$

where C is independent of h, H, and δ.

As in [7], we prove Theorem 3 using a result for the conforming elements if we can construct a conforming space that is isomorphic to $N_h(\Gamma)$. For this, let $U_{h/2}(\Gamma)$ denote the restriction of $U_{h/2}$ to $\cup_{i=1}^n \partial\Omega_i$. Then, define $\mathcal{I}_h : N_h(\Gamma) \to U_{h/2}(\Gamma)$ as in (2.9). Namely, if $v \in N_h(\Gamma)$ and $(x, y) \in \Gamma \cup \partial\Omega$,

$$\begin{aligned} \mathcal{I}_h v(x,y) &= 0 && \text{if } (x,y) \in \partial\Omega, \\ \mathcal{I}_h v(x,y) &= v(x,y) && \text{if } (x,y) \text{ is a midpoint of an edge of } E \in \mathcal{E}_h, \\ \mathcal{I}_h v(x,y) &= \tfrac{1}{\mathcal{N}_1} \textstyle\sum_j v(\bar{x}_j, \bar{y}_j) && \text{if } (x,y) \text{ is a vertex of } E \in \mathcal{E}_h, \end{aligned} \tag{3.6}$$

where $\mathcal{N}_1$ is the number of the adjacent midpoints $(\bar{x}_j, \bar{y}_j)$ on $\Gamma \cap \bar{E}$, $E \in \mathcal{E}_h$, to (x, y). With this definition we have the following lemma [7]. Thus Theorem 3 can be proven from the same result [6] on the conforming elements.

Lemma 4. *Let $\mathcal{I}_h : N_h(\Gamma) \to U_{h/2}(\Gamma)$ be defined as in* (3.6). *Then there are constants C_1 and C_2 independent of mesh parameters such that for all $v \in N_h(\Gamma)$*

$$C_1 b(v, v) \le |\mathcal{I}_h v|_{1/2,\partial\Omega}^2 \le C_2 b(v, v),$$

where

$$|\mathcal{I}_h v|^2_{1/2,\partial\Omega} = \int_{\partial\Omega}\int_{\partial\Omega} \frac{|v(s)-v(t)|^2}{|s-t|^2} dsdt.$$

Mandel's balancing domain decomposition. The balancing domain decomposition algorithm [10] provides an efficient preconditioner for the iterative solution of the reduced problem (3.1), and is defined in terms of two sets of matrices $\{D_i\}_{i=1}^n$ and $\{Y_i\}_{i=1}^n$. The matrices D_i form a decomposition of unity on $N_h(\Gamma)$:

$$\sum_{i=1}^n L_i D_i L_i^t v = v, \quad \forall v \in N_h(\Gamma),$$

where L_i is the canonical inclusion operator $L_i : N_h(\Gamma_i) \to N_h(\Gamma)$ given by setting the elements in $N_h(\Gamma_i)$ to be zero at the nodes outside Γ_i. For each i, let n_i =dim($N_h(\Gamma_i)$) and Y_i be an $n_i \times l_i$ matrix of full column rank with $0 \le l_i \le n_i$ such that

$$\text{Null } Z^{(i)} \subset \text{Range } Y_i.$$

The simplest choice for D_i is the diagonal matrix with diagonal elements equal to the reciprocal of the number of substructures with which the degree of freedom is associated [10]. We now define a "coarse space" $N_h(\Gamma^H) \subset N_h(\Gamma)$ by

$$N_h(\Gamma^H) = \{v \in N_h(\Gamma) : v = \sum_{i=1}^n L_i D_i \varphi,\ \varphi \in \text{Range } Y_i\}.$$

As in [10], $v \in N_h(\Gamma)$ is said *balanced* if

$$Y_i^t D_i^t L_i^t v = 0, \quad i = 1, \cdots, n.$$

The following balancing domain decomposition algorithm yields a preconditioner $\mathcal{M}$ for the system (2.2).

Balancing algorithm. Given $v \in N_h(\Gamma)$, calculate $\mathcal{M}^{-1}v$ as follows:

(1) Balance the original residual by solving the following problem for $\theta_i \in \Re^{l_i}$:

$$Y_i^t D_i^t L_i^t \left(v - Z \sum_{j=1}^n L_j D_j Y_j \theta_j \right) = 0, \quad i = 1, \cdots, n; \tag{3.7}$$

(2) Set

$$r = v - Z \sum_{j=1}^n L_j D_j Y_j \theta_j, \quad r_i = D_i^t L_i^t r, \quad i = 1, \cdots, n;$$

(3) Find any solution $v_i \in N_h(\Gamma_i)$

$$Z^{(i)} v_i = r_i, \quad i = 1, \cdots, n;$$

(4) Balance the residual by solving the equation for $\mu_i \in \Re^{l_i}$:

$$Y_i^t D_i^t L_i^t \left(v - Z \sum_{j=1}^n L_j D_j (v_j + Y_j \mu_j) \right) = 0, \quad i = 1, \cdots, n; \tag{3.8}$$

(5) Average the result on the interfaces by

$$\mathcal{M}^{-1} v = \sum_{j=1}^n L_j D_j (v_j + Y_j \mu_j).$$

The case where some $l_j = 0$ is allowed. In this case Y_j as well as the block unknowns θ_j and μ_j are void and the jth block equation is taken out of (3.7) and (3.8).

The convergence property of this balancing algorithm can be obtained from the corresponding result [10] for the conforming elements using the result in Lemma 4.

Theorem 5. *There is a constant C independent of h and H such that the condition number $c(\mathcal{M}^{-1}Z)$ of the matrix $\mathcal{M}^{-1}Z$ satisfies*

$$c(\mathcal{M}^{-1}Z) \leq C(1 + \log(H/h))^2.$$

4. Two-Level Additive Schwarz Method on Rectangles

For $0 < h < 1$, let $\mathcal{E}_h$ be a triangulation of Ω into rectangles of size h, and define the rotated Q_1-nonconforming finite element space

$$\begin{aligned} N_h = \Big\{ v \in L^2(\Omega) : v|_E = a_E^1 + a_E^2 x + a_E^3 y + a_E^4(x^2 - y^2),\ a_E^i \in \Re,\ \forall E \in \mathcal{E}_h; \\ \text{if } E_1 \text{ and } E_2 \text{ share an edge } e, \text{ then } \int_e \xi|_{\partial E_1} ds = \int_e \xi|_{\partial E_2} ds; \\ \text{and } \textstyle\int_{\partial E \cap \Gamma} \xi|_\Gamma ds = 0 \Big\}. \end{aligned}$$

The analysis and results in the last two sections hold for this element [4]. Also, due to the equivalence between nonconforming and mixed methods [1], [11–13], all the analysis and results apply to the mixed methods for second-order elliptic problems [1–4].

References

1. Z. Chen, East-West J. Numer. Math., **4**, 1–33 (1996).
2. Z. Chen and R. Ewing, in: Domain Decomposition Methods in Sciences and Engineering, R. Glowinski et al., (Eds.), Wiley-Interscience-Europe (1997) pp. 213–220.
3. Z. Chen, R. E. Ewing, Y. Kuznetsov, R. Lazarov and S. Maliassov, J. Numer. Lin. Alg. Appl., **30**, 427–453 (1996).
4. Z. Chen, R. E. Ewing and R. Lazarov, Math. Comp., **65**, 467–490 (1996).
5. P. G. Ciarlet, The Finite Element Method for Elliptic Problems, North-Holland, Amsterdam (1978).
6. M. Dryja and O. Widlund, SIAM J. Sci. Comput., **15**, 604–620 (1994).
7. L. Cowsar, in: Proc. Sixth Copper Mountain Conference on Multigrid Methods, N. Melson et al., (Eds.), NASA Conference Publication 3224, Part 1 (1993) pp. 93–109.
8. M. Sarkis, in: Proc. Sixth Copper Mountain Conference on Multigrid Methods, N. Melson et al., (Eds.), NASA Conference Publication 3224, Part 2 (1993) pp. 544–566.
9. B. Smith, SIAM J. Sci. Stat. Comput., **13**, 364–378 (1992).
10. J. Mandel, Commun. Appl. Math., **9**, 233–241 (1993).
11. T. Arbogast and Z. Chen, Math. Comp., **64**, 943–972 (1995).

12. D. N. Arnold and F. Brezzi, RAIRO Model. Math. Anal. Numer., **19**, 7–32 (1985).

13. Z. Chen, RAIRO Math. Model. Numer. Anal., **27**, 9–34 (1993).

9th Int. Coll. on Differential Equations, pp. 85-90
D. Bainov (Ed)

Local Mesh Refinement for Degenerate Two-Phase Incompressible Flow Problems

ZHANGXIN J. CHEN
Department of Mathematics, Southern Methodist University, Dallas, Texas 75275-0156, USA
RICHARD E. EWING
Institute for Scientific Computation, Texas A&M University, College Station, TX 77843-3404, USA

Abstract: In this paper we consider finite element methods for a degenerate elliptic-parabolic partial differential system which describes the flow of two incompressible, immiscible fluids in porous media. Local mesh refinement techniques for solving this degenerate system are discussed.
Key Words and Phrases: *Porous medium, degenerate elliptic-parabolic system, finite element, local mesh refinement.*

1. Two-Phase Flow Equations

In this paper we consider finite element methods for the flow of two incompressible, immiscible fluids in porous media $\Omega \subset \Re^d$, $d \leq 3$:

$$\begin{aligned} -\nabla \cdot \{\kappa(\lambda(s)\nabla p + \gamma_1'(s))\} &= q(s), \\ \phi\partial_t s - \nabla \cdot \{\kappa(\nabla\theta + \lambda_w(s)\nabla p + \gamma_2'(s))\} &= q_w(s), \end{aligned} \tag{1.1}$$

where ϕ and κ are the porosity and absolute permeability of the porous system, s is the (reduced) saturation of the wetting phase (e.g. water), λ is the total mobility, p and θ are the so called global pressure and complementary pressure respectively, λ_w is the wetting phase mobility, γ_1' and γ_2' are the given functions of s, and q and q_w are the sink/source terms. s is related to θ through

$$s = \mathcal{S}(\theta), \tag{1.2}$$

where $\mathcal{S}(x,\theta)$ is a strictly increasing function of θ. For the derivation of (1.1) from the usual two-phase flow equations, refer to [1, 2]. The pressure equation is given by the first equation of (1.1), while the saturation equation is described by the second equation. They determine the main unknowns p, s, and θ. With the following division of the boundary Γ of Ω:

$$\Gamma = \Gamma_1^p \cup \Gamma_2^p = \Gamma_1^\theta \cup \Gamma_2^\theta, \quad \emptyset = \Gamma_1^p \cap \Gamma_2^p = \Gamma_1^\theta \cap \Gamma_2^\theta,$$

the boundary conditions are specified by

$$\begin{aligned} -\kappa(\lambda(s)\nabla p + \gamma_1'(s)) \cdot \nu &= \varphi_1(x,t), & (x,t) &\in \Gamma_1^p \times J, \\ p &= \varphi_2(x,t), & (x,t) &\in \Gamma_2^p \times J, \\ -\kappa(\nabla\theta + \lambda_w(s)\nabla p + \gamma_2'(s)) \cdot \nu &= \varphi_3(x,t), & (x,t) &\in \Gamma_1^\theta \times J, \\ \theta &= \varphi_4(x,t), & (x,t) &\in \Gamma_2^\theta \times J, \end{aligned} \tag{1.3}$$

where the φ_i are given functions, $J = (0, T]$ $(T > 0)$, and ν is the outer unit normal to Γ. The initial condition is given by

$$s(x,0) = s_0(x), \quad x \in \Omega. \tag{1.4}$$

The usual Sobolev spaces $W^{l,\pi}(\Omega)$ with the norm $\|\cdot\|_{W^{l,\pi}(\Omega)}$ will be used, where l is a nonnegative integer and $0 \le \pi \le \infty$. When $\pi = 2$, we simply write $H^l(\Omega) = W^{l,2}(\Omega)$. When $l = 0$, we have $L^2(\Omega) = H^0(\Omega)$. Below $(\cdot,\cdot)_Q$ denotes the $L^2(Q)$ inner product (or sometimes the duality pairing); Q is omitted if $Q = \Omega$. Also, set $\Omega_T = \Omega \times J$.

For $d = 2$ or 3, let

$$H(\text{div}, \Omega) = \{v \in (L^2(\Omega))^d : \nabla \cdot v \in L^2(\Omega)\}.$$

Also, set

$$\begin{aligned} V(g) &= \{v \in H(\text{div},\Omega) : v \cdot \nu = g \text{ on } \Gamma_1^p\}, \\ W &= \{v \in L^2(\Omega) : \text{ if } \Gamma_2^p = \emptyset, \text{ then } \textstyle\int_\Omega v dx = 0\}, \\ M(g) &= \{v \in H^1(\Omega) : v = g \text{ on } \Gamma_2^\theta\}. \end{aligned}$$

Below M^* indicates the dual of $M(0)$.

We introduce the total velocity

$$u = -\kappa(\lambda(s)\nabla p + \gamma_1'(s)),$$

and the weak form for the system (1.1)–(1.4): find $u \in L^\infty(J; V(\varphi_1))$, $p \in L^\infty(J; W)$, and $\theta \in L^2(J; M(\varphi_4))$ such that

(1.5)

$$\begin{aligned} s &= S(\theta), \quad \phi\partial_t s \in L^2(J; M^*), \\ &\quad 0 \le \theta(x,t) \le \theta^*(x) \\ &\quad \text{a.e. on } \Omega_T, \\ (\nabla \cdot u, v) &= (q(s), v) \quad \forall v \in L^\infty(J; W), \\ (a(s)u, v) - (p, \nabla \cdot v) &= (\gamma_1(s), v) - (\varphi_2, v \cdot \nu)_{\Gamma_2^p} \\ &\quad \forall v \in L^\infty(J; V(0)), \\ \textstyle\int_0^t(\phi\partial_t s, v)d\tau + \int_0^t(\kappa\{\nabla\theta + \gamma_2(s)\} - f_w(s)u, \nabla v)d\tau &= \textstyle\int_0^t(q_w(s), v)d\tau \\ &\quad - \textstyle\int_0^t(\varphi_3, v)_{\Gamma_1^\theta}d\tau \\ &\quad \forall v \in L^2(0,t; M(0)),\ t \in J, \\ s(\cdot, 0) &= s_0, \end{aligned}$$

where

$$\begin{aligned} a(s) &= (\kappa\lambda(s))^{-1}, & \gamma_1(s) &= -\gamma_1'(s)\lambda^{-1}(s), \\ f_w(s) &= \lambda_w(s)/\lambda(s), & \gamma_2(s) &= \gamma_2'(s) - f_w(s)\gamma_1'(s). \end{aligned}$$

Under physically reasonable assumptions on the data [1], the system (1.1)–(1.4) has a weak solution in the sense of (1.5). Furthermore, the solution satisfies

$$\begin{aligned} &\|\partial_t s\|_{L^2(J;M^*)} + \|s\|_{L^2(\Omega_T)} + \|\theta\|_{L^2(J;H^1(\Omega))} \\ &\quad + \|u\|_{L^\infty(J;H(\text{div},\Omega))} + \|p\|_{L^\infty(J;L^2(\Omega))} \le C, \end{aligned}$$

where C is a positive constant and the $H(\text{div},\Omega)$-norm is given by

$$\|u\|_{H(\text{div},\Omega)} = \{\|u\|^2_{L^2(\Omega)} + \|\nabla \cdot u\|^2_{L^2(\Omega)}\}^{1/2}.$$

2. Finite Element Methods

Let Ω be a convex polygonal domain. For $0 < h_p < 1$ and $0 < h < 1$, let T_{h_p} and T_h be regular partitions into elements, say, simplexes, rectangular parallelepipeds, and/or prisms. In both partitions, each exterior edge or face has either Dirichlet or Neumann boundary conditions imposed on it, but not both. We assume neither the quasi-uniformity nor the acuteness of both T_{h_p} and T_h. Thus local mesh refinements discussed later can be allowed here.

Let $M_h \subset H^1(\Omega)$ be a standard C^0-finite element space associated with T_h such that

$$\inf_{v_h \in M_h} \|v - v_h\|_{H^1(\Omega)} \leq Ch\|v\|_{H^2(\Omega)}. \tag{2.1}$$

Here and below C (with or without a subscript) indicates a generic constant independent of h, which will probably take on different values in different occurrences. In this paper we only consider lowest-order C^0-finite elements such that (2.1) is satisfied. Due to lacking in regularity on the solution, no improvements in the asymptotic convergence rate result from taking higher-order finite element spaces. Now, for $g \in C^0(\bar{\Omega})$ define

$$M_h(g) = \{v \in M_h : v = I_h g \text{ on } \Gamma_2^\theta\},$$

where $I_h g$ indicates the interpolant of g in M_h.

We now introduce mixed finite element spaces. Without loss of generality, we consider the case of $\varphi_1 \equiv 0$ in the analysis below; otherwise, φ_1 can be incorporated into the differential equation, or the mixed finite element method can be handled by introducing the space of Lagrange multipliers [3]. Thus, set $V = V(0)$.

Associated with the partition T_{h_p}, let $V_h \times W_h = V_{h_p} \times W_{h_p} \subset V \times W$ be the Raviart-Thomas-Nedelec [4, 5], the Brezzi-Douglas-Fortin-Marini [6], the Brezzi-Douglas-Marini [7] (if $d = 2$), the Brezzi-Douglas-Durán-Fortin [8] (if $d = 3$), or the Chen-Douglas [9] mixed finite element space of index such that the approximation properties below are satisfied:

$$\begin{aligned}
&\inf_{v_h \in V_h} \|v - v_h\|_{L^2(\Omega)} \leq Ch_p^l \|v\|_{H^l(\Omega)}, && 0 \leq l \leq k^* + 1,\\
&\inf_{v_h \in V_h} \|\nabla \cdot (v - v_h)\|_{L^2(\Omega)} \leq Ch_p^l \|\nabla \cdot v\|_{H^l(\Omega)}, && 0 \leq l \leq k^{**},\\
&\inf_{v_h \in W_h} \|v - v_h\|_{L^2(\Omega)} \leq Ch_p^l \|v\|_{H^l(\Omega)}, && 0 \leq l \leq k^{**},
\end{aligned}$$

where $k^{**} = k^* + 1$ for the first two spaces, $k^{**} = k^*$ for the second two spaces, and both cases are included in the last space. Again, we consider the lowest-order mixed finite elements. In the case of $k^{**} = k^* + 1$, we take $k^* = 0$; in the case of $k^{**} = k^*$, we take $k^* = 1$.

For $t \in \bar{J}$ a semidiscrete approximation is defined as follows: find $(u_h(\cdot, t), p_h(\cdot, t)) \in V_h \times W_h$ such that

$$\begin{aligned}
(\nabla \cdot u_h, v) &= (q(s_h), v) && \forall v \in W_h,\\
(a(s_h)u_h, v) - (p_h, \nabla \cdot v) &= (\gamma_1(s_h), v) - (\varphi_2, v \cdot \nu)_{\Gamma_2^p} && \forall v \in V_h,
\end{aligned} \tag{2.2}$$

where $s_h = \mathcal{S}(\theta_h)$ and for each $t \in \bar{J}$, $\theta_h(\cdot, t) \in M_h(\varphi_4)$ satisfies

$$\begin{aligned}
&\textstyle\int_0^t (\phi \partial_t s_h, v) d\tau + \int_0^t (\kappa\{\nabla \theta_h + \gamma_2(s_h)\} - f_w(s_h)u_h, \nabla v) d\tau\\
&\quad = \textstyle\int_0^t (q_w(s_h), v) d\tau - \int_0^t (\varphi_3, v)_{\Gamma_1^\theta} d\tau \quad \forall v \in M_h(0),\ t \in J.
\end{aligned} \tag{2.3}$$

The initial datum is defined by

$$s_{0,h} = P_h s_0, \tag{2.4}$$

where P_h indicates the L^2-projection into M_h.

Note that if bases are introduced in V_h, W_h, and M_h, equations (2.2)–(2.4) can be written as a nonlinear system of ordinary differential equations for s_h (after substituting (2.2) into (2.3)). This nonlinear system can be shown to have a unique solution from the fundamental theorem of ordinary differential equations [10]. Error analysis has been given in [2] under physically reasonable assumptions on the data.

3. Characteristic-Based Methods

It is known [11] that an accurate approximation to the total velocity u is important; in (2.2), the mixed finite element method is used. Since the physical transport dominates the diffusive effects in incompressible flow, a more efficient approximation procedure should be used to solve the saturation equation. Standard finite difference and finite element methods produce numerical solutions with excessive oscillations, while upwinding and stabilized versions of these methods tend to generate solutions with nonphysical dispersion. Although conventional Eulerian-Lagrangian methods can overcome these difficulties, they fail to conserve mass. Here we very briefly review the Eulerian-Lagrangian localized adjoint method (ELLAM), which can accurately and efficiently solve advection-dominated problems in a mass-conservative manner.

To fix the ideas, we present it for the saturation equation in one space dimension

$$\begin{aligned}
(\phi s)_t + (Vs - Ds_x)_x &= q_w, & (x,t) &\in (a,b)\times J,\\
Vs - Ds_x &= g_a(t), & x = a,\ t &\in J,\\
-Ds_x &= g_b(t), & x = b,\ t &\in J,\\
s(x,0) &= g_0(x), & x &\in [a,b].
\end{aligned} \tag{3.1}$$

An extension to several space dimensions is straightforward. Also, in (3.1) we consider inflow flux and outflow Neumann boundary conditions; the ELLAM formalism can handle other types of boundary conditions [12]. Finally, for notational simplicity we consider uniform partitions in space and time below.

For positive integers I and $\mathcal{N}$, we introduce the partitions

$$\begin{aligned}
h &= (b-a)/I, & x_i &= a + ih, & i &= 0,1,\dots,I,\\
\Delta t &= T/\mathcal{N}, & t^n &= n\Delta t, & n &= 0,1,\dots,\mathcal{N}.
\end{aligned}$$

For any space-time test function v that vanishes outside $[a,b]\times(t^n,t^{n+1}]$ and is possibly discontinuous in time at t^n, a space-time weak formulation of (3.1) reads as follows:

$$\begin{aligned}
&\int_a^b \phi(x,t^{n+1})s(x,t^{n+1})v(x,t^{n+1})dx + \int_a^b\int_{t^n}^{t^{n+1}} Ds_x v_x dt dx\\
&\quad + \int_{t^n}^{t^{n+1}} (Vs - Ds_x)\, v|_a^b dt - \int_a^b\int_{t^n}^{t^{n+1}} s\phi\left(v_t + V^\phi v_x\right) dt dx\\
&= \int_a^b \phi(x,t^n)s(x,t^n)v(x,t_+^n)dx + \int_a^b\int_{t^n}^{t^{n+1}} q_w v dt dx,
\end{aligned} \tag{3.2}$$

where $v(x,t_+^n) = \lim_{t\to t_+^n} v(x,t)$ and $V^\phi = V/\phi$.

The ELLAM scheme below suggests that the test function v should be chosen from the solution space of the formal homogeneous adjoint problem of (3.1)

$$-\phi v_t - Vv_x - Dv_{xx} = 0.$$

This solution space is infinite-dimensional. For a numerical procedure, only a finite number of test functions are needed. Different choices of these functions lead to different classes of approximation methods. One choice is to solve the adjoint problem using the following operator-splitting technique:

$$\phi v_t + V v_x = 0 \quad \text{and} \quad D v_{xx} = 0. \tag{3.3}$$

This implies that v is constant along the characteristic in the direction (ϕ, V) or $(1, V^{\phi})$, so the last term in the left-hand side of (3.2) vanishes. In general, we cannot track the characteristic exactly. However, the residual should be small if the test function is constant along an approximate characteristic. We can approximate the characteristic via the backward Euler method. Also, it follows from (3.3) that v can be chosen as standard hat functions in space. Substituting this type of v into (3.2) and carrying out algebraic manipulations, we derive the ELLAM scheme for the resulting weak formulation. Error analysis of this scheme under physically reasonable assumptions on the data will be presented in a forthcoming paper.

4. Local Mesh Refinement

There are different types of self-adaptive mesh refinement techniques. For problems with fixed wells, faults, and pinchouts, fixed local mesh refinements have proven to be effective. With these fully local methods, an arbitrary level of refinement can be used at an arbitrary region in space. This technique requires a special data structure for setting up matrices as well as special algorithms for efficient solution. For problems where the local behavior along moving fluid interfaces is crucial to the flow process, the local mesh refinement must be capable of moving with the fluids. This type of refinement and the associated data structure are generally much more complex than fixed local refinement.

The complexity of the data structure and the associated solution procedures make many of the true local refinement techniques inefficient for large-scale porous medium flow problems. If different meshes are used for each time-step in a large-scale problem, the overhead associated with the data structure and solution procedures can easily dominate the overall computation time. For this reason, different techniques which do not require complex data structures or mesh regeneration at each time-step are desirable.

A second type of fixed local refinement is a patch-type refinement, where patches are located in regions of interest and fixed refinement is carried out on the patches. This patch refinement technique is an attractive alternative to truly local refinement. This technique does not need a complex data structure, but does involve the idea of passing information from one grid to another. The idea behind a local patch refinement is to select a patch that includes most of the critical behavior requiring better resolution, and use a special, possibly uniform, refinement within this patch. If a uniform or logically rectangular fine mesh is applied in the patch, very fast solution procedures can be utilized locally in this region using boundary data from the original coarse grid. This local refinement is much easier to vectorize than truly local refinement. Also, each patch can be sent to a different processor to attain parallelism in the algorithms [13, 14, 15].

The difficult problem associated with the patch-type refinement is the communication of the solution between the fine and coarse meshes. The use of local mesh refinement in large-scale porous medium problems often destroys the vectorization and efficient solution capabilities of numerical algorithms. Local patch refinement techniques combined with domain decomposition iterative solution procedures have proven to be very effective for developing accurate and efficient local mesh refinement in the context of large-scale porous medium simulation problems [16]. The patch-type refinement technique coupled with the domain decomposition solution procedure for the present two-phase problem under reasonable assumptions on the physical data is being investigated by the authors.

Acknowledgements

This work is supported in part by National Science Foundation grant DMS-9626179.

References

1. Z. Chen, SMU Math. Report 97–10, Dallas, Texas.
2. Z. Chen and R. Ewing, Numer. Math., (submitted).
3. Z. Chen and R. Ewing, SIAM J. Numer. Anal., **34**, 2228–2253 (1997).
4. J. Nedelec, Numer. Math., **35**, 315–341 (1980).
5. P. Raviart and J. Thomas, Lecture Notes in Mathematics, Vol. 606, Springer, Berlin, 292–315 (1977).
6. F. Brezzi, J. Douglas, Jr., M. Fortin and L. Marini, RAIRO Modèl. Math. Anal. Numér., **21**, 581–604 (1987).
7. F. Brezzi, J. Douglas, Jr. and L. Marini, Numer. Math., **47**, 217–235 (1985).
8. F. Brezzi, J. Douglas, Jr., R. Durán and M. Fortin, Numer. Math., **51**, 237–250 (1987).
9. Z. Chen and J. Douglas, Jr., Calcolo, **26**, 135–148 (1989).
10. Z. Chen, M. Espedal and R. Ewing, Appl. Math., **40**, 203–226 (1995).
11. J. Douglas, Jr., R. E. Ewing and M. Wheeler, RAIRO Anal. Numér., **17**, 17–33 (1983).
12. M. A. Celia, T. F. Russell, I. Herrera and R. E. Ewing, Advances in Water Resources, **13**, 187–206 (1990).
13. M. Espedal and R. Ewing, Comp. Methods Appl. Mech. Engrg., **64**, 113–135 (1987).
14. R. Ewing, Comp. Methods Appl. Mech. Engrg., **55**, 89–103 (1986).
15. R. Ewing, B. Boyett, D. Babu and R. Heinemann, SPE 18413, Houston, TX, (1989).
16. J. H. Bramble, R. E. Ewing, J. E. Pasciak and A. H. Schatz, Comp. Meth. Appl. Mech. Eng., **67**, 149–159 (1988).

9th Int. Coll. on Differential Equations, pp. 91-98
D. Bainov (Ed)

The Existence of Infinitely Many Solutions of a Nonlinear Wave Equation

Q-HEUNG CHOI
Department of Mathematics, Inha University, Incheon 402-751, Korea
TACKSUN JUNG
Department of Mathematics, Kunsan N. University, Kunsan 573-701, Korea

Abstract: We investigate a relation between multiplicity of solutions and source terms of jumping problem in a wave equation. We show that the jumping problem has infinitely many solutions when the source term is positive multiple of the positive eigenfunction.
Key Words and Phrases: *Infinitly many solutions, Dirichlet boundary, eigenfunction.*
1991 Mathematics Subject Classification. 35B10, 35L20.

1. Introduction

We investigate the existence of solutions $u(x,t)$ for a piecewise linear perturbation of the one-dimensional wave operator $u_{tt} - u_{xx}$ under Dirichlet boundary condition on the interval $(-\frac{\pi}{2}, \frac{\pi}{2})$ and periodic condition on the variable t,

$$u_{tt} - u_{xx} + g(u) = f(x,t) \quad \text{in} \quad (-\frac{\pi}{2}, \frac{\pi}{2}) \times R, \tag{1.1}$$

$$u(\pm\frac{\pi}{2}, t) = 0, \tag{1.2}$$

$$u \text{ is } \pi - \text{periodic in } t \text{ and even in } x, \tag{1.3}$$

where we assume that $g(u) = bu^+ - au^-$. We let L the wave operator, $Lu = u_{tt} - u_{xx}$. Then the eigenvalue problem for $u(x,t)$

$$Lu = \lambda u \quad \text{in} \quad (-\frac{\pi}{2}, \frac{\pi}{2}) \times R$$

with (1.2) and (1.3), has infinitely many eigenvalues

$$\lambda_{mn} = (2n+1)^2 - 4m^2 \qquad (m, n = 0, 1, 2, \cdots)$$

and corresponding normalized eigenfunctions $\phi_{mn}, \psi_{mn} (m, n \geq 0)$ given by

$$\begin{aligned}
\phi_{0n} &= \frac{\sqrt{2}}{\pi} \cos(2n+1)x \text{ for } n \geq 0, \\
\phi_{mn} &= \frac{2}{\pi} \cos 2mt \cos(2n+1)x \quad \text{for} \quad m > 0, n \geq 0, \\
\psi_{mn} &= \frac{2}{\pi} \sin 2mt \cos(2n+1)x \quad \text{for} \quad m > 0, n \geq 0.
\end{aligned}$$

We note that all eigenvalues in the interval $(-9,9)$ are given by

$$\lambda_{21} = -7 < \lambda_{10} = -3 < \lambda_{00} = 1 < \lambda_{11} = 5.$$

Let Q be the square $(-\frac{\pi}{2}, \frac{\pi}{2}) \times (-\frac{\pi}{2}, \frac{\pi}{2})$ and H the Hilbert space defined by

$$H = \{u \in L^2(Q) : u \ \ is \ \ even \ \ in \ \ x\}.$$

Then the set of eigenfunctions $\{\phi_{mn}, \psi_{mn}\}$ is an orthonormal base in H. Hence equation (1.1) with (1.2) and (1.3) is equivalent to

$$Lu + bu^+ - au^- = f \quad \text{in} \quad H. \tag{1.4}$$

In [1] the authors investigate multiplicity of solutions of (1.4) when the nonlinearity $-(bu^+ - au^-)$ crosses the eigenvalue λ_{10} and the source term f is a multiple of the positive eigenfunction ϕ_{00}.

Our concern is to investigate a relation between multiplicity of solutions and source terms of jumping problem (1.4) when the nonlinearity $-(bu^+ - au^-)$ crosses the eigenvalue λ_{10} and the source term f is generated by three eigenfunctions $\phi_{00}, \phi_{10}, \psi_{10}$.

Let $-1 < \theta \leq 1$ and $\phi_\theta = \theta\phi_{10} + \sqrt{1-\theta^2}\psi_{10}$, which is an eigenfunction of L corresponding to λ_{10}. Let H_θ be the subspace of H defined by

$$H_\theta = span\{\{\phi_{mn}, \psi_{mn} : \phi_{mn} \neq \phi_{10}, \psi_{mn} \neq \psi_{10}\} \cup \{\phi_\theta\}\}.$$

In Section 2, we suppose that the nonlinearity $-(bu^+ - au^-)$ crosses an eigenvalue λ_{10} and the source term f is generated by ϕ_{00} and ϕ_θ and we investigate the existence of solutions of the equation

$$Lu + bu^+ - au^- = f \quad \text{in} \quad H_\theta. \tag{1.4$'$}$$

In Section 3, we reveal a relation between multiplicity of solutions and source terms in equation (1.4′) when f belongs to the two dimensional space $V_\theta = span\{\phi_{00}, \phi_\theta\}$. In Section 4, we reveal a relation between multiplicity of solutions and source terms in equation (1.4) when f belongs to the three dimensional space $V_\theta = span\{\phi_{00}, \phi_{10}, \psi_{10}\}$. We also show that (1.4) has infinitely many solutions in H when the source term is positive multiple of the positive eigenfunction ϕ_{00}.

2. A Variational Reduction Method

We suppose that $-1 < a < 3$ and $3 < b < 7$. Under this assumption, we have a concern with a relation between multiplicity of solutions in H_θ and source terms of a nonlinear wave equation

$$Lu + bu^+ - au^- = f \quad \text{in} \quad H_\theta. \tag{2.1}$$

Here we suppose that f is generated by two eigenfunctions ϕ_{00} and ϕ_θ.

Let V_θ be the two dimensional subspace of H_θ spanned by $\{\phi_{00}, \phi_\theta\}$ and W be the orthogonal complement of V_θ in H_θ. Let P_θ be an orthogonal projection H_θ onto V_θ. Then every element $u \in H_\theta$ is expressed by $u = v + w$, where $v = P_\theta u$, $w = (I - P_\theta)u$. Hence equation (2.1) is equivalent to a system

$$Lw + (I - P_\theta)(b(v+w)^+ - a(v+w)^-) = 0, \tag{2.2}$$

$$Lv + P_\theta(b(v+w)^+ - a(v+w)^-) = s_1\phi_{00} + s_2\phi_\theta. \tag{2.3}$$

LEMMA 2.1. *For fixed $v \in V_\theta$,* (2.2) *has a unique solution $w = w_\theta(v)$. Furthermore, $w_\theta(v)$ is Lipschitz continuous (with respect to L^2 norm) in terms of v.*

The proof of the lemma is similar to that of Lemma 1.1 of [2]. From now, for simplicity, we denote $w(v) = w_\theta(v)$.

By Lemma 2.1, the study of multiplicity of solutions of (2.1) is reduced to the study of multiplicity of solutions of an equivalent problem

$$Lv + P_\theta(b(v+w(v))^+ - a(v+w(v))^-) = s_1\phi_{00} + s_2\phi_\theta \tag{2.4}$$

defined on the two dimensional subspace V_θ spanned by $\{\phi_{00}, \phi_\theta\}$.

If $v \geq 0$ or $v \leq 0$, then $w(v) \equiv 0$. For example, let us take $v \geq 0$ and $w(v) = 0$. Then equation (2.2) reduces to

$$L0 + (I - P_\theta)(bv^+ - av^-) = 0$$

which is satisfied because $v^+ = v$, $v^- = 0$ and $(I - P_\theta)v = 0$, since $v \in V_\theta$.

Since the subspace V is spanned by $\{\phi_{00}, \phi_\theta\}$ and $\phi_{00}(x,t) > 0$ in Q, there exists a cone C_1 defined by

$$C_1 = \{v = c_1\phi_{00} + c_2\phi_\theta \mid c_1 \geq 0, K_1(\theta)c_1 \leq c_2 \leq K_2(\theta)c_1\}$$

for some $K_1(\theta) < 0, K_2(\theta) > 0$ so that $v \geq 0$ for all $v \in C_1$ and a cone C_3 defined by

$$C_3 = \{v = c_1\phi_{00} + c_2\phi_\theta \mid c_1 \leq 0, K_2(\theta)c_1 \leq c_2 \leq K_1(\theta)c_1\}$$

so that $v \leq 0$ for all $v \in C_3$.

We define a map $\Phi_\theta : V_\theta \to V_\theta$ given by

$$\Phi_\theta(v) = Lv + P_\theta(b(v+w(v))^+ - a(v+w(v))^-), \quad v \in V_\theta. \tag{2.5}$$

Then Φ_θ is continuous on V_θ, since $w(v)$ is continuous on V_θ and hence we have the following lemma (cf. Lemma 2.2 of [2]).

LEMMA 2.2. *$\Phi_\theta(cv) = c\Phi_\theta(v)$ for $c \geq 0$ and $v \in V_\theta$.*

Lemma 2.2 implies that Φ maps a cone with vertex 0 onto a cone with vertex 0. We define the cones C_2, C_4 as follows

$$C_2 = \left\{v = c_1\phi_{00} + c_2\phi_\theta \mid c_2 \geq 0, \frac{1}{K_1(\theta)}c_2 \leq c_1 \leq \frac{1}{K_2(\theta)}c_2\right\}$$

$$C_4 = \left\{v = c_1\phi_{00} + c_2\phi_\theta \mid c_2 \leq 0, \frac{1}{K_2(\theta)}c_2 \leq c_1 \leq \frac{1}{K_1(\theta)}c_2\right\}.$$

Then the union of $C_i (1 \leq i \leq 4)$ is V_θ. We have the images of the cones C_1, C_3 under Φ_θ as follows

$$R_1 = \left\{d_1\phi_{00} + d_2\phi_\theta \mid d_1 \geq 0, K_1(\theta)\left(\frac{b+\lambda_{10}}{b+\lambda_{00}}\right)d_1 \leq d_2 \leq K_2(\theta)\left(\frac{b+\lambda_{10}}{b+\lambda_{00}}\right)d_1\right\},$$

$$R_3 = \left\{d_1\phi_{00} + d_2\phi_\theta \mid d_1 \leq 0,\ K_1(\theta)\frac{\lambda_{10}+a}{\lambda_{00}+a}d_1 \leq d_2 \leq K_2(\theta)\frac{\lambda_{10}+a}{\lambda_{00}+a}d_1\right\}.$$

We note that R_1 is in the right half plane of V_θ and R_3 is in the left half plane of it. The restrictions $\Phi_\theta|_{C_1} : C_1 \to R_1$ and $\Phi_\theta|_{C_3} : C_3 \to R_3$ are bijective.

THEOREM 2.1. *(i) If f belongs to R_1, then equation (2.1) has a positive solution and no negative solution. (ii) If f belongs to R_3, then equation (2.1) has a negative solution and no positive solution.*

Lemma 2.2 means that the images $\Phi_\theta(C_2)$ and $\Phi(C_4)$ are the cones in the plane V_θ. Before we investigate the images $\Phi_\theta(C_2)$ and $\Phi_\theta(C_4)$, we set

$$R_2' = \left\{ d_1\phi_{00} + d_2\phi_{10} \mid d_2 \geq 0, \frac{1}{K_2(\theta)} \frac{\lambda_{00}+a}{\lambda_{10}+a} d_2 \leq d_1 \leq \frac{1}{K_2(\theta)} \frac{b+\lambda_{00}}{b+\lambda_{10}} d_2 \right\},$$

$$R_4' = \left\{ d_1\phi_{00} + d_2\phi_{10} \mid d_2 \leq 0, \frac{1}{K_1(\theta)} \frac{\lambda_{00}+a}{\lambda_{10}+a} d_2 \leq d_1 \leq \frac{1}{K_1(\theta)} \frac{b+\lambda_{00}}{b+\lambda_{10}} d_2 \right\}.$$

Then the union of four cones R_1, R_2', R_3, R_4' is also the space V_θ.

To investigate a relation between multiplicity of solutions and source terms in the nonlinear wave equation

$$Lu + bu^+ - au^- = f \quad \text{in} \quad H_\theta, \tag{2.6}$$

we consider the restrictions $\Phi_\theta|_{C_i} (1 \leq i \leq 4)$ of Φ to the cones C_i. Let $\Phi_{\theta i} = \Phi_\theta|_{C_i}$, i.e.,

$$\Phi_{\theta i} : C_i \to V_\theta.$$

For $i = 1, 3$, the image of $\Phi_{\theta i}$ is R_i and $\Phi_{\theta i} : C_i \to R_i$ is bijective.

From now on, our goal is to find the image of C_i under $\Phi_{\theta i}$ for $i = 2, 4$. Suppose that γ is a simple path in C_2 without meeting the origin, and end points (initial and terminal) of γ lie on the boundary ray of C_2 and they are on each other boundary ray. Then the image of one end point of γ under Φ_θ is on the ray $c_1(b+\lambda_{00})\phi_{00} + K_2(\theta)c_1(b+\lambda_{10})\phi_\theta, c_1 \geq 0$ (a boundary ray of R_1) and the image of the other end point of γ under Φ_θ is on the ray $-c_1(\lambda_{00}+a)\phi_{00} + K_1(\theta)c_1(\lambda_{10}+a)\phi_\theta, c_1 \geq 0$ (a boundary ray of R_3). Since Φ_θ is continuous, $\Phi_\theta(\gamma)$ is a path in V_θ. By Lemma 1.2, $\Phi_\theta(\gamma)$ does not meet the origin. Hence the path $\Phi_\theta(\gamma)$ meets all rays (starting from the origin) in $R_1 \cup R_4'$ or all rays (starting from the origin) in $R_2' \cup R_3$.

Therefore it follows from Lemma 1.2 that the image $\Phi_\theta(C_2)$ of C_2 contains one of sets $R_1 \cup R_4'$ and $R_2' \cup R_3$.

Similarly, we have that the image $\Phi_\theta(C_4)$ of C_4 contains one of sets $R_1 \cup R_2'$ and $R_4' \cup R_3$.

LEMMA 2.3. *Let A be one of the sets $R_1 \cup R_4'$ and $R_2' \cup R_3$ such that it is contained in $\Phi_\theta(C_2)$. Let γ be any simple path in A with end points on ∂C_2, where each ray (starting from the origin) in A intersect only one point of γ. Then the inverse image $\Phi_{\theta 2}^{-1}(\gamma)$ of γ is a simple path in C_2 with end points on ∂C_2, where any ray (starting from the origin) in C_2 intersects only one point of this path.*

Similarly, we have the following lemma.

LEMMA 2.3$'$. *Let A be one of the sets $R_1 \cup R_2'$ and $R_4' \cup R_3$ such that it is contained in $\Phi_\theta(C_4)$. Let γ be any simple path in A with end points on ∂A, where each ray (starting from the origin) in A intersect only one point of γ. Then the inverse image $\Phi_{\theta 4}^{-1}(\gamma)$ of γ is a simple path in C_4 with end points on ∂C_4, where any ray (starting from the origin) in C_4 intersects only one point of this path.*

With Lemma 2.3 and Lemma 2.3′, we have the following theorem, which is very important to investigate a relation between the multiplicity of solutions and source terms in a nonlinear wave equation.

THEOREM 2.2. *For $i = 2,4$, if we let $\Phi_{\theta i}(C_i) = R_i$, then R_2 is one of sets $R_1 \cup R_4'$, $R_2' \cup R_3$ and R_4 is one of sets $R_3 \cup R_4'$, $R_1 \cup R_2'$. Furthermore, for each $1 \le i \le 4$, the restriction $\Phi_{\theta i}$ maps C_i onto R_i. In particular, $\Phi_{\theta 1}$ and $\Phi_{\theta 3}$ are bijective.*

3. Multiplicity of Solutions and Source Terms in H_θ

In this section we reveal the relation between multiplicity of solutions and source terms in the nonlinear wave equation (2.1). Now we remember the map $\Phi_\theta : V_\theta \to V_\theta$ given by

$$\Phi_\theta(v) = Lv + P_\theta(b(v + w(v))^+ - a(v + w(v))^-), \quad v \in V_\theta,$$

where $-1 < a < 3 < b < 7$, $w(v)$ is a solution of (2.2), and V_θ is the two-dimensional subspace of H_θ spanned by two eigenfunctions ϕ_{00}, ϕ_θ. The map Φ_θ is continuous on V_θ, since $w(v)$ is continuous on V_θ.

For $f \in V_\theta$, we establish an *a priori* bound for solutions of

$$Lv + P_\theta(b(v + z(v))^+ - a(v + z(v))^-) = f \quad \text{in} \quad V_\theta. \tag{3.1}$$

LEMMA 3.1. *Let $C = \{(a,b) : \frac{1}{\sqrt{b+1}} + \frac{1}{\sqrt{a+1}} = 1\}$. Let $k(\ge 16)$ be fixed and $f \in V_\theta$ with $\|f\| = k$. Let $\alpha, \beta, \epsilon > 0$ be given. Let $3 + \alpha < b < 7 - \alpha$, $-1 + \beta < a < 3 - \beta$ satisfy the condition $\frac{1}{\sqrt{b+1}} + \frac{1}{\sqrt{a+1}} \ne 1$ and $dist((a,b), C) \ge \epsilon$. Then there exists $R_0 > 0$ (depending only on k and α, β, ϵ) such that the solutions of (3.1) satisfy $\|v\| < R_0$.*

LEMMA 3.2. *Let $-1 < a < 3$, $-1 < b < 7$ satisfy*

$$\frac{1}{\sqrt{b+1}} + \frac{1}{\sqrt{a+1}} < 1. \tag{3.2}$$

Let $k(\ge b+1)$ be fixed and $f \in V_\theta$ with $\|f\| = k$. Then we have

$$d(v - L^{-1}(f - P_\theta(b(v + z(v))^+ - a(v + z(v))^-)), B_R, 0) = 1$$

for all $R \ge R_0$.

LEMMA 3.3. *Let $-1 < a < 3 < b < 7$ satisfy the condition (3.2) and $f = (b+1)\phi_{00}$. Then equation (3.1) has a positive solution in* Int C_1, *at least one sign changing solution in* Int C_2, *and at least one sign changing solution in* Int C_4.

With Theorem 2.2, Lemma 3.3, we get the following.

LEMMA 3.4. *Let $-1 < a < 3 < b < 7$ satisfy the condition (3.2). For $1 \le i \le 4$, let $\Phi_\theta(C_i) = R_i$. Then $R_2 = R_1 \cup R_4'$ and $R_4 = R_1 \cup R_2'$, where R_2', R_4' are the same cones as in section 2.*

Proof. It follows from Lemma 3.3 that $R_2 \cap R_1 \ne \phi$. Since R_2 is one of sets $R_1 \cup R_4'$, $R_3 \cup R_2'$ (Theorem 2.2), the image R_2 of C_2 under Φ must be $R_1 \cup R_4'$.

On the other hand, it follows from Lemma 2.6 that $R_4 \cap R_1 \ne \phi$. Since R_4 is one of sets $R_1 \cup R_2'$, $R_3 \cup R_4'$ (Theorem 2.2), the image R_4 of C_4 under Φ_θ must be $R_1 \cup R_2'$. ∎

If a solution of (2.4) is in C_1, then it is positive. If a solution of (2.4) is in C_3, then it is negative. If a solution of (2.4) is in Int $(C_2 \cup C_4)$, then it has both signs. Therefore we have the main theorem of this paper with aid of Theorem 2.1, Theorem 2.2, and Lemma 3.4.

THEOREM 3.1. *Let $-1 < a < 3 < b < 7$ satisfies the condition (3.2). Then we have the followings.*
(i) If $f \in$ Int R_1, then equation (2.1) has a positive solution and at least two sign changing solutions in H_θ.
(ii) If $f \in \partial R_1$, then equation (2.1) has a positive solution and at least one sign changing solution in H_θ.
(iii) If $f \in$ Int $R_i'(i = 2, 4)$, then equation (2.1) has at least one sign changing solution in H_θ.
(iv) If $f \in$ Int R_3, then equation (2.1) has only the negative solution.
(v) If $f \in \partial R_3$, then equation (2.1) has a negative solution.

4. A Note on the Existence of Infinitely Many Solutions in H

We suppose that $-1 < a < 3$ and $3 < b < 7$. Under this codition, we investigate the existence of solutions in H of a nonlinear wave equation

$$Lu + bu^+ - au^- = f \quad \text{in} \quad H \tag{4.1}$$

in the weak sence; u is a solutions of (4.1) iff $(Lu + bu^+ - au^-, h) = (f, h)$ holds for all $h \in H$. Here we suppose that f is generated by three eigenfunctions ϕ_{00}, ϕ_{10}, ψ_{10}. We reveal a relation between multiplicity of solutions in H and source terms of equation (4.1).

Let V be the subspace of H spanned by three eigenfunctions ϕ_{00}, ϕ_{10}, ψ_{10}. Let $f \in V$. Then $f \in V_\theta$ for some θ with $-1 < \theta \le 1$ and hence f belongs to some cone R_i in V_θ (defined in section 3).

LEMMA 4.1. *Let $f \in V_\theta$. If u_θ is a solution in H_θ of (2.1), then it becomes a weak solution in H of (4.1).*

Proof. If $h \in H$, then it is expressed by $h = h_\theta + h_\theta^\perp$, where $h_\theta \in H_\theta$, $h_\theta^\perp \in H_\theta^\perp$. Here $H_\theta^\perp$ is an orthogonal compliment of H_θ in H. Since $Lu_\theta + bu_\theta^+ - au_\theta^- - f \in H_\theta$, for all $h \in H$ we have

$$\begin{aligned}
&(Lu_\theta + bu_\theta^+ - au_\theta^- - f, h)\\
&= (Lu_\theta + bu_\theta^+ - au_\theta^- - f, h_\theta + h_\theta^\perp)\\
&= (Lu_\theta + bu_\theta^+ - au_\theta^- - f, h_\theta) + (Lu_\theta + bu_\theta^+ - au_\theta^- - f, h_\theta^\perp)\\
&= 0.
\end{aligned}$$

∎

Let P be an orthogonal projection H onto V. Then every element $u \in H$ is expressed by $u = v + w$, where $v = Pu$, $w = (I - P)u$. Hence equation (4.1) is equivalent to a system

$$Lw + (I - P)(b(v + w)^+ - a(v + w)^-) = 0, \tag{4.2}$$

$$Lv + P(b(v + w)^+ - a(v + w)^-) = f. \tag{4.3}$$

For fixed $v \in V$, (4.2) has a unique solution $w = w(v)$ (cf. Lemma 2.1). We note that if $v \in V_\theta$ then $Lw + (I - P)(b(v+w)^+ - a(v+w)^-) = Lw + (I - P_\theta)(b(v+w)^+ - a(v+w)^-)$

$= 0$ and hence $w(v) = w_\theta(v)$ for $v \in V_\theta$. Furthermore, $w(v)$ is Lipschitz continuous (with respect to L^2 norm) in terms of v. Hence the multiplicity of solutions of (4.1) is equal to that of (4.3). We investigate the multiplicity of solutions of

$$Lv + P(b(v + w(v))^+ - a(v + w(v))^-) = f. \tag{4.4}$$

Since $\phi_{00} > 0$ in Q, there is a cone $\Gamma_1 \subset V$ such that $u > 0$ for all $u \in \Gamma_1$ and a cone $\Gamma_3 \subset V$ such that $u < 0$ for all $u \in \Gamma_3$. Let $\Gamma_2 = V \backslash (\Gamma_1 \cup \Gamma_3)$. Then every element $u \in \Gamma_2$ has both sign.

We define a map $\Phi : V \to V$ given by $\Phi(v) = Lv + P(b(v + w(v))^+ - a(v + w(v))^-)$. Then $\Phi(v) = \Phi_\theta(v)$ for $v \in V_\theta$. Let $\Sigma_1 = \Phi(\Gamma_1)$, $\Sigma_3 = \Phi(\Gamma_3)$, $\Sigma_2 = V \backslash (\Sigma_1 \cup \Sigma_3)$. Then for any θ $(-1 < \theta \leq 1)$ $\Sigma_1 \cap V_\theta = R_1$, $\Sigma_3 \cap V_\theta = R_3$, $\Sigma_2 \cap V_\theta = \mathrm{Int}(C_2 \cup C_4)$ in V_θ. Hence we have the following theorem.

THEOREM 4.1. *Let $-1 < a < 3 < b < 7$ satisfy condition (3.2). Then we have:*
(i) If $f \in \mathrm{Int}\Sigma_1$, then equation (4.1) has a positive solution and at least two sign changing solutions in H.
(ii) If $f \in \partial\Sigma_1$, then equation (4.1) has a positive solution and at least one sign changing solution in H.
(iii) If $f \in \Sigma_2$, then equation (4.1) has at least one sign changing solution in H.
(iv) If $f \in \mathrm{Int}\Sigma_3$, then equation (4.1) has only the negative solution in H.
(v) If $f \in \partial\Sigma_3$, then equation (4.1) has a negative solution in H.

In particular, for the case $f = s\phi_{00} (s > 0)$ we have

THEOREM 4.2. *Let $-1 < a < 3 < b < 7$ satisfy condition (3.2). For $f = s\phi_{00} (s > 0)$ equation (4.1) has infinitely many solutions in H.*

Proof. Let $s > 0$. For any $\theta(-1 < \theta \leq 1)$ and $f = s\phi_{00}(s > 0)$

$$Lu + P(bu^+ - au^-) = s\phi_{00} \text{ in } H_\theta \tag{4.5}$$

has a positive solution and at least two sign changing solutions in H_θ. By Lemma 3.4, there are solutions $u_{2\theta}$, $u_{4\theta}$ of (4.5) such that $P_\theta(u_{2\theta}) \in C_2$ and $P_\theta(u_{4\theta}) \in C_4$. If $\theta_1 \neq \theta_2 (-1 < \theta_1, \theta_2 \leq 1)$, then $u_{2\theta_1} \neq u_{2\theta_2}$ and $u_{4\theta_1} \neq u_{4\theta_2}$. This proves the theorem. ∎

Acknowledgements

The research of the first author is supported in part by GARC-KOSEF and Inha University Research Fund.

The research of the second author is supported in part by BSRI Program BSRI-97-1436.

References

1. Q.H. Choi and T. Jung, J. Differential Equations, **117**, 390-410 (1995).

2. Q.H. Choi, S. Chun and T. Jung, Studia Math., **120**, 259-270 (1996).

3. A. Ambrosetti and G. Prodi, A Primer of Nonlinear Analysis, Cambridge, University Press, Cambridge Studies in Advanced Math., **No. 34** (1993).

4. A.C. Lazer and P.J. McKenna, J. Math. Anal. Appl., **107**, 371-395 (1985).

5. A.C. Lazer and P.J. McKenna, Comm.in P.D.E., **11**, 1653-1676 (1986).

6. K. Schmitt, Rocky Mountain Math. J., **16**, 481-496 (1986).

9th Int. Coll. on Differential Equations, pp. 99-106
D. Bainov (Ed)

The Approximation of Continuous Periodic Functions by Means of Extrapolation Techniques for the De La Valle'e Poussin Operator

FRANCESCO COSTABILE AND MARIA ITALIA GUALTIERI
Dipartimento di Matematica, Università della Calabria, Italy
E-mail: Costabil@unical.it

STEFANO SERRA
Dipartimento di Energetica, Università di Firenze, Italy
E-mail: Serra@mail.dm.unipi.it

Abstract: We are concerned with the approximation of continuous 2π-periodic functions f by using trigonometric polynomials. It is well known that the De La Vallée Poussin integral produces a trigonometric polynomial approximation $V_n(f)$ which converges uniformly to the function f. Unfortunately, this technique is very slow and consequently is not employed for computational purposes. In this paper first we give a trigonometric version of the Taylor formula and then we prove an asymptotic expansion formula of $V_n(f)$ which, joint with classical extrapolation procedures, allows one to approximate f fastly.
Key Words and Phrases: *De La Vallée Poussin operator, extrapolation.*
1991 Mathematics Subject Classification. 65D15, 65B05.

1. Introduction

In this paper, we consider the use of De La Vallée Poussin integrals [1]

$$V_n(f) = \frac{(2n)!!}{(2n-1)!!}\,\frac{1}{2\pi}\int_{-\pi}^{\pi} f(t)\,\cos^{2n}\frac{(t-x)}{2}\,dt \tag{1.1}$$

in order to approximate a given function belonging to the linear space of the continuous 2π functions $C_{2\pi}$.

The theorem of Korovkin (see [2], pages 14-17) asserts that the convergence of positive linear operators generically defined in

$$C_\theta = \{f : \Re \to \Re : \ f \text{ is continuous}, \ f(x+\theta) = f(x), \ x \in \Re\}$$

on itself, to the identity, depends only on its convergence on three test functions. Moreover, the related rate of convergence depends also on the specific convergence speed with regard to these test functions. We point out that the operator sequence defined by $\{V_n(\cdot)\}_n$ (see (1.1)) has the previous feature [1] and, consequently, it gives a *theoretical* scheme to approximate the given function f.

Unfortunately, the magnitude of the convergence is computationally not satisfactory and improves with the smoothness of the functions up to $C^2_{2\pi}$, whereas it does not improve with smoother class of functions. More precisely, if $f \in C^2_{2\pi}$ we find [1]

$$\|V_n(f) - f\|_\infty = O(n^{-1}) \tag{1.2}$$

which is a result analogous to that one of Voronowskaja [3] for the Bernstein polynomial approximation.

As shown in [4], to improve the order of the approximation we must consider non-positive operators. Actually, in [4] the authors also define special linear combinations of few terms $V_n(f)$ with different choices of n and they prove approximation results finer than (1.2).

In the sequel, we fully generalize this idea by following two main stages (see also [5, 6, 7] for other operators):

- We give a trigonometric ("sine") version of Taylor's formula.
- We demonstrate an asymptotic expansion for $V_n(f)$ by proving that, under some regularity conditions, there exist constant values $S_i(f,x)$ independent of n such that the following relation holds

$$V_n(f) = f(x) + \sum_{i=1}^{k} S_i(f,x)h^i + z_k(f,x,h)o(h^k).$$

  Here $h = n^{-1}$ and z_k is bounded with the parameter h, for any choice of $x \in I = [-\pi,\pi]$.
- We apply an estrapolation algorithm and, by virtue of the previos expansion, we obtain an error of order h^k for the estrapolated expression (of course, as remarked in [4], the new operator is a nonpositive one).

Observe that the resulting expression is a linear combination of the De La Vallée Poussin terms; actually, by using this procedure we can easily obtain also the linear combinations considered in [4]. Since we are interested in computational algorithms, we do not calculate the terms $V_n(f)$, but we use an extended trapezoidal rule in order to evaluate the given integral. In addition, by calling $T_N(f_n)$ the quantity obtained by the extended trapezoidal rule applied to $f_n(t) = f(t)\cos^{2n}\dfrac{(t-x)}{2}$, we proved [8, 6] another asymptotic expansion for $T_N(f_n)$. More precisely, putting $H = N^{-1}$ and t_n the constant defined in equation (2.2), we have

$$t_n T_N(f_n) = f(x) + \sum_{i=1}^{k} S_i(f,x)h^i + z_k(f,x,h)o(h^k) + w_k(f,x,H)H^{2k}$$

with z_k and w_k bounded with regard to h and H, respectively. We notice that, in this way, we require the knowledge of the function f only on a uniform grid, since we apply the extrapolation methods, not to the theoretic function $V_n(f)$, but to the values $t_n T_N(f_n)$.

The paper is organized as follows. In section 2, first we give a trigonometric version of the Taylor formula and then we prove the asymptotic expansion for $V_n(f)$.

2. Main Results

2.1. Preliminaries

We start by recalling some simple preliminary lemmas and definitions.

Definition 2.1 *If $f \in C_{2\pi}$, then we define as the De La Vallée-Poussin integral the trigonometric polynomial of degree n given by*

$$V_n(x) = \frac{(2n)!!}{(2n-1)!!}\frac{1}{2\pi}\int_{-\pi}^{\pi} f(t)\cos^{2n}\frac{(t-x)}{2}\,dt \tag{2.1}$$

where the symbol $n!!$ denotes the product of all the positive natural numbers not exceeding n and even or uneven according as n is even or uneven.

Lemma 2.1 *For any natural constant n, the equation*

$$\int_0^{\pi/2} \cos^{2n} t\,dt = \frac{(2n-1)!!}{(2n)!!}\frac{\pi}{2} \tag{2.2}$$

is verified. We call t_n^{-1} the quantity in (2.2).

Lemma 2.2 *If $f \in C_{2\pi}$ then*

$$V_n(x) = \frac{(2n)!!}{(2n-1)!!}\frac{1}{\pi}\int_0^{\pi/2} [f(x+2t) + f(x-2t)]\cos^{2n} t\,dt. \tag{2.3}$$

Proof The proof is obtained by exploiting the substitutions $t = x + u$ and $u = 2t$ in (2.1) and by using the periodicity of f. •

Theorem 2.1 *For all $x \in \Re$*

$$\lim_{n\to\infty} V_n(x) = f(x). \tag{2.4}$$

Proof We can simply apply the Korovkin theorem. Indeed, the definition of the De La Vallée Poussin integrals and the former lemmas are used to fulfil the assumption of the Korovkin theorem (for details see also [1]). •

Now we give a trigonometric version of Taylor expansion which is similar (but not identical) to the one considered by Bleiman and Stark [9].

Lemma 2.3 *If $f \in C_{2\pi}^k$, then we may deduce the following asymptotic expansion*

$$f(x+t) = c_0(x) + c_1(x)\sin(t) + c_2(x)\sin^2(t) + \cdots + c_k(x)\sin^k(t) + \alpha_k(x,t)\sin^k(t) \tag{2.5}$$

with $\alpha_k(x,t)$ defined and continuous for $t \in (-\pi,\pi)$ and vanishing when t goes to 0. More precisely, we have $c_0(x) = f(x)$, $c_1(x) = f'(x)$ and

$$c_{j+1}(x) = ((j+1)!)^{-1}\left(f^{(j+1)}(x) - \sum_{l=1}^{j} c_l(x)d(l,j)\right),$$

where

$$d(l,j) = (j+1)!\sum_{(2i_1+1)+\ldots+(2i_l+1)=j+1} \alpha_{i_1}\cdots\alpha_{i_l}$$

and $\sin(t) = \sum_{k=0}^{\infty}\alpha_k t^{2k+1}$.

Proof By continuity we have

$$\lim_{t\to 0} f(x+t) = f(x),$$

while the term $c_1(x)$ is computed as follows

$$c_1(x) = \lim_{t\to 0}\frac{f(x+t)-f(x)}{\sin(t)} = \lim_{t\to 0}\frac{f'(x+t)}{\cos(t)} = f'(x)$$

by using the L'Hospital rule. For $1 \le j < k$ the other coefficients are calculated, recursively, in the subsequent manner:

$$\begin{aligned} c_{j+1}(x) &= \lim_{t\to 0}\frac{f(x+t)-\sum_{l=0}^{j} c_l(x)\sin^l(t)}{\sin^{j+1}(t)} \qquad (2.6)\\ &= \lim_{t\to 0}\frac{f^{(j+1)}(x+t)-\frac{d^{j+1}}{dt^{j+1}}\left[\sum_{l=0}^{j} c_l(x)\sin^l(t)\right]}{\frac{d^{j+1}}{dt^{j+1}}\sin^{j+1}(t)} \end{aligned}$$

with $c_0(x) = f(x)$.
From the equality

$$\sin^{j+1}(t) = t^{j+1} + t^{j+2}\sum_{l=0}^{\infty} z_j t^j$$

we obtain

$$\left.\frac{d^{j+1}}{dt^{j+1}}\sin^{j+1}(t)\right|_{t=0} = (j+1)!.$$

Then

$$c_{j+1}(x) = \frac{f^{(j+1)}(x+t)-\sum_{l=1}^{j} c_l(x)\frac{d^{j+1}}{dt^{j+1}}\sin^l(t)\Big|_{t=0}}{(j+1)!}.$$

In order to calculate

$$\left.\frac{d^{j+1}}{dt^{j+1}}\sin^l(t)\right|_{t=0}$$

we notice that the considered function is analytic and, therefore, we need only the $(j+1)$-th coefficient of the Taylor expansion of $\sin^l(t)$. Of course, by making use of the well-known expansion of $\sin(t)$ and by computing the formal l-th power of this infinite sum, we recover the desired coefficient which is equal to

$$(j+1)! \sum_{(2i_1+1)+\ldots+(2i_l+1)=j+1} \alpha_{i_1}\cdots\alpha_{i_l}.$$

Here α_k indicates the constant value of the term t^k in the Taylor expansion of $\sin t$, i.e.,

$$\sin t = \alpha_0 t + \alpha_1 t^3 + \ldots$$

•

From lemma 2.3, $\forall f \in C_{2\pi}^{2k}$ and setting $S = f(x+2t) + f(x-2t)$, we find

$$\begin{aligned} S &= f(x) + \textstyle\sum_{j=1}^{2k} c_j(x)\sin^j(2t) + \\ &\quad + f(x) + \textstyle\sum_{j=1}^{2k} c_j(x)\sin^j(-2t) + \bar{\alpha}_k(x,t)\sin^{2k}(2t) \\ &= 2f(x) + \textstyle\sum_{j=1}^{2k} c_j(x)\sin^j(2t) + \\ &\quad + \textstyle\sum_{j=1}^{2k} c_j(x)(-1)^j\sin^j(2t) + \bar{\alpha}_k(x,t)\sin^{2k}(2t) \\ &= 2f(x) + 2\textstyle\sum_{j=1}^{k} c_{2j}(x)\sin^{2j}(2t) + \bar{\alpha}_k(x,t)\sin^{2k}(2t). \end{aligned}$$

Therefore, for any $t \in (-\pi/2, \pi/2)$, we obtain

$$f(x+2t) + f(x-2t) = 2\sum_{j=0}^{k} c_{2j}(x)\sin^{2j}(2t) + \bar{\alpha}_k(x,t)\sin^{2k}(2t) \qquad (2.7)$$

with $c_0(x) = f(x)$, and $\bar{\alpha}_k(x,t) \to 0$ when $t \to 0$.

2.2. The asymptotic expansion of $V_n(f)$

In this subsection we prove the main result which allows one to use extrapolation algorithms in order to speed up the convergence of linear combinations of the De La Vallée terms.

Theorem 2.2 *If $f \in C_{2\pi}^{2k}$ then the following expansion holds true*

$$V_n(f) = f(x) + S_1(f,x)h + S_2(f,x)h^2 + \ldots + S_k(f,x)h^k + z_k(f,x,h)o(h^k) \qquad (2.8)$$

with $S_i(f,x)$ independent of $h = \frac{1}{n}$ and $z_k(f,x,h)$ bounded with h, $\forall x \in I$.

Proof By recalling (2.2) and (2.3) we have

$$V_n(f) = t_n \int_0^{\pi/2} [f(x+2t) + f(x-2t)] \cos^{2n}(t)\,dt$$

and then

$$\begin{aligned} V_n(f) &= t_n \textstyle\int_0^{\pi/2} \left[2\sum_{j=0}^{k} c_{2j}(x)\sin^{2j}(2t)\,dt + \bar{\alpha}_k(x,t)\sin^{2k}(2t)\right] \times \\ &\quad \times \cos^{2n}(t)\,dt = \\ &= 2t_n \left\{\textstyle\sum_{j=0}^{k} c_{2j}(x) \int_0^{\pi/2} \sin^{2j}(2t)\cos^{2n}(t)\,dt\right\} + \\ &\quad + t_n \textstyle\int_0^{\pi/2} \bar{\alpha}_k(x,t)\sin^{2k}(2t)\cos^{2n}(t)\,dt = \\ &= 2t_n \textstyle\sum_{j=0}^{k} c_{2j}(x) \int_0^{\pi/2} 2^{2j}\sin^{2j}(t)\cos^{2j}(t)\cos^{2n}(t)\,dt + \\ &\quad + 2t_n E_{n,k}(x,h) = \\ &= 2t_n \textstyle\sum_{j=0}^{k} 2^{2j} c_{2j}(x) \int_0^{\pi/2} \cos^{2(j+n)}(t)\,(1-\cos^2(t))^j\,dt + \\ &\quad + 2t_n E_{n,k}(x,h) = \\ &= 2t_n \textstyle\sum_{j=0}^{k} 2^{2j} c_{2j}(x) \int_0^{\pi/2} \cos^{2(j+n)}(t) \times \\ &\quad \times \left[\textstyle\sum_{l=0}^{j} \binom{j}{l} (-1)^l \cos^{2l}(t)\right] dt + 2t_n E_{n,k}(x,h) = \\ &= 2t_n \left\{\textstyle\sum_{j=0}^{k} 2^{2j} c_{2j}(x) \sum_{l=0}^{j} \binom{j}{l} (-1)^l \int_0^{\pi/2} \cos^{2(j+n+l)}(t)\,dt\right\} \\ &\quad + 2t_n E_{n,k}(x,h). \end{aligned}$$

Therefore, by using lemma 2.1, we obtain

$$\begin{aligned} V_n(f) &= 2\,t_n \sum_{j=0}^k 2^{2j}\, c_{2j}(x) \sum_{l=0}^j \binom{j}{l} (-1)^l \tfrac{(2(j+n+l)-1)!!}{(2(j+n+l))!!}\, \tfrac{\pi}{2} + \\ &\quad +2\, t_n E_{n,k}(x,h) = \\ &= \sum_{j=0}^k 2^{2j}\, c_{2j}(x) \sum_{l=0}^j \binom{j}{l} (-1)^l \times \\ &\quad \times \frac{(2n+2j+2l-1)(2n+2j+2l-3)\dots(2n+1)}{(2n+2j+2l)(2n+2j+2l-2)\dots(2n+2)} + \\ &\quad +\bar{E}_{n,k}(x,h). \end{aligned}$$

In conclusion, we have found

$$V_n(x,f) = \sum_{j=0}^k 2^{2j} c_{2j}(x)\theta_j(h) + \bar{E}_{n,k}(x,h) \tag{2.9}$$

with

$$\theta_j(h) = \sum_{l=0}^j \binom{j}{l} (-1)^l \frac{(2n+2j+2l-1)(2n+2j+2l-3)\dots(2n+1)}{(2n+2j+2l)(2n+2j+2l-2)\dots(2n+2)} \tag{2.10}$$

and

$$\bar{E}_{n,k}(x,h) = t_n \int_0^{\pi/2} \bar{\alpha}_k(x,t) \sin^{2k}(2t)\, \cos^{2n}(t)\, dt.$$

We would like to prove

$$\theta_j(h) = h^j \left(\beta_0 + h\beta_1 + h^2\beta_2 + \dots\right).$$

We set

$$\sigma_{n,j} = \frac{(2n+2j+2l-1)(2n+2j+2l-3)\dots(2n+1)}{(2n+2j+2l)(2n+2j+2l-2)\dots(2n+2)}$$

and then, by virtue of lemma 2.5 in [8], we find

$$\sigma_{n,j} = 1 + \frac{p_1(l)}{n} + \frac{p_2(l)}{n^2} + \dots \tag{2.11}$$

where $p_i(l)$ is a polynomial with l of degree i and coefficients depending only on the parameter j. In such a way

$$\theta_j(h) = \sum_{l=0}^j \binom{j}{l} (-1)^l \left(1 + \frac{p_1(l)}{n} + \dots \frac{p_j(l)}{n^j} + \dots\right)$$

and, since

$$\sum_{l=0}^j \binom{j}{l} (-1)^l l^k = 0 \qquad 0 \le k \le j-1,$$

we conclude

$$\theta_j(h) = \frac{c_j}{n^j} + \frac{c_{j+1}}{n^{j+1}} + \dots$$

Now we have to express explicitly the term $\bar{E}_{n,k}(x,h)$. From its expression

$$\bar{E}_{n,k}(x,h) = t_n \int_0^{\pi/2} \bar{\alpha}_k(x,t) \sin^{2k}(2t)\, \cos^{2n}(t)\, dt$$

and by choosing a value $\delta \in \left(0, \frac{\pi}{2}\right)$ we deduce

$$\begin{aligned}\bar{E}_{n,k}(x,h) &= t_n \int_0^\delta \bar{\alpha}_k(x,t)\sin^{2k}(2t)\cos^{2n}(t)\,dt+ \\ &\quad +t_n \int_\delta^{\pi/2} \bar{\alpha}_k(x,t)\sin^{2k}(2t)\cos^{2n}(t)\,dt.\end{aligned}$$

Since $\delta < \pi/2$, Lemma 2.4 and equation (2.7), we have that $\bar{\alpha}_k(x,t)$ is continuous on $[0,\delta]$. Therefore we apply the mean value theorem and we find

$$\begin{aligned}\bar{E}_{n,k}(x,h) &= t_n\bar{\alpha}_k(x,t_{k,n,x}) \int_0^\delta \sin^{2k}(2t)\cos^{2n}(t)\,dt+ \\ &\quad +t_n \int_\delta^{\pi/2} \bar{\alpha}_k(x,t)\sin^{2k}(2t)\cos^{2n}(t)\,dt\end{aligned}$$

where $t_{k,n,x} \le \delta$ and then there exists a constant $M_1(\delta)$ such that

$$|\bar{\alpha}_k(x,t_{k,n,x})| \le M_1(\delta)$$

with $M_1(\delta) \to 0$ when $\delta \to 0$. Moreover

$$\begin{aligned}|\bar{E}_{n,k}(x,h)| &\le t_n M_1(\delta) \int_0^\delta \sin^{2k}(2t)\cos^{2n}(t)\,dt+ \\ &\quad +t_n \int_\delta^{\pi/2} |\bar{\alpha}_k(x,t)\sin^{2k}(2t)|\cos^{2n}(t)\,dt.\end{aligned}$$

Now, by invoking the equation (2.7), we obtain that $|\bar{\alpha}_k(x,t)\sin^{2k}(2t)|$ is bounded by another constant M_2 and finally

$$|\bar{E}_{n,k}(x,h)| \le t_n M_1(\delta)\int_0^\delta \sin^{2k+2}(2t)\cos^{2n}(t)\,dt + t_n M_2\cos^{2n}(\delta)\,dt.$$

By virtue of the definition of the quantities $\theta_j(h)$, we arrive to

$$|\bar{E}_{n,k}(x,h)| \le M_1(\delta)\theta_k(h) + M_2 c^{2n}$$

with $c < 1$.

Consequently, by exploiting again lemma 2.1, we obtain that, for any $\epsilon > 0$, there exists a positive δ such that

$$|n^k \bar{E}_{n,k}(x,h)| \le \epsilon + M_2 n^k c^{2n}. \tag{2.12}$$

Of course, the previous expression can be rewritten as follows $\bar{E}_{n,k}(x,h) = z_k(x,f,h)o(h^k)$ where the quantity $z_k(x,f,h)$ is bounded with h for any choice of $x \in I$ (observe that relation (2.12) for $k = 1$ was proved in the theorem at page 212 of [1]). •

Now we apply a polynomial procedure of extrapolation to the terms $V_n(f)$ by obtaining, as a very special case, the nonpositive operators found in [4].

Theorem 2.3 *Let $f \in C_{2\pi}^{2k}$, let $\{n_0 = n < n_1 < n_2 < \cdots\}$ be an increasing sequence of positive integers and let us define $h_i = n_i^{-1}$. We set $\{X_q^{(i)}\}$ a sequence of values constructed in the following way*

$$\begin{cases} X_0^{(i)} := V_{n_i}(f), \\ X_q^{(i)} := \frac{h_{i+q}X_{q-1}^{(i)} - h_i X_{q-1}^{(i+1)}}{h_{i+q}-h_i}, \quad q = 1,2,\ldots,k-1. \end{cases}$$

Then, for i fixed positive number, we have

$$\lim_{h_i \to 0} X_q^{(i)} = f(x), \tag{2.13}$$

while the error takes the form

$$X_q^{(i)} - f(x) = h_i h_{i+1}\cdots h_q(-1)^q(S_{q+1}(f,x) + O(h_i)). \tag{2.14}$$

Proof It is a straightforward consequence of the application of the theory developed in Chapters 3.3 and 3.4 [10] to the expansion in Theorem 2.2. •

Observe that, if we choose the Romberg sequence $n_i = n\rho^i$ with $\rho = 2$ and we apply the considered procedure by assuming $q = 1, 2$, we obtain

$$X_1^{(0)} = 2V_{2n}(f) - V_n(f), \quad (n_0 = n),$$

$$X_2^{(0)} = \frac{8}{3}V_{4n}(f) - 2V_{2n}(f) + \frac{1}{3}V_n(f), \quad (n_0 = n).$$

It is worth noting that this two operators have been considered as linear combinations of the De La Vallée Poussin terms in [4]; according to the previous theorem, if the function f is regular enough then we have an error asymptotic to $\frac{1}{n^2}$ and $\frac{1}{n^3}$, respectively. Finally, we remark that by means of this extrapolation procedure, we can construct, in a general and uniform way, several examples of linear combinations of the terms $V_n(f)$ which well approximate the given function f.

Since we are interested in computational methods and the trigonometric polynomials $V_n(f)$ are defined by an integral, in [8] we considered the use of an integration rule joint with the asymptotic expansion in Theorem 2.2 and with the use of polynomial extrapolation. The numerical experiments performed in [8, 6] confirmed the adherence with the theoretical analysis.

References

1. I.P. Natanson, Constructive Function Theory, Vol 1. Frederick Ungar Publishing Co., New York (1964).

2. P.P. Korovkin, Linear Operators and Approximation Theory, Hindustan Publishing Co., Delhi (1960).

3. E. Voronowskaja, C.R. Acad. Sci. USSR, 79–85 (1932).

4. Z. Dizian and G. Freud, J. Approx. Th., **12**, 23–31 (1974).

5. F. Costabile, M.I. Gualtieri and S. Serra, BIT, **36-4**, 676–687 (1996).

6. F. Costabile, M.I. Gualtieri and S. Serra, in: Proc. 7th Int. Coll. on Diff. Equat., Plovdiv 1996, D. Bainov (Ed.), VSP, Utrecht (1997) pp. 67–74.

7. F. Costabile, M.I. Gualtieri and S. Serra, Rend. Circolo Mat. Palermo, Serie II, **52**, 345–355 (1998).

8. F. Costabile, M.I. Gualtieri and S. Serra, Technical Report 18, LAN - Dept. Math., University of Calabria (1996).

9. G. Bleiman and E. Stark, in: Proc. Conf. on Approx. and Function Spaces, Gdansk (1979), pp. 149–162.

10. J. Stoer and R. Bulirsh, Introduction to Numerical Analysis, Vol. 1. Springer-Verlag, Berlin (1970).

11. P.L. Butzer, Can. J. Math., **5**, 559–567 (1953).

12. R. De Vore and G. Lorentz, Constructive Approximation, Springer-Verlag, Berlin (1991).

9th Int. Coll. on Differential Equations, pp. 107-112
D. Bainov (Ed)

Variational Inequalities, Equivalent Optimization Problems and Associated Lagrangean Function

PATRIZIA DANIELE
Dipartimento di Matematica, Università di Catania,
Viale A. Doria 6, I-95125 Catania, Italy
daniele@dipmat.unict.it

Abstract: We associate to a Variational Inequality an equivalent optimization problem by means of a suitable *Gap* function using the separation theory and a Lagrangean function whose saddle point is related to the solution of the Variational Inequality.
Key Words and Phrases: *Variational Inequality, Gap function, Lagrangean function, Separation Theory, Saddle point.*

1. Introduction

We study the Variational Inequality (see [1] for more details)

$$\text{(1.1)} \qquad \text{"Find } \; H \in \mathbb{K} : \int_0^T \langle C(t, H(t)), F(t) - H(t) \rangle \, dt \quad \forall F \in \mathbb{K}\text{"}$$

where $\mathbb{K} = \{F(t) \in L^2(0, T; \mathbb{R}^m) : F_r(t) \geq 0 \quad r = 1, \dots, m \quad \text{a.e. in} \quad [0, T],$

$$\text{(1.2)} \qquad \sum_{r=1}^{m} \varphi_{jr} F_r(t) = \overline{\rho}_j \quad j = 1, 2, \dots, l \quad \text{a.e. in} \quad [0, T]\},$$

and, by means of the separation theory, we associate to it the function

$$G(H) = \min_{\lambda \in C^*} \max_{F \in L^2(0,T;\mathbb{R}^m)} \left[\int_0^T \langle C(t, H(t)), H(t) - F(t) \rangle \, dt + \right.$$

$$+ \sum_{i \in I^0} \lambda_i \int_0^T \omega_i(H(t), t, F(t)) \psi_i(H(t), t, F(t)) \, dt$$

$$\left. + \sum_{i \in I^+} \lambda_i \int_0^T \omega_i(H(t), t, F(t)) \psi_i(H(t), t, F(t)) \, dt \right],$$

where ω_i are functions of the kind

$$\omega_i(H(t), t, F(t)) = \left[\sum_{r=1}^{m} \varphi_{jr} F_r(t) - \overline{\rho}_i(t) \right] c_i \quad i \in I^0 = \{1, 2, \dots, l\}$$

with $c_i \neq 0$ and $\omega_i(H(t), t, F(t)) = -c_i \min\{F_i(t), 0\} \quad i \in I^+ = \{1, 2, \dots, m\}$ with $c_i > 0$.

G(H) is a Gap function, that is $G(H) \geq 0 \quad \forall H \in \mathbb{K}$ and $G(H) = 0$ if and only if H is solution of the Variational Inequality. In such a way we have associated to the Variational Inequality an equivalent minimum problem.

Moreover we are able to associate to the Variational Inequality a Lagrangean function

$$L(F,u,v) = \Psi(F) + \sum_{i=1}^{m} u_i \int_0^T F_i(t)\,dt + \sum_{j=1}^{l} v_j \Big(\sum_{R_r \in \Re_j} \int_0^T F_r(t)\,dt - \overline{\rho}_j \Big),$$

where $\Psi(F) = \sup_{\Phi \in \mathbb{K}} \int_0^T \langle C(t,F(t)), F(t) - \Phi(t) \rangle\, dt$.

This recent formulation gives the advantage to deal the solutions of the Variational Inequality as saddle points of the Lagrangean function.

2. The Static Model and the Dynamic Model

First of all we will recall the model of equilibrium flows in a network in the static case and next in the evolution case. The notations and the definitions we need are the following ones:

$L = \{P_1, P_2, \dots, P_q\}$ denotes the set of nodes;
$N = \{a_1, a_2, \dots, a_n\}$ the set of l inks;
$W = \{w_1, w_2, \dots, w_l\}$ the set of Origin-Destination pairs;
R_r $r = 1, \dots, m$ the paths connecting the O/D pairs;
$\mathcal{R}_j = \{R_r : R_r \text{ connects the O/D pair } w_j\} \quad j = 1, \dots, l$;
$\mathcal{R} = \bigcup_{j=1}^{l} \mathcal{R}_j$;
$F' = (F_1, F_2, \dots, F_m)$ the path-flow vector (the apex denotes the transposition);
$C'(F) = (C_1(F), C_2(F), \dots, C_m(F))$ the path-cost vector;
$\rho' = (\rho_1, \rho_2, \dots, \rho_l)$ the vector travel demand between the O/D pairs w_j;
$\varphi = \{\varphi_{jr}\}_{\substack{j=1,\dots,l \\ r=1,\dots,m}} \qquad \varphi_{jr} = \begin{cases} 1 & \text{if } R_r \in \mathcal{R}_j \\ 0 & \text{otherwise;} \end{cases}$ (φ is the pair-path incidence matrix);
$\Delta = \{\delta_{ir}\}_{\substack{i=1,\dots,n \\ r=1,\dots,m}} \qquad \delta_{ir} = \begin{cases} 1 & \text{if } a_i \in R_r \\ 0 & \text{otherwis e;} \end{cases}$ (Δ is the link-path incidence matrix).

If

$$c(f) = (c_1(f), \dots, c_n(f)) \quad f = (f_1, \dots, f_n)$$

are the link–flow cost vector and the link–flow vector, it results

$$f = \Delta F, \quad C(F) = \Delta' c(\Delta F).$$

We assume that the traffic demand is satisfied, that is the so called *traffic conservation law* holds:

$$\sum_{r=1}^{m} \varphi_{jr} F_r = \rho_j \quad j = 1, \dots, l.$$

Then, if we denote by $\mathbb{K}$ the set of feasible path–flow vectors, it will be given by

$$\mathbb{K} = \{F \in \mathbb{R}^m : F_r \geq 0 \quad r = 1, \dots, m, \quad \sum_{r=1}^{m} \varphi_{jr} F_r = \rho_j \quad j = 1, \dots, l\}.$$

If we adopt as an equilibrium condition the well known user's equilibrium principle due to Wardrop, we obtain

"A vector $H \in \mathbb{K}$ will be an equilibrium vector in the static case if

$$\forall \mathcal{R}_j, \quad \forall R_r,\ R_s \in \mathcal{R}_j \quad \textit{if it results} \quad C_r(H) > C_s(H) \tag{2.1}$$

then $H_r = 0$". In the year 1979 it was proved by M.S. Smith (see [2], [1], [3]), that

"H is an equilibrium solution to the traffic equilibrium problem in Wardrop sense if and only if

$$H \in \mathbb{K} \quad \textit{is such that} \quad \langle C(H), F - H\rangle \geq 0 \quad \forall F \in \mathbb{K}\text{"} \tag{2.2}$$

(see also [4] for more informations).

Now let us present the evolution case. Let us suppose that the travel demands ρ_j, $j = 1, \dots, l$, evolve in the time

$$\rho_j : [0,T] \to \mathbb{R}^+ \quad j = 1, \dots, l$$

and, hence, the flows and the costs in turn become functions of the time $F(t)$, $C(t, F(t))$ with

$$F : [0,T] \to \mathbb{R}^m_+ \qquad C : [0,T] \times \mathbb{R}^m_+ \to \mathbb{R}^m_+.$$

For technical reasons, the functional setting for the flow trajectories is $L^2(0,T;\mathbb{R}^m)$. Then the set of feasible flows is given by

$$\mathbb{K} := \Bigg\{ F(t) \in L^2(0,T;\mathbb{R}^m) : F_r(t) \geq 0 \quad r = 1, \dots, m \quad \text{a.e. in} \quad [0,T],$$

$$\sum_{r=1}^{m} \varphi_{jr} F_r(t) = \rho_j(t) \quad j = 1, \dots, l \quad \text{a.e. in} \quad [0,T] \Bigg\}.$$

The equilibrium definition becomes (see [3])

Definition 2.1. $H \in \mathbb{K}$ is an equilibrium flow if and only if

$$\forall \mathcal{R}_j, \quad \forall R_r,\ R_s \in \mathcal{R}_j \quad \text{a.e. in}\ [0,T] \quad \text{if it results} \quad C_r(t, H(t)) > C_s(t, H(t))$$

then

$$H_r(t) = 0.$$

It is possible to show (see [3]) that $H \in \mathbb{K}$ is an equilibrium solution in the sense of Def. 2.1 if and only if

$$\int_0^T \langle C(t, H(t)), F(t) - H(t)\rangle\, dt \geq 0 \quad \forall F(t) \in \mathbb{K}.$$

In [3] theorems 3 and 4 give some sufficient conditions to ensure the existence of a solution.

3. The V. I. and the Gap Function

The aim of this section is to extend to (1.1)-(1.2) the image space approach defined for constrained extremum problems (see [5]) and for Variational Inequalities (see [5], [2]) in order to obtain the *gap* function. To this end, let us set

$$I^0 := \{1, \dots, l\}, \quad I^+ := \{1, \dots, m\}, \quad I := \{1, \dots, l, l+1, \dots, l+m\},$$

and let us consider the set $C := 0_l \times \mathbb{R}^m_+$. Then, setting

$$\Psi_i(H,F) := \left\{ \sum_{r=1}^{m} \varphi_{ir} F_r(t) - \rho_j \quad \text{a.e. in} \quad [0,T] \right\} \quad i \in I^0,$$

and

$$\Psi_i(H,F) := \{F_i(t) \quad \text{a.e. in} \quad [0,T]\} \quad i \in I^+,$$

we can write:

$$\mathbb{K} = \{F \in L^2(0,T;\mathbb{R}^m) : \Psi(H,F) \subseteq C\}.$$

Now, $\forall H$ we consider the multifunction

$$M : L^2(0,T;\mathbb{R}^m) \to \mathbb{R} \times \mathbb{R}^{m+l}$$

defined by:

$$M(H,F) := \left\{ (u,v) \in \mathbb{R} \times \mathbb{R}^{m+l} : u = \int_0^T \langle C(t,H(t)), H(t) - F(t) \rangle \, dt, \right.$$

$$\left. v = \Psi(H,F) \right\}.$$

$M(H,F)$ is a subset of a finite dimensional space, whereas the vector H belongs to an infinite dimensional space $L^2(0,T;\mathbb{R}^m)$; we try to overcome this difficulty. To this end, we remark that $H \in L^2(0,T;\mathbb{R}^m)$ is a solution to (1.1) iff the system (in the unknown F)

$$\int_0^T \langle C(t,H(t)), H(t) - F(t) \rangle \, dt > 0, \quad \Psi(H,F) \subseteq C, \quad F \in L^2(0,T;\mathbb{R}^m)$$

is impossible.
Then, for each $H, F \in L^2(0,T;\mathbb{R}^m)$, denoted by $\Omega(H,F)$ the set of vectors $\omega := (\omega_i(H(t), t, F(t)),\ i \in I)$ such that

$$\omega_i : L^2(0,T;\mathbb{R}^m) \times [0,T] \times L^2(0,T;\mathbb{R}^m) \to \mathbb{R}$$

and whose elements $\omega_i(H,F)$ $i \in I^0$ are not identically zero on $[0,T]$, whereas $\omega_i(H,F) \geq 0$ $i \in I^+$, $\forall H, F$, let us consider the vector function

$$\Phi : L^2(0,T;\mathbb{R}^m) \times \mathbb{R}^{1+l+m} \times \Omega \to \mathbb{R}^{1+l+m},$$

defined $\forall F \in L^2(0,T;\mathbb{R}^m)$ by:

$$\Phi(H; M(H,F); \omega(H,F)) :=$$

$$\left(\int_0^T \langle C(t,H(t)), H(t) - F(t) \rangle \, dt, \int_0^T \omega_i(H(t),t,F(t)) \cdot \Psi_i(H(t),t,F(t)) \, dt, \quad i \in I \right)$$

and set

$$\begin{aligned} \mathcal{H} := \{ \quad & (u, v_1, \ldots, v_l, v_{l+1}, \ldots, v_{l+m}) \in \mathbb{R}^{1+l+m} : u > 0, \\ & v_i = 0 \quad i \in I^0, \ v_{l+i} \geq 0 \quad i \in I^+\}. \end{aligned}$$

We ask Φ to be such that $\forall H \in L^2(0,T;\mathbb{R}^m)$

(3.1) $$M(H,F) \subseteq \mathcal{H} \Leftrightarrow \Phi(H; M(H,F); \omega(H,F)) \in \mathcal{H} \ \forall \omega(H,F) \in \Omega(H,F).$$

If we set

$$\mathcal{K}(H,\omega) := \{\Phi(H; M(H, L^2(0,T;\mathbb{R}^m)); \omega(H,F)), \quad H \in L^2(0,T;\mathbb{R}^m),$$
$$\omega(H,F) \in \Omega(H,F)\},$$

then, obviously H is a solution to (1.1) iff there exists a vector $\omega(H,F)$ such that

$$\mathcal{H} \cap \mathcal{K}(H; \omega(H,F)) = \emptyset,$$

as it easily follows taking into account (3.1) and thus we have obtained that the solution exists if and only if the two sets $\mathcal{H}$ and $\mathcal{K}$ are disjointed.
Following the separation theory, there exist separation functions and we may consider, for example, the following function:

$$\Psi(H,F,\lambda,\omega) := \int_0^T \langle C(t,H(t)), H(t) - F(t)\rangle\, dt +$$
$$+ \sum_{i\in I} \lambda_i \int_0^T \omega_i(H(t),t,F(t)) \cdot \Psi_i(H(t),t,F(t))\, dt,$$

where $\lambda_i \in C^* := \{\lambda \in \mathbb{R}^{m+l} : \lambda_r \geq 0,\ r = 1,\dots,m\}$ (C^* is the positive polar of C) as a particular case of weak separation function (see [4], [5], [6]).

It is possible to prove that the function

$$G(H) = \min_{\lambda\in C^*} \max_{F\in L^2(0,T;\mathbb{R}^m)} \left[\int_0^T \langle C(t,H(t)), H(t) - F(t)\rangle\, dt + \right.$$
$$\left. + \sum_{i\in I} \lambda_i \int_0^T \omega_i(H(t),t,F(t)) \cdot \Psi_i(H(t),t,F(H))dt \right]$$

is a *gap* function, that is, $G(H) \geq 0 \quad \forall H \in \mathbb{K}$ and $G(H) = 0$ iff H is a solution to (1.1).

Actually the following theorem holds:

Theorem 3.1. *Let us assume that there exist*

$$F^0 \in L^2(0,T;\mathbb{R}^m) \quad \textit{and} \quad \lambda^0 \in C^* \quad \textit{such that}$$

$$G(H) = \int_0^T \langle C(t,H(t)), H(t) - F^0(t)\rangle\, dt +$$
$$+ \sum_{i\in I^0} \lambda_i^0 \int_0^T \omega_i(H(t),t,F^0(t)) \cdot \left(\sum_{r=1}^m \varphi_{ir} F_r^0(t) - \rho_i(t) \right) dt +$$
$$+ \sum_{i\in I^+} \lambda_i^0 \int_0^T \omega_i(H(t),t,F^0(t)) \cdot F_i^0(t) dt.$$

Then $G(H) \geq 0 \quad \forall H \in L^2(0,T;\mathbb{R}^m)$ and, if $G(H) = 0$, then $H \in \mathbb{K}$ and H is a solution to the Variational Inequality (1.1).

Moreover it easy to show that there exist ω_i such that (3.1) is verified. For example, we may choose:

$$\omega_i(H(t),t,F(t)) = \left[\sum_{r=1}^m \varphi_{ir} F_r(t) - \rho_i(t) \right] c_i \quad i \in I^0,$$

where $c_i \neq 0$ and $\omega_i(H(t),t,F(t)) = -c_i \min\{F_i(t), 0\} \quad i \in I^+$, where $c_i > 0$.

4. The V. I. and the Lagrangean Function

Now we want to find a Lagrangean Function which can be associated to the Variational Inequality (1.1) such that a solution of the Variational Inequality can be expressed by means of the saddle point of such a function.

Such a Lagrangean Function is the following:

$$L(F,u,v) = \Psi(F) + \sum_{i=1}^{m} u_i \int_0^T F_i(t)\,dt + \sum_{j=1}^{l} v_j \Big(\sum_{R_r \in \mathfrak{R}_j} \int_0^T F_r(t)\,dt - \rho_j \Big),$$

where

$$\Psi(F) = \sup_{\Phi \in \mathbf{K}} \int_0^T \langle C(t,F(t)), F(t) - \Phi(t) \rangle \,dt.$$

It is possible to prove that if H is a solution to the Variational Inequality (1.1), then we may associate to H a point $(\overline{u},\overline{v}) \in C^*$ with $C^* = \{(u,v) \in \mathbb{R}^{m+l} : u_i \leq 0,\ i = 1,\dots,m\}$ such that $(H,\overline{F},\overline{u},\overline{v})$ is saddle point of L, then H is solution of the Variational Inequality. In such a formulation the dual variables (u,v) do not depend on the time, which can be an advantage from a computational point of view. The proof is obtained by taking into account some procedures followed to determine the common Lagrangean Function for the most expressive result is the following. Let us consider the Dual Problem

$$\text{(4.1)} \quad \max_{(u,v)\in C^*} \inf_{F \in L^2(0,T;\mathbb{R}^m)} \Big[\Psi(F) + \sum_{i=1}^{m} u_i \int_0^T F_i(t)\,dt + \sum_{j=1}^{l} v_j \Big(\sum_{R_r \in \mathfrak{R}_j} \int_0^T F_r(t)\,dt - \rho_j \Big) \Big],$$

we have

Theorem 4.1. *Let H be a solution of the Variational Inequality and $(\overline{u},\overline{v})$ a solution of the Dual problem (4.1), then $(H,\overline{u},\overline{v})$ is a saddle point of the Lagrangean Function L. Viceversa, a saddle point $(H,\overline{u},\overline{v})$ of the Lagrangean Function L is such that H is solution of the Variational Inequality and $(\overline{u},\overline{v})$ is solution of the Dual problem (4.1).*

References

1. A. Maugeri, Rend. Circ. Mat. Palermo, **48**, 179–184 (1997).

2. A. Maugeri, in: Recent Developments in Theory and Algorithms, F. Giannessi and A. Maugeri (Eds.), Plenum Publishing, New York (1995) 195–211.

3. A. Maugeri, in: New Trends in Mathematical Programming, F. Giannessi et al. (Eds.), The Netherlands, (1998) 191–202.

4. F. Giannessi, Le Matematiche, **49**, 243–247 (1994).

5. F. Giannessi, in: Separation of Sets and Gap Functions for Quasi–Variational Inequalities, F. Giannessi and A. Maugeri (Eds.), Plenum Publishing Co., New York, (1995) 101-121.

6. J. Jahn, Introduction to the Theory of Nonlinear Optimization, Springer–Verlag Berlin Heidelberg (1996).

9th Int. Coll. on Differential Equations, pp. 113–118
D. Bainov (Ed)

Studying the Numerical Properties of Solvers for Systems of Nonlinear Equations

DEBORAH DENT AND MARCIN PAPRZYCKI
School of Mathematical Sciences, University of Southern Mississippi, Hattiesburg, MS 39406-5106, USA
{m.paprzycki, deborah.dent}@usm.edu
ANNA KUCABA-PIĘTAL
Department of Aeronautical Engineering, Technical University of Rzeszów, Poland
anpietal@ewa.prz.rzeszow.pl

Abstract: Solution of systems of nonlinear equations is a relatively complicated problem for which a number of different approaches have been proposed. The aim of this note is to attempt at a comparison between some of them when applied to a number of test problems. A discussion of the properties of the solvers and the interrelationship between the solvers and the test problems will be presented.
Key Words and Phrases: *Systems of linear algebraic equations, solution methods, performance comparison.*

1. Introduction

The solution of systems of algebraic equations has a well-developed mathematical and computational theory when the equations are linear, or when a single nonlinear equation is to be solved [1]. The situation is much more complicated for systems of nonlinear algebraic equations. Both the mathematical theory and computational practice are far from a complete understanding of the solution process. When systems of linear equations are considered, libraries of subprograms solving standard problems have been developed. These libraries may not provide the most efficient way of solving certain special problems, but they should be robust enough to effectively solve most problems. In addition, sets of test cases have been developed which can be used as benchmarks against which the newly developed software is to be tested. These benchmarks allow for establishing the quality of the new approach and a comparison with existing software targeting the same problem. No such development can be observed when the solution of nonlinear systems is concerned.

The aim of this note is to report on the initial steps in the direction of building a library of nonlinear solvers and compiling a set of test problems. These test problems are used to compare the performance of existing nonlinear solvers. In section 2 we briefly describe the software that we have performed experiments. Section 3 presents and discusses the results of our numerical experiments. Directions for future research are discussed.

2. Algorithms for the Solution of the System of Nonlinear Algebraic Equations

This section contains a brief summary of algorithms behind nonlinear solvers that we have performed experiments (in all cases the references cited should be consulted for the details). We assume that a system of n nonlinear algebraic equations $F(x) = 0$ is to be solved.

Newton's Method

The Newton's Method for a system of equations is a natural extension of the Newton's Method for a single equation [2]. Let us assume that the function G is defined by

$$G(\underline{x}) = \underline{x} - J(\underline{x})^{-1}F(\underline{x}),$$

and the functional iteration procedure is: select starting vector x_0 and generate a series of vectors

$$\underline{x}_k = G(\underline{x}_{k-1}) = \underline{x}_{k-1} - J(\underline{x}_{k-1})^{-1}F(\underline{x}_{k-1})$$

where $J(x)$ is the Jacobian matrix.

The convergence rate for this method is fast, but the success of the method depends on a good starting vector x_0.

Brown's Method

Brown's Method is a modification of Newton's Method [3]. Here, we replace the Jacobian matrix with its difference quotient approximation. For each iteration, only one function from the system at a time is evaluated (function f_i for use with f_{i+1}). A successive substitution scheme is being used for treatment of f_i. As in case of Newton's Method, the convergence is fast, but it requires a good starting vector x_0.

Secant (Broyden's Method)

Broyden's Method belongs to the general category of quasi-Newton methods [4]. Here, matrix B_k is updated at each iterate so that the new approximation B_{k+1} satisfies the general quasi-Newton equation:

$$B_{k+1}(\underline{x}_{k+1} - \underline{x}_k) = F(\underline{x}_{k+1}) - F(\underline{x}_k).$$

Thus given an initial matrix B_0 (e.g. the finite-difference approximation to the Jacobian matrix), the subsequent matrices are generated by:

$$B_{k+1} = B_k + \frac{(\underline{y}_k - B\underline{s}_k)\underline{s}_k^T}{\|\underline{s}_k\|_{21}^2}$$

where $s_k = x_k + 1 - x_k$, $y_k = F(x_{k+1}) - F(x_k)$.

The convergence of Broyden's Method is fast, but as in the case of Newton's Method it requires a good choice of the initial vector x_0.

Steepest Descent Method

In the Steepest Descent Method the problem of solving a system of nonlinear algebraic equations is transformed into a minimization problem [5]. A relative minimum x^* of the function $f(x, y)$ with known partial derivatives $g = \frac{\partial f}{\partial x}$, $h = \frac{\partial f}{\partial x}$ is located starting from a given initial guess (x_0, y_0). Using a given step size c a sequence of steps is generated according to:

$$\underline{x}_n = \underline{x}_n - \frac{c\underline{g}}{\sqrt{\underline{g}^2 + \underline{h}^2}},$$

$$\underline{y}_n = \underline{y}_n - \frac{c\underline{h}}{\sqrt{\underline{g}^2 + \underline{h}^2}}.$$

The stopping criteria is met when the relative minimum x^* is located and

$$\sqrt{\underline{g}^2 + \underline{h}^2} < \varepsilon.$$

The convergence rate of the steepest descent method is slow, but it works for any choice of starting vectors.

Trust Region Method

In the Trust Region Method we replace the Jacobian matrix with its approximation [6]. Then we calculate

$$\min\{\|f(\underline{x}_k) + B_k\delta\| : \quad \|D_k\delta\|_2 <= \Delta_k\},$$

where D_k is a scaling matrix and Δ_k is the trust region radius. The solution to this minimization problem is an approximate solution to the original problem. Stopping criteria is met when

$$p_k = \frac{\|f(\underline{x}_k)\| - \|f(\underline{x}_k + \delta_k)\|}{\|f(\underline{x}_k)\| - \|f(\underline{x}_k + B_k\delta_k)\|}$$

is smaller than some constant σ_0 (typically .0001). Otherwise, we decrease the radius of the trust region and re-solve the minimization problem.

The convergence rate of this method is slow, but it can use an arbitrary starting solution vector.

Experimental Results

We have experimented with the algorithms described above. We used two implementations of the Brown's method, our own implementation (based on [7]) and the SOS algorithm from [8]. Two hybrid methods, an algorithm by Kucaba-Piętal [9, 10] and the HYBRDI algorithm (a combination of Trust Region, Steepest Descent, and Newton's methods) from [8]. All algorithms were Fortran-based implementations and the experiments were run in double precision on a PC with a Pentium Pro 200 MHz processor. For the numerical tests we have used 17 problems found at [11, 12, 13]. These problems come from the three collections of test problems (with complete source code) for the solution of systems of nonlinear algebraic equations that we were able to locate. While some of the problems come from the real applications, others are artificially generated with properties not typical for real life applications. The problems used were (see the references for the complete formulation):

1. Rosenbrock's function [11]	2. Powell singular function [11]
3. Powell badly scaled function [11]	4. Wood function [11]
5. Helical valley function [11]	6. Watson function [11]
7. Chebyquad function [11]	8. Brown almost-linear function [11]
9. Discrete boundary value function [11]	10. Discrete integral equation function [11]
11. Trigonometric function [11]	12. Variably dimensioned function [11]
13. Broyden tridiagonal function [11]	14. Broyden banded function [11]
15. Exponential/Sine Function [12]	16. The Freudenstein-Roth function [13]
17. Semiconductor Boundary Condition [12]	

Table 1: Comparison of performance of nonlinear solvers for the test problems.

		QUASI_A		HYBRD		SOS		BROWN	
#	n	IT	FC	IT	FC	IT	FC	IT	FC
1	2	8	33	8	22	1	10	5	20
2	4	*85*	164	*83*	108	*11*	168	*18*	238
3	2	87	266	87	181	13	55	14	65
4	4	52	142	52	94	*16*	188	*820*	11466
5	3	14	38	14	27	19	174	8	63
6	6	38	127	38	96	23	648	36	987
6	9	57	180	57	182	8	486	30	1629
7	5	11	24	11	17	3	50	10	180
7	6	10	34	10	25	7	111	12	297
7	7	11	27	11	20	3	84	20	665
7	8	*30*	153	*30*	120	*14*	660	*88*	3884
7	9	19	57	19	43	8	486	18	918
8	10	8	37	8	31	5	280	9	520
8	30	*1*	110	*1*	85	*8*	2130	*10*	4455
8	40	*1*	130	*1*	105	*9*	4500	*10*	7740
9	10	6	20	6	16	2	140	6	325
10	1	6	11	6	7	2	5	5	8
10	10	6	20	6	16	2	140	5	260
11	10	*30*	160	*30*	130	*4*	270	8	455
12	10	27	75	27	60	*4*	585	*9999*	9999
13	10	11	27	11	21	3	205	8	455
14	10	20	41	20	30	4	270	7	390
15	2	12	22	12	15	*1*	32	4	29
16	6	3	11	3	9	1	54	5	108
17	2	1	7	1	5	*1*	126	5	36

Table 1 summarizes the results obtained for the test problems for the four codes tested. Here, IT denotes the number of iterations, FC the number of times the function was evaluated. ***9999*** denotes lack of convergence while results ***italicized bold and underlined*** represent the lack of solution due to the lack of significant improvement over the previous five Jacobian evaluations. Some of the problems have been run varying the number of equations. In each case a number of starting points has been tried and the best results are reported. All problems converged for some starting point except #2 (for $n = 2$) and #7 (for $n = 8$). In addition to runs reported in the table, problems 6–14 were run for $n = 10, 20, 40, 80$, and 100 with multiple initial starting points. All converged except #6, #7, and #12 for $n = 40, 80$ and 100. Problems #15 and #17 for SOS ended because the Jacobian matrix was singular.

From the results presented in the table and the discussion above it is clear that the difficulty of the system increases with the number of equations. The system either becomes unsolvable or the solution requires more iterations (and thus function evaluations). The numbers also show that the hybrid methods require much less function evaluation (while they require more irerations) than the implementations of the Brown's method (which require fewer iterations, but substantially more function evaluations).

In all experiments we have observed extreme sensitivity of the solvers to the selection of the starting vector x_0. This becomes more of a problem as the number of equations increases.

At this moment it is difficult to assess which of the test examples should be kept when a library of test cases will be compiled. The tests used here clearly cover a wide spectrum of functions but they clearly do not exhaust the possibilities arising in practical engineering applications. It should also be stressed that in real-life engineering applications systems of 100's of nonlinear algebraic equations have to be solved for the method to have a real predictive power (see for instance [9, 10]). Only few of the examples found can be even extended to this many equations.

Conclusions and Future Work

In this note we have presented an initial report of our comparisons between solvers for systems of nonlinear algebraic equations. We have found that methods based on similar algorithms behave similarly and the implementation details have only a minimal impact on the performance. All methods, regardless of their underlying algorithm, showed high sensitivity to the starting vector and this sensitivity increased as the number of equations in the system increased. We were not able to come to firm conclusions on the quality of the collection of test sets.

In the near future we plan to proceed in three directions. First, we will expand the test set by including additional recently located test problems. We will experiment with the described above codes on these sets. We will also use the Homotopy and Continuation methods on all test problems. Second, we will use the experimental data to attempt at evaluating the test set and to develop a library of test problems. Finally, we plan to investigate various methods for finding the starting vectors.

Acknowledgements

Work of Deborah Dent was sponsored from the Long Term Training Program, US Army Corps of Engineers, Waterways Experiment Station, Vicksburg, Mississippi, USA.

References

1. W.C. Rheinboldt, Methods for Solving Systems of Nonlinear Equations, Society for Industrial and Applied Mathematics, Philadelphia (1998).

2. G.H. Hostetter, M.S. Santina and P.D'Carpio-Montalvo, Analytical, Numerical and Computational Methods for Science and Engineering, Prentice Hall, New Jersey (1991).

3. K.M. Brown, SIAM J. Num. Anal., **6**, 560-569 (1969).

4. http://www.mcs.anl.gov/otc/Guide/OptWeb/continuous/unconstrained/nonlineareq/broydn.html

5. R.L. Burden and J.D. Faires, Numerical Analysis, PWS-Kent Publishing Company, Boston (1993).

6. http://www.mcs.anl.gov/otc/Guide/OptWeb/continuous/unconstrained/nonlineareq/trls.hml

7. W. Prosnak, Introduction to Fluid Mechanics, (in Polish), Ossolineum, Wrocław (1993).

8. http://www.netlib.org/liblist.html

9. A. Kucaba-Piętal and L. Laudański, in: Proceedings of I Symposium on Dynamics of Mechanical Systems, Warsaw (1994).

10. A. Kucaba-Piętal and L. Laudański, Zeszyty Naukowe Politechniki Śląskiej, Seria Mechanika, **121**, 173 (1995).

11. J.J. More, B.S. Garbow and K.E. Hillstrom, ACM Trans. Math. Software, **7**, 17-41 (1981).

12. U.N. Weimann, A Family of Newton Codes for Systems of Highly Nonlinear Equations, ZIB Technical Report TR-91-10, Berlin (1991), http://weyl.zib-berlin.de/Documents/tr91-10/section1.1.4.1.html.

13. J.J. More, A Collection of Nonlinear Model Problems, Preprint MCS-P60-0289, Mathematics and Computer Science Division, Argonne National Laboratory (1989).

9th Int. Coll. on Differential Equations, pp. 119-126
D. Bainov (Ed)

Superlinear Convergence of Asynchronous Waveform Relaxation Methods for Nonlinear ODEs

M. EL-KYAL, J.-C. MIELLOU AND J. BAHI
UMR CNRS 6623, IUT de Belfort, BP. 527, 90016 Belfort cedex, France

Abstract: In this paper, we prove the superlinear convergence of asynchronous waveform relaxation (AWR) methods applied to a system of nonlinear ordinary differential equations (ODE's). This study is based on an asynchronous convergence theorem introduced by Bertsekas and Tsitsiklis.
Key Words and Phrases: *Asynchronous algorithms, waveform relaxation, superlinear convergence, nonlinear ODEs.*

1. Introduction

We consider an asynchronous waveform relaxation (AWR) method applied to a system of nonlinear ordinary differential equations (ODE's).
Waveform relaxation is an iterative solution technique for initial-value problems (IVP's) consisting of (ODE's) [1] or differential algebraic equations (DAE's) [2]. The method was introduced by Lelarasmee [1] to overcome the deficiencies of standard time-stepping schemes when applied to IVP's that model the behavior of very large-scale electrical networks. It consists in decomposing the original problem at the differential equation level, the large initial-value system is then solved iteratively, by solving the smaller subsystems among different processors. Every processor is then responsible for computing the update to a subsystem of differential equation.
The main characteristic of an asynchronous mode is that the local algorithm do not have to wait at predetermined messages to become available. We allow some processors to communicate more frequently than others, and we allow the communication delays to be substantial and unpredictable.
Our goal is to prove the superlinear convergence of AWR methods, this study is based on an asynchronous convergence theorem introduced by Bertsekas and Tsitsiklis [3]. Note that synchronous algorithms in the computer science sense are particular cases of our formulation of asynchronous methods.

2. Preliminaries

2.1. Asynchronous Algorithms

Asynchronous algorithms are used in the mathematical modelling of the parallel treatment of problems taking into account interaction processes.

Let B_i, $i \in \{1, ..., \alpha\}$ be Banach spaces equipped with the norms $|| \, . \, ||_i$.
Consider $B = \prod_{i=1}^{\alpha} B_i$ equipped with the norm

$$|| \, . \, ||_\gamma = \max_{1 \leq l \leq \alpha} \frac{|| \, . \, ||_l}{\gamma_l}.$$

Define
$J = \{J(p)\}_{p \in \mathbb{N}}$ a sequence of nonempty subsets of $\{1, ..., \alpha\}$
$S = \{\rho_1(p), ..., \rho_\alpha(p)\}_{p \in \mathbb{N}}$ a sequence of $\mathbb{N}^\alpha$ such that:
$(h1)$ $\forall i \in \{1, ..., \alpha\}$, the subset $\{p \in \mathbb{N}, i \in J(p)\}$ is infinite
$(h2)$ $\forall i \in \{1, ..., \alpha\}, \forall p \in \mathbb{N}, \rho_i(p) \leq p$
$(h3)$ $\forall i \in \{1, ..., \alpha\}, \lim_{p \to \infty} \rho_i(p) = \infty$
The asynchronous algorithm associated to the operator $T = (T_1, ..., T_\alpha)$ and denoted (T, u^0, J, S) is defined by, see [4]

$$\left\{ \begin{array}{l} Given\ u^0 = (u_1^0, ..., u_\alpha^0) \\ \quad for\ p = 1, 2... \\ \quad\quad for\ l = 1, ..., \alpha \\ \quad\quad u_l^{p+1} = \left\{ \begin{array}{r} T_l(u^{1,\rho_1(p)}, ..., u^{\alpha,\rho_\alpha(p)})\ if\ l \in J(p) \\ u_l^p\ if\ l \notin J(p) \end{array} \right. \end{array} \right. \tag{1}$$

$-$ $J(p)$ is the set of components to be updated at step p
$-\rho_l(p)$ is the delay due to l^{th} processor when it computes the lth block at the pth iteration.
If we take $\rho_i(p) = p$ then (1) describes synchronous algorithms.
If we take

$$\left\{ \begin{array}{l} \rho_i(p) = p\ for\ p = 1, 2, ...\ and\ i = 1, 2, ... \\ J(p) = \{1, 2..., \alpha\}\ \ for\ p = 1, 2, ... \end{array} \right.$$

then (1) describes the algorithm of Jacobi.
If we take

$$\left\{ \begin{array}{l} \rho_i(p) = p\ for\ p = 1, 2, ...\ and\ i = 1, 2, ... \\ J(p) = 1 + p(mod\ \alpha)\ for\ p = 1, 2, ... \end{array} \right.$$

then (1) describes the algorithm of Gauss Seidel.

2.2. Convergence Theorem

We suppose that there is a sequence of nonempty sets $\{\mathcal{Y}^p\}_{p \in \mathbb{N}}$ with

$$... \subset \mathcal{Y}^{p+1} \subset \mathcal{Y}^p \subset ... \mathcal{Y}^0 \subset B$$

satisfying the following three conditions:

1. (Synchronous Convergence) We have

$$T(x) \in \mathcal{Y}^{p+1}, \quad \forall x \in \mathcal{Y}^p \tag{2}$$

   Furthermore, if $\{x^p\}_{p \in \mathbb{N}}$ is a sequence such that $x^p \in \mathcal{Y}^p$ for every p, then every limit point of $\{x^p\}_{p \in \mathbb{N}}$ is a fixed point of T.

2. (Box Condition) For every p, there exists sets $\mathcal{Y}_i^p \subset B_i$ such that

$$\mathcal{Y}^p = \mathcal{Y}_1^p \times ... \times \mathcal{Y}_i^p \times ... \times \mathcal{Y}_\alpha^p \tag{3}$$

3. there exists u^* in B, a fixed point of T, such that

$$\bigcap_{p=0}^{+\infty} \mathcal{Y}^p = \{u^*\} \tag{4}$$

Theorem 1 *Let T be a mapping from $D(T) \subset B$ in B and suppose that conditions (2), (3) and (4) hold, and the initial solutiòn estimate $u^0 = (u_1^0, ..., u_\alpha^0)$ belongs to the set $D(T)$, then every asynchronous algorithm (T, u^0, J, S) associated to T converges to the fixed point u^* of T.*

Proof. See [3].

3. An Asynchronous WR Method Applied to Nonlinear Ordinary Differential Systems

We consider a nonlinear ordinary differential system of the form :

$$\begin{cases} x'(t) = F(x(t), t) \\ x(0) = x_0 \end{cases} \tag{5}$$

where $t \geq 0$, $F = (F_1, ..., F_\alpha)^t : \prod_{i=1}^{i=\alpha} \mathbb{R}^{n_i} \times [0, +\infty[\to \prod_{i=1}^{i=\alpha} \mathbb{R}^{n_i}$,
$x_0 = (x_{0_1}, ..., x_{0_\alpha})^t \in \prod_{i=1}^{i=\alpha} \mathbb{R}^{n_i}$ the initial condition,
$x(t) = (x_1(t), ..., x_\alpha(t))^t \in \prod_{i=1}^{i=\alpha} \mathbb{R}^{n_i}$ the solution vector at t and x' the derivative of x with respect to t where:

1. $x \in C = \prod_{i=1}^{i=\alpha} C_i = \prod_{i=1}^{i=\alpha} C\left([0, +\infty[, \mathbb{R}^{n_i}\right)$
2. For all $i \in \{1, ..., \alpha\}$ and $t \in [0, +\infty[$, we will assume the following Lipschitz condition :

$$|F_i(x(t), t) - F_i(\bar{x}(t), t|_i \leq L_1 \, |x(t) - \bar{x}(t)|_\infty . \tag{6}$$

where for all $i \in \{1, ..., \alpha\}$, $\mathbb{R}^{n_i}$ is equipped with the norm:

$$|a_i|_i = \max_{1 \leq j \leq n_i} |a_{i,j}| \,, \; a_i = (a_{i,1}, ..., a_{i,n_i})^t \tag{7}$$

and the space $\prod_{i=1}^{i=\alpha} \mathbb{R}^{n_i}$ is equipped with the norm:

$$|a|_\infty = \max_{1 \leq i \leq \alpha} |a_i|_i \tag{8}$$

For all $i \in \{1, ..., \alpha\}$, let the space C_i equipped with the norm:

$$\|x\|_{i,K} = \max_{t \geq 0} e^{-Kt} \, |x_i(t)|_i \qquad \text{where } K > L_1 \text{ (L_1 is defined by (6))}$$

and the space C equipped with the norm $\|x\|_K = \max_{1 \leq i \leq \alpha} \|x_i\|_{i,K}$,
we denote by $\|x_i\|_{i,K,t} = \sup_{0 \leq \tau \leq t} e^{-K\tau} \, |x_i(\tau)|_i$
and by $\|x\|_{K,t} = \max_{1 \leq i \leq \alpha} \|x_i\|_{i,K,t}$,
we denote by:

$$\rho_p(t) = \begin{cases} \left(\frac{KL_1}{K-L_1}\right)^p \frac{t^p}{p!} \; \textit{if} \; t < t_p = \frac{K-L_1}{KL_1} (p!)^{\frac{1}{p}} , \\ 1 \; \textit{if} \; t \geq t_p. \end{cases}$$

Remark 2 *By using the formula of Stirling, we have:*
$p! \simeq p^p e^{-p}\sqrt{2\pi p} \Rightarrow (p!)^{\frac{1}{p}} < p$, *then for all* $t < t_p$, $\frac{KL_1}{K-L_1}\frac{t}{p} < 1$,
furthermore, $t_{p+1} \geq t_p$, *indeed*

$$\begin{aligned}
\ln \frac{t_{p+1}}{t_p} &= \frac{1}{p+1}\sum_{i=1}^{p} \ln i + \frac{\ln (p+1)}{p+1} - \frac{1}{p}\sum_{i=1}^{p} \ln i \\
&= \frac{1}{p+1}\left(\ln (p+1) - \frac{1}{p}\sum_{i=1}^{p} \ln i\right) \\
&\geq \frac{1}{p+1}\left(\ln (p+1) - \frac{1}{p} p \ln (p+1)\right) = 0
\end{aligned}$$

4. Superlinear Convergence of Asynchronous Waveform Relaxation

Define a sequence of nonempty sets $\{\mathcal{Y}^p\}_{p\in\mathbb{N}}$ in C where for $p = 0, 1, \ldots$

$$\mathcal{Y}^p = \prod_{i=1}^{i=\alpha} \mathcal{Y}_i^p$$

so that for all $i \in \{1, \ldots, \alpha\}$, the set $\mathcal{Y}_i^p$ is defined as follows :

$$\mathcal{Y}_i^p = \left\{x_i \in C_i / \|x_i - x_i^*\|_{i,K,t} \leq \rho_p(t) \left\|x^0 - x^*\right\|_K \forall t \geq 0 \text{ and } x_i(0) = x_{0_i}\right\}$$

Lemma 3 $\forall p \in \mathbb{N}$, $\mathcal{Y}^{p+1} \subset \mathcal{Y}^p$.

Proof. Let $i \in \{1, \ldots, \alpha\}$, in order to prove that

$$\mathcal{Y}_i^{p+1} \subset \mathcal{Y}_i^p$$

we will prove that

$$\rho_{p+1}(t) \leq \rho_p(t).$$

Let $t \geq 0$,
(a) if $t < t_p$ then

$$\begin{aligned}
\rho_{p+1}(t) &= \left(\frac{KL_1}{K-L_1}\right)^{p+1} \frac{t^{p+1}}{(p+1)!} \\
&= \frac{KL_1}{K-L_1}\frac{t}{p+1}\left(\frac{KL_1}{K-L_1}\right)^{p} \frac{t^{p}}{p!} \\
&= \frac{KL_1}{K-L_1}\frac{t}{p+1}\rho_p(t) \leq \frac{KL_1}{K-L_1}\frac{t}{p}\rho_p(t) \\
&\leq \rho_p(t) \quad (\textit{see remark } 1)
\end{aligned}$$

(b) if $t_p \leq t < t_{p+1}$ then

$$\rho_{p+1}(t) = \left(\frac{KL_1}{K-L_1}\right)^{p+1} \frac{t^{p+1}}{(p+1)!} \text{ and } \rho_p(t) = 1$$

but $t < t_{p+1}$ and by the remark 1, we have

$$\left(\frac{KL_1}{K-L_1}\right)^{p+1} \frac{t^{p+1}}{(p+1)!} < 1$$

then

$$\rho_{p+1}(t) \leq \rho_p(t)$$

(c) if $t \geq t_{p+1}$ then

$$\rho_{p+1}(t) = 1 = \rho_p(t)$$

so for all $t \geq 0$, $\rho_{p+1}(t) \leq \rho_p(t)$.
Thus for all $i \in \{1, ..., \alpha\}$,

$$\mathcal{Y}_i^{p+1} \subset \mathcal{Y}_i^p$$

and hence

$$\mathcal{Y}^{p+1} \subset \mathcal{Y}^p \quad \square$$

Let's define a fixed point mapping $\tilde{T}$ as fallows :

$$\begin{aligned} \tilde{T} : D\left(\tilde{T}\right) \rightarrow C \\ u \rightarrow x \end{aligned} \tag{9}$$

such that for $i \in \{1, ..., \alpha\}$, x_i is the solution of the system

$$\begin{cases} x_i'(t) = F_i\left(u_1(t), ..., u_{i-1}(t), x_i(t), u_{i+1}(t), ..., u_\alpha(t), t\right), \forall t \geq 0 \\ x_i(0) = x_{0_i} \end{cases}$$

and for all $p = 0, 1, ...$, let's define a mapping T^p as follows :

$$\begin{aligned} T^p : D\left(T^p\right) \rightarrow C \\ u \rightarrow x \end{aligned} \tag{10}$$

with

$$\begin{cases} x(t) = u(t) \quad if \ t \geq t_{p+1}, \\ x(t) = v(t) \quad if \ t < t_{p+1} \quad where \quad v = \tilde{T}(u). \end{cases}$$

Note that for all $u \in \mathcal{Y}^0$, there exists some p in $\mathbb{N}$ such that $u \in \mathcal{Y}^p \ominus \mathcal{Y}^{p+1}$.
To the problem 5, we associate the fixed point mapping defined as follows :

$$\begin{aligned} T : \mathcal{Y}^0 \rightarrow C \\ u \rightarrow x \end{aligned} \tag{11}$$

such that $x(t) = T^p(u)$ where $u \in \mathcal{Y}^p \ominus \mathcal{Y}^{p+1}$.

Theorem 4 *Any asynchronous algorithm (T, u^0, J, S) associated to T converges to the fixed point u^* of T.*

Proof. 1) By Lemma 1, we have $\mathcal{Y}^{p+1} \subset \mathcal{Y}^p$ for all $p = 0, 1, ...$
2) By construction we have $\mathcal{Y}^p = \mathcal{Y}_1^p \times ... \times \mathcal{Y}_\alpha^p, \quad \forall p \in \mathbb{N}$
3) Let's prove that $T\left(\mathcal{Y}^p\right) \subset \mathcal{Y}^{p+1}$ for all $p = 0, 1, ...$
For $u \in \mathcal{Y}^p$ and $x = T(u)$.

(a) If $t < t_{p+1}$ then:
$x_i'(t) = F_i(u_1(t), ..., u_{i-1}(t), x_i(t), u_{i+1}(t), ...u_\alpha(t), t)$

$$\begin{aligned}|x_i(t) - x_i^\star(t)|_i = &\left|\int_0^t F_i(u_1(s), ..., u_{i-1}(s), x_i(s), u_{i+1}(s), ...u_\alpha(s), s)\right.\\ &\left.-F_i(x^\star(s), s)\, ds\right|_i\\ \leq &\int_0^t \left|F_i(u_1(s), ..., u_{i-1}(s), x_i(s), u_{i+1}(s), ...u_\alpha(s), s)\right.\\ &\left.-F_i\left(x_1^\star(s), ..., x_{i-1}^\star(s), x_i^\star(s), x_{i+1}^\star(s), ..., x_\alpha^\star(s), s\right)\right|_i ds\\ \leq &\int_0^t \left|F_i(u_1(s), ..., u_{i-1}(s), x_i(s), u_{i+1}(s), ...u_\alpha(s), s)\right.\\ &\left.- F_i\left(x_1^\star(s), ..., x_{i-1}^\star(s), x_i(s), x_{i+1}^\star(s), ..., x_\alpha^\star(s), s\right)\right|_i ds\\ &+\int_0^t \left|F_i\left(x_1^\star(s), ..., x_{i-1}^\star(s), x_i(s), x_{i+1}^\star(s), ..., x_\alpha^\star(s), s\right)\right.\\ &\left.-F_i(x^\star(s), s)\right|_i ds\\ \leq &\int_0^t L_1 |u(s) - x^\star(s)|_\infty ds + \int_0^t L_1 |x_i(s) - x_i^\star(s)|_i ds\end{aligned}$$

then

$$\begin{aligned}e^{-Kt} |x_i(t) - x_i^\star(t)|_i \leq &\underbrace{\int_0^t L_1 e^{-K(t-s)} e^{-Ks} |u(s) - x^\star(s)|_\infty ds}_{(1)}\\ &+ \underbrace{\int_0^t L_1 e^{-K(t-s)} e^{-Ks} |x_i(s) - x_i^\star(s)|_i ds}_{(2)}\end{aligned}$$

We have

$$\begin{aligned}(1) &= \textstyle\int_0^t L_1 e^{-K(t-s)} e^{-Ks} \max\limits_{1\leq i\leq\alpha} |u_i(s) - x_i^\star(s)|_i ds\\ &= \textstyle\int_0^t L_1 e^{-K(t-s)} \max\limits_{1\leq i\leq\alpha} e^{-Ks} |u_i(s) - x_i^\star(s)|_i ds\\ &\leq \textstyle\int_0^t L_1 e^{-K(t-s)} \max\limits_{1\leq i\leq\alpha} \left(\sup\limits_{0\leq r\leq s} e^{-Kr} |u_i(r) - x_i^\star(r)|_i\right) ds\\ &\leq \textstyle\int_0^t L_1 e^{-K(t-s)} \max\limits_{1\leq i\leq\alpha} \|u_i - x_i^\star\|_{i,K,s} ds\\ &\leq \textstyle\int_0^t L_1 e^{-K(t-s)} \rho_p(s) \|x^0 - x^\star\|_K ds\end{aligned}$$

(a1) if $t < t_p$ then $\rho_p(s) = \left(\frac{KL_1}{K_1 - L_1}\right)^p \frac{s^p}{p!}$.

$$\begin{aligned}(1) &\leq L_1 \int_0^t e^{-K(t-s)} \left(\frac{KL_1}{K_1 - L_1}\right)^p \frac{s^p}{p!} \left\|x^0 - x^\star\right\|_K ds\\ &\leq \frac{L_1}{p!} \left(\frac{KL_1}{K - L_1}\right)^p \int_0^t s^p ds. \left\|x^0 - x^\star\right\|_K\\ &\leq L_1 \left(\frac{KL_1}{K - L_1}\right)^p \frac{t^{p+1}}{(p+1)!}. \left\|x^0 - x^\star\right\|_K\end{aligned}$$

(a2) if $t_p \le t < t_{p+1}$ then

$$\begin{aligned}
(1) \le & L_1 \int_0^{t_p} e^{-K(t-s)} \rho_p(s) \left\| x^0 - x^* \right\|_K ds + L_1 \int_{t_p}^{t} e^{-K(t-s)} \rho_p(s) \left\| x^0 - x^* \right\|_K ds \\
\le & \frac{L_1}{p!} \left(\frac{KL_1}{K - L_1} \right)^p \int_0^{t_p} s^p ds \left\| x^0 - x^* \right\|_K + L_1 \int_{t_p}^{t} e^{-K(t-s)} \left\| x^0 - x^* \right\|_K ds \\
\le & \frac{L_1}{p!} \left(\frac{KL_1}{K - L_1} \right)^p \int_0^{t_p} s^p ds \left\| x^0 - x^* \right\|_K + \frac{L_1}{p!} \left(\frac{KL_1}{K - L_1} \right)^p \int_{t_p}^{t} s^p ds \left\| x^0 - x^* \right\|_K \\
\le & L_1 \left(\frac{KL_1}{K - L_1} \right)^p \frac{t^{p+1}}{(p+1)!} \left\| x^0 - x^* \right\|_K .
\end{aligned}$$

Moreover

$$\begin{aligned}
\textstyle\int_0^t L_1 e^{-K(t-s)} e^{-Ks} \left| x_i(s) - x_i^*(s) \right|_i ds & \le \textstyle\int_0^t L_1 e^{-K(t-s)} e^{-Ks} ds . \sup_{0 \le s \le t} \left| x_i(s) - x_i^*(s) \right|_i \\
& \le L_1 \left(\tfrac{1 - e^{-Kt}}{K} \right) \left\| x_i - x_i^* \right\|_{i,K,t} \\
& \le \tfrac{L_1}{K} \left\| x_i - x_i^* \right\|_{i,K,t}
\end{aligned}$$

so

$$e^{-Kt} \left| x_i(t) - x_i^*(t) \right|_i \le L_1 \left(\frac{KL_1}{K - L_1} \right)^p \frac{t^{p+1}}{(p+1)!} \left\| x^0 - x^* \right\|_K + \frac{L_1}{K} \left\| x_i - x_i^* \right\|_{i,K,t}$$

then

$$\left\| x_i - x_i^* \right\|_{i,K,t} \le \rho_{p+1}(t) \left\| x^0 - x^* \right\|_K$$

(b) if $t \ge t_{p+1}$ then

$$\begin{aligned}
& x_i(t) = u_i(t) \qquad \mathit{for}\ i = 1, ..., \alpha \\
& \Rightarrow \left| x_i(t) - x_i^*(t) \right|_i = \left| u_i(t) - x_i^*(t) \right|_i \\
& \Rightarrow e^{-Kt} \left| x_i(t) - x_i^*(t) \right|_i = e^{-Kt} \left| u_i(t) - x_i^*(t) \right|_i
\end{aligned}$$

but:

$$\begin{aligned}
& \sup_{0 \le r \le t} e^{-Kr} \left| x_i(r) - x_i^*(r) \right|_i \\
& = \max \left(\sup_{0 \le r < t_p} e^{-Kr} \left| x_i(r) - x_i^*(r) \right|_i , \sup_{t_p \le r \le t} e^{-Kr} \left| u_i(r) - x_i^*(r) \right|_i \right) \\
& \le \max \left(\left\| x_i - x_i^* \right\|_i , \sup_{t_p \le r \le t} e^{-Kr} \left| u_i(r) - x_i^*(r) \right|_i \right) \\
& \le \max \left(\left\| x^0 - x^* \right\|_K , \rho_p(t) \left\| x^0 - x^* \right\|_K \right) \\
& \le \left\| x^0 - x^* \right\|_K = \rho_{p+1}(t) \left\| x^0 - x^* \right\|_K
\end{aligned}$$

so:

$$x_i \in \mathcal{Y}_i^{p+1}$$

hence

$$x \in \prod_{i=1}^{i=\alpha} \mathcal{Y}_i^{p+1}$$

then the asynchronous algorithm (T, u^0, J, S) associated to T converges superlinearly to the fixed point u^* of T. □

References

1. E.Lelarasmee, A.E.Ruehli and A.L.Sangiovanni-Vincentelli, IEEE Trans. on CAD, **1**, 131–145 (1982).

2. J. Bahi, E. Griepentrog and J.-C. Miellou, SIAM Journal on Numerical Analysis, **23**, 1969–1980 (1996).

3. D. P. Bertsekas and J. N. Tsitsiklis, Parallel and Distributed Computation, Englewood Cliffs, Prentice Hall, New Jersey (1989).

4. M. N. El Tarazi, Numer. Math., **39**, 325-340 (1982).

9th Int. Coll. on Differential Equations, pp. 127–132
D. Bainov (Ed)

A Discretization Scheme and Error Estimate for First-Order Systems and Elliptic Problems

RICHARD E. EWING
Institute for Scientific Computation, Texas A&M University, College Station, TX 77843-3404, USA
JIAN SHEN
DynCorp/Information & Engineering Technology, 12750 Fair Lakes Circle, Fairfax, VA 22033, USA

Abstract: A discretization scheme applicable to the direct approximation of the velocity variable for either a first-order system or second-order elliptic equation is proposed in this paper. The scheme is motivated by mixed finite element methods and extends the cell-centered finite difference and the finite volume element methods to problems with discontinuous coefficients and various boundary conditions. Error estimates of order $O(h^2)$ in certain discrete L^2-norms for both the velocity and the pressure variables can be obtained, even in the presence of discontinuous coefficients.

1. Introduction

Accurate numerical methods for fluid velocity in porous media are important in reservoir simulation [1, 2]. Unlike the poorly approximated derivatives of standard finite elements methods, mixed finite elements methods approximate the flux terms directly [2] pp. 311. In this article, we consider the approximate solutions of the elliptic system

$$-\nabla \cdot (\mathcal{K}\nabla p) = f, \qquad \text{in } \Omega, \tag{1.1}$$

where Ω is an open, bounded domain in $\mathbf{R}^2$, and $\mathcal{K} = \text{diag}\,(a, b)$ is a diagonal matrix function such that

$$0 < \tau_0 \leq a(x,y),\ b(x,y) \leq \tau_1, \qquad (x,y) \in \Omega, \tag{1.2}$$

for some positive real numbers τ_0 and τ_1. We are concerned with numerical schemes that approximate the smooth physical velocity variable

$$\mathbf{u} = -\mathcal{K}\nabla p, \tag{1.3}$$

with high accuracy. It is known that a mixed formulation for (1.1) seeks $(\mathbf{u}, p) \in H(\text{div}\,, \Omega) \times L^2(\Omega) \equiv \mathbf{V} \times W$ such that

$$\begin{aligned} \mathcal{K}^{-1}\mathbf{u} + \nabla p &= 0, \qquad \text{in } \Omega, \\ \nabla \cdot \mathbf{u} &= f, \qquad \text{in } \Omega, \end{aligned} \tag{1.4}$$

with appropriate boundary conditions. Here $H(\mathrm{div}, \Omega)$ is the space of two-dimensional vectors whose components and divergence are in $L^2(\Omega)$.

Based on system (1.4), numerical schemes such as the mixed finite element method (cf., [3, 4, 5, 6, 7, 8, 9]) and the cell-centered finite difference method (cf., [10, 11]) are known to provide adequate approximations for the velocity variable in the case of smooth coefficients. In particular, the cell-centered finite difference scheme has advantages over other schemes in that the algebraic linear system can be solved by eliminating the velocity variable with a diagonal matrix inversion. We refer to [12, 13, 10] for discussions of efficiency in solving the corresponding linear algebraic system.

Our objective in this paper is to propose a discretization scheme based on (1.4) which covers the mixed finite element method, the finite volume element method, and the cell-centered finite difference method. The scheme is designed for problems with piecewise smooth coefficients. More precisely, we might assume that the coefficient $\mathcal{K}$ is smooth only on each cell. Optimal and superconvergence error estimates are established for both the velocity $\mathbf{u}$ and the pressure p either with smooth coefficients or with smooth jumps across elements. Our scheme and the corresponding error estimate can be extended to problems in $\mathbf{R}^3$ without any difficulty.

The major consideration of this paper is the development of a linearlized family of discretization scheme for either first-order systems or second-order elliptic equations. The motivation comes from the physical mass conservation interpretation of the second equation in (1.4), the continuity of the pressure [1, 2] and the fact that mixed finite element approximation maintains mass balance in each element by imposing continuity of the velocity at the element boundaries in the normal direction [8]. Therefore, given (1.4), the mixed finite element, cell-centered finite difference, or finite volume approximations can provide a velocity that is continuous in the normal direction on any interface inside the domain if it corresponds to a cell interface. Thus, the optimal and superconvergence error estimates can be derived [14] under a local smoothness assumption of the solution as long as any potential interface of the coefficient discontinuity can be put on cell interfaces We provide analytical and quantitative support for the use of mixed finite element type of approximations in petroleum reservoir simulation [2] pp.187.

Moreover, the finite volume method [15] considered in this paper for discretizing (1.4) can be viewed as a particular case of our scheme. In other words, we provide superconvergence error estimates for Cai and McCormick's method [15]. In [14], we also extend the superconvergence results of [11] to inhomogeneous Dirichlet boundary conditions. In [16], Falk and Osborn compared several kinds of rough coefficient problems and approximations of different finite element methods. Their analysis shows the importance of regularity in assessing when a mixed method will be effective. In [14], we tried to relax the restrictions on solutions as much as we can.

The rest of this paper is organized as follows. In §2, we set notations and review the Raviart-Thomas finite element space of the lowest order [8]. In §3, we present a discretization scheme for (1.4) with a homogeneous Neumann boundary condition. The finite volume element method [15] is described in this context in §4.

2. Preliminaries and Notation

For simplicity, let $\bar{\Omega} = [0,1] \times [0,1]$, and Δ_x and Δ_y be partitions of $I = [0,1]$ given by

$$\Delta_x : \qquad 0 = x_{\frac{1}{2}} < x_{\frac{3}{2}} < \cdots < x_{N_x+\frac{1}{2}} = 1,$$

$$\Delta_y : \qquad 0 = y_{\frac{1}{2}} < y_{\frac{3}{2}} < \cdots < y_{N_y+\frac{1}{2}} = 1.$$

Denote by $\mathcal{T}$ the partition $\Delta_x \times \Delta_y$; i.e., $\mathcal{T}$ is composed of elements of the form $e_{ij} = \left(x_{i-\frac{1}{2}}, x_{i+\frac{1}{2}}\right) \times \left(y_{j-\frac{1}{2}}, y_{j+\frac{1}{2}}\right)$, for $i = 1, \cdots N_x$ and $j = 1, \cdots N_y$. Let

$$h_i = x_{i+\frac{1}{2}} - x_{i-\frac{1}{2}}, \qquad h'_j = y_{j+\frac{1}{2}} - y_{j-\frac{1}{2}},$$

and

$$x_i = x_{i-\frac{1}{2}} + h_i/2, \qquad y_j = y_{j-\frac{1}{2}} + h'_j/2,$$

for $i = 1, \cdots, N_x$ and $j = 1, \cdots, N_y$. Denote the mesh size of the partition $\mathcal{T}$ by h. Assume that the partition $\mathcal{T}$ is quasi-regular.

The Raviart-Thomas element of lowest order associated with the rectangular partition contains piecewise constant functions for the pressure unknown. The velocity is approximated by piecewise polynomials for which the x-component is a continuous piecewise linear polynomial in the x variable and constant in the y-direction, and the y-component is defined analogously with the roles of x and y interchanged denoted by $\mathbf{V}_h = V_{1,h} \times V_{2,h}$ and W_h the finite element spaces for the velocity and pressure, respectively. With an abuse of notation, by $\mathbf{v} = (v^x, v^y)$ we understand either a vector-valued function or a vector. Its vector form is defined by $v^x_{i-\frac{1}{2},j} = v^x\left(x_{i-\frac{1}{2}}, y_j\right)$; the vector v^y is defined similarly. This observation also applies to functions in W_h; i.e., any function $w \in W_h$ is understood to be either a function or a vector $w = \{w_{ij}\}$ defined by $w_{ij} = w(x_i, y_j)$.

Let $(\cdot, \cdot)_h$ be an inner product on $\mathbf{V}_h$ defined by

$$(\mathbf{u}, \mathbf{v})_h = (u^x, v^x)_1 + (u^y, v^y)_2, \tag{2.1}$$

where $(\cdot, \cdot)_i$ are inner products on $V_{i,h}$ given by

$$(u^x, v^x)_1 = \sum_{j=1}^{N_y} \sum_{i=0}^{N_x} u^x_{i+\frac{1}{2},j} v^x_{i+\frac{1}{2},j}, \qquad (u^y, v^y)_2 = \sum_{i=1}^{N_x} \sum_{j=0}^{N_y} u^y_{i,j+\frac{1}{2}} v^y_{i,j+\frac{1}{2}}.$$

Let $(\cdot, \cdot)_0$ be an inner product on W_h such that

$$(w, p)_0 = \sum_{j=1}^{N_y} \sum_{i=1}^{N_x} w_{ij} p_{ij}. \tag{2.2}$$

If B is an operator (matrix) from W_h to $\mathbf{V}_h$, by B^t we mean the adjoint (transpose) of B defined by the following equation:

$$(Bw, \mathbf{v})_h = (w, B^t \mathbf{v})_0, \qquad w \in W_h, \ \mathbf{v} \in \mathbf{V}_h. \tag{2.3}$$

Finally, denote by $W^m_s(\Omega)$ the usual Sobolev spaces on Ω with norms and semi-norms defined, respectively, by

$$\|f\|_{m,s,\Omega} = \left(\sum_{k=0}^{m} |f|^2_{k,s,\Omega}\right)^{\frac{1}{2}}, \qquad |f|_{k,s,\Omega} = \left(\int_\Omega \sum_{|\beta|=k} |\partial^\beta f|^s d\Omega\right)^{\frac{1}{s}}.$$

For simplicity, let $H^m(\Omega) = W^m_2(\Omega)$ with norm $\|\cdot\|_m \equiv \|\cdot\|_{m,2,\Omega}$.

3. The Numerical Discretization Scheme

We first discuss the treatment for the homogeneous Neumann boundary value problem; the algorithm and analysis can be extended to the non-homogeneous case without any difficulty.

Let A and B be two matrices defined on $\mathbf{V}_h$ and W_h which approximate the operator $\mathcal{K}^{-1}$ and the gradient operator $-\nabla$, respectively. Note that the vector-valued functions in $\mathbf{V}_h$ have a vanishing boundary value in the normal component, and W_h consists of piecewise constants with a vanishing mean value. A numerical scheme for discretizing (1.4) is then described by seeking a pair $(\mathbf{U}, P) \in \mathbf{V}_h \times W_h$ such that

$$\begin{aligned} A\mathbf{U} - BP &= 0, \\ B^t\mathbf{U} &= f_h, \end{aligned} \tag{3.1}$$

where f_h is an appropriately-defined vector form of the right-hand side function $f(x)$. In this paper, we choose f_h such that

$$(f_h)_{ij} = \int_{e_{ij}} f\,dx\,dy.$$

Note that the operator A returns a vector in $\mathbf{V}_h$ and so does the operator B.

Now we move toward the definition of A and B. Let α be a real number less than $\frac{1}{4}$. Recall that $\mathcal{K} = \text{diag}\,(a, b)$. For any $U^x \in V_{1,h}$, define $A_x(\alpha)U^x \in V_{1,h}$ by

$$(A_x(\alpha)U^x)_{i+\frac{1}{2},j} = t^x_{i-\frac{1}{2},j}U^x_{i-\frac{1}{2},j} + d^x_{i+\frac{1}{2},j}U^x_{i+\frac{1}{2},j} + t^x_{i+\frac{1}{2},j}U^x_{i+\frac{3}{2},j}, \tag{3.2}$$

where

$$\begin{aligned} t^x_{i-\frac{1}{2},j} &= \alpha\frac{h_i h'_j}{a_{ij}}, \\ d^x_{i+\frac{1}{2},j} &= h'_j\left(\frac{1}{2} - 2\alpha\right)\left(\frac{h_i}{a^-_{i+\frac{1}{2},j}} + \frac{h_{i+1}}{a^+_{i+\frac{1}{2},j}}\right) + t^x_{i-\frac{1}{2},j} + t^x_{i+\frac{1}{2},j}. \end{aligned}$$

Similarly, for any $U^y \in V_{2,h}$, we define $A_y(\alpha)U^y \in V_{2,h}$ by

$$(A_y(\alpha)U^y)_{i,j+\frac{1}{2}} = t^y_{i,j-\frac{1}{2}}U^y_{i,j-\frac{1}{2}} + d^y_{i,j+\frac{1}{2}}U^y_{i,j+\frac{1}{2}} + t^y_{i,j+\frac{1}{2}}U^y_{i,j+\frac{3}{2}}, \tag{3.3}$$

where

$$\begin{aligned} t^y_{i,j-\frac{1}{2}} &= \alpha\frac{h'_j h_i}{b_{ij}}, \\ d^y_{i,j+\frac{1}{2}} &= h_i\left(\frac{1}{2} - 2\alpha\right)\left(\frac{h'_j}{b^-_{i,j+\frac{1}{2}}} + \frac{h'_{j+1}}{b^+_{i,j+\frac{1}{2}}}\right) + t^y_{i,j-\frac{1}{2}} + t^y_{i,j+\frac{1}{2}}. \end{aligned}$$

For any $w \in W_h$, define $Bw = (B_x w, B_y w)^t \in \mathbf{V}_h$, where

$$(B_x w)_{i+\frac{1}{2},j} = h'_j(w_{ij} - w_{i+1,j}), \qquad (B_y w)_{i,j+\frac{1}{2}} = h_i(w_{ij} - w_{i,j+1}). \tag{3.4}$$

Lemma 0.1 *For any $w \in W_h$ and $\mathbf{v} \in \mathbf{V}_h$, the following is valid*

$$(Bw, \mathbf{v})_h = (\nabla\cdot\, \mathbf{v}, w)_\Omega,$$

where $(\cdot,\cdot)_\Omega$ denotes the standard L^2 inner product over Ω.

PROOF: See [14].

For simplicity, we introduce two bilinear forms defined, respectively, as follows:

$$A(\mathbf{u},\mathbf{v}) = (A_x u^x, v^x)_1 + (A_y u^y, v^y)_2, \qquad \mathbf{u},\mathbf{v} \in \mathbf{V}_h,$$

$$B(\mathbf{v},w) = (Bw,\mathbf{v})_h, \qquad \mathbf{v} \in \mathbf{V}_h,\ w \in W_h.$$

The matrix A is symmetric and positive definite for any $\alpha < \frac{1}{4}$. Let $|||\mathbf{v}|||_A = A(\mathbf{v},\mathbf{v})^{\frac{1}{2}}$ be the A-norm on $\mathbf{V}_h$. Since by Lemma 0.1 the $B(\cdot,\cdot)$-form satisfies the inf-sup condition of Brezzi [17], then the discretized problem (3.1) has a unique solution for any $\alpha < \frac{1}{4}$.

Remark 0.1 *The algorithm (3.1) involves a parameter α. It should be noted that the choice of $\alpha = \frac{1}{6}$ yields the standard mixed finite element method by using a specific numerical integration: Simpson's rule in the x-variable and the mid-point rule in the y-variable in evaluating $(a^{-1}u^x, v^x)$. The finite volume element method using lowest-order rectangular R-T elements [8] coincides with the case $\alpha = \frac{1}{8}$ (see §4 and [15]). The choice of $\alpha = 0$ yields the cell-centered finite difference scheme in the mixed form (cf., [10, 11] if the coefficient is smooth).*

4. The Finite Volume Element Method

A finite volume element method can be applied to discretize the system (1.4), which turns out to be a particular case $\alpha = \frac{1}{8}$ of the scheme proposed in §3. For simplicity, we consider equation (1.4) with a unit coefficient. From the second equation of (1.4), we have

$$\int_{e_{i,j}} \nabla \cdot \mathbf{u}\, de = \int_{e_{i,j}} f\, de, \tag{8.1}$$

for each element $e_{i,j}$. Applying the divergence theorem yields

$$\begin{aligned}\int_{e_{i,j}} \nabla \cdot \mathbf{u}\, de = \int_{\partial e_{i,j}} \mathbf{u}\cdot\mathbf{n}\, ds = &\int_{y_{j-\frac{1}{2}}}^{y_{j+\frac{1}{2}}} \left[u^x\left(x_{i+\frac{1}{2}}, y\right) - u^x\left(x_{i-\frac{1}{2}}, y\right)\right] dy \\ &+ \int_{x_{i-\frac{1}{2}}}^{x_{i+\frac{1}{2}}} \left[u^y\left(x, y_{j+\frac{1}{2}}\right) - u^y\left(x, y_{j-\frac{1}{2}}\right)\right] dx.\end{aligned} \tag{8.2}$$

Replace the exact solution $\mathbf{u}$ by an approximation $\mathbf{U} \in \mathbf{V}_h$. It follows from (8.1) and (8.2) that

$$(B^t\mathbf{U})_{i,j} = (f_h)_{i,j}. \tag{8.3}$$

To discretize the first equation of (1.4), we define finite volumes for the velocity by

$$V_{i+\frac{1}{2},j} = [x_i, x_{i+1}] \times \left[y_{j-\frac{1}{2}}, y_{j+\frac{1}{2}}\right].$$

Thus, it follows from the first equation of (1.4) that

$$\int_{V_{i+\frac{1}{2},j}} u^x de = -\int_{V_{i+\frac{1}{2},j}} \partial_x p\, de = \int_{y_{j-\frac{1}{2}}}^{y_{j+\frac{1}{2}}} [p(x_i, y) - p(x_{i+1}, y)] dy. \tag{8.4}$$

Let $(\mathbf{U}; P)$ be the approximation of $(\mathbf{u}; p)$ in $\mathbf{V}_h \times W_h$. We have

$$\int_{V_{i+\frac{1}{2},j}} U^x de = h_j'(P_{i,j} - P_{i+1,j}),$$

which is equivalent to

$$A_x\left(\frac{1}{8}\right)U^x - B_xP = 0. \tag{8.5}$$

Similarly, we have

$$A_y\left(\frac{1}{8}\right)U^y - B_yP = 0. \tag{8.6}$$

Combining (8.3) with (8.5) and (8.6) yields an algorithm from the finite volume element method. The discretized system clearly coincides with the one defined in §3 with $\alpha = \frac{1}{8}$.

Acknowledgement

This work was supported by the Department of Energy contract #DE-FG05-92ER25143 and by the National Science Foundation grant #DMS-9626179.

References

1. J. Bear, Dynamics of Fluids in Porous Media, Dover Pub. Inc. (1972).
2. G. Chavent and J. Jaffre, Mathematical Models and Finite Elements for Reservoir Simulation, Elsevier Science Publishers B.V. (1986).
3. M.B. Allen, R.E. Ewing and J.V. Koebbe, Num. Meth. for PDE's, **3**, 195 (1983).
4. B.L. Darlow, R.E. Ewing and M.F. Wheeler, Soc. Pet. Eng. J., **4**, 391 (1984).
5. J. Douglas, Jr., R.E. Ewing and M.F. Wheeler, R.A.I.R.O. Anal. Numer., **17**, 17 (1983).
6. R.E. Ewing, R.D. Lazarov and J. Wang, SIAM J. Numer. Anal., **28**, 1015 (1991).
7. R.S. Falk and J.E. Osborn, R.A.I.R.O. Anal. Numer., **14**, 249 (1980).
8. P.A. Raviart and J.M. Thomas, in: Mathematics Aspects of the Finite Element Method, Lecture Notes in Mathematics, **606**, Springer-Verlag, 292 (1977).
9. J.E. Robert and J.M. Thomas, Mixed and Hybrid Methods, Handbook of Numerical Analysis, **II**: Finite Element Methods, (Part 1), P.G. Ciarlet and J.L. Lions, (Eds.), North-Holland, Amsterdam (1989).
10. J. Shen, Mixed Finite Element Methods: Analysis and Computational Aspects, Ph.D. Thesis, The University of Wyoming (1992).
11. A. Weiser and M.F. Wheeler, SIAM J. Numer. Anal., **25**, 351 (1988).
12. J.H. Bramble, R.E. Ewing, J.E. Pasciak and J. Shen, Advances in Computational Mathematics, (1996).
13. R.E. Ewing and J. Shen, in: The Proceedings of the 6th Copper Mountain Conference on Multigrid Methods, NASA Conference Publication 3224, April (1993).
14. R.E. Ewing and J. Shen, Numerical Methods for Partial Differential Equations, (submitted).
15. Z. Cai and S.F. McCormick, SIAM J. Numer. Anal. **27**, 636 (1990).
16. R.S. Falk and J.E. Osborn, Math. Comp., **62**, 1 (1994).
17. F. Brezzi, R.A.I.R.O. Anal. Numer., **2**, 129 (1974).

9th Int. Coll. on Differential Equations, pp. 133–138
D. Bainov (Ed)

Stability Estimates for a Boundary Identification Problem in a Thin Domain

DARIO FASINO
Dipartimento di Matematica e Informatica, Università di Udine, Udine, Italy

GABRIELE INGLESE
Istituto di Analisi Globale e Applicazioni, CNR, Firenze, Italy

Abstract: We collect in this paper some new results concerning stability of a boundary identification problem related to the Laplace's equation in a thin domain. We prove a logarithmic type stability result that takes into account the regularity of the unknown boundary; we characterize the derivative with respect to the unknown boundary of the solution to the direct problem; and we propose a new stability result in which the contribution of the thickness of the domain and the error on the data are separated.
Key Words and Phrases: *Boundary identification problem, Laplace's equation, stability.*

Introduction

For $0 < a < 1$, let $\Omega^\theta \subset \mathbb{R}^2$ be the domain

$$\Omega^\theta = \{(x,y) : 0 < x < 1, 0 < y < a + \theta(x)\}.$$

We will consider functions θ satisfying some regularity and admissibility conditions, e.g., for some $k \geq 1$,

$$\theta(x) \in C^k(0,1), \quad \theta(0) = \theta(1) = 0, \quad \theta'(0) = \theta'(1) = 0, \quad \max_x |\theta(x)| \leq m \ll a.$$

We call *Problem $D(\theta)$* the following Boundary Value Problem, subject to suitable regularity conditions on the data:

$$\begin{aligned} \Delta u &= 0 \qquad \text{in } \Omega^\theta \\ u_y(x,0) &= \phi(x) \\ u_n(x, a+\theta(x)) &= 0. \end{aligned}$$

Here u_n means the normal derivative of u to the boundary of Ω^θ, and ϕ is assigned. In the vertical sides of the domain, $x = 0$ and $x = 1$, either Dirichlet or Neumann homogeneous boundary conditions can be considered; in the latter case, to enforce uniqueness, one should constrain u, e.g., to have zero mean in Ω^θ. In this paper we will consider Neumann conditions, for consistency with previous works; however, the treatment of Dirichlet conditions requires only minor changes, e.g., the substitution of some cosine series with analogous sine series.

The inverse problem of recovering θ from the knowledge of the data $\phi(x)$ and $u(x,0)$ is a boundary identification problem that will be called *Problem I* in what follows. Problem I is a particular case (due to the simple underlying geometry of the domain of interest) of a more general boundary identification problem that received considerable interest, see e.g., [1, 2, 3, 4, 5]. In this note we address the question of stability of Problem I, i.e., the continuous dependence of θ from the data ϕ and $u(x,0)$. Alessandrini proves in [6] that logarithmic type stability results are the best possible that can be achieved for boundary identification problems like our. Moreover, Beretta and Vessella obtained in [5] a particular logarithmic type continuous dependence of an unknown boundary from Cauchy data and for a generic domain, under suitable a priori regularity assumptions.

In this note we derive new stability results for Problem I, making explicit use of its particular geometry. Throughout this paper, we always assume that θ and ϕ are sufficiently regular, so that u is a *classical* solution of Problem $D(\theta)$: $u \in C^2(\bar{\Omega}^\theta)$. Because of this assumption, our results are not optimal; optimal results require a Sobolev space setting, taking into account the right regularity classes of data and solutions. In fact, results in this paper are in a preliminar form and are kept with a minimum of notation; we are planning to improve their form in a further paper.

Logarithmic Stability from Fourier Analysis

For any $f \in L^1(0,1)$ let

$$F_k(f) = \frac{1}{\pi}\int_0^1 \cos(k\pi x) f(x)\, dx$$

be the kth cos-Fourier coefficient of f. Consider the Fourier series of θ', the derivative of θ:

$$\theta'(x) = \sum_{k=1}^{\infty} t_k \cos(k\pi x), \qquad t_k = F_k(\theta').$$

Due to the regularity and admissibility assumptions, the above series is uniformly convergent. Let θ'_n be its nth partial sum; if $\theta \in C^k(0,1)$ with $k \geq 1$ then by Jackson's theorem there exists a constant C such that in the L^2-norm it holds (see [7, Thm 13.3.1]):

$$\|\theta' - \theta'_n\| \leq C n^{-k}. \tag{1}$$

Let u_1 and u_2 be the solutions of Problems $D(\theta_1)$ and $D(\theta_2)$ respectively. We assume a perfect knowledge of the trace $u_i(x,0)$, and address the problem of deriving stability estimates for the quantity $\|\theta_1 - \theta_2\|$.

We borrow [2, Thm. 4.1] adapting it to our notations:

Theorem 1 *There exist constants C and $\hat{a}$, depending only on ϕ, a and m, such that, for all $n > 1$*

$$|F_n(\theta'_1) - F_n(\theta'_2)| \leq C\|u_1(x,0) - u_2(x,0)\|_\infty n \cosh(n\pi\hat{a}).$$

The main result of this section is the following:

Theorem 2 *Let $\|u_1(x,0) - u_2(x,0)\|_\infty \leq \epsilon$. Under the given assumptions on θ, there exists a constant C depending only on ϕ, a and m, such that*

$$\|\theta_1 - \theta_2\|_\infty \leq C\left|\frac{1}{\log \epsilon}\right|^k.$$

Proof. Let $\theta'_{i,n}$ be the nth finite Fourier sum of θ'_i, for $i = 1, 2$. For all $n \geq 1$ we have from (1) and Theorem 1

$$\begin{aligned} \|\theta'_1 - \theta'_2\| &\leq \|\theta'_1 - \theta'_{1,n}\| + \|\theta'_2 - \theta'_{2,n}\| + \|\theta'_{1,n} - \theta'_{2,n}\| \\ &\leq C_1 n^{-k} + C_2 n^{3/2} \cosh(n\pi\hat{a})\epsilon. \end{aligned}$$

We estimate the minimum of the right hand size by taking $n(\epsilon) = c|\log \epsilon|$ for some $c > 0$. Hence,

$$\begin{aligned} \|\theta'_1 - \theta'_2\| &\leq C_1 c^{-k} |\log \epsilon|^{-k} + C_2 c^{3/2} |\log \epsilon|^{-3/2} \epsilon^{c\pi\hat{a}+1} \\ &\leq C |\log \epsilon|^{-k}. \end{aligned}$$

Finally, observe that $\theta_1(0) = \theta_2(0)$, hence by Cauchy–Schwarz inequality

$$\|\theta_1 - \theta_2\|_\infty \leq \int_0^1 |\theta'_1(t) - \theta'_2(t)|\, dt \leq \|\theta'_1 - \theta'_2\|,$$

completing the proof. □

Derivative with Respect to the Domain

In this section we prove that, under our regularity assumptions, Problem $D(\theta)$ can be differentiated in zero with respect to θ, and its derivative can be expressed as classical solution of a BVP.

With a little abuse of notation, we denote $D(\theta)$ the *forward map* of our problem: Let

$$D : C_0^3(0,1) \mapsto C^2(\bar{\Omega}^\theta), \qquad D(\theta) = u.$$

Moreover, let θ be fixed, let $u^0 = D(0)$ be the solution of the direct problem in the reference domain $\Omega^0 = (0,1) \times (0,a)$ and let $u^h = D(h\theta)$ for $h > 0$. The *local* derivative of D computed in zero in the direction θ is defined as [8]

$$\frac{\partial D}{\partial \theta}(0)\theta = \lim_{h \to 0} \frac{u^h - u^0}{h}.$$

The limit is defined in Ω^0, as for any point $x \in \Omega^0$, for sufficiently small h it holds $x \in \Omega^{h\theta}$. In the next theorem, we prove that the latter limit exists and give a closed form for it. The proof of the following theorem is similar to that of the results in [8, Sect. 6].

Theorem 3 *Under the stated regularity assumptions on θ, there exists a function $\tilde{u} \in C^2(\bar{\Omega}^0)$ such that $\tilde{u} = (\partial D/\partial\theta)(0)\theta$. The function $\tilde{u}$ is the unique solution of*

$$\begin{aligned} \Delta\tilde{u} &= 0 && \text{in } \Omega^0 \\ \tilde{u}_x(0,y) = \tilde{u}_x(1,y) &= 0 && \text{for } y \in (0,a) \\ \tilde{u}_y(x,0) &= 0 && \\ \int_{\Omega^0} \tilde{u} &= 0 && \end{aligned}$$

and

$$\tilde{u}_y(x,a) = (\theta(x) u_x^0(x,a))_x. \tag{2}$$

Proof. Let $\Theta : (0,1) \times \mathbb{R} \mapsto (0,1) \times \mathbb{R}$ be defined as

$$\Theta(x,y) = (0, \frac{y}{a}\theta(x)).$$

Observe that for small h the map $I + h\Theta$ is a diffeomorphism, and $\Omega^{h\theta} = (I + h\Theta)\Omega^0$. Introduce the functions $v^h = u^h \circ (I + h\Theta)$:

$$v^h(x,y) = u^h(x, y + h\frac{y}{a}\theta(x)).$$

For all h it holds $v^h \in C^2(\bar{\Omega}^0)$. We can characterize v^h as the solution of a particular BVP: Let M^h be the inverse matrix of the jacobian of the map $I + h\Theta$,

$$M^h = \begin{pmatrix} 1 & 0 \\ h\frac{y}{a}\theta'(x) & 1 + \frac{h}{a}\theta(x) \end{pmatrix}^{-1}.$$

The matrix M^h is well defined for $|h| \leq 1$ since $\|\theta\|_\infty \ll a$ by hypotesis, and has the same regularity of θ' there. Define on $\partial\Omega^0$ the function ψ such that $\psi = \phi$ for $y = 0$, $\psi = 0$ else. Changing variables in the relations defining u^h, it follows that v^h is the unique solution in $C^2(\bar{\Omega}^0)$ of the following elliptic selfadjoint BVP:

$$\begin{aligned} \nabla \cdot M^h(M^h)^t\nabla(v^h) &= 0 \qquad \text{in } \Omega^0 & (3)\\ n \cdot (M^h)^t\nabla v^h &= \psi \qquad \text{on } \partial\Omega^0 & (4) \end{aligned}$$

(n is the normal to $\partial\Omega^0$), subject to

$$\int_{\Omega^0}(1 + \frac{h}{a}\theta(x))^{-1}v(x,y) = 0. \qquad (5)$$

Indeed, existence and uniqueness follow from analogous properties of u^h, via the relationship $v^h = u^h \circ (I + h\Theta)$.

Observe that the maps on the left hand sides of equations (3—5), as functions defined in $(-1,1) \times C^2(\bar{\Omega}^0)$, are differentiable in a neighborhood of $(0, v^0)$. Then the theorem of differentiability of implicit functions, applied to (3—5), leads us to the conclusion that the map $(I + h\Theta) \circ D : C^k(0,1) \mapsto C^2(\bar{\Omega}^0)$, whose value in θ is v^θ, is differentiable at 0. In particular, there exists a function $\tilde{v} \in C^2(\bar{\Omega}^0)$ such that

$$\tilde{v} = \lim_{h\to 0} \frac{v^h - v^0}{h}. \qquad (6)$$

As a consequence, also the map $D : \theta \mapsto u^\theta$ is differentiable at 0 in the direction θ. This proves existence (and uniqueness, incidentally) of $\tilde{u}$.

By construction, $\tilde{u}$ is obviously harmonic in Ω^0, and all boundary conditions except (2) are trivial. We prove (2) now. From the boundary conditions of Problem $D(\theta)$ it follows

$$u_y^h(x, a + h\theta(x)) - h\theta'(x)u_x^h(x, a + h\theta(x)) = 0,$$

hence

$$v_y^h(x,a) - h\theta'(x)v_x^h(x,a) = 0.$$

Then, from (6) and $v_y^0(x,a) = 0$,

$$\tilde{v}_y(x,a) = \lim_{h\to 0} \theta'(x)v_x^h(x,a) = \theta'(x)v_x^0(x,a).$$

From [8, Lemma 1.2] we obtain

$$\begin{aligned}\tilde{u}(x,y) &= \tilde{v}(x,y) - \Theta(x,y)\cdot\nabla v^0(x,y)\\ &= \tilde{v}(x,y) - \frac{y}{a}\theta(x)v_y^0(x,y).\end{aligned}$$

Finally, since $v^0 = u^0$ and $v_y^0(x,a) = 0$, we have

$$\begin{aligned}\tilde{u}_y(x,a) &= \tilde{v}_y(x,a) - [\theta(x)v_y^0(x,a)/a + \theta(x)v_{yy}^0(x,a)]\\ &= \theta'(x)v_x^0(x,a) - \theta(x)v_{yy}^0(x,a)\\ &= \theta'(x)u_x^0(x,a) + \theta(x)u_{xx}^0(x,a)\\ &= (\theta(x)u_x^0(x,a))_x.\end{aligned}$$

□

As an immediate consequence we have:

Theorem 4 *For any $(x,y)\in\Omega^0$ and for sufficiently small h it holds:*

$$u^h(x,y) = u^0(x,y) + h\tilde{u}(x,y) + o(h).$$

Stability from Thin Plate Approximation

In the paper [9] the expression

$$\theta_{\mathrm{TPA}}(x) = a\frac{u_x^0(x,0) - u_x^\theta(x,0)}{u_x^0(x,0)}, \tag{7}$$

called *Thin Plate Approximation,* is proposed as an approximation to θ. This expression is derived by formal techniques, assuming $a \ll 1$, hence the reader is left without an indication of its accuracy. However, there is some numerical evidence that for small a and θ the quality of the Thin Plate Approximation is rather good.

In this section, we derive an error bound for θ_{TPA}, and use it to derive a stability result for Problem I of a completely new kind. Since we are interested in letting $a \to 0$ here, we consider the scaled functions $\tilde{\theta} = a^{-1}\theta$. Observe that $\tilde{\theta}$ is actually a *relative* perturbation of the upper boundary of Ω^0. Indeed, $a+\theta(x) = a(1+\tilde{\theta}(x))$, hence $\Omega^\theta = \Omega^{a\tilde{\theta}}$. Accordingly, we suppose the bound $\|\tilde{\theta}\|_\infty \le \tilde{m}$.

Theorem 5 *For sufficiently small a it holds uniformly in x*

$$\tilde{\theta}(x) = \frac{u_x^0(x,0) - u_x^{a\tilde{\theta}}(x,0)}{u_x^0(x,0)} + O(a),$$

hence

$$\theta(x) = a\frac{u_x^0(x,0) - u_x^\theta(x,0)}{u_x^0(x,0)} + O(a^2).$$

Proof. Let $u^h = D(h\tilde{\theta})$, for small h, so that $u^a = D(\theta) = u^\theta$. If a is sufficiently small, we can invoke Theorem 4 and write

$$u^a(x,y) = u^0(x,y) + a\tilde{u}(x,y) + o(a),$$

where $\tilde{u} = (\partial D/\partial\theta)(0)\theta$. Hence

$$\begin{aligned} u^a_{xx}(x,y) &= u^0_{xx}(x,y) + a\tilde{u}_{xx}(x,y) + o(a) \\ &= u^0_{xx}(x,y) - a\tilde{u}_{yy}(x,y) + o(a) \\ &= u^0_{xx}(x,y) - \tilde{u}_y(x,a) + O(a), \end{aligned}$$

using the approximation $a\tilde{u}_{yy}(x,y) = \tilde{u}_y(x,a) - \tilde{u}_y(x,0) + O(a)$, so that

$$u^0_{xx}(x,y) - u^a_{xx}(x,y) = (\tilde{\theta}(x)u^0_x(x,a))_x + O(a).$$

Since $u^0_x(0,y) = u^a_x(0,y) = 0$, integrating with respect to x we obtain

$$u^0_x(x,y) - u^a_x(x,y) = \tilde{\theta}(x)u^0_x(x,a) + O(a).$$

Finally, use the approximation $u^0_x(x,a) = u^0_x(x,0) + O(a)$. Letting $y = 0$ the claim is proved. □

In the next theorem we restrict our data $u(x,0)$ to a compact set, using an a priori information on the second derivative, and derive a stability bound for Problem I in which the contribution of the thickness of the domain and the error on the data are separated.

Theorem 6 *Let $\|u_1(x,0) - u_2(x,0)\|_\infty \le \epsilon$ and assume $\|u_{i,xx}(x,0)\| \le E$, for $i = 1,2$. If there exists a positive constant μ such that $|u^0_x(x,0)| \ge \mu > 0$ for $x \in \mathcal{I} \subseteq (0,1)$, then for sufficiently small a and ϵ there exist constants C_1 and C_2, depending on μ, ϕ, E and $\tilde{m}$ but not on a and ϵ, such that*

$$\sup_{x\in\mathcal{I}} |\theta_1(x) - \theta_2(x)| \le C_1a^2 + C_2\epsilon^{1/2}.$$

Proof. For simplicity, put $\mathcal{I} = (0,1)$. From the hypoteses,

$$\|u_{1,x}(x,0) - u_{2,x}(x,0)\|_\infty \le 20E^{3/2}\epsilon^{1/2},$$

see e.g., [10]. Let $\theta_{i,\mathrm{TPA}}$ be the Thin Plate Approximation of θ_i, for $i = 1,2$. From (7) and Theorem 5 we obtain

$$\begin{aligned} \|\theta_1 - \theta_2\|_\infty &\le \|\theta_1 - \theta_{1,\mathrm{TPA}}\|_\infty + \|\theta_2 - \theta_{2,\mathrm{TPA}}\|_\infty + \|\theta_{1,\mathrm{TPA}} - \theta_{2,\mathrm{TPA}}\|_\infty \\ &\le C_1a^{-2} + 20E^{3/2}\mu^{-1}\epsilon^{1/2}, \end{aligned}$$

whence the thesis. □

References

1. S. Andrieux, A. Ben Abda and M. Jaoua, C. R. Acad. Sci. Paris, **316**, 429–434 (1993).
2. N.D. Aparicio and M.K. Pidcock, Inverse Problems, **12**, 565–577 (1996).
3. H.T. Banks and F. Kojima, Quarterly of Appl. Math., **48**, 273–293 (1989).
4. H.T. Banks, F. Kojima and W.P. Winfree, Inverse Problems, **6**, 897–921 (1990).
5. E. Beretta and S. Vessella, SIAM J. Math. Anal., (in press) (1998).
6. G. Alessandrini, Inverse Problems, **13**, 887–897 (1997).
7. P.J. Davis, Interpolation and Approximation, Dover, New York (1975).
8. J. Simon, Numer. Funct. Anal. and Opt., **2**, 649–687 (1980).
9. P. Kaup, F. Santosa and M. Vogelius, Inverse Problems, **12**, 279–293 (1996).
10. G. Alessandrini, Applicable Analysis, **11**, 45–59 (1980).

9th Int. Coll. on Differential Equations, pp. 139-144
D. Bainov (Ed)

Partial Hölder Continuity Results for Solution of non Linear non Variational Elliptic Systems with Limit Controlled Growth

LUISA FATTORUSSO AND GIOVANNA IDONE
Università degli Studi di Reggio Calabria, Dipartimento di Ingegneria Elettronica and Matematica Applicata, via Graziella, Località Feo di Vito, Reggio Calabria, Italy

Abstract: Let Ω be a bounded open subset of R^n, $n > 4$, of class C^2.

Let $u \in H^2(\Omega)$ a solution of elliptic non linear non variational system

$$a(x, u, Du, H(u)) = b(x, u, Du)$$

where $a(x, u, \mu, \xi)$ and $b(x, u, \mu)$ are vectors in R^N, $N \geq 1$, measurable in x, continuous in (u, μ, ξ) and (u, μ) respectively.

We prove that if $b(x, u, \mu)$ has limit controlled growth, if $a(x, u, \mu, \xi)$ is of class C^1 in ξ and statisfies the Campanato condition (A) and, together with $\frac{\partial a}{\partial \xi}$, certain continuity assumptions, then the vertor Du is partially Holder continuous for every exponent $\alpha < 1 - \frac{n}{p}$.

1. Introduction

In this work we study the partial Hölder continuity for solutions of second order non linear non variational elliptic systems of type

$$a(x, u, Du, H(u)) = b(x, u, Du) \tag{1.1}$$

with limit controlled growth. This result is similar to the one proved in the case of strictly controlled growth by L. Fattorusso and G. Idone [1].

Let $u \in H^2(\Omega)$ a solution of elliptic non linear non variational system (1.1) where $a(x, u, \mu, \xi)$ and $b(x, u, \mu)$ are vectors of R^N, $N \geq 1$, measurable in x, continuous in (u, μ, ξ) and (u, μ) respectively, satisfying the conditions:

(1.2) $a(x, u, \mu, 0) = 0$

(1.3) $a(x, u, \mu, \xi)$ is of class C^1 in ξ with derivatives $\frac{\partial a}{\partial \xi_{ij}}$ (we have set $\frac{\partial a(x,u,\mu,\xi)}{\partial \xi_{ij}} = \left\{\frac{\partial a^h(x,u,\mu,\xi)}{\partial \xi_{ij}^k}\right\}$, $h, k = 1, \ldots, N$) uniformly continuous and bounded in $\Omega \times R^N \times R^{nN} \times R^{n^2N}$,

(1.4) there exists a constant c such that, $\forall u \in R^N$, $\forall \mu \in R^{nN}$ and for almost every $x \in \Omega$ we have

$$\| b(x, u, \mu) \| \leq c \left\{ f(x) + \| u \|^{\alpha} + \| \mu \|^{\beta} \right\}$$

with $f \in L^2(\Omega)$ and with $\alpha = \frac{n}{n-4}$ and $\beta = \frac{n}{n-2}$

(A) there exist three positive constants $\bar{\alpha}, \bar{\gamma}$, and $\bar{\delta}$, with $\bar{\gamma}+\bar{\delta}<1$, such that, $\forall u \in R^N$, $\forall \mu \in R^{nN}$, $\forall \tau, \eta \in R^{n^2N}$ and for almost every $x \in \Omega$ we have

$$\left\| \sum_{i=1}^{n} \tau_{ii} - \bar{\alpha}\left[a(x,u,\mu,\tau+\eta) - a(x,u,\mu,\eta)\right]\right\|^2 \leq \bar{\gamma} \| \tau \|^2 + \bar{\delta}\left\|\sum_{i=1}^{n}\tau_{ii}\right\|^2$$

(B) there exists a non negative function $\omega(t)$, defined for $t \geq 0$, continuous, bounded, concave, non decreasing with $\omega(0)=0$ such that $\forall x, y \in \Omega, \forall u, v \in R^N, \forall \mu, \bar{\mu} \in R^{nN}$ and $\forall \xi, \tau \in R^{n^2N}$

$$\| a(x,u,\mu,\xi) - a(y,u,\bar{\mu},\xi) \| \leq \omega\left(d^2(x,y) + \| u-v \|^2 + \| \mu - \bar{\mu} \|^2\right) \cdot \| \xi \|,$$

$$\left\|\frac{\partial a(x,u,\mu,\xi)}{\partial \xi} - \frac{\partial a(x,u,\mu,\tau)}{\partial \xi}\right\| \leq \omega\left(\|\xi-\tau\|^2\right) \text{ (here } \frac{\partial a(x,u,\mu,\eta)}{\partial \xi} = \left\{\frac{\partial a(x,u,\mu,\eta)}{\partial \xi_{ij}}\right\} i,j=1,\dots,n).$$

Partial Hölder continuity results for solutions of the system (1.1) had been obtained in [1] in the strictly controlled growth case.

In this work we obtain the following partial Hölder continuity result for solutions of (1.1) in the limit controlled growth case:

Theorem 1.1. *If $u \in H^2(\Omega, R^N)$ is a solution to the system (1.1) and if the assumptions (1.3) (1.4) with $f \in L^p(\Omega), p > n$ hold, then there exists a set $\mathcal{B}_0$, closed in Ω (In particular $m(\mathcal{B}_0)=0$), with*

$$\mathcal{B}_2 \subset \mathcal{B}_0 \subset \mathcal{B}_1 \subset \mathcal{B}_2$$

such that

$$Du \in C^{0,\alpha}\left(\Omega \backslash \mathcal{B}_0, R^{nN}\right), \qquad \forall \alpha < 1 - \frac{n}{p}.$$

In order to demonstrate the theorem mentioned, above we need to prove the following L^p local regularity theorem for solutions of (1.1). It serves an auxiliary role, but it has indeed an interest in its own right.

Theorem 1.2. *If $u \in H^2(\Omega, R^N)$ is a solution to the system (1.1) and if (1.2), (1.4) with $f \in L^p(\Omega), p > 2$, and (A) hold, then $u \in H^{2,p}_{loc}(\Omega, R^N)$ and there exists $\sigma_0(u)$ such that $\forall B(x^0,\sigma) \subset B(x^0, 2\sigma) \subset\subset \Omega$, with $\sigma < \sigma_0$ one has:*

$$(1.5) \qquad \fint_{B(x^0,\sigma)} \left(\| u \|^{\frac{2n}{n-4}} + \| Du \|^{2^*} + \| H(u) \|^2 \right)^{p/2} dx \leq$$

$$\leq k \left\{ \left[\fint_{B(x^0,2\sigma)} \left(\| u \|^{\frac{2n}{n-4}} + \| Du \|^{2^*} + \| H(u) \|^2 \right) dx \right]^{p/2} + \fint_{B(x^0,2\sigma)} |f|^p \, dx \right\}$$

where k does not depend on σ.

2. Local L^p-Regularity of the Matrix $H(u)$

Proof of the Theorem 1.2. Let $u \in H^2(\Omega, R^N)$ be a solution in Ω to the system (1.1), being $a(x,u,\mu,\xi)$ and $b(x,u,\mu)$ vectors of R^N satisfying assumptions (A), (1.2) and (1.4) with $f \in L^p, p > 2$.

From the estimate (3.2) of lemma (3.1) of [2],

$$(2.1) \qquad \int_{B(x^0,\sigma)} \| H(u) \|^2 \, dx \leq c\sigma^{-2} \int_{B(x^0,2\sigma)} \| Du - (Du)_{2\sigma} \|^2 \, dx +$$

$$+c\int_{B(x^0,2\sigma)}\| b(x,u,Du)\|^2\,dx$$

and by the theorem of Poincaré, we have

$$\int_{B(x^0,2\sigma)}\| Du-(Du)_{2\sigma}\|^2\,dx \le c\left(\int_{B(x^0,2\sigma)}\| H(u)^{\frac{2n}{n+2}}\|^2\,dx\right)^{\frac{n+2}{n}}. \tag{2.2}$$

From the assumption (1.4) one has:

$$\int_{B(x^0,2\sigma)}\| b(x,u,Du)\|^2\,dx \le c\int_{B(x^0,2\sigma)}|f(x)|^2\,dx+\int_{B(x^0,2\sigma)}\| u\|^{\frac{2n}{n-4}}\,dx+ \tag{2.3}$$

$$+\int_{B(x^0,2\sigma)}\| Du\|^{\frac{2n}{n-2}}\,dx.$$

Now, observing that

$$\int_{B(x^0,2\sigma)}\| Du\|^{2^*}\,dx \le$$
$$\le c\int_{B(x^0,2\sigma)}\| Du-(Du)_{2\sigma}\|^{2^*}\,dx+\int_{B(x^0,2\sigma)}\|(Du)_{2\sigma}\|^{2^*}\,dx \le$$
$$\le c\int_{B(x^0,2\sigma)}\| Du-(Du)_{2\sigma}\|^{2^*}\,dx+\sigma^{n(1-2^*)}\left(\int_{B(x^0,2\sigma)}\| Du\|\right)^{2^*}dx \le$$
$$\le c\int_{B(x^0,2\sigma)}\| Du-(Du)_{2\sigma}\|^{2^*}\,dx+\sigma^{-2}\left[\int_{B(x^0,2\sigma)}\| Du\|^{2^*\frac{n}{n+2}}\,dx\right]^{\frac{n+2}{n}}$$

and taking into account the (3.19) of page 21 of [3], we have

$$\int_{B(x^0,2\sigma)}\| Du\|^{2^*}\,dx \le$$
$$\le\left(\int_{B(x^0,2\sigma)}\| H(u)\|\right)^{2^*/2}dx+$$
$$+\sigma^{-2}\left[\int_{B(x^0,2\sigma)}\| Du\|^{2^*\frac{n}{n+2}}\,dx\right]^{\frac{n+2}{n}}.$$

Now from the absolute - continuity of the integral, we have that, $\forall\lambda$ there exists $\sigma(\mu,\lambda)$ such that if $\sigma<\sigma_1$

$$\int_{B(x^0,2\sigma)}\| Du\|^{2^*}\,dx \le \frac{\lambda}{3}\left(\int_{B(x^0,2\sigma)}\| H(u)\|^2\,dx\right)+ \tag{2.4}$$

$$+\sigma^{-2}\left[\int_{B(x^0,2\sigma)}\| Du\|^{2^*\frac{n}{n+2}}\,dx\right]^{\frac{n+2}{n}}.$$

Moreover being

$$\int_{B(x^0,2\sigma)}\| u\|^{\frac{2n}{n-4}}\,dx \le c\sigma^n\|P\|^{\frac{2n}{n-4}}+\int_{B(x^0,2\sigma)}\| u-P\|^{\frac{2n}{n-4}}\,dx$$

where $P\equiv(P_1,P_2,\ldots,P_N)$ is the polynomial vector of degree ≤ 1 such that

$$\int_{B(x^0,\sigma)}D^\alpha(u-P)dx=0 \qquad \forall\alpha,\ |\alpha|\le 1$$

we have:

$$\sigma^n \|P\|^{\frac{2n}{n-4}} \leq \sigma^{-2}\left[\int_{B(x^0,2\sigma)} \| u \|^{\frac{2n}{n-4}\frac{n}{n+2}}\, dx\right]^{\frac{n+2}{n}} + c(u)\int_{B(x^0,2\sigma)} \| Du \|^{2^*}\, dx$$

and

$$\int_{B(x^0,2\sigma)} \| u - P \|^{\frac{2n}{n-4}}\, dx \leq c\left(\int_{B(x^0,2\sigma)} \| H(u) \|^2\, dx\right)^{\frac{n}{n-4}}.$$

Hence

$$\int_{B(x^0,2\sigma)} \| u \|^{\frac{2n}{n-4}}\, dx \leq c\,\sigma^{-2}\left[\int_{B(x^0,2\sigma)} \| u \|^{\frac{2n}{n-4}\frac{n}{n+2}}\, dx\right]^{\frac{n+2}{n}} + \tag{2.5}$$

$$+c(u)\int_{B(x^0,2\sigma)} \| Du \|^{2^*}\, dx + c\left(\int_{B(x^0,2\sigma)} \| H(u) \|^2\, dx\right)^{\frac{n}{n-4}}.$$

From (2.1), taking into account (2.2) (2.3) (2.4) and (2.5) it follows

$$\int_{B(x^0,2\sigma)}\|H(u)\|^2 dx \leq c\,\sigma^{-2}\left[\int_{B(x^0,2\sigma)}\left(\|u\|^{\frac{2n}{n-4}} + \|Du\|^{2^*} + \|H(u)\|^2\right)^{\frac{n}{n+2}} dx\right]^{\frac{n+2}{n}} + \tag{2.6}$$

$$+\frac{\lambda}{3}\left(\int_{B(x^0,2\sigma)}\|H(u)\|^2 dx\right) + \int_{B(x^0,2\sigma)}| f(x) |^2 dx + c(u)\int_{B(x^0,2\sigma)}\|Du\|^{2^*} dx +$$

$$+c\left(\int_{B(x^0,2\sigma)}\|H(u)\|^2 dx\right)^{\frac{n}{n-4}}.$$

Now, from the absolute continuity of the integral, we have that $\forall\lambda > 0\ \exists\sigma_2(u,\lambda)$ such that, if $\sigma < \sigma_2$,

$$c\left(\int_{B(x^0,2\sigma)}\|H(u)\|^2 dx\right)^{\frac{n}{n-4}} < \frac{\lambda}{3}.$$

From this and from (2.4) (2.6)

$$\int_{B(x^0,2\sigma)}\|H(u)\|^2 dx \leq c\left\{\sigma^{-2}\left[\int_{B(x^0,2\sigma)}\left(\|u\|^{\frac{2n}{n-4}} + \|Du\|^{2^*} + \|H(u)\|^2\right)^{\frac{n}{n+2}} dx\right]^{\frac{n+2}{n}} + \right. \tag{2.7}$$

$$\left. +\lambda\int_{B(x^0,2\sigma)}\|H(u)\|^2 dx + \int_{B(x^0,2\sigma)}| f(x) |^2 dx\right\}.$$

Adding member to member (2.4), (2.5) and (2.7)

$$\int_{B(x^0,2\sigma)}\|u\|^{\frac{2n}{n-4}} dx + \int_{B(x^0,2\sigma)}\|Du\|^{2^*} dx + \int_{B(x^0,2\sigma)}\|H(u)\|^2 dx \leq \tag{2.8}$$

$$\leq c\left\{\sigma^{-2}\left[\int_{B(x^0,2\sigma)}\left(\|u\|^{\frac{2n}{n-4}} + \|Du\|^{2^*} + \|H(u)\|^2\right)^{\frac{n}{n+2}} dx\right]^{\frac{n+2}{n}} + \right.$$

$$\left. +\lambda\left(\int_{B(x^0,2\sigma)}\|H(u)\|^2 dx\right) + \int_{B(x^0,2\sigma)}|f(x)|^2 dx\right\}.$$

Now, setting

$$U(x) = \left\{\|u\|^{\frac{2n}{n-4}} + \|Du\|^{2^*} + \|H(u)\|^2 dx\right\}^{\frac{n}{n+2}}$$
$$G(x) = \{f(x)\}^{\frac{n}{n+2}}$$

the inequality (2.8), if $\sigma < \sigma_0(u, \lambda)$, can be written in the following form:

$$\fint_{B(x^0,\sigma)} U^{\frac{n}{n+2}} dx \leq$$
$$\leq c \left\{ \left(\fint_{B(x^0,2\sigma)} U dx \right)^{\frac{n}{n+2}} + \lambda \fint_{B(x^0,2\sigma)} U^{\frac{n}{n+2}} dx + \int_{B(x^0,2\sigma)} G^{\frac{n}{n+2}} dx \right\}.$$

From this, using a lemma of Gehring - Giaquinta - Modica (see lemma 4.1 page 125 of [3]), written for $r = \frac{n+2}{n}$, $s = p\,\frac{n+2}{n}$ where $p > 2$, we deduce that there exists $\lambda_0(r, s)$ such that $\forall \lambda < \lambda_0$ $\exists \varepsilon > 0$ for which $U \in L^1_{loc}(\Omega)$, $\forall t \in [r, r + \varepsilon)$ and

$$\text{(2.9)} \qquad \left(\int_{B(x^0,\sigma)} U^t dx \right)^{\frac{1}{t}} \leq k \left\{ \left(\fint_{B(x^0,2\sigma)} U^r dx \right)^{\frac{1}{r}} + \left(\fint_{B(x^0,2\sigma)} G^t dx \right)^{\frac{1}{t}} \right\}.$$

Setting in (2.9) $t = p\,\frac{n+2}{2n}$ it follows (1.5).

3. Partial Hölder Continuity of the Vector Du

With the same technique used in [1], we have the following lemmas:

Lemma 3.1. *If $u \in H^2(\Omega, R^N)$ is a solution to the system (1.1) and if the assumptions (1.3) (1.4) hold, then, $\forall B(x^0, \sigma) \subset\subset \Omega$, with $\sigma < 2$, $\forall \tau \in (0,1)$ and $\forall \varepsilon \in \left(0, (n-2)\left(1 - \frac{2}{p}\right)\right)$, it results:*

$$\text{(3.1)} \qquad \Phi(u, x^0, \tau\sigma) \leq A\,\Phi(u, x^0, \sigma) \left\{ \tau^\lambda + \sigma^{2(1-\frac{2}{p})} + \left[\omega(c\,\sigma^{2-n}\,\Phi(u, x^0, \sigma)\right]^{1-\frac{2}{p}} + \right.$$
$$\left. + \left[\omega\left(\fint_{B(x^0,\sigma)} \|H(u) - (H(u))_\sigma\|^2 d\sigma \right)\right]^{1-\frac{2}{p}} \right\}$$

where $\lambda = n\left(1 - \frac{2}{p}\right) - \varepsilon$ *and*

$$\text{(3.2)} \qquad \Phi(u, x^0, \sigma) = \sigma^\xi + \int_{B(x^0,\sigma)} \left[\|u\|^{\frac{2n}{n-4}} + \|Du\|^{2^*} + \|H(u)\|^2\right] dx$$

with $\xi = n\left(1 - \frac{2}{p}\right)$.

Let us set:

$$\mathcal{B}_1 = \left\{ x \in \Omega : \lim_{\sigma \to 0}{}'' \fint_{B(x^0,\sigma)} \|H(u) - (H(u))_\sigma\|^2 dy > 0 \right\}$$

$$\mathcal{B}_2 = \left\{ x \in \Omega : \lim_{\sigma \to 0}{}' \sigma^{2-n}\,\Phi(u, x^0, \sigma) > 0 \right\}.$$

By a known property of the Lebesgue integral we have

$$mis\ \mathcal{B}_1 = 0$$

and taking into account a theorem of Giusti ([3] p.142) we obtain

$$\mathcal{H}_{n-2}\left(\mathcal{B}_2\right)=0$$

where $\mathcal{H}_\gamma$ is the γ-dimensional Hausdorff measure.

Hence the set $\mathcal{B}_1 \cup \mathcal{B}_2$ has measure zero.

Now, reasoning exactly as in theorem 5.1 of [4], it is easy to prove:

Lemma 3.2. *If $u \in H^2(\Omega, R^N)$ is a solution to the system (1.1), if the assumptions (1.3) (1.4) hold, then, for every fixed $\varepsilon \in \left(0, 1-\frac{n}{p}\right)$, it is possible to associate to every $x^0 \in \Omega \backslash \mathcal{B}_1 \cup \mathcal{B}_2$ a ball $B\left(x^0, R_{x^0}\right) \subset \Omega \backslash \mathcal{B}_2$ and a positive number σ_ε such that, $\forall t \in (0,1)$ and $\forall y \in B\left(x^0, R_{x^0}\right)$*

$$\Phi\left(u, y, t\sigma_\varepsilon\right) \leq (1+A)\, t^{\,n(1-\frac{2}{p})-2\varepsilon}\, \Phi\left(u, y, \sigma_\varepsilon\right)$$

and hence (see [4])

$$H(u) \in L^{2, n(1-\frac{2}{p})-2\varepsilon}\left(B\left(x^0, R_{x^0}\right), R^{n^2 N}\right)$$

$$Du \in \mathcal{L}^{2, n(1-\frac{2}{p})-2\varepsilon+2}\left(B\left(x^0, R_{x^0}\right), R^{nN}\right).$$

From Lemma (3.2) the theorem 1.1 easily follows.

References

1. L. Fattorusso and G. Idone, Rend. Sem. Mat. Padova., (in print).
2. S. Campanato, Rend. Matem. Serie VII., **10**, 531–549 (1990).
3. S. Campanato, Sistemi Ellittici in Forma di Divergenza: Regolarità All'Interno, Quaderni scuola Normale Sup. Pisa (1980).
4. S. Campanato, Rend. Sem. Mat. e Fis. Milano, **L** (1982).

9th Int. Coll. on Differential Equations, pp. 145-154
D. Bainov (Ed)

Applications of the Lyapunov Direct Method for Stability Investigations of Some Functional Differential Equations

P. FERGOLA
Dip. di Matematica e Applicazioni, Univ. degli Studi di Napoli "Federico II", Compl. Universitario Monte S. Angelo, Via Cintia, 80126 Napoli, Italy

V.B. KOLMANOVSKII
Moscow State Inst. of Electronics and Math. (Technical Univ.), Bol. Vuzovskii, 3/12, Moscow, 109028, Russia

Abstract: A chemostat model has been considered where two populations compete for two critical nutrients in presence of delayed nutrient recycling processes. New asymptotic stability results have been proved due to the construction of suitable Lyapunov functionals. The adopted, constructive procedure seems to be quite general for a class of FDE's.
Key Words and Phrases: *Chemostat model, delayed nutrient recycling, Lyapunov functionals, stability.*
1991 Mathematics Subject Classification. 34K20, 92D25.

Introduction

Lyapunov Direct Method is among the most powerful methods in the stability theory of ordinary differential equations. In addition, general theorems of this method can be generalized for other types of equations, e.g., partial differential equations, stochastic equations, functional differential equations.

With respect to FDE basic obstacle to apply the Lyapunov method is connected with the difficulties to construct Lyapunov functionals for the concrete equations. Moreover, these difficulties are so huge that some authors compare the process of construction of the Lyapunov functionals with an art or even have doubt about possibilities to apply it.

Nevertheless, many examples of very complex Lyapunov functionals can be found in FDE's literature. In particular, important applications, including delayed processes, both of nutrient recycling and growth of populations (see, for instance, [1]–[4]), can be found in the theory of chemostat models.

Chemostat models are often used to study the growth of microorganisms populations in aquatic ecosystems. These models can describe well enough some of the main features of the growing of species receiving nutrient inputs from streams or for nutrient regeneration.

Usually in these models strong approximations are made by neglecting, for instance, the influence on the growth of the light and/or temperature fluctuations. Due to the level of the adopted approximation and to that of time scaling, these models can be mathematically represented by autonomous and nonautonomous systems of ODE, by PDE systems and by FDE systems.

Autonomous systems refer to an ideal situation in which the nutrient is uniformly distributed and the growth of microorganisms naturally occurs by supposing that the parameters of the environment are constants.

Nonautonomous systems of ODE, still keeping the space independence hypothesis, include time fluctuations of parameters due to seasonal variations.

PDE models include the space dependence of the unknowns, which will depend on the time as well as on the space variables.

In the papers of the last groups (FDE systems) delay terms are introduced in the systems in order to take into account the delays in the growth of populations and in nutrient recycling process coming from the bacterial decomposition of dead biomass.

Recently in these models the influence of always present toxicant substances has been considered ([5]-[8]).

In this paper we begin by analyzing the form of Lyapunov-Krasowskii functional used in [9]. We observe that the functional can be splitted into two parts taking care, respectively, of the instantaneous and delayed terms in the equations system. We points out that such a structure of the functional can be obtained by following a formal procedure (algorithm) consisting of four steps and using essentially Lyapunov functions for auxiliary specially constructed ODE's.

Proposed firstly for FDE's, the procedure appears to be formally the same for neutral type, stochastic FDE's and Volterra type difference equations and their applications [10].

The plan of the paper is as follows. In Section 1, by introducing model and notations, we recall an asymptotic stability result obtained in [5] for a positive equilibrium of a two-populations chemostat model with two recyclable nutrients.

In Section 2 we give two examples of Lyapunov functionals related to the same system of Section 1 and obtained by following a formal constructive procedure here suggested.

Some final remarks (Section 3) conclude the paper.

1. Model, Notation and Known Stability Results

In [9] a chemostat-like model has been considered where two populations compete for two critical nutrients. By supposing that delayed processes of nutrient recycling are present, the system can be written in the form:

$$\begin{cases} \dot{R} = D(R^{(0)} - R) - \sum_{i=1}^{2} \frac{c_i}{y_{R_i}} U_i(R,S) N_i + \sum_{i=1}^{2} b_{R_i} \gamma_i I_i \\ \dot{S} = D(S^{(0)} - S) - \sum_{i=1}^{2} \frac{c_i}{y_{S_i}} U_i(R,S) N_i + \sum_{i=1}^{2} b_{S_i} \gamma_i I_i \\ \dot{N}_i = N_i[c_i U_i(R,S) - (D + \gamma_i) - \delta_{ii} N_i] \end{cases} \tag{1.1}$$

where $i = 1, 2$

$$U_i(R,S) = \begin{cases} \dfrac{RS}{K_{S_i} R + RS + K_{R_i} S} & \text{if} \quad (R,S) \in R^2 - \{(0,0)\} \\ 0 & \text{if} \quad (R,S) = (0,0) \end{cases}$$

and

$$I_i = \int_0^{+\infty} f_i(s) N_i(t-s) ds$$

with the delay kernels f_i non-negative, bounded functions for $s \in [0, +\infty)$ such that

$$\int_0^{+\infty} f_i(s)ds = 1, \qquad \int_0^{+\infty} s f_i(s)ds < +\infty, \qquad \int_0^{+\infty} s^2 f_i(s)ds < +\infty. \tag{1.2}$$

Furthermore,

- R and S are the concentrations of the two limiting nutrients.
- $R^{(0)} > 0$ and $S^{(0)} > 0$ are the constant input concentrations of the two resources.
- $D > 0$ is the constant input and output flow rate of medium containing R and S.
- N_i are measures of the two species in competition.
- y_{R_i} and y_{S_i} are the constant yields of i-th species per unit resource R and S respectively consumed ($0 \leq y_{R_i}, y_{S_i} \leq 1$).
- $K_{R_i} > 0$ and $K_{S_i} > 0$ are the half saturation constants for the i-th species on resource R and S.
- $c_i > 0$ is the maximum specific growth of i-th species.
- $b_{R_i}, b_{S_i} \in (0, 1)$ are the rate of the nutrient R and S respectively recycled after the death of the i-th species.
- $\gamma_i > 0$ is the death rate of the i-th species.
- $\delta_{ii} > 0$ is the intraspecific competition coefficient of the i-th biotic species.

Denote by $BC(-\infty, 0]$ the space of the continuous and bounded functions $\phi : (-\infty, 0] \to R^4$ and $\|\phi\| = \sup_\theta |\phi(\theta)|$, $\theta \in (-\infty, 0]$, $|\cdot|$ being any norm in R^4. Indicating by H a positive constant, it has been set

$$Q_H = \{\phi \in BC(-\infty, 0], \ \|\phi\| \leq H\}, \qquad R^4_{+0} = \{u \in R^4, \quad u_i \geq 0, \ i = 1, 2, 3, 4\}.$$

In [9] we prove that under the conditions

$$\begin{cases} \dfrac{1}{y_{R_i}} = \dfrac{1}{y_{S_i}} \equiv \dfrac{1}{y_i}, \quad b_{R_i} = b_{S_i} \equiv b_i \\ \dfrac{1}{\delta_{11}}\left[\dfrac{c_1}{1 + K_{S_1} + K_{R_1}} - (D + \gamma_1)\right] = \dfrac{1}{\delta_{22}}\left[\dfrac{c_2}{1 + K_{S_2} + K_{R_2}} - (D + \gamma_2)\right] \\ \dfrac{2a}{\delta_{11}}\dfrac{c_1(D + \gamma_1)}{1 + K_{S_1} + K_{R_1}} + \dfrac{b}{\delta_{11}}(D + \gamma_1) \leq \left[\dfrac{ac_1^2}{\delta_{11}^2(1 + K_{S_1} + K_{R_1})} + \dfrac{a}{\delta_{11}^2}(D + \gamma_1)^2 + \right. \\ \qquad\qquad \left. + \dfrac{bc_1}{2\delta_{11}(1 + K_{S_1} + K_{R_1})}\right] \end{cases} \tag{1.3}$$

where

$$a = \sum_{i=1,2} \frac{\delta_{ii}}{y_i}, \qquad b = \sum_{i=1,2} \left[b_i \gamma_i + \frac{D + \gamma_i}{y_i}\right] \tag{1.4}$$

system (1.1) admits the positive equilibrium

$$E_{++++} = (R^+, S^+, N_1^+, N_2^+) = (c, c, n, n), \quad c \geq 0, \quad n \geq 0. \tag{1.5}$$

The asymptotic stability of a positive equilibrium of type (1.5) has been also proved in [9] with the following

Theorem 1.1. *Under the conditions*

$$i) \qquad \frac{1}{y_{S_1}} > \frac{b_{R_1}+b_{S_1}}{2\delta_{11}}U^+_{1/S}, \qquad \frac{1}{y_{S_2}} > \frac{b_{R_2}+b_{S_2}}{2\delta_{22}}U^+_{2/S} \tag{1.6}$$

where $U^+_{i/S} = \left(\frac{\partial U_i}{\partial S}\right)_{R=R^+,\ S=S^+}$ *a positive equilibrium of type* (1.5), *if it exists, is asymptotically stable provided that the wash-out D is such that the following inequalities are satisfied:*

$$ii) \qquad -a_{22}(w_1R_1 - Y_1^R)^2 - a_{11}(w_2R_2 - Y_2^R)^2 + (b_{R_1}\gamma_1 + b_{R_2}\gamma_2)a_{11}a_{22} \le 2a_{11}a_{22}R_3$$

$$iii) \qquad S_4 \ge \frac{1}{2}\left[(b_{S_1}\gamma_1 + b_{S_2}\gamma_2) + \frac{(-S_R+R_S)^2 a_{11}a_{22}}{|\Delta_3|}\right].$$

The proof of this theorem can be shortly resumed as follows.
By means of a change of variables involving (1.5), system (1.1) is transformed into the system

$$\begin{cases} \dot{x}_i = (x_i+N_i^+)[c_i\xi_i(x_3,x_4) - \delta_{ii}x_i], \quad i=1,2 \\ \dot{x}_3 = -Dx_3 - \sum_{i=1}^{2}\frac{c_i}{y_{R_i}}U_ix_i - \sum_{i=1}^{2}\frac{c_i}{y_{R_i}}\xi_i(x_3,x_4)N_i^+ \\ \qquad + \sum_{i=1}^{2} b_{R_i}\gamma_i\int_0^{+\infty} f_i(s)x_i(t-s)ds \\ \dot{x}_4 = -Dx_4 - \sum_{i=1}^{2}\frac{c_i}{y_{S_i}}U_ix_i - \sum_{i=1}^{2}\frac{c_i}{y_{S_i}}\xi_i(x_3,x_4)N_i^+ \\ \qquad + \sum_{i=1}^{2} b_{S_i}\gamma_i\int_0^{+\infty} f_i(s)x_i(t-s)ds \end{cases} \tag{1.7}$$

where

$$\xi_i(x_3,x_4) = U_i(R^+ + x_3, S^+ + x_4) - U_i^+, \qquad U_i^+ = U_i(R^+,S^+).$$

The associated linearized system of (1.7) is

$$\begin{cases} \dot{x}_1 = -\delta_{11}N_1^+x_1 + R_1x_3 + S_1x_4 \\ \dot{x}_2 = -\delta_{22}N_2^+x_2 + R_2x_3 + S_2x_4 \\ \dot{x}_3 = -Y_1^Rx_1 - Y_2^Rx_2 - R_3x_3 - S_Rx_4 + \sum_{i=1}^{2} b_{R_i}\gamma_i\int_0^{+\infty} f_i(s)x_i(t-s)ds \\ \dot{x}_4 = -Y_1^Sx_1 - Y_2^Sx_2 - R_Sx_3 - S_4x_4 + \sum_{i=1}^{2} b_{S_i}\gamma_i\int_0^{+\infty} f_i(s)x_i(t-s)ds \end{cases} \tag{1.8}$$

where

$$R_1 = c_1N_1^+U^+_{1/R} \qquad S_1 = c_1N_1^+U^+_{1/S}$$

$$R_2 = c_2N_2^+U^+_{2/R} \qquad S_2 = c_2N_2^+U^+_{2/S}$$

$$R_3 = D + \frac{R_1}{y_{R_1}} + \frac{R_2}{y_{R_2}} \qquad S_4 = D + \frac{S_1}{y_{S_1}} + \frac{S_2}{y_{S_2}}$$

$$R_S = \frac{R_1}{y_{S_1}} + \frac{R_2}{y_{S_2}} \qquad S_R = \frac{S_1}{y_{R_1}} + \frac{S_2}{y_{R_2}}$$

$$Y_1^R = \frac{c_1}{y_{R_1}} U_1^+ \qquad Y_2^R = \frac{c_2}{y_{R_2}} U_2^+$$

$$Y_1^S = \frac{c_1}{y_{S_1}} U_1^+ \qquad Y_2^S = \frac{c_2}{y_{S_2}} U_2^+.$$

Finally, it is shown that the functional

$$\begin{aligned} V(\underline{x_t}) &= w_1 x_1^2 + w_2 x_2^2 + x_3^2 + x_4^2 + \sum_{i=1}^{2} b_{R_i}\gamma_i \int_0^{+\infty} f_i(s) \int_{t-s}^{t} x_i^2(u)duds \\ &+ \sum_{i=1}^{2} b_{S_i}\gamma_i \int_0^{+\infty} f_i(s) \int_{t-s}^{t} x_i^2(u)duds \end{aligned} \tag{1.9}$$

satisfies all the conditions required by the extension to the FDE's theory of the Lyapunov theorem on asymptotic stability.

We point out that, due to the form (1.9) of the Lyapunov–Krasovskii functional, the conditions required from Theorem 1.1 do not involve the delay kernels connected to the recycling processes of nutrients. In the next Section we show how different stability conditions can be obtained by taking advantage from delay terms.

2. Procedures to Construct Lyapunov Functionals

We can observe that the functional (1.9) can be splitted into two parts. First of them corresponds to the instantaneous part of the system, the second one takes into account the influence of delay.

Analogously, we represent right hand side of system (1.1) as a sum of two terms as follows:

$$\dot{x} = Lx + \int_0^{+\infty} K(s)x(t-s)ds \tag{2.1}$$

where

$$L = \begin{pmatrix} -\delta_{11}N_1^+ & 0 & R_1 & S_1 \\ 0 & -\delta_{22}N_2^+ & R_2 & S_2 \\ -Y_1^R & -Y_2^R & -R_3 & -S_R \\ -Y_1^S & -Y_2^S & -R_S & -S_4 \end{pmatrix}$$

and

$$K(s) = \begin{pmatrix} 0 & 0 & 0 & 0 \\ 0 & 0 & 0 & 0 \\ b_{R_1}\gamma_1 f_1(s) & b_{R_2}\gamma_2 f_2(s) & 0 & 0 \\ b_{S_1}\gamma_1 f_1(s) & b_{S_2}\gamma_2 f_2(s) & 0 & 0 \end{pmatrix}.$$

Suppose that the linear system $\dot{x} = Lx$ is asymptotically stable. Then a positive definite matrix P there exists such that $L^T P + PL = -I$, where I is the identity matrix. Let us consider the Lyapunov functional

$$V_1(t) = x^T(t)Px(t).$$

Calculation of the derivative of $V_1(t)$ along the solutions of system (2.1) yields

$$\begin{aligned}
\dot{V}_1(t) &= \dot{x}^T(t)Px(t) + x^T(t)P\dot{x}(t) \\
&= \left(Lx(t) + \int_0^{+\infty} K(s)x(t-s)ds\right)^T Px(t) + x^T(t)P\Big(Lx(t) + \int_0^{+\infty} K(s)x(t-s)ds\Big) \\
&= \left(x^T(t)L^T + \left(\int_0^{+\infty} K(s)x(t-s)ds\right)^T\right) Px(t) \\
&\quad + x^T(t)P\left(Lx(t) + \int_0^{+\infty} K(s)x(t-s)ds\right) \\
&= x^T(t)L^TPx(t) + x^T(t)PLx(t) + \left(\int_0^{+\infty} K(s)x(t-s)ds\right)^T Px(t) \\
&\quad + x^T(t)P\int_0^{+\infty} K(s)x(t-s)ds = \\
&= x^T(t)(L^TP + PL)x(t) + 2x^T(t)P\int_0^{+\infty} K(s)x(t-s)ds \\
&= -x^T(t)x(t) + 2x^T(t)P\int_0^{+\infty} K(s)x(t-s)ds.
\end{aligned}$$

If $|K(s)|$ denotes the norm of matrix $K(s)$ compatible with the Euclidean norm

$$|x| = \sqrt{\sum_{i=1}^{n} x_i^2}$$

we have

$$\begin{aligned}
\dot{V}_1 &\leq -|x(t)|^2 + 2|x^T(t)P\int_0^{+\infty} K(s)x(t-s)ds| \\
&\leq -|x(t)|^2 + 2|x(t)||P|\int_0^{+\infty} |K(s)||x(t-s)|ds \\
&\leq -|x(t)|^2 + |P|\int_0^{+\infty} |K(s)|(|x(t)|^2 + |x(t-s)|^2)ds \\
&\leq -|x(t)|^2 + |P|\int_0^{+\infty} |K(s)|ds|x(t)|^2 + |P|\int_0^{+\infty} |K(s)||x(t-s)|^2ds.
\end{aligned}$$

Now we consider the functional

$$V_2 = |P|\int_0^{+\infty} |K(s)|\int_{t-s}^{t} |x(u)|^2duds$$

and we set

$$V = V_1 + V_2.$$

The derivative of V along the solutions of system (1.9) satisfies

$$\begin{aligned}\dot{V} &\leq -|x(t)|^2 + 2|P| \int_0^{+\infty} |K(s)|ds|x(t)|^2 \\ &= -(1 - 2|P| \int_0^{+\infty} |K(s)|ds)|x(t)|^2.\end{aligned}$$

Therefore we get

Theorem 2.1. *Let us suppose that system $x' = Lx$ is asymptotically stable. Then, if the inequality*

$$|P| \int_0^{+\infty} |K(s)|ds < \frac{1}{2} \tag{2.2}$$

holds, the positive equilibrium (1.5) is locally asymptotically stable.

Now let us use another type of transformation of equation (1.1). We remark that

$$\begin{aligned}\frac{d}{dt} \int_0^{+\infty} K(s)ds \int_{t-s}^{t} x(\tau)d\tau &= \int_0^{+\infty} K(s)ds \frac{d}{dt} \int_{t-s}^{t} x(\tau)d\tau \\ &= \int_0^{+\infty} K(s)ds(-x(t-s) + x(t)) \\ &= x(t) \int_0^{+\infty} K(s)ds - \int_0^{+\infty} K(s)x(t-s)ds.\end{aligned}$$

Therefore

$$\dot{x} = Lx + x \int_0^{+\infty} K(s)ds - \frac{d}{dt} \int_0^{+\infty} K(s)ds \int_{t-s}^{t} x(\tau)d\tau.$$

Put

$$B = \int_0^{+\infty} K(s)ds$$

as a result we have

$$\frac{d}{dt}\Big(x + \int_0^{+\infty} K(s)ds \int_{t-s}^{t} x(\tau)d\tau\Big) = (L + B)x. \tag{2.3}$$

Furthermore, if we set

$$B_1 = L + B,$$

$$z = x + \int_0^{+\infty} K(s)ds \int_{t-s}^{t} x(\tau)d\tau \tag{2.4}$$

we can write (2.3) in the form

$$\dot{z} = B_1 x. \tag{2.5}$$

Let us consider the functional V_1

$$V_1 = z^T P z. \tag{2.6}$$

We have, by virtue of $(2.4)_1$, $(2.4)_2$ and (2.5)

$$\begin{aligned}
\dot{V}_1 &= \dot{z}^T P z + z^T P \dot{z} = x^T B_1 P z + z^T P B_1 \dot{x} \\
&= x^T B_1^T P x + x^T B_1^T P \int_0^{+\infty} K(s) ds \int_{t-s}^{t} x(\tau) d\tau \\
&\quad + x^T P B_1 x + \left(\int_0^{+\infty} K(s) ds \int_{t-s}^{t} x(\tau) d\tau \right)^T P B_1 x \\
&= x^T (B_1^T P + P B_1) x + x^T B_1^T P \int_0^{+\infty} K(s) ds \int_{t-s}^{t} x(\tau) d\tau \\
&\quad + \left(\int_0^{+\infty} K(s) ds \int_{t-s}^{t} x(\tau) d\tau \right)^T P B_1 x.
\end{aligned}$$

Now, suppose that P is a positive definite matrix such that

$$B_1^T P + P B_1 = -I.$$

Then we have

$$\dot{V}_1 \le -x^2 + |x^T B_1^T P \int \int_0^{+\infty} K(s) ds \int_{t-s}^{t} x(\tau) d\tau| + |\int_0^{+\infty} K(s) ds \int_{t-s}^{t} x(\tau) d\tau P B_1 x|.$$

Moreover

$$\begin{aligned}
|x^T B_1 P \int_0^{+\infty} K(s) ds \int_{t-s}^{t} x(\tau) d\tau| &\le |B_1||P||x| \int_0^{+\infty} |K(s)| ds \int_{t-s}^{t} |x(\tau)| d\tau \\
&\le |B_1||P| \int_0^{+\infty} |K(s)| ds \int_{t-s}^{t} \frac{1}{2} (|x(t)|^2 + |x(\tau)|^2) d\tau \\
&\le \frac{1}{2} |B_1||P| \int_0^{+\infty} |K(s)| ds \left(s|x|^2 + \int_{t-s}^{t} |x(\tau)|^2 d\tau \right)
\end{aligned}$$

and similarly

$$\left| \int_0^{+\infty} K(s) ds \int_{t-s}^{t} x(\tau) d\tau P B_1 x \right| \le \frac{1}{2} |B_1||P| \int_0^{+\infty} |K(s)| ds \left(\int_{t-s}^{t} |x(\tau)|^2 d\tau + s|x|^2 \right).$$

Hence

$$\dot{V}_1 \le -x^2 + |B_1||P| \Big(|x|^2 \int_0^{+\infty} s|K(s)| ds + \int_0^{+\infty} |K(s)| \int_{t-s}^{t} |x(\tau)|^2 d\tau ds \Big).$$

As a functional V_2 we can take

$$V_2 = |B_1||P| \int_0^{+\infty} |K(s)| ds \int_{t-s}^{t} d\tau \int_{\tau}^{t} |x(\alpha)|^2 d\alpha.$$

Then

$$\dot{V}_2 = |B_1||P||x(t)|^2 \int_0^{+\infty} s|K(s)|ds - |B_1||P| \int_0^{+\infty} |K(s)|ds \int_{t-s}^{t} |x(\tau)|^2 d\tau.$$

As a result we have that the derivative of the function $V = V_1 + V_2$ along the solutions of system (2.1) satisfies

$$\dot{V} = \dot{V}_1 + \dot{V}_2 \leq -x^2(t) + 2|B_1||P||x^2(t)| \int_0^{+\infty} sK(s)ds = -x^2 \left(1 - 2|B_1||P| \int_0^{+\infty} s|K(s)|ds\right).$$

The last inequality leads us to state the following asymptotic stability result.

Theorem 2.2. *Let us suppose that a definite positive matrix P there exists such that*

$$B_1^T P + PB_1 = -I.$$

Then, if the inequalities

$$|B_1||P| \int_0^{+\infty} s|K(s)|ds < \frac{1}{2}, \qquad \int_0^{+\infty} s|K(s)|ds < 1 \tag{2.7}$$

are satisfied, the positive equilibrium (1.5) *is locally asymptotically stable.*

Conclusion. We have shown some possibilities to investigate stability of chemostat by constructing different Lyapunov functionals which, generally speaking, could lead us to various parts of stability domain. For example comparison of stability conditions (2.2) and (2.7) shows that conditions (2.2) are more suitable for the kernels $K(s)$ larger at infinity whereas (2.7) can be easier to be applied for the kernels $K(s)$ larger near the origin. Further applications of the same procedure could enlarge the stability domains obtained in the space of parameters.

Acknowledgements

This research has been partially supported by Gruppo Nazionale per la Fisica Matematica of the C.N.R. and it has been performed according to a bilateral cooperation research project between University of Naples "Federico II" and Technical University of Moscow.

References

1. E. Beretta, G.I. Bischi and F. Solimano, J. Math. Biol., **28**, 99–111 (1990).
2. G.S.K. Wolkowicz, H. Xia and S. Ruan, SIAM J. Appl. Math., **57**, 1281–1310 (1997).
3. E. Beretta, P. Fergola and C. Tenneriello, J. Biol. Systems, **3**, 483–494 (1995).
4. X. He and S. Ruan, (in press).
5. P. Fergola, C. Tenneriello, Z. Ma and X. Wen, in: Proc. Twelfth European Meeting on "Cybernetics and System Research '94", Vienna, April 5–8 (1994).

6. P. Fergola, C. Tenneriello, Z. Ma and X. Wen, in: Proc. VIII Int. Conf. on "Waves and Stability on Continuous Media", Palermo (1995).

7. P. Fergola, B. Buonomo and C. Ruggieri, Mathematical and Computer Modelling, **24**, (1996).

8. P. Fergola, C. Tenneriello, Z. Ma and X. Wen, Dynamical Systems and Applications, **6**, (1997).

9. P. Fergola, S. Rionero and C. Tenneriello, in: Proc. VII Int. Conf. on "Waves and Stability on Continuous Media", Bologna, World Scientific (1993).

10. V.B. Kolmanovskii and A.D. Myshkis, Applied Theory of Functional Differential Equations, Kluwer Academic Publishers (1992).

9th Int. Coll. on Differential Equations, pp. 155-162
D. Bainov (Ed)

On Global Weak Solutions of a Nonlinear Evolution Equation in Noncylindrical Domain

J. FERREIRA
Departamento de Matemática, Universidade Estadual de Maringá, Av. Colombo 5790-Zona 7, 87020-900 Maringá- PR, Brazil

M.A. ROJAS-MEDAR
DMA-IMECC-UNICAMP, C.P. 6065, 13081-970, Campinas-SP, Brazil

Abstract: In this paper we establish the existence of global weak solutions of the mixed problem for the nonlinear evolution equation in noncylindrical domains.

1. Introduction

Let Q be a noncylindrical domain of $R^n \times]0,T[$ with lateral boundary Σ, $T > 0$ is an arbitrary real number, Δ is the Laplace operator, $u' = \frac{\partial u}{\partial t}$ and Ω be a bounded open set in R^n with smooth boundary Γ such that

$$Q \subset \Omega \times]0,T[.$$

In Q consider the following mixed nonlinear hyperbolic problem:

$$|u_t|^\rho u_{tt} - \Delta u - \Delta u_{tt} + G(u) = f \text{ in } Q, \qquad (1)$$

$$u = 0 \text{ on } \Sigma, \qquad (2)$$

$$u(0) = u^0 \text{ and } u_t(0) = u^1 \qquad (3)$$

where $\rho \geq 2$ is a real number and $G(s)$ is a continuous function satisfying the conditions

$$0 \leq H(s) \leq C\,|s|^{\gamma+2} \text{ where } H(s) = \int_0^s G(\xi)d\xi \qquad (4)$$

$$|G(s)| \leq C\,|s|^{\gamma+1}, \gamma \in R \qquad (5)$$

such that $0 < \gamma \leq \frac{2}{n-2}$ if $n > 2$, $0 < \gamma < \infty$ if $n = 1$ or $n = 2$ and C is a positive constant.

The equation (1) is interesting not only from the point of view of the general theory of PDE's but also due to various applications in Mathematical Physics.

The most famous representative of this class is the equation

$$u'' - \Delta u - \Delta u'' = f \qquad (6)$$

which models in the study of the extensional vibrations of thin rods, see Love [1] for details. Moreover, the equation

$$u'' - \Delta u - M\left(\int_\Omega |\nabla u|^2 dx\right)\Delta u'' = f \tag{7}$$

with $M(\lambda) = \lambda_0$, $\lambda_0 = \int_\Omega \phi^2 dx$, where ϕ is the torsion-function arises in the study of the torsional vibrations of thin rods, see Love [1].

A great number of papers dealt with the equation (1) in cylindrical domains, but very few of them are devoted to weak solutions in noncylindrical domains. It seemed for us worthwhile to study this problem in the present paper.

The existence of global weak solutions to problem (1) in a cylinder domain was studied in [2]. Some papers dealt also with classical solutions to problem (7) in a cylinder, see [3], [4].

Most results on hyperbolic problems in noncylindrical domains are concerned with the existence and decay of weak solutions, see [5], [6], [7] and to equations of a variable type in noncylindrical domains, see [8].

In this paper we establish the mixed problem for (1)-(3) and obtain existence of weak solutions under the hypothesis that the domain is monotone increasing. Our proof is based upon the penalty method introduced by Lions in [5], the Faedo-Galerkin method and compactness argument with some technical idea. We observe that the uniqueness solution is open problem with respect to the problem (1)-(3).

The paper is organized as follows:

2. Notations and assumptions.
3. Existence of weak solutions.

2. Notations and Assumptions

We shall simply denote by $|u|$ the standard $L^2(\Omega)$ norm of u , several classical functional spaces, such that, the L^p and Sobolev spaces, will be used in the sequel. For their definitions and main properties we refer to Adams [9]. We denote the $H^1(\Omega)$ norm by $\|\cdot\|$ and by $a(.,.)$ the classical Dirichlet form.

Let Ω be an open and bounded subset of R^n with smooth boundary Γ, $Q \subset \Omega\times]0,T[$ an open noncylindrical domain. We will use the following notations:

$$\begin{aligned} \Omega_s &= Q \cap \{t = s\} \text{ for } s > 0,\ \Omega_0 = \operatorname{int}(\overline{Q} \cap \{t = 0\}) \\ \Gamma_s &= \partial\Omega_s\ ,\ \Sigma = \bigcup_{0 \le s < T} \Gamma_s \text{ and } \partial Q = \overline{\Omega_0} \cup \Sigma \text{ is the boundary of } Q. \end{aligned}$$

Of course, $\Omega_0 \neq \emptyset$.

Our assumptions on Q are:

$(H.1)$ Ω_t is monotone increasing, that is, $\Omega_t^* \subset \Omega_s^*$ if $t < s$, where Ω_t^* is the projection of Ω_t in the hyperplane $t = 0$.

$(H.2)$ For each $t \in]0,T[$, Ω_t has the following property of regularity: if $u \in H_0^1(\Omega)$ and $u = 0$ a.e. in $\Omega\backslash\Omega_t^*$, then the restriction of u to Ω_t^* belongs to $H_0^1(\Omega_t^*)$.

In order to simplify the notation we will identify Ω_t^* with Ω_t.

We define $L^q(0,T;L^p(\Omega_t))$ as the space of functions $w \in L^q(0,T;L^p(\Omega))$ such that $w(x,t) = 0$ a.e. in $\Omega\backslash\Omega_t$ and for almost all $t \in]0,T[$. When $1 \le q < \infty$ we consider the norm

$$\|w\|_{L^q(0,T;L^p(\Omega_t))} = \left[\int_0^T \|w(t)\|^q_{L^p(\Omega_t)}\, dt\right]^{\frac{1}{q}},$$

which agrees with $\|w\|_{L^q(0,T;L^p(\Omega))}$. For the case $q = \infty$ we consider

$$\|w\|_{L^\infty(0,T;L^p(\Omega_t))} = ess \sup_{0<t<T} \|w(t)\|_{L^p(\Omega_t)} .$$

We observe that $L^q(0,T;L^p(\Omega_t))$ is a closed subspace of $L^q(0,T;L^p(\Omega))$ for $1 \le q \le \infty$. In the same way we define $L^q(0,T;H_0^1(\Omega_t))$ as the space of functions $w \in L^q(0,T;H_0^1(\Omega))$ such that $w(x,t) = 0$ a.e. in $\Omega\backslash\Omega_t$ and for almost all in $t \in]0,T[$ with the norm:

$$\|w\|_{L^q(0,T;H_0^1(\Omega_t))} = \left[\int_0^T \|w(t)\|^q_{H_0^1(\Omega_t)}\, dt\right]^{\frac{1}{q}} , \ 1 \le q < \infty ,$$

and

$$\|w\|_{L^\infty(0,T;H_0^1(\Omega_t))} = ess \sup_{0<t<T} \|w(t)\|_{H_0^1(\Omega_t)} ,$$

it follows by $(H.2)$ that these norms agree with the norms in $L^q(0,T;H_0^1(\Omega))$ for $1 \le q \le \infty$.

We also have that $L^q(0,T;H_0^1(\Omega_t))$ is a closed subspace of $L^q(0,T;H_0^1(\Omega))$.

Finally in this section we give the following Lemma, which will be needed for the proof of an existence of weak solutions.

Lemma 2.1: Let $w : \Omega\times]0,T[\to R$ be Lebesgue measurable. Then $w(x,t) = 0$ a.e. in $\Omega\backslash\Omega_t$ for almost all $t \in]0,T[$ if and only if $w = 0$ a.e. in $\Omega\times]0,T[\backslash Q$.

Proof. See [7].

3. Existence of Weak Solutions

Theorem 3.1: Let $u^0 \in H_0^1(\Omega_0) \cap L^p(\Omega_0)$, $u^1 \in H_0^1(\Omega_0)$, $f \in L^2(0,T;L^2(\Omega_t))$ and $p = \rho + 2$. Assume that G satisfies the conditions (4)-(5). Then under the assumptions $(H.1)$ and $(H.2)$, there exists a function u defined in Q such that

$$u \in L^\infty(0,T;H_0^1(\Omega_t) \cap L^p(\Omega_t)) \tag{8}$$

$$u' \in L^\infty(0,T;H_0^1(\Omega_t) \cap L^p(\Omega_t)) \tag{9}$$

$$u'' \in L^2(0,T;H_0^1(\Omega_t)) \tag{10}$$

and for all $\theta \in D(]0,T[)$ the identity holds

$$-\frac{1}{\rho+1}\int_0^T (|u'|^\rho u', v)\theta' dt + \int_0^T a(u,v)\theta dt + \int_0^T a(u'',v)\theta dt \\ + \int_0^T (G(u),\ v)\theta dt = \int_0^T (f,\ v)\,\theta dt \tag{11}$$

where v is an arbitrary function from $H_0^1(\Omega_t) \cap L^p(\Omega_t)$ and

$$u(0) = u^0,\ u'(0) = u^1 \text{ in } \Omega_0. \tag{12}$$

Proof of Theorem 3.1. We observe that by (8), (9) and (10) the initial conditions make sense. To prove the theorem we consider $\tilde{u}^0$, $\tilde{u}^1$ extensions of u^0, u^1 by zero outside of Ω_0, and $M \in L^\infty(\Omega\times]0,T[)$ defined by

$$M(x,t) = \begin{cases} 1 \text{ in } \Omega\times]0,T[\backslash(Q \cup \Omega_0 \times \{0\}) \\ 0 \text{ in } Q \cup \Omega_0 \times \{0\}. \end{cases}$$

It is clear that $\tilde{u}^0 \in H_0^1(\Omega) \cap L^p(\Omega)$, $\tilde{u}^1 \in H_0^1(\Omega)$.

For each $\epsilon > 0$, we consider the problem:

$$\left| \begin{array}{c} |u'_\epsilon|^\rho u''_\epsilon + \epsilon u''_\epsilon - \Delta u_\epsilon - \Delta u''_\epsilon + \frac{1}{\epsilon} M u'_\epsilon + G(u_\epsilon) = f \text{ in } \Omega \times]0,T[\\ u_\epsilon(0) = \tilde{u}^0 \text{ in } \Omega \\ u'_\epsilon(0) = \tilde{u}^1 \text{ in } \Omega. \end{array} \right.$$

The proof of Theorem 3.1 will be a consequence of the following:

Lemma 3.2: For each $\epsilon > 0$ there exists a function u_ϵ, defined in $\Omega \times]0,T[$ such that:

$$u_\epsilon \in L^\infty(0,T; H_0^1(\Omega) \cap L^p(\Omega)) \tag{13}$$

$$u'_\epsilon \in L^\infty(0,T; H_0^1(\Omega) \cap L^p(\Omega)) \tag{14}$$

$$u''_\epsilon \in L^2(0,T; H_0^1(\Omega)) \tag{15}$$

and for all $\theta \in D(]0,T[)$ the identity holds

$$\begin{aligned} & -\frac{1}{\rho+1} \int_0^T (|u'_\epsilon|^\rho u'_\epsilon, v)\theta' dt + \int_0^T (\epsilon u''_\epsilon, v)\theta dt + \int_0^T a(u_\epsilon, v)\theta dt \\ & + \int_0^T a(u''_\epsilon, v)\theta dt + \int_0^T (\frac{1}{\epsilon} M u'_\epsilon, v)\theta dt + \int_0^T (G(u_\epsilon), v)\theta dt \\ = & \int_0^T (f, v)\theta dt \end{aligned} \tag{16}$$

where v is an arbitrary function from $H_0^1(\Omega) \cap L^p(\Omega)$ and

$$u_\epsilon(0) = \tilde{u}^0, \; u'_\epsilon(0) = \tilde{u}^1 \text{ in } \Omega. \tag{17}$$

Proof of Lemma 3.2. Let $(w_\nu)_{\nu \in N}$ be a basis of $H_0^1(\Omega) \cap L^p(\Omega)$ and $V_m = [w_1, ..., w_m]$ the subspace generated by the m first vectors of the basis (w_ν). For each $\epsilon > 0$, we determine the penalized approximate solutions $u_{\epsilon m} : [0, t_{\epsilon m}[\to V_m$ as solutions of the following system:

$$\begin{aligned} \Big((|u'_{\epsilon m}|^\rho + \epsilon) u''_{\epsilon m}, v\Big) + a(u_{\epsilon m}, v) + a(u''_{\epsilon m}, v) + (\frac{1}{\epsilon} M u'_{\epsilon m}, v) + (G(u_{\epsilon m}), v) \\ = (f, v), \text{ for } v \in V_m \end{aligned} \tag{18}$$

$$u_{\epsilon m}(0) = u_{0m} \to \tilde{u}^0 \text{ strongly in } H_0^1(\Omega) \cap L^p(\Omega), \; u_{0m} \in V_m \tag{19}$$

$$u'_{\epsilon m}(0) = u_{1m} \to \tilde{u}^1 \text{ strongly in } H_0^1(\Omega), \; u_{1m} \in V_m. \tag{20}$$

The system (19)-(21) has solutions in $[0, t_{\epsilon m}[$, $0 < t_{\epsilon m} < T$. The extension of the solution to the whole interval $[0,T]$ is a consequence of the first estimate.

First Estimate: In fact, setting $v = u'_{\epsilon m}$ in (18), using the Cauchy-Schwarz inequality, integrating from 0 to t, $t < t_{\epsilon m}$, after of use Gronwall's Lemma we have

$$\begin{aligned} & \frac{1}{p} \|u'_{\epsilon m}\|^p_{L^p(\Omega)} + \frac{1}{2} \left|\sqrt{\epsilon} u'_{\epsilon m}\right|^2 + \frac{1}{2} \|u_{\epsilon m}\|^2 + \frac{1}{2} \|u'_{\epsilon m}\|^2 \\ & + \frac{1}{\epsilon} \int_0^t \int_\Omega |M(u'_{\epsilon m})|^2 dx ds + \int_\Omega H(u_{\epsilon m}) dx \leq C. \end{aligned} \tag{21}$$

Here C is a positive constant independent of $m \in N$, ϵ and t.

Observe that the estimate (21) implies that there exists a subsequence of $(u_{\epsilon m})$, which we still denote by $(u_{\epsilon m})$, aand a function u_ϵ such that

$$\begin{aligned} u_{\epsilon m} &\rightarrow u_\epsilon \text{ , weakly-star in } L^\infty(0,T;H_0^1(\Omega)\cap L^p(\Omega)) \\ u'_{\epsilon m} &\rightarrow u'_\epsilon \text{ , weakly-star in } L^\infty(0,T;H_0^1(\Omega)\cap L^p(\Omega)) \\ \frac{1}{\sqrt{\epsilon}} M u'_{\epsilon m} &\rightarrow \frac{1}{\sqrt{\epsilon}} M u'_\epsilon \text{ , weakly in } L^2(0,T;L^2(\Omega)). \end{aligned} \tag{22}$$

Second Estimate: The estimate (21) is not sufficient to pass to the limits in the approximate equation. To obtain more estimates, we need, first of all, to estimate $\frac{1}{\epsilon}\int_0^t\int_\Omega(M(s)u'_{\epsilon m}(s), u''_{\epsilon m}(s))ds$ in the $L^2(\Omega)$ norm.

In fact, let

$$\phi_\mu(x,t) = \mu \int_t^{t+\frac{1}{\mu}} M(x,s)ds.$$

We can prove that $\phi_\mu \in C([0,T[;L^\infty(\Omega))$, ϕ_μ is differentiable with respect to t and

$$\frac{\partial}{\partial t}\phi_\mu(x,t) \leq 0 \text{ for } t \in [0,T[, \text{ a.e. in } \Omega,$$

being the derivative in distributional sense of ϕ_μ with respect to t agree with $\frac{\partial}{\partial t}\phi_\mu$. Moreover, we have:

$$\begin{aligned} \frac{\partial}{\partial t}\phi_\mu &\in L^\infty(0,T;L^\infty(\Omega)) \\ |\phi_\mu(x,t)| &\leq 1, \text{ for } (x,t) \in \Omega \times [0,T[\text{ and } \mu > 0 \\ \phi_\mu(x,t) &\rightarrow M(x,t), \text{ for } t \in]0,T[\text{ a.e. in } \Omega, \text{ when } \mu \rightarrow \infty \\ \int_0^t (\phi_\mu(s)w(s), w'(s))ds &= \frac{1}{2}(\phi_\mu(t)w(t), w(t)) - \frac{1}{2}(\phi_\mu(0)w(0), w(0)) \\ &\quad -\frac{1}{2}\int_0^t (\phi'_\mu(s)w(s), w(s))ds \end{aligned}$$

for $w \in L^\infty(0,T;H_0^1(\Omega))$ such that $w' \in L^\infty(0,T;L^2(\Omega))$.

When we take $w = u'_{\epsilon m}$ in the last equality and then make $\mu \rightarrow \infty$, we obtain by the above results for ϕ_μ that:

$$\int_0^t (M(s)u'_{\epsilon m}(s), u''_{\epsilon m}(s))ds \geq \frac{1}{2}\left|M(t)u'_{\epsilon m}(t)\right|^2 - \frac{1}{2}\left|M(0)u_{1m}\right|^2. \tag{23}$$

From (20) we also obtain

$$\frac{1}{\epsilon}M(0)u_{1m} \rightarrow 0 \text{ strongly in } L^2(\Omega) \text{ when } m \rightarrow \infty. \tag{24}$$

Taking $v = u''_{\epsilon m}(t)$ in (18) and integrating from 0 to t, we have

$$\begin{aligned} &\int_0^t \left[(|u'_{\epsilon m}|^\rho u''_{\epsilon m}, u''_{\epsilon m}) + \epsilon\left|u''_{\epsilon m}\right|^2 + \left\|u''_{\epsilon m}\right\|^2\right] ds + \frac{1}{2}\left\|u_{\epsilon m}\right\|^2 \\ &+\frac{1}{\epsilon}\int_0^t (M(s)u'_{\epsilon m}(s), u''_{\epsilon m}(s))ds \leq C + \frac{3\delta}{2}\int_0^t \left\|u''_{\epsilon m}(s)\right\|^2 ds. \end{aligned} \tag{25}$$

Using (23) in the last inequality we have the following estimate:

$$\int_0^t \left[(|u'_{\epsilon m}|^\rho u''_{\epsilon m}, u''_{\epsilon m}) + \epsilon |u''_{\epsilon m}|^2 + (1 - \frac{3\delta}{2}) \|u''_{\epsilon m}\|^2 \right] ds + \frac{1}{2} \|u_{\epsilon m}\|^2 \\ + \frac{1}{2\epsilon} |M(t) u'_{\epsilon m}(t)|^2 - \frac{1}{2\epsilon} |M(0) u_{1m}|^2 \leq C. \tag{26}$$

Here C is a positive constant independent of $m \in N$, ϵ and t.

Now, choosing δ such that $0 < \delta < \frac{2}{3}$, we obtain from (26) that there exists a subsequence of $(u_{\epsilon m})$, which we still denote by $(u_{\epsilon m})$, and a function u_ϵ such that

$$\begin{aligned} u''_{\epsilon m} &\to u''_\epsilon \text{ weakly in } L^2(0,T; H_0^1(\Omega)) \\ \sqrt{\epsilon} u''_{\epsilon m} &\to \sqrt{\epsilon} u''_\epsilon \text{ weakly in } L^2(0,T; L^2(\Omega)). \end{aligned} \tag{27}$$

To prove the convergence of nonlinear term of (18), first we show that they are bounded in $L^2(0,T;L^2(\Omega))$ and then using $(22)_1$, $(22)_2$, compactness arguments (see Lions [5]), Ch.I, Lemma 1.3, we conclude that:

$$G(u_{\epsilon m}) \to G(u_\epsilon) \text{ weakly in } L^2(0,T; L^2(\Omega)). \tag{28}$$

Therefore, by the same argument used to obtain (28), we conclude that

$$|u'_{\epsilon m}|^\rho u'_{\epsilon m} \to |u'_\epsilon|^\rho u'_\epsilon \text{ weakly in } L^{p'}(\Omega \times]0,T[) \text{ where } \frac{1}{p} + \frac{1}{p'} = 1. \tag{29}$$

Now , the next step is the limit passage. But, once these latter convergence have been established, there is a standard procedure.

Proof of Theorem 3.1: Observe that the estimates early obtained are independent of ϵ. Therefore, by the same argument used to obtain u_ϵ from $u_{\epsilon m}$, which is the solution of (18)-(20) we can pass to the limit when ϵ goes to zero in u_ϵ, or a subsequence, obtaining a function u independent of ϵ and m, such that:

$$\left| \begin{array}{c} u_\epsilon \to u \text{ weakly-star in } L^\infty(0,T; H_0^1(\Omega) \cap L^p(\Omega)) \\ u'_\epsilon \to u' \text{ weakly-star in } L^\infty(0,T; H_0^1(\Omega) \cap L^p(\Omega)) \\ u''_\epsilon \to u'' \text{ weakly in } L^2(0,T; H_0^1(\Omega)) \\ \sqrt{\epsilon} u''_\epsilon \to 0 \text{ weakly in } L^2(0,T; L^2(\Omega)) \\ G(u_\epsilon) \to G(u) \text{ weakly in } L^2(0,T; L^2(\Omega)) \\ |u'_\epsilon|^\rho u'_\epsilon \to |u'|^\rho u' \text{ weakly in } L^{p'}(0,T; L^{p'}(\Omega)) \text{ where } \frac{1}{p} + \frac{1}{p'} = 1. \end{array} \right. \tag{30}$$

From (22) we obtain that

$$\int_0^T \int_\Omega |M(u'_\epsilon)|^2 \, dx dt \leq C\epsilon,$$

therefore,

$$M u'_\epsilon \to 0 \text{ strongly in } L^2(0,T; L^2(\Omega)) \tag{31}$$

as $\epsilon \to 0$.

Using $(30)_1$, $(30)_3$ and compactness arguments Lions [5] we have

$$u'_\epsilon \to u' \text{ strongly in } L^2(0,T; L^2(\Omega))$$

and therefore

$$Mu'_\epsilon \to Mu' \text{ strongly in } L^2(0,T;L^2(\Omega)). \tag{32}$$

From (31) and (32) we conclude that

$$Mu' = 0 \text{ a.e. in } \Omega\times]0,T[. \tag{33}$$

But, $M = 1$ in $\Omega\times]0,T[\backslash Q$, therefore $u' = 0$ a.e. in $\Omega\times]0,T[\backslash Q$. Applying the Lemma 2.1, we obtain that $u' = 0$ a.e. in $\Omega\backslash\Omega_t$ for almost all $t \in]0,T[$. As $u' \in L^\infty(0,T;H_0^1(\Omega)\cap L^p(\Omega))$ we have that

$$u' \in L^\infty(0,T;H_0^1(\Omega_t)\cap L^p(\Omega_t)).$$

By the same argument it can be shown (8) and (10).

Now, we consider the equation of Lemma 3.2:

$$\begin{aligned} -\frac{1}{\rho+1}\int_0^T(|u'_\epsilon|^\rho u'_\epsilon, v)\theta' dt + \int_0^T(\epsilon u''_\epsilon, v)\theta dt + \int_0^T a(u_\epsilon, v)\theta dt \\ -\int_0^T a(u''_\epsilon, v)\theta dt + \int_0^T (G(u_\epsilon), v)\theta dt + \int_0^T(\frac{1}{\epsilon}Mu'_\epsilon, v)\theta dt \\ = \int_0^T (f,v)\theta dt \text{ ,} \forall v \in H_0^1(\Omega)\cap L^p(\Omega) \text{ and } \forall\theta \in D(]0,T[). \end{aligned} \tag{34}$$

Therefore, restricting the equation (34) on Q, we obtain by (30) and by definition of M, letting $\epsilon \to 0$, the identity (11). In order to check up that $u(0) = u^0$ it is sufficient to use $(30)_1$ and $(30)_2$ followed by an integration by parts.

By the same argument, using $(33)_3$, it can be shown that $u'(0) = u^1$. Thus the proof of Theorem 3.1 is completed.

Acknowledgements

The first author (J. Ferreira) is partially supported by CAPES-Brasília/Brazil under grant BEX2480/95-6 when him visited the University of Pisa, Italy, in a Post-Doctoral Program during 1996/1997. The second author (M.A. Rojas-Medar) is supported by UEM-Brazil as a Visiting Professor at the State University of Maringá, a grant 300116/93-4 from CNPq-Brazil and FAPESP-Brazil under grant 1997/3711-0.

References

1. A.H. Love, A Treatise on the Mathematical Theory of Elasticity, Dover, New York (1944).

2. J. Ferreira and D.C. Pereira, Bol. Soc. Par. Mat., **11**, 70-90 (1990).

3. Y. Ebihara and D.C. Pereira, Int. J. Math. and Math. Sci., **12**, 29-38 (1989).

4. J.E.M. Rivera, Int. J. Math. and Math. Sci., **14**, 731-736 (1991).

5. J.L. Lions, Quelques Métodes de Résolution des Problèmes aux Limites non Linéaires, Dunod, Paris (1969).

6. M. Nakao and T. Narazaki, Math. Rep., **XI-2**, 117-161 (1987).

7. T.N. Rabello, Int. J. Math. and Math. Sci., **17**, 561-570 (1994).

8. J. Ferreira, Rend. Cir. Mat. Palermo, **II, XLIV**, 135-146 (1995).

9. R. Adams, Sobolev Spaces, Acad. Press, New York (1975).

10. J. Ferreira and N.A. Lar'kin, Port. Mat., **53**, 381-395 (1996).

11. J. Ferreira and D.C. Pereira, Int. J. Math. and Math. Sci., **15**, 543-552 (1992).

12. L.A. Medeiros, Arch. Rational Mech. Anal., 47-58 (1972).

9th Int. Coll. on Differential Equations, pp. 163–168
D. Bainov (Ed)

Lifting Theory and its Application to Quantum Amplifier Processes

SHIGERU FURUICHI AND KUNIO OSHIMA
Science University of Tokyo in Yamaguchi, Japan
ICHIRO HOFUKU
Tokyo Metropolitan College of Technology, Japan

Abstract: A state change on physical processes is generally described by a concept of channel, especially quantum mechanical channel in quantum system. However, in many cases it is difficult to formulate this quantum mechanical channel. In such cases the mathematical concept "lifting" is useful to get the quantum mechanical channel. In this paper, we will review how to formulate the quantum mechanical channel for two cases through the concept of lifting.
Key Words and Phrases: *quantum mechanical channel, lifting, Jaynes-Cummings model, quantum amplifier processes.*

1. Introduction

In mathematical physics, the dynamics of system is described by the concept of state [1]. This concept of state is a kind of the mathematical representation, for instance, there are vector, probability distribution and density operator. In the course of information transmission or more generally, in some physical processes, an initial state changes to another state called final state under the external or internal effect of the system. Therefore, to investigate this state change from an initial state to a final state is one of the main method in natural science. Especially, in communication theory the study of the state change is indispensable and this state change is described by channeling transformation.

In quantum system, some entropy-type functions such as von Neumann entropy and quantum mutual entropy [2] are defined for a quantum state, density operator which is a special case of trace class operator, so we should describe physical processes as the state change. For this end, the concept of "quantum mechanical channel" was introduced in [2]. This quantum mechanical channel can uniformly treat many physical transformations as a state change. But in many cases, it is difficult to formulate the quantum mechanical channel. In such cases, the concept of lifting [3] is very useful. By using the concept of lifting, we can formulate the quantum mechanical channel, and we can describe various physical processes such as optical communication processes and open system model and so on. In this paper, we will apply this lifting method to Jaynes-Cummings model and quantum amplifier processes in order to formulate the quantum mechanical channel.

This paper is organized by the following content. In section 2, the mathematical concepts, "quantum mechanical channel and lifting" in quantum information theory are briefly reviewed. In section 3, the lifting expression for Jaynes-Cummings model is shown [4]. In section 4, the lifting expression for quantum amplifier processes is shown [5].

2. Quantum Mechanical Channel and Lifting

A quantum mechanical channel expresses a state change from an input system to an output one, corresponding to a channel in classical Shannon theory. Now, we will construct a quantum mechanical channel. Let $\mathcal{H}_k$ be separable complex Hilbert spaces. And let $B(\mathcal{H}_k)$ be the set of all bounded linear operators on the Hilbert spaces $\mathcal{H}_k$;

$$B(\mathcal{H}_k) \equiv \{A; \|Ax\| \leq M \|x\| (\exists M \geq 0, \forall x \in \mathcal{H}_k)\} \tag{2.1}$$

and let the set of all density operators on the Hilbert spaces $\mathcal{H}_k$ be

$$\mathfrak{S}(\mathcal{H}_k) \equiv \{\rho \in B(\mathcal{H}_k); \rho = \rho^*, \rho \geq 0, tr\rho = 1\}. \tag{2.2}$$

In quantum system an element of $B(\mathcal{H}_k)$ is considered as an observable and an element of $\mathfrak{S}(\mathcal{H}_k)$ is considered as a state of the system. The mathematical definition of the quantum mechanical channel is as follows:

Definition 2.1 *[2, 6] A mapping $\Lambda^* : \mathfrak{S}(\mathcal{H}_1) \longrightarrow \mathfrak{S}(\mathcal{H}_2)$ is called a completely positive channel (channel for short) if its dual map $\Lambda : B(\mathcal{H}_2) \longrightarrow B(\mathcal{H}_1)$ satisfies the completely positivity:*

$$\sum_{i,j}^{n} B_i^* \Lambda(A_i^* A_j) B_j \geq 0 \quad for \quad {}^\forall B_i \in B(H_1), {}^\forall A_j \in B(H_2), {}^\forall n \in N. \tag{2.3}$$

Here the dual map of Λ is defined as

$$tr\Lambda^*(\rho) A = tr\rho\Lambda(A) \quad for \quad {}^\forall \rho \in \mathfrak{S}(\mathcal{H}_1), {}^\forall A \in B(\mathcal{H}_2). \tag{2.4}$$

It is well known that most of physical transformations satisfy the above condition [2]. The concept of "channel" is deeply related to that of "lifting" introduced in [3].

Definition 2.2 *Let $\mathfrak{S}_0, \mathfrak{S}_1$ be the set of all states on $\mathcal{H}_0, \mathcal{H}_1$ respectively. A lifting from $\mathfrak{S}_0$ to $\mathfrak{S}_0 \otimes \mathfrak{S}_1$ is a continuous map*

$$\mathcal{E}^* : \mathfrak{S}_0 \to \mathfrak{S}_0 \otimes \mathfrak{S}_1. \tag{2.5}$$

The concept "lifting" has been introduced as a common expression of the compound state and transition expectation and it generalizes some physical processes such as reduction dynamics [7]. Moreover, once we can obtain the lifting expression, we can get a channel in the following way :

$$\Lambda^*\rho = tr_{\mathcal{K}}\mathcal{E}^*\rho \tag{2.6}$$

where $tr_{\mathcal{K}}$ is the partial trace;

$$\langle\Phi_1, tr_{\mathcal{K}}\, Q\, \Phi_2\rangle = \sum_n \langle\Phi_1 \otimes \Psi_n, Q\, \Phi_2 \otimes \Psi_n\rangle, \quad \text{for} \quad {}^\forall Q \in B(\mathcal{H} \otimes \mathcal{K}) \tag{2.7}$$

where Φ_1, Φ_2 are arbitarily elements of the Hilbert space $\mathcal{H}$ and $\{\Psi_n\}$ is a CONS of the Hilbert space $\mathcal{K}$.

Example 2.3 *[5] (Lifting expression for quantum attenuation processes)*
The mapping of quantum attenuation processes is given by an isometric operator, $V : \mathcal{H} \longrightarrow \mathcal{H} \otimes \mathcal{K}$ such that;

$$V|\theta\rangle \equiv |\alpha\theta\rangle \otimes |\beta\theta\rangle.$$

Then lifting $\mathcal{E}^* : \mathfrak{S}(\mathcal{H}) \to \mathfrak{S}(\mathcal{H} \otimes \mathcal{K})$, and the channel $\Lambda^* : \mathfrak{S}(\mathcal{H}) \to \mathfrak{S}(\mathcal{K})$ are given as follows:

$$\mathcal{E}^*\rho = V\rho V^*.$$
$$\Lambda^*\rho = tr_{\mathcal{K}}\mathcal{E}^*\rho.$$

3. Lifting Expression for Jaynes-Cummings Model

The resonant Jaynes-Cummings model [8, 9, 10] Hamiltonian can be expressed by using rotating-wave approximation in the following form:

$$H = H_A + H_F + H_I,\ H_A = \frac{1}{2}\hbar\omega_0\sigma_z\ H_F = \hbar\omega_0 a^* a,\ H_I = \hbar g(a \otimes \sigma^+ + a^* \otimes \sigma^-), \quad (3.1)$$

where g is a coupling constant, $\sigma^\pm$ are the pseudo-spin operators of two-level atom, σ_z is the Pauli spin operator, a is the annihilation operator of a photon and a^* is the creation operator of a photon.

In order to derive information quantities such as von Neumann and quantum mutual entropy, we have to give the quantum mechanical channel for the JCM. Now, we suppose that the Hilbert spaces $\mathcal{H}_A$ and $\mathcal{H}_F$ describe an atomic system and a field system, respectively. We also suppose that the initial states of $\mathcal{H}_A$ and $\mathcal{H}_F$ are represented by $\rho \in \mathfrak{S}_A$ and $\omega \in \mathfrak{S}_F$, respectively. We then suppose that the initial state of the atom is a superposition of the upper level and the lower level:

$$\rho = \lambda_0 E_0 + \lambda_1 E_1 \in \mathfrak{S}_A, \quad (3.2)$$

where, $E_0 = |1\rangle\langle 1|$, $E_1 = |2\rangle\langle 2|$, $\lambda_0 + \lambda_1 = 1$. We also suppose that the field state is the coherent state:

$$\omega = |\theta\rangle\langle\theta| \in \mathfrak{S}_F,\ |\theta\rangle = \exp\left[-\frac{1}{2}|\theta|^2\right]\sum_l \frac{\theta^l}{\sqrt{l!}}|l\rangle. \quad (3.3)$$

Here we set the lifting $\mathcal{E}_t^*$ and the *time-dependent* quantum mechanical channel Λ_t^* which describes the time evolution of the atom for the JCM as follows:

$$\begin{aligned} \Lambda_t^* &: \mathfrak{S}_A \longrightarrow \mathfrak{S}_A, \\ \mathcal{E}_t^* &: \mathfrak{S}_A \longrightarrow \mathfrak{S}_A \otimes \mathfrak{S}_F. \end{aligned}$$

Then, the lifting is given by

$$\mathcal{E}_t^*\rho = U_t(\rho \otimes \omega)U_t^*,$$

which describes the final state $\mathcal{E}_t^*\rho \in \mathfrak{S}_A \otimes \mathfrak{S}_F$ after the interaction between the atom and the photons at the time t. The quantum mechanical channel Λ_t^* is written by this litfing $\mathcal{E}_t^*$ as follows:

$$\Lambda_t^*\rho = tr_{\mathcal{H}_F}\mathcal{E}_t^*\rho = tr_{\mathcal{H}_F}U_t(\rho \otimes \omega)U_t^*,$$

which represents the final state of the atom at the time t. From the Hamiltonian form in (3.1), the following communication relation holds [11]:

$$[H_A + H_F, H_I] = 0. \quad (3.4)$$

Therefore the time evolution of the system is determined by

$$U_t = \exp(-it(H_A + H_F)/\hbar)\exp(-itH_I/\hbar).$$

Take a *dressed state*

$$\left|\Phi_j^{(n)}\right\rangle = \frac{1}{\sqrt{2}}\left(|2 \otimes n\rangle + (-1)^j|1 \otimes n+1\rangle\right), (j = 0, 1).$$

Then the following eigen-equation holds for $j = 0, 1$

$$(H_A + H_F)\left|\Phi_j^{(n)}\right\rangle = \hbar\omega_0\left(n + \frac{1}{2}\right)\left|\Phi_j^{(n)}\right\rangle. \quad (3.5)$$

Moreover, since we have

$$\left\langle \Phi_0^{(n)} \middle| \Phi_1^{(n)} \right\rangle = \left\langle \Phi_1^{(n)} \middle| \Phi_0^{(n)} \right\rangle = 0, \left\| \Phi_j^{(n)} \right\| = 1 \ (j = 0, 1), \tag{3.6}$$

(3.5) can be written as

$$H_A + H_F = \sum_{n=0}^{\infty} \left(H_A^{(n)} + H_F^{(n)} \right) = \sum_{n=0}^{\infty} \sum_{j=0}^{1} \hbar\omega_0 \left(n + \frac{1}{2} \right) \left| \Phi_j^{(n)} \right\rangle \left\langle \Phi_j^{(n)} \right|. \tag{3.7}$$

Therefore,

$$\exp\left(-it\left(H_A + H_F\right)/\hbar\right) = \sum_{n=0}^{\infty} \sum_{j=0}^{1} \exp\left[-it\omega_0 \left(n + \frac{1}{2} \right) \right] \left| \Phi_j^{(n)} \right\rangle \left\langle \Phi_j^{(n)} \right|. \tag{3.8}$$

Also the following eigen-equation holds for $j = 0, 1$

$$H_I \left| \Phi_j^{(n)} \right\rangle = (-1)^j \hbar\Omega \left| \Phi_j^{(n)} \right\rangle, \tag{3.9}$$

where $\Omega = g\sqrt{n+1}$ is called the Rabi frequency. From (3.6), (3.9) can be written as

$$H_I = \sum_{n=0}^{\infty} H_I^{(n)} = \sum_{n=0}^{\infty} \sum_{j=0}^{1} (-1)^j \hbar\Omega \left| \Phi_j^{(n)} \right\rangle \left\langle \Phi_j^{(n)} \right|, \tag{3.10}$$

so that

$$\exp\left(-itH_I/\hbar\right) = \sum_{n=0}^{\infty} \sum_{j=0}^{1} \exp\left[-it(-1)^j \Omega \right] \left| \Phi_j^{(n)} \right\rangle \left\langle \Phi_j^{(n)} \right|. \tag{3.11}$$

From (3.8) and (3.11), we obtain

$$U_t = \exp\left(-itH/\hbar\right) = \sum_{n=0}^{\infty} \sum_{j=0}^{1} E_{n,j} \left| \Phi_j^{(n)} \right\rangle \left\langle \Phi_j^{(n)} \right|, \tag{3.12}$$

where $E_{n,j} = \exp\left[-it \left\{ \omega_0 \left(n + \frac{1}{2} \right) + (-1)^j \Omega \right\} \right]$. From this unitary operator, the final state of the atom of the JCM is given by

$$\begin{aligned} \Lambda_t^* \rho &= tr_{\mathcal{H}_F} \mathcal{E}_t^* \rho = tr_{\mathcal{H}_F} U_t \left(\rho \otimes \omega \right) U_t^* \\ &= \sum_{m,n=0}^{\infty} \sum_{i,j=0}^{1} E_{n,j} E_{m,i}^* \left\langle \Phi_j^{(n)} \middle| \rho \otimes \omega \middle| \Phi_i^{(m)} \right\rangle tr_{\mathcal{H}_F} \left| \Phi_j^{(n)} \right\rangle \left\langle \Phi_i^{(m)} \right|. \end{aligned}$$

Moreover, from some simple computations, we obtain the quantum mechanical channel such as

$$\Lambda_t^* \rho = \left(\lambda_0 \tilde{c}_1(t) + \lambda_1 \tilde{s}_0(t) \right) E_0 + \left(\lambda_0 \tilde{s}_1(t) + \lambda_1 \tilde{c}_0(t) \right) E_1, \tag{3.13}$$

where, for $i = 0, 1$, $\tilde{c}_i(t)$ and $\tilde{s}_i(t)$ are given by

$$\tilde{c}_i(t) = \sum_n P(n+i) \cos^2 \Omega t, \ \tilde{s}_i(t) = \sum_n P(n+i) \sin^2 \Omega t, \tag{3.14}$$

where, $P(n)$ is the Poisson distribution

$$P(n) = \exp\left[-|\theta|^2 \right] \frac{|\theta|^{2n}}{n!}.$$

This final state is a reduced state for the atom after the interaction with the field. By using the above quantum mechanical channel, we can compute some information quantities such as von Neumann entropy and quantum mutual entropy, and then these quantities can give a new interpretation to Jaynes-Cummings model from the information theoretical point of view [4, 12].

4. Lifting Expression for Quantum Amplifier Processes

In this section, a procedure to get a lifting expression for a quantum amplifier process is briefly discussed.

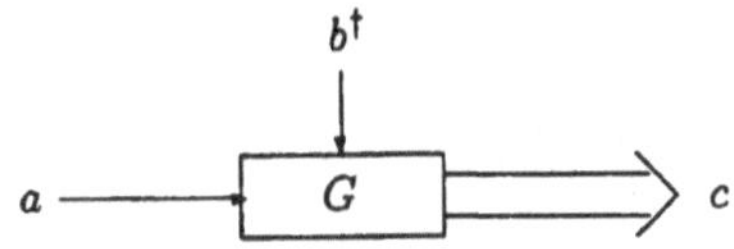

Fig.1 Quantum Amplifier Process

In many references on quantum optics [5, 11], the expression on quantum amplifier process is given by a change of operator as follows:

$$c = \sqrt{G}a + \sqrt{G-1}b^{\dagger} \tag{4.1}$$

where a is an annihilation operator on Hilbert space $\mathcal{H}$ and $b^{\dagger}$ is a creation operator on another Hilbert space $\mathcal{K}$. G is the gain of this amplifier.

More precisely, (4.1) is written by the annihilation operator c on tensor product Hilbert space $\mathcal{H} \otimes \mathcal{K}$ as

$$c = \mu a \otimes 1 + \nu 1 \otimes b^{\dagger} \tag{4.2}$$

where $\mu = \sqrt{G}, \nu = \sqrt{G-1}, (G > 1)$. These three operators a, b and c represent an annihilation operators of an input system, a control system and an output system in quantum amplifier processes, respectively. We have to solve the following eigen-equation, to see the state change due to the above operator c;

$$c\,|\gamma\rangle = \gamma\,|\gamma\rangle \tag{4.3}$$

where $|\gamma\rangle$ is an eigenvector of c on a tensor Hilbert space $\mathcal{H} \otimes \mathcal{K}$. We obtain the following inner product $\langle \alpha \otimes \beta \mid \gamma \rangle$ from this eigen-equation;

$$\langle \alpha \otimes \beta \mid \gamma \rangle = \frac{1}{\mu} \exp\left[-\frac{|\alpha|^2}{2} - \frac{|\beta|^2}{2} - \frac{|\gamma|^2}{2} - \frac{\nu}{\mu}\alpha^{*}\beta^{*} + \frac{1}{\mu}\alpha^{*}\gamma \right] \tag{4.4}$$

where $|\alpha\rangle$ and $|\beta\rangle$ are coherent state vectors in $\mathcal{H}$ and $\mathcal{K}$, respectively. Here a coherent state vector is the eigenvector of an annihilation operator, namely;

$$a|\alpha\rangle = \alpha|\alpha\rangle, b|\beta\rangle = \beta|\beta\rangle.$$

In the process to obtain the above inner product, we apply the following formula;

$$\frac{1}{\pi}\int d^2w \exp\{-\mid w \mid^2 + aw + bw^{*} + cw^2 + dw^{*2}\} = \frac{1}{\sqrt{1-4cd}} \exp\left\{\frac{a^2d + ab + b^2c}{1-4cd}\right\}. \tag{4.5}$$

See [5] for the detailed derivation of (4.4).

Then a quantum expression for a linear amplifier can be written as :

$$V_G\,|\alpha\rangle = |\gamma\rangle\,(= |\alpha \otimes \beta; \mu, \nu\rangle)\,. \tag{4.6}$$

Therefore, the lifting $\mathcal{E}^*$ and the quantum mechanical channel Λ^* are given by the followings;

$$\begin{aligned} \mathcal{E}^*\rho &= V_G \rho V_G^* \\ \Lambda^*\rho &\equiv tr_{\mathcal{K}} \mathcal{E}^*\rho. \end{aligned} \tag{4.7}$$

This expression was applied to the rigorous derivation of some efficiencies such as error probability [13] and SNR [5], quantum mutual entropy [14] and quantum capacity [15].

5. Conclusion

As we have seen, we concretely gave the quantum mechanical channel which represents a state change for Jaynes-Cummings model and quantum amplifier processes through the concept of lifting.

We are now trying to uniformly treat quantum information by the concept of lifting. For physical phenomena such as communication processes and irreversible processes, giving the lifting expression, we can obtain the quantum mechanical channel which describes the state change. For this channel, we compute the information quantities such as von Neumann entropy and quantum mutual entropy, and then we would like to give a new interpretation form the information theoretical point of view.

References

1. M. Ohya and D. Petz, Quantum Entropy and its Use, Springer-Verlag (1993).
2. M. Ohya, IEEE Transactions of Information Theory, **IT-29**, 770-774 (1983).
3. L. Accardi and M. Ohya, J. Appl. Math. Optim., (in press).
4. S. Furuichi and M. Ohya, Quantum Mutual Entropy for Jaynes-Cummings Model, (preprint).
5. M. Ohya and H. Suyari, Rep. Math. Phys., **36**, 403-420 (1995).
6. M. Ohya, Rep. Math. Phys., **27**, 19-47 (1989).
7. E.B. Davies, Quantum Theory of Open System, Academic Press (1976).
8. E.T. Jaynes and F.W. Cummings, Proc. IEEE, **51**, 89 (1963).
9. H.-I. Yoo and J.H. Eberly, Phys. Rep., **118**, 239-337 (1985).
10. B.W. Shore and P.L. Knight, Journal of Modern Optics, **40**, 1195-1238 (1993).
11. D.F. Walls and G.J. Milburn, Quantum Optics, Springer-Verlag (1994).
12. S. Furuichi, H. Suyari, K. Oshima and M. Ohya, Japanese Journal, (in press).
13. M. Ohya, H. Suyari and S. Furuichi, Electronics and Communications in Japan, **81**, 809-817 (1998).
14. S. Furuichi, M. Ohya and H. Suyari, in: Computation of Mutual Entropy in Quantum Amplifer Processes, Quantum Communication and Measurement, Hirota et al. (Ed.), Plenum Press, New York (1997) 147–155.
15. S. Furuichi, M. Ohya and K. Oshima, Journal of Nonlinear Analysis B., (in press).

9th Int. Coll. on Differential Equations, pp. 169-176
D. Bainov (Ed)

Tangential Oblique Derivative Problem for Nonlinear Elliptic Discontinuous Operators in the Plane

SOFIA GIUFFRÈ
Department of Mathematics, University of Catania,
Viale A. Doria 6, 95125, Catania, Italy

Abstract: Strong solvability in Sobolev space $W^{2,q}(\Omega)$ is proved for the regular oblique derivative problem for a class of nonlinear discontinuous differential operators in the plane with quadratic growth. We assume the nonlinear operator to satisfy Carathèodory's condition and a suitable ellipticity condition. Moreover it is allowed the field l to be tangential to the boundary at a finite number of points.
Key Words and Phrases: *Nonlinear Discontiuous Elliptic Operators, Oblique Derivative Problem.*

1. Introduction

Let $\mathcal{A}(x, u, Du, H(u))$ be a Carathèodory's function and satisfy an ellipticity condition (see (A)). Our aim is to study the following regular oblique derivative problem

$$(1.1) \qquad \begin{cases} \mathcal{A}(x,u,Du,H(u)) = f(x,u,Du) & \text{a.e. } in\ \Omega \subset R^2 \\ \frac{\partial u}{\partial l} + \sigma\, u = 0 & \text{on } \partial\Omega \end{cases}$$

where the number σ is greater than zero and l is a unit vector field C^1-piecewice on $\partial\Omega$. Under some geometrical conditions we will prove that there exists a number $\tilde{q} > 2$, such that, for all $q \in]2, \tilde{q}[$, the problem (1.1) admits solution belonging to the Sobolev space $W^{2,q}(\Omega)$.

The oblique derivative problem has been very well studied in the case of sufficiently smooth Hölder continuous data. In the case of discontinuous operators the oblique derivative problem is less studied. The linear planar problem $\mathcal{L}u = \sum_{i,j=1}^{n} a_{ij}(x)\frac{\partial^2 u}{\partial x_i \partial x_j}$ has been well studied by G.Talenti ([1]), who established a $W^{2,2}(\Omega)$ solvability, assuming $a_{ij}(x)$ to be only bounded and measurable functions. In the multidimensional case the strong solvability of the linear oblique derivative problem for operator $\mathcal{L}u = \sum_{i,j=1}^{n} a_{ij}(x)\frac{\partial^2 u}{\partial x_i \partial x_j} + \sum_{i=1}^{n} b_i(x)\frac{\partial u}{\partial x_i} + c(x)u$ is studied by C.Miranda ([2]), M.Chicco ([3]), G.Viola ([4]) if $a_{ij} \in W^{1,n}(\Omega)$ and by M.Chicco ([5]) and F.Nicolosi ([6]) if a_{ij} are measurable and satisfy the Cordes condition.

Recently, these results has been improved by G.Di Fazio and D.K.Palagachev ([7]), who assumed the coefficients of the elliptic differential operator to belong to the space VMO of

functions with vanishing mean oscillations (infact $a_{ij} \in C^0(\overline{\Omega})$, $a_{ij} \in W^{1,n}(\Omega)$ imply $a_{ij} \in VMO$). Their results have been successfully applied to the study of oblique derivative problem for quasilinear elliptic operators with VMO principal coefficients and general boundary operator (see [8] and [9]).
On the other hand recently the planar oblique derivative problem has been studied for discontinuous nonlinear operator $\mathcal{A}(x,\xi) : \Omega \times R^4 \to R$ and linear free term $f(x)$ by S.Giuffrè ([10], [11], [12]), under the above-mentioned ellipticity condition (A) for the Carathèodory function $\mathcal{A}(x,\xi)$.

2. Main Results

Let $\Omega \subset R^2$ be a bounded, convex and open set with C^2-smooth boundary $\partial\Omega$. We assume that $\partial\Omega$ is a closed curve and let

$$\begin{cases} x_1 = x_1(\varphi), \\ x_2 = x_2(\varphi) \end{cases}$$

be the normal parameterization of $\partial\Omega$, with φ being a curvilinear parameter, $\varphi \in [0, L]$. On the boundary $\partial\Omega$ is defined a unit vector field $l = (Y_1(\varphi), Y_2(\varphi))$, such that $Y_1(L) = Y_1(0)$; $Y_2(L) = Y_2(0)$. Setting $n = (X_1, X_2)$ for the unit outward normal to $\partial\Omega$, and denoting by θ the angle between the normal n and the vector l, we assume that

$$\cos\theta = \sum_{i=1}^{2} X_i(\varphi)Y_i(\varphi) \geq 0 \tag{2.1}$$

with $\cos\theta = 0$ at a finite number of points $\varphi_j \in]0, L[$, $j = 1, .., n$, with $\varphi_1 < \varphi_2 < ... < \varphi_n$.
Futhermore we suppose

$$\begin{cases} Y_i(\varphi) \in C^1([\varphi_{j-1}, \varphi_j]), \quad i = 1, 2, \ j = 1, .., n. \\ \lim_{\varphi \to \varphi_j^-} Y_i(\varphi) = -\lim_{\varphi \to \varphi_j^+} Y_i(\varphi). \end{cases} \tag{2.2}$$

If $u(x)$ is a real valued function defined on Ω, we denote by $H(u) = \{D_{ij}u\}_{i,j=1,2}$ its Hessian matrix. We are aimed at the investigation of the strong solvability of the regular oblique derivative problem

$$\begin{cases} \mathcal{A}(x, u, Du, H(u)) = f(x, u, Du) & \text{a.e. } in\ \Omega \\ \frac{\partial u}{\partial l} + \sigma\, u = 0 & \text{on } \partial\Omega \end{cases} \tag{2.3}$$

where the number σ is greater than zero.
As usual by strong solution $u \in W^{2,q}(\Omega)$ we mean a twice weakly differentiable function with L^q summable derivatives. Troughout the paper we shall suppose that the functions $\mathcal{A}(x, z, p, \xi)$, $f(x, z, p)$ are real valued functions which satisfy Carathèodory's condition, i.e. they are measurable in $x \in \Omega$ for all $z \in R$, $p \in R^2$, $\xi \in R^4$ and continuous in the other variables for almost all $x \in \Omega$. Denoting by κ the mean curvature of $\partial\Omega$, we impose the following requirements:
the function $\mathcal{A}(x, z, p, \xi)$ satisfies the condition
(A) there exist α, γ, $\delta > 0$, $\gamma + \delta < 1$, such that, for almost all $x \in \Omega$, for all $z \in R$, $p \in R^2$, ξ, $\tau \in R^4$, one has

$$\left|\sum_{i=1}^{2} \xi_{ii} - \alpha\left[\mathcal{A}(x, z, p, \xi + \tau) - \mathcal{A}(x, z, p, \tau)\right]\right| \leq \gamma \|\xi\| + \delta \left|\sum_{i=1}^{2} \xi_{ii}\right|$$

$$\mathcal{A}(x,z,p,0)=0;$$

moreover

$$(B)\qquad \begin{cases} \kappa \le \dfrac{d\theta}{d\varphi} \le 0, \quad \forall\varphi \in [\varphi_{j-1},\varphi_j], \ \ j=1,\dots,n \\ \dfrac{d\theta(\varphi_j^-)}{d\varphi} < 0; \ \ \dfrac{d\theta(\varphi_j^+)}{d\varphi} < 0. \end{cases}$$

As it concerns the function $f(x,z,p)$ we assume

$$(2.4)\qquad |f(x,z,p)| \le f_1(|z|)\left[f_2(x)+|p|^2\right]$$

for almost all $x \in \Omega$, for all $(z,p) \in R \times R^2$, where $f_1 \in C^0(R^+)$ is positive, monotone nondecreasing function and $f_2 \in L^r(\Omega)$, $r > 2$, is positive and

$$(2.5)\qquad (sign\ z) f(x,z,p) \le 2\sqrt{\frac{g(x)\det|a_{ij}|}{h(p)}}$$

for a.a. $x \in \Omega$, $|z| \ge M = const > 0$, $p \in R^2$, where $a_{ij}(x,z,p,\xi) = \int_0^1 \frac{\partial\mathcal{A}}{\partial\xi_{ij}}(x,z,p,s\xi)ds \in L^\infty(\Omega \times R \times R^2 \times R^4)$, $g(x) \in L^1(\Omega)$ and $h \in L^1_{loc}(R^2)$ are positive functions such that $\int_\Omega g(x)dx < \int_{R^2} h(p)dp$. Let us note that Campanato's (A) condition ensures that the derivatives $\frac{\partial\mathcal{A}}{\partial\xi_{ij}}(x,z,p,\xi)$ exist almost everywhere and they are essentially bounded. We are in position to formulate our existence result.

Theorem 2.1. *Let the conditions* A), B), (2.1), (2.2), (2.4), (2.5) *be fulfilled. Then there exists a number* $\bar{q} > 2$, *such that, for each* $q \in]2,\bar{q}[$ *and for each* $\sigma > 0$, *the problem* (2.3) *is solvable in the Sobolev space* $W^{2,q}(\Omega)$.

Remark 2.1. As it is known from the classical theory (cf. [13]), the linear oblique derivative problem in two dimensions is always a non-degenerate one if the field l is nowhere zero. That is, the problem has a finite index, i.e., finite dimensional kernel and cokernel. On the other hand in our situation, since the vector field l is tangential to $\partial\Omega$ at a finite number of points, the kernel of the linear problem is nontrivial. This difficulty is avoidable by prescribing the values of the solution u at the points of tangency (cf. [13]). This way a new problem arises having a trivial kernel and therefore solvable for any data. Concerning our problem, from condition (2.2) we get

$$\lim_{\varphi\to\varphi_j^-}\left(\frac{\partial u}{\partial l}+\sigma u\right) = A + \sigma u(\varphi_j) = 0$$

$$\lim_{\varphi\to\varphi_j^+}\left(\frac{\partial u}{\partial l}+\sigma u\right) = -A + \sigma u(\varphi_j) = 0.$$

Then it follows

$$(2.6)\qquad A = \frac{\partial u}{\partial\varphi}(\varphi_j) = 0, \quad u(\varphi_j) = 0.$$

Therefore, the values of u at the points of tangency are known from the boundary condition and, hence, it follows the uniquely solvability of the problem for the Laplacian with the boundary condition as in (2.3), that we will use in some stage of the proof.

3. Preliminary Results

We recall here some auxiliary results, which will be useful in that follows. Let us start with the following Miranda-Talenti type estimate.

Lemma 3.1. *Under the assumptions* (2.1), (2.2) and B), *for every* $u \in C^2(\overline{\Omega})$ *such that* $\frac{\partial u}{\partial l} + \sigma u = 0$ *on* $\partial\Omega$, *it results*

$$\|H(u)\|_{L^2(\Omega)} \le \|\Delta u\|_{L^2(\Omega)}\,. \tag{3.1}$$

Proof. We will prove the Lemma supposing $\cos\theta = 0$ in the unique point $\varphi = \varphi_0$. Let us set $p_i = \frac{\partial u}{\partial x_i}$, $i = 1,2$, $p_{ij} = \frac{\partial^2 u}{\partial x_i \partial x_j}$, $i,j = 1,2$. By the identity

$$\sum_{i,k=1}^{2} p_{ik}^2 + \sum_{i,k=1}^{2} (p_{ii}p_{kk} - p_{ik}^2) = (\Delta u)^2$$

in order to obtain the estimate (3.1), we have to prove

$$\int_\Omega (p_{11}p_{22} - p_{12}^2)dx \ge 0.$$

We obtain (see [12])

$$\int_\Omega (p_{11}p_{22}-p_{12}^2)dx = \frac{1}{2}\int_0^{\varphi_0}(p_1^2+p_2^2)\left(\frac{d\theta}{d\varphi}-\kappa\right)d\varphi + \frac{1}{2}\int_{\varphi_0}^{L}(p_1^2+p_2^2)\left(\frac{d\theta}{d\varphi}-\kappa\right)d\varphi + \sigma\int_0^L \frac{du}{d\varphi}c(\varphi)d\varphi.$$

Let us observe that

$$\int_0^L \frac{du}{d\varphi}\, c(\varphi)\, d\varphi = \lim_{\varepsilon\to 0}\left[\int_0^{\varphi_0-\varepsilon} \frac{du}{d\varphi}\, c(\varphi)\, d\varphi \; + \; \int_{\varphi_0+\varepsilon}^{L} \frac{du}{d\varphi}\, c(\varphi)\, d\varphi\right].$$

In the intervals $[0, \varphi_0 - \varepsilon]$, $[\varphi_0 + \varepsilon, L]$ we have $\cos\theta > 0$, then from the system

$$\begin{cases} p_1 Y_1 + p_2 Y_2 = -\sigma u, \\ -p_1 X_2 + p_2 X_1 = \dfrac{du}{d\varphi} \end{cases}$$

we get

$$\frac{du}{d\varphi}c(\varphi) = \frac{1}{\cos\theta}\left(\frac{du}{d\varphi}\right)^2 + \sigma u \tan\theta \frac{du}{d\varphi}.$$

Then

$$\int_0^L \frac{du}{d\varphi}\, c(\varphi)\, d\varphi = \lim_{\varepsilon\to 0}\Big[\int_0^{\varphi_0-\varepsilon} \frac{1}{\cos\theta}\left(\frac{du}{d\varphi}\right)^2 d\varphi \; + \; \frac{\sigma}{2}\int_0^{\varphi_0-\varepsilon} \tan\theta \frac{du^2}{d\varphi} d\varphi +$$

$$+ \int_{\varphi_0+\varepsilon}^{L} \frac{1}{\cos\theta}\left(\frac{du}{d\varphi}\right)^2 d\varphi + \frac{\sigma}{2}\int_{\varphi_0+\varepsilon}^{L} \tan\theta \frac{du^2}{d\varphi} d\varphi\Big].$$

Taking into account Remark 2.1, we have for all $\varphi \in [0, \varphi_0]$ and for a suitable point $\overline{\varphi} \in]\varphi, \varphi_0[$

$$\frac{du(\varphi)}{d\varphi} = \frac{du(\varphi)}{d\varphi} - \frac{du(\varphi_0)}{d\varphi} = (\varphi - \varphi_0)\frac{d^2u(\overline{\varphi})}{d\varphi^2}.$$

Bearing in mind condition (B)

$$\lim_{\varphi\to\varphi_0^-}\left|\frac{1}{\cos\theta}\left(\frac{du}{d\varphi}\right)^2\right| \le M \lim_{\varphi\to\varphi_0^-}\frac{(\varphi-\varphi_0)^2}{\cos\theta} = M \lim_{\varphi\to\varphi_0^-}\left|\frac{\theta(\varphi)-\theta(\varphi_0)}{\cos\theta}\right| \frac{|\varphi-\varphi_0|}{\left|\frac{\theta(\varphi)-\theta(\varphi_0)}{\varphi-\varphi_0}\right|} = 0$$

and then

$$\lim_{\varepsilon\to 0}\int_0^{\varphi_0-\varepsilon}\frac{1}{\cos\theta}\left(\frac{du}{d\varphi}\right)^2 d\varphi = \int_0^{\varphi_0}\frac{1}{\cos\theta}\left(\frac{du}{d\varphi}\right)^2 d\varphi.$$

Similarly we get

$$\lim_{\varepsilon\to 0}\int_{\varphi_0+\varepsilon}^{L}\frac{1}{\cos\theta}\left(\frac{du}{d\varphi}\right)^2 d\varphi = \int_{\varphi_0}^{L}\frac{1}{\cos\theta}\left(\frac{du}{d\varphi}\right)^2 d\varphi.$$

On the other hand

$$\int_0^{\varphi_0-\varepsilon}\frac{du^2}{d\varphi}\tan\theta d\varphi = \left[u^2(\varphi)\tan\theta(\varphi)\right]_0^{\varphi_0-\varepsilon} - \int_0^{\varphi_0-\varepsilon}\frac{u^2(\varphi)}{\cos^2\theta}\frac{d\theta}{d\varphi}d\varphi.$$

As above, we have for all $\varphi \in [0,\varphi_0]$ and for a suitable $\varphi^* \in]\varphi,\varphi_0[$

$$u(\varphi) = u(\varphi) - u(\varphi_0) = (\varphi-\varphi_0)\frac{du(\varphi^*)}{d\varphi}$$

and

$$\lim_{\varphi\to\varphi_0^-}\left|\frac{u(\varphi)}{\cos\theta}\right| \le M \lim_{\varphi\to\varphi_0^-}\left|\frac{\varphi-\varphi_0}{\cos\theta}\right| = M \lim_{\varphi\to\varphi_0^-}\left|\frac{\theta(\varphi)-\theta(\varphi_0)}{\cos\theta}\right| \frac{1}{\left|\frac{\theta(\varphi)-\theta(\varphi_0)}{\varphi-\varphi_0}\right|} = k$$

Then

$$\lim_{\varepsilon\to 0}\int_0^{\varphi_0-\varepsilon}\frac{u^2(\varphi)}{\cos^2\theta}\frac{d\theta}{d\varphi}d\varphi = \int_0^{\varphi_0}\frac{u^2(\varphi)}{\cos^2\theta}\frac{d\theta}{d\varphi}d\varphi$$

and similarly

$$\lim_{\varepsilon\to 0}\int_{\varphi_0+\varepsilon}^{L}\frac{u^2(\varphi)}{\cos^2\theta}\frac{d\theta}{d\varphi}d\varphi = \int_{\varphi_0}^{L}\frac{u^2(\varphi)}{\cos^2\theta}\frac{d\theta}{d\varphi}d\varphi.$$

Finally, being

$$\lim_{\varepsilon\to 0}u^2(\varphi_0-\varepsilon)\tan(\varphi_0-\varepsilon) \le k \lim_{\varepsilon\to 0}\frac{\varepsilon^2}{\cos(\varphi_0-\varepsilon)} = 0$$

and taking into account that $Y_i(0) = Y_i(L)$, $X_i(0) = X_i(L)$, $i = 1,2$ implies $\tan\theta(L) = \tan\theta(0)$, we get

$$(3.2)\quad \int_\Omega (p_{11}p_{22}-p_{12}^2)dx = \frac{1}{2}\int_0^{\varphi_0}\left(p_1^2+p_2^2\right)\left(\frac{d\theta}{d\varphi}-\kappa\right)d\varphi + \frac{1}{2}\int_{\varphi_0}^{L}\left(p_1^2+p_2^2\right)\left(\frac{d\theta}{d\varphi}-\kappa\right)d\varphi$$

$$+\sigma\int_0^{\varphi_0}\frac{1}{\cos\theta}\left(\frac{du}{d\varphi}\right)^2 d\varphi + \sigma\int_{\varphi_0}^{L}\frac{1}{\cos\theta}\left(\frac{du}{d\varphi}\right)^2 d\varphi - \frac{\sigma^2}{2}\int_0^{\varphi_0}\frac{u^2}{\cos^2\theta}\frac{d\theta}{d\varphi}d\varphi - \frac{\sigma^2}{2}\int_{\varphi_0}^{L}\frac{u^2}{\cos^2\theta}\frac{d\theta}{d\varphi}d\varphi.$$

Then by the conditions (2.1), (B), we obtain the estimate (3.1).

Let us observe that the estimate (3.1) holds true also for a more extended class of functions W, where W is the closure in $W^{2,2}(\Omega)$ of the set of functions u belonging to $C^2(\overline{\Omega})$ such that $\frac{\partial u}{\partial l} + \sigma u = 0$ on $\partial\Omega$.

Moreover let us note that in the case $\sigma = 0$ and $\cos\theta > 0$ the equality (3.2) was proved by G.Talenti (see [1]), who provides also some examples from which the necessity of the condition $\frac{d\theta}{d\varphi} \geq \kappa$ can be derived.

In the general case $q \in [2, +\infty)$ the following result holds :

Lemma 3.2. *For each $u \in W^{2,q}(\Omega)$ such that $\frac{\partial u}{\partial l} + \sigma u = 0$ on $\partial\Omega$, we have*

$$\|H(u)\|_{L^q(\Omega)} \leq c(q)\, \|\Delta u\|_{L^q(\Omega)} \tag{3.3}$$

where $c(q) : [2, +\infty) \to [1, +\infty)$ is a continuous function at $q = 2$ and $c(2) = 1$.

We will need also the following existence result related to a Caratheòdory operator $\mathcal{A}(x, \xi)$, which is easily derived from [10] and [12]:

Theorem 3.1. *Under the assumptions* (2.1), (2.2), A), B), *there exists a number $q_0 > 2$, such that $\forall q \in [2, q_0[$, $\forall f \in L^q(\Omega)$, $\forall \sigma > 0$, the problem*

$$\begin{cases} \mathcal{A}(x, H(u)) = f(x) & a.e.\ in\ \Omega \\ \frac{\partial u}{\partial l} + \sigma\, u = 0 & on\ \partial\Omega \end{cases} \tag{3.4}$$

admits a unique solution $u \in W^{2,q}(\Omega)$ and it results:

$$\|H(u)\|_{L^q(\Omega)} \leq \frac{\alpha}{1 - (\gamma c(q) + \delta)}\, \|f(x)\|_{L^q(\Omega)}\,. \tag{3.5}$$

We note that q_0 is such that, for each $q \in [2, q_0[$, it results $1 - (\gamma c(q) + \delta) > 0$.

4. Proof of the Theorems

In the present paper the main tool in proving Theorem 2.1 is Leray-Shauder fixed point theorem. For this let us fix $v \in W^{1,2q}(\Omega)$, $q < r$ and consider for fixed $\tau \in [0, 1]$

$$\begin{cases} \mathcal{A}(x, v, Dv, H(u)) = \tau f(x, v, Dv) & \text{a.e. } in\ \Omega \\ \frac{\partial u}{\partial l} + \sigma\, u = 0 & \text{on } \partial\Omega \end{cases} \tag{4.1}$$

by means of Theorem 3.1, there exists a number $q_0 > 2$ such that, if we set $\bar{q} = min\,\{q_0, r\}$, for each $q \in]2, \bar{q}[$, the problem (4.1) admits a unique solution $u \in W^{2,q}(\Omega)$. So we have defined an operator

$$T : W^{1,2q}(\Omega) \times [0, 1] \to W^{2,q}(\Omega) \hookrightarrow W^{1,2q}(\Omega)$$

where the image $u = T(v, \tau)$ is the unique solution of (4.1). It is easily seen that each fixed point of the mapping $T(\cdot, 1)$ is a solution of (2.3). The condition $\mathcal{A}(x, z, p, 0) = 0$, as required above, and uniqueness of solution of (4.1) show that $T(v, 0) = 0$ for all $v \in W^{1,2q}(\Omega)$. Taking into account Theorem 3.1, the continuity and the compactness of the mapping T follows as in [14] and [15]. To assert the existence of a fixed point of $T(\cdot, 1)$ it remains to prove the a priori estimate

$$\|u\|_{W^{1,2q}(\Omega)} \leq C \tag{4.2}$$

with a constant C independent of u and τ, for each solution u of the problem

$$(4.3) \qquad \begin{cases} \mathcal{A}(x,u,Du,H(u)) = \tau f(x,u,Du) & \text{a.e. } in\ \Omega \\ \frac{\partial u}{\partial l} + \sigma\, u = 0 & \text{on } \partial\Omega. \end{cases}$$

The first step in deriving this estimate is to obtain a bound for $\|u\|_{L^\infty}$ and it follows easily from a variant of Aleksandrov-Bakelman-Pucci maximum principle proved by N.S.Trudinger ([16]) (see [8]).

Lemma 4.1. *Suppose conditions* A), (2.5) *to be fulfilled. Then*

$$(4.4) \qquad \|u\|_{L^\infty(\Omega)} \le M + \left\{ diam\Omega + sup_{\partial\Omega}\ \frac{1}{\sigma} \right\} R$$

for each solutions $u \in W^{2,q}(\Omega)$ *of the problem* (2.3), *where* R *is such that*

$$\int_{B_R(0)} h(p)dp = \int_\Omega g(x)dx$$

and $B_R(0)$ *is a ball with center at the origin and radius* R.

The second step is devoted to a priori estimate for $L^{2q}(\Omega)$ norm of the gradient Du. Using essentially the condition (A), which enables to linearize the equation in a suitable manner, and, then, applying a topological approach due to H.Amann and M.Crandall ([17]) (cf. [15]), we derive

Theorem 4.1. *Suppose conditions* A), B), (2.1), (2.2), (2.4) *to be fulfilled. Then there exists a constant*

$$C = C(\alpha, \gamma, \delta, q, \partial\Omega, f_1, f_2, \|u\|_{L^\infty(\Omega)})$$

such that, for each $q \in]2, \tilde{q}[$,

$$(4.5) \qquad \|Du\|_{L^{2q}(\Omega)} \le C$$

for each solution $u \in W^{2,q}(\Omega)$ *of the problem* (2.3).

From (4.4) and (4.5) we obtain the estimate (4.2) and then the existence of the solution of the problem (2.3).

References

1. G.Talenti, Boll. Un. Mat. Ital. (3), 22, 505-526 (1967).
2. C. Miranda, Ann. Mat. Pura Appl., 63, 353-386 (1963).
3. M. Chicco, Ann. Mat. Pura Appl., (4), 112, 241-259 (1977); errata, ibid. (4), 130, 399-401 (1982).
4. G. Viola, Boll. Un. Mat. Ital. (6), 3-B, 397-411 (1984).
5. M. Chicco, Rend. Ist. Mat. Univ. Trieste, 4, 85-94 (1972).
6. F. Nicolosi, Rend. Circ. Mat. Palermo, 33, 351-368 (1984).
7. G. Di Fazio and D.K. Palagachev, Comment. Math. Univ. Carolinae, 37, 537-556 (1996).

8. G. Di Fazio and D. K. Palagachev, Bull. Austral. Math. Soc., 53, 501-513 (1996).

9. A. Maugeri and D.K. Palagachev, Forum Math., 10, no.4, 393-405 (1998).

10. S. Giuffrè, in: Proc. 8th Int. Coll. on Differential Equations, D. Bainov (Ed.), VSP, Utrecht, 187-193 (1998).

11. S. Giuffrè, Comm. Appl. Anal., 2, no.4, 585-594 (1998).

12. S. Giuffrè, C.R. Acad. Sci. Paris, t. 325, Sèrie I, 1081-1086 (1997).

13. P.R. Popivanov and D.K. Palagachev, The Degenerate Oblique Derivative Problem for Elliptic and Parabolic Equations, Akad. Verlag, Berlin (1997).

14. D. K. Palagachev, Le Matematiche, XLVIII, 311-321 (1993).

15. L. Softova, Z. Anal. Anwen., 17, no.1, 57-66 (1998).

16. N.S.Trudinger, Non Linear Second Order Elliptic Equations, Lecture Notes of Math. Inst. of Nankai Univ., Tianjin, China (1986).

17. H. Amann and M. Crandall, Indiana Univ. Math. J., 27, 779-790 (1978).

9th Int. Coll. on Differential Equations, pp. 177-183
D. Bainov (Ed)

Life Span of Solutions for Non Linear Cauchy Problem

DANIEL GOURDIN
Université Paris 6, U.F.R 920 - U.M.R 9994, 4 Place Jussieu, 75252 Paris, France
MUSTAPHA MECHAB
Université Djilali Liabès, B.P 89, 22000 Sidi Bel Abbès, Algérie

Abstract: We study the global non linear Cauchy problem with initial data which have small speed of variation, in the space of holomorphic functions with respect to t, belonging to Gevrey classes with respect to the other variables, without hyperbolicity.

1. Notations and Results

Let $m,\ n \in \mathbb{N}^*$, $t \in \mathbb{C}$, $x = (x_1, \dots, x_n) \in \mathbb{R}^n$. B a part of

$$\{(j, \alpha) \in \mathbb{N} \times \mathbb{N}^n;\ j + |\alpha| \leq m,\ j < m \text{ and } \alpha \neq 0\}$$

with $|\alpha| = \sum_{i=1}^{n} \alpha_i$ the length of α, r the cardinal of B, $D^B u = (D^\sigma u)_{\sigma \in B}$, $D_t^{-1} u$ the primitive of u with respect to t vanishing for $t = 0$, $d = \inf\left\{\frac{m-j}{|\alpha|},\ (j, \alpha) \in B\right\}$, Ω an open neighbourhood of $0 \in \mathbb{R}^n$ such that $\lambda x \in \Omega$ for all $x \in \Omega$ and $\lambda \in [0, 1]$.
Denote $\mathcal{U}_T = \{t \in \mathbb{C};\ |t| \leq T\}$ and $\Omega_T = \mathcal{U}_T \times \Omega$ $(T > 0)$, $\mathbf{G}^d(\Omega)$ the Gevrey class of functions $\varphi \in \mathcal{C}^\infty(\Omega, \mathbb{R})$ with the property:

$$\exists C > 0,\ \forall \alpha \in \mathbb{N}^n,\ \forall x \in \Omega, \qquad |D^\alpha \varphi(x)| \leq C^{|\alpha|+1} (\alpha!)^d$$

which is equivalent to existence of constants, written $([\varphi]_0, [\varphi]_1)$, such that

$$\forall \alpha \in \mathbb{N}^n,\ \forall x \in \Omega, \qquad |D_x{}^\alpha \varphi(x)| \leq [\varphi]_0 . ([\varphi]_1)^{|\alpha|} (\alpha!)^d$$

$\mathbf{G}^{\omega,d}(\mathcal{U}_T \times \Omega)$ is the algebra of functions $u : \mathcal{U}_T \times \Omega \longrightarrow \mathbb{C}$, with real values in $(\mathcal{U}_T \cap \mathbb{R}) \times \Omega$ admitting continuous derivatives D_x^α upon $\mathcal{U}_T \times \Omega$, $\forall \alpha \in \mathbb{N}^n$, holomorphic with respect to t, and such that

$$\exists\, C > 0,\ \forall \alpha \in \mathbb{N}^n,\ \forall t \in \mathcal{U}_T,\ \forall x \in \Omega, \quad |D_x^\alpha u(t, x)| \leq C^{|\alpha|+1} (|\alpha|!)^d$$

For $\varphi_j \in \mathbf{G}^d(\Omega)$, $(j = 1, \dots, m-1)$, and $\varepsilon > 0$ small, we consider the non linear Cauchy problem

$$\begin{cases} D_t^m u(t, x) = f(t, D^B u(t, x))\ , \\ D_t^j u(0, x) = \varphi_j(\varepsilon . x), \qquad \forall\, j = 0, \dots, m-1 \end{cases} \tag{1}$$

We assume that $f(t, 0) = 0$ and we denote $\varphi_j^\varepsilon(x) = \varphi_j(\varepsilon . x)$.
We get the following results

Theorem 1 *For any holomorphic function f in a neighbourhood of $\mathcal{U}_T \times \{0\} \subset \mathbb{C} \times \mathbb{C}^r$ and $\varphi_j \in \mathbf{G}^d(\Omega)$, $(j = 0, \ldots, m-1)$, there exists $\varepsilon_0 > 0$ such that for all $\varepsilon < \varepsilon_0$, the Cauchy problem (1) has a unique solution u_ε in $\mathbf{G}^{\omega,d}(\mathcal{U}_T \times \Omega)$.*
If Ω is bounded, the solution of (1) with null data is stable, that is if $\varphi_j(0) = 0$ $(j = 0, \ldots, m-1)$ $(u_\varepsilon)_\varepsilon \underset{\varepsilon \to 0}{\longrightarrow} 0$ in $\mathbf{G}^{\omega,d}(\mathcal{U}_T \times \Omega)$ for the topology of uniform convergence of the holomorphic functions in t.

When f does not depend of t, we have:

Theorem 2 *For any holomorphic function f in a neighbourhood of $0 \in \mathbb{C}^r$ and $\varphi_j \in \mathbf{G}^d(\Omega)$, $(j = 0, \ldots, m-1)$ and for all $\varepsilon > 0$, the Cauchy problem (1) has a unique solution $u_\varepsilon \in \mathbf{G}^{\omega,d}(\mathcal{U}_T \times \Omega)$ with asymptotic estimate of life span:*

$$T_\varepsilon \geq \mu \left(\frac{1}{\varepsilon}\right)^s, \quad \text{with} \quad s = \inf\left\{\frac{|\alpha|}{m-k}, \ (k, \alpha) \in B\right\}.$$

2. Sketch of the Proofs

2.1. Reduction of the Problem

Denote the successive change of unknowns.

$$\begin{aligned} v(t,x) &= u(t,x) - \sum_{j=0}^{m-1} \frac{1}{j!} t^j \varphi_j^\varepsilon(x) && (2)\\ w(t,x) &= D_t^m v(t,x) && (3)\\ \rho(t,x) &= w(t,x) - f(0, (D_x^\alpha \varphi_j^\varepsilon(x))_{(k,\alpha)\in B}) \rho(t,x) = w(t,x) - \psi_\varepsilon(x) && (4)\\ & && (5) \end{aligned}$$

As $(k,\alpha) \in B \Longrightarrow k < m$ then

$$D^B u = D^B \left[D_t^{-m}(\rho + t^m \psi_\varepsilon(x) + \sum_{j=0}^{m-1} \frac{1}{j!} t^j \varphi_j^\varepsilon(x) \right]$$

and (1) is equivalent to the equation

$$\rho = f(t, D^B \left[D_t^{-m}(\rho + t^m \psi_\varepsilon(x) + \sum_{j=0}^{m-1} \frac{1}{j!} t^j \varphi_j^\varepsilon(x) \right] (t,x) - \psi_\varepsilon(x) \qquad (6)$$

with $\rho(0,x) = 0$ for all $x \in \Omega$.
We consider the application $\mathcal{L} : \ u \mapsto \mathcal{L}u$ defined by

$$\mathcal{L}\rho(t,x) = f(t, D^B \left[D_t^{-m}(\rho + t^m \psi_\varepsilon(x) + \sum_{j=0}^{m-1} \frac{1}{j!} t^j \varphi_j^\varepsilon(x) \right] (t,x) - \psi_\varepsilon(x) \qquad (7)$$

Then resolve (1) becomes equivalent to research fixed points of $\mathcal{L}$.
We shall prove that $\mathcal{L}$ is a strict contraction with Banach algebras $\mathbf{G}^{\omega,d}_{T,\varsigma}(\Omega_T)$ defined in [1] and [2] with $\xi = 1$.

2.2. The Banach Algebras $\mathrm{G}^{\omega,d}_{T,\zeta}(\Omega_T)$

Recall briefly the definition of Banach algebras $\mathrm{G}^{\omega,d}_{T,\zeta}(\Omega_T)$ ([1], [2]).

Let be $\theta(t) = \sum_{j=0}^{+\infty} \frac{t^j}{(j+1)^2}$ the Lax function with the property, for a constant K:

$$\theta^2(t) \ll K\theta(t)$$

with the following meaning of majorizing series $\ll$.
$v = \sum_\alpha v_\alpha x^\alpha$ and $u = \sum_\alpha u_\alpha x^\alpha$ where $v_\alpha \in \mathbb{C}^n$ and $u_\alpha \geq 0$ verify $v \ll u$ iff

$$|v_\alpha| \leq u_\alpha \qquad (\forall \alpha)$$

So $\phi_T(t) = \frac{1}{K}\theta\left(\frac{t}{T}\right)$ satisfies:

$$(\phi_T(t))^2 \ll \phi_T(t).$$

And for all $\zeta = (\zeta_1, \dots, \zeta_n) \in \left(\mathbb{R}^*_+\right)^n$, $|t| < T$, we consider

$$\Phi^{\omega,d}_{T,\zeta} = \Phi^{\omega,d}_{T,\zeta}(t,x) = \sum_{k\in\mathbb{N}} \frac{(\zeta.x)^k}{k!}(k!)^{d-1} D^k \phi_T(t)$$

Definition 1 *$u \in \mathrm{G}^{\omega,d}_{T,\zeta}(\Omega_T)$ if $u : \mathcal{U}_T \times \Omega \longrightarrow \mathbb{C}$ is real on $(\mathcal{U}_T \cap \mathbb{R}) \times \Omega$, admits derivatives for each order of x, continuous on $\mathcal{U}_T \times \Omega$ and holomorphic with respect to t such that*

$$\exists C > 0,\ \forall \alpha \in \mathbb{N}^n,\ \forall x \in \Omega,\quad D^\alpha_x u(t,x) \ll C\,\zeta^\alpha\,|\alpha|!^{d-1}\,D^{|\alpha|}\phi_T(t). \tag{8}$$

We shall denote also

$$u \ll C.\Phi^{\omega,d}_{T,\zeta} \tag{9}$$

So we have $\left(\Phi^{\omega,d}_{T,\zeta}\right)^2 \ll \Phi^{\omega,d}_{T,\zeta}$ and $\mathrm{G}^{\omega,d}_{T,\zeta}(\Omega_T)$ with norm $u \mapsto \|u\| = \inf\left\{C > 0;\ u \ll C\Phi^{\omega,d}_{T,\zeta}\right\}$ is a Banach algebra.

Proposition 1

1. $\mathrm{G}^{\omega,d}_{T,\zeta}(\Omega_T) \subset \mathrm{G}^{\omega,d}(\Omega_T)$.
2. *Let $u \in \mathrm{G}^{\omega,d}(\Omega_T)$ such that $u(0,.) = 0$, $\zeta \in \left(\mathbb{R}^*_+\right)^n$ and $C > 0$, then there exists $T_0 > 0$ such that for all $T' \in]0, T_0[$, $u \in \mathrm{G}^{\omega,d}_{T',\zeta}(\Omega_{T'})$ and $\|u\| \leq C$.*
3. *If $u \in \mathrm{G}^{\omega,d}_{T,\zeta}(\Omega_T)$ satisfies $\|u\| < R$, then $\frac{R}{R-u} \in \mathrm{G}^{\omega,d}_{T,\zeta}(\Omega_T)$ and*

$$\left\|\frac{R}{R-u}\right\| \leq \left(K + \frac{\|u\|}{R - \|u\|}\right)$$

Proposition 2 *For all $(k,\alpha) \in \mathbb{Z} \times \mathbb{N}^n$ with $k + d|\alpha| \leq 0$, there exist $C_{k,\alpha} > 0$ such that:*

$$\forall T > 0,\ \forall u \in \mathrm{G}^{\omega,d}_{T,\zeta}(\Omega_T),\qquad \left\|D^k_t D^\alpha_x u\right\| \leq C_{k,\alpha}\,\zeta^\alpha T^{-k-|\alpha|}\,\|u\|$$

$$\forall T > 0,\ \forall u,\ u' \in \mathrm{G}^{\omega,d}_{T,\zeta}(\Omega_T),\qquad \left\|D^k_x D^\alpha_y u - D^k_x D^\alpha_y u'\right\| \leq C_{k,\alpha}\zeta^\alpha T^{-|k|-|\alpha|}\|u - u'\|$$

2.3. Preliminaries

Lemma 1 *For all $\alpha \in \mathbb{N}^n$, there exists $C_{|\alpha|} > 0$ such that for all $\varphi \in \mathbf{G}^d(\Omega)$, $D_x^\alpha \varphi \in \mathbf{G}^d(\Omega)$ and we choose*

$$([D_x^\alpha \varphi]_0, [D_x^\alpha \varphi]_1) = \left(\left[(C_{|\alpha|})^d [\varphi]_1\right]^{|\alpha|} [\varphi]_0 \,,\ [(C_{|\alpha|})^d]^{|\alpha|}.[\varphi]_1\right)$$

This result is implied by $\lim_{|\beta| \to +\infty} \frac{log(|\beta| + j)}{|\beta| + 1} = 0, \forall j \in \mathbb{N}$.
From lemma 1 we get

Lemma 2 *Let $\varphi \in \mathbf{G}^d(\Omega)$, $T > 0$ and $\alpha \in \mathbb{N}^n$, then for all $\zeta \geq \left([(C_{|\alpha|})^d]^{|\alpha|}[\varphi]_1 Te^2, \ldots , [(C_{|\alpha|})^d]^{|\alpha|}[\varphi]_1 Te^2\right)$, $D_x^\alpha \varphi \in \mathbf{G}^{\omega,d}_{T,\zeta}(\Omega_T)$ and*

$$\|D_x^\alpha \varphi\| \leq K.[\varphi]_0 \left[(C_{|\alpha|})^d.[\varphi]_1\right]^{|\alpha|} \tag{10}$$

and from proposition 2 we get

Lemma 3 *Let $k \in \mathbb{N}$, $\varphi \in \mathbf{G}^d(\Omega)$, then $t^k\varphi \in \mathbf{G}^{\omega,d}_{T,\zeta}(\Omega_T)$ for all $T > 0$ and we have for all $\zeta \geq \zeta^0 = (T.e^2.[\varphi]_1, \ldots , T.e^2.[\varphi]_1)$,*

$$\|t^k \varphi\| \leq K.C_{k,0}.T^k.[\varphi]_0$$

Further more

Lemma 4 *There exists constants C_B and $C > 0$, such that for $\varphi_j \in \mathbf{G}^d(\Omega)$ $(j = 0, 1, \ldots , m - 1)$:*

$$\forall\, T > 0,\ \forall \sigma = (k, \alpha) \in B,\ D^\sigma \left[\sum_{j=0}^{m-1} \frac{t^j}{j!}\varphi_j\right] \in \mathbf{G}^{\omega,d}_{T,\zeta^0}(\Omega_T)$$

$$\left\|D^\sigma \left[\sum_{j=0}^{m-1} \frac{t^j}{j!}\varphi_j\right]\right\| \leq C. \sum_{j=k}^{m-1} T^{j-k} C_{0,\varphi}. \left[C_{1,\varphi}\right]^{|\alpha|} \tag{11}$$

with $\zeta^0 = (C_B.T.C_{1,\varphi}, \ldots , C_B.T.C_{1,\varphi})$, $C_{0,\varphi} = \sup_{j=0,\ldots,m-1} [\varphi_j]_0$, $C_{1,\varphi} = \sup_{j=0,\ldots,m-1} [\varphi_j]_1$.

2.4. $\mathcal{L}$ is a Contraction

Let $\eta > 1$ and $T,\ R > 0$ such that f holomorphic and bounded in

$$\Delta = \Delta_{T,R,\eta} = \{(t, z) \in \mathbb{C} \times \mathbb{C}^n,\ |t| < \eta T \text{ and } |z_\sigma| < \eta R,\ \forall \sigma \in B\}$$

With Cauchy inequalities we can find a constant $C = C_{(f,T,R,\eta)} > 0$ such that

$$\zeta \in (\mathbb{R}_+^*)^n, \qquad f(t, z) \ll C_{(f,T,R,\eta)} \frac{\eta R}{\eta R - 1.z} \Phi^{\omega,d}_{T,\zeta}(t, x) \tag{12}$$

Proposition 3 *For $f(t,z)$ holomorphic function in a neighbourhood of $\mathcal{U}_T \times \{0\} \subset \mathbb{C} \times \mathbb{C}^r$, there exists two constants $K_f > 0$ and $\varepsilon > 0$ such that for all functions $\varphi_j \in \mathbf{G}^d(\Omega)$, $j = 0, \ldots, m-1$, verifying $C_{1,\varphi} < \varepsilon$ we have:*

$$\forall\ a \geq K_f, \qquad \mathcal{L}(\mathcal{B}(0,a)) \subset \mathcal{B}(0,a)$$

where $\mathcal{B}(0,a)$ is the closed ball with center 0 and radius a in $\mathbf{G}^{\omega,d}_{T,\zeta^0}(\Omega_T)$ and $\zeta^0 = (C_B.T.C_{1,\varphi}, \ldots, C_B.T.C_{1,\varphi})$.

Proof:
Let $a > 0$ and $u \in \mathcal{B}(0,a) \subset \mathbf{G}^{\omega,d}_{T,\zeta^0}(\Omega_T)$. Denote

$$z_\sigma = z_{(k,\alpha)} = D_t^k D_x^\alpha \left[D_t^{-m} u + \sum_{j=0}^{m-1} \frac{t^j}{j!} \varphi_j \right]$$

(for simplicity, we suppose $\psi_\varepsilon = 0$).
With proposition 2 we get

$$\left\| D_t^k D_x^\alpha [D_t^{-m} u] \right\| \leq C_{k-m,\alpha}\, \zeta^\alpha\, T^{(m-k)-|\alpha|}\, \|u\| \tag{13}$$

As $f(t,0) = 0$, we may write

$$\mathcal{L}(u) = f(t,z) = \sum_{\sigma \in B} f_\sigma(t,z).z_\sigma \tag{14}$$

where $f_\sigma(t,z)$ are holomorphic functions in $\Delta_{T,\eta,R}$.
With small η, we can suppose these functions bounded in $\Delta_{T,R,\eta}$, and from (12) we get

$$\forall\ (t,z) \in \Delta_{T,\eta,R}, \quad f_\sigma(t,z) \ll C_{f_\sigma,T,R,\eta} \frac{\eta R}{\eta R - 1.z} \Phi^{\omega,d}_{T,\zeta}(t,x) \tag{15}$$

With inequalities (11), (13) and the expression of ζ^0 we get

$$\|z_\sigma\| \leq C.[C_{1,\varphi}]^{|\alpha|} T^{m-k} \|u\| + C. \sum_{j=k}^{m-1} T^{j-k} C_{0,\varphi} [C_{1,\varphi}]^{|\alpha|} \tag{16}$$

As $\alpha \neq 0$, for a and T fixed and small $C_{1,\varphi}$ we have $\|z\| \leq \eta.R$; thanks to proposition 1, from (15) we get

$$\|f_\sigma(t,z)\| \leq C_{(f_\sigma,T,\eta,R)}(K + \mathcal{E}\,(C_{1,\varphi})) \tag{17}$$

with $\lim_{s \to 0} \mathcal{E}(s) = 0$.
$\mathbf{G}^{\omega,d}_{T,\zeta}(\Omega_T)$ is a Banach algebra, these inequalities used in (14) lead to

$$\begin{aligned} \|\mathcal{L}(u)\| \quad &\leq \quad \sum_{(k,\alpha) \in B} C_{(f_\sigma,T,\eta,R)}(K + \mathcal{E}\,(C_{1,\varphi})) \\ &\qquad \left[C_{k-m,\alpha}.[C_{1,\varphi}]^{|\alpha|} T^{m-k} \|u\| + C. \sum_{j=k}^{m-1} T^{j-k} C_{0,\varphi} [C_{1,\varphi}]^{|\alpha|} \right] \\ &\leq \quad \sum_{\sigma \in B} C_{f_\sigma,T,\eta,R}(K + \mathcal{E}\,(C_{1,\varphi}))\, [\mathcal{E}\,(C_{1,\varphi})\, \|u\| + \mathcal{E}\,(C_{1,\varphi})] \end{aligned} \tag{18}$$

For $C_{1,\varphi}$ small we can write (18) in the following way

$$\forall\ u \in \mathcal{B}(0,a) \subset \mathbf{G}^{\omega,d}_{T,\zeta}(\Omega_T), \quad \|\mathcal{L}(u)\| \leq C_0 \|u\| + K_f \tag{19}$$

where $0 < C_0 < 1$ and $K_f > 0$, so for $C_0 = \frac{1}{2}$ and small $C_{1,\varphi}$ we have

$$\forall\, a \geq 2K_f, \qquad \mathcal{L}(\mathcal{B}(0,a)) \subset \mathcal{B}(0,a) \tag{20}$$

the proposition is proved. ◇

The next result has analogous proof.

Proposition 4 *Let $f(t,z)$ be a holomorphic function in a neighbourhood of $\mathcal{U}_T \times \{0\} \subset \mathbb{C} \times \mathbb{C}^r$, then ther exists $\varepsilon > 0$ such that for all $a > 0$ and all functions $\varphi_j \in \mathbf{G}^d(\Omega)$, $j = 0, \ldots, m-1$, with $C_{1,\varphi} < \varepsilon$, we can find constant $C \in]0,1[$ such that*

$$\forall u,\; u' \in \mathcal{B}(0,a) \subset \mathbf{G}^{\omega,d}_{T,\zeta^0}(\Omega_T), \qquad \|\mathcal{L}u - \mathcal{L}u'\| \leq C\|u - u'\|$$

2.5. Proof of Theorem 1

We use the following lemma.

Lemma 5 *Let $\varphi \in \mathbf{G}^d(\Omega)$, then for all $\varepsilon \in]0,1]$, the function $x \longrightarrow \varphi_\varepsilon(x) = \varphi(\varepsilon.x)$ belongs to $\mathbf{G}^d(\Omega)$ and we can choose: $(C_{0,\varphi_\varepsilon}, C_{1,\varphi_\varepsilon}) = (C_{0,\varphi}\,,\ \varepsilon C_{1,\varphi})$.*

By lemma 5 we have $C_{1,\varphi^\varepsilon} = \varepsilon C_{1,\varphi} \underset{\varepsilon\to 0}{\longrightarrow} 0$.
Let T and $R > 0$ and $f(t,z)$ be a holomorphic function in a neighbourhood of $\mathcal{U}_T \times \{0\}$. With propositions 3 and 4 and lemma 5, we see that for small $\varepsilon > 0$, there exists ζ such that $\mathcal{L}$ is a strict contraction of a ball $\mathcal{B}(0,a) \subset \mathbf{G}^{\omega,d}_{T,\zeta}(\Omega_T)$; so $\mathcal{L}$ has a unique fixed point. From proposition 1 we can assert that this fixed point is in $\mathbf{G}^{\omega,d}(\Omega_T)$; so the problem (1) admits a solution in $\mathbf{G}^{\omega,d}(\Omega_T)$.

For uniqueness we use the second assertion of proposition 1.

Proof of stability.

Using inequalities (18) and lemma 5 we have:

$$\|u_\varepsilon\| \;\leq\; \sum_{(k,\alpha)\in B} C.(K + \mathcal{E}(\varepsilon C_{1,\varphi})).\varepsilon^{|\alpha|}{C_{1,\varphi}}^{|\alpha|}\left\{2K_f + C_{0,\varphi^\varepsilon}\right\} + C.C_{0,\varphi^\varepsilon} \tag{21}$$

in $\mathbf{G}^{\omega,d}_{T,\zeta_\varepsilon}(\Omega_T)$ with u_ε solution of problem (1) and $\zeta_\varepsilon = (C_B T C_{1,\varphi^\varepsilon}, \ldots, C_B T C_{1,\varphi^\varepsilon})$.
$\varphi_j(0) = 0$ implies $\lim_{\varepsilon\to 0}\left[\varphi_j^\varepsilon\right]_0 = 0$. So the right hand member χ_ε of (21) admits 0 for limit when $\varepsilon \longrightarrow 0$.
Furthermore (21) can be expressed with the majorizing functions and we get

$$\forall \alpha \in \mathbb{N}^n,\; \forall x \in \Omega, \qquad D^\alpha_x u_\varepsilon(t,x) \ll \chi_\varepsilon.(\zeta_\varepsilon)^\alpha(|\alpha|!)^{d-1} D^{|\alpha|}\phi_T(t) \tag{22}$$

and $\left|D^{|\alpha|}\phi_T(t)\right| \leq C_T^{|\alpha|+1}|\alpha|!$, so

$$\forall\, \alpha \in \mathbb{N}^n,\; \forall\, x \in \Omega, \qquad D^\alpha_x u_\varepsilon(t,x) \ll (C_\varepsilon)^{|\alpha|+1}(|\alpha|!)^d$$

with $\lim_{\varepsilon\to 0} C_\varepsilon = 0$. Theorem 1 is proved. ◇

2.6. Proof of Theorem 2

We use (18) and lemma 5 and we get

$$C_f[K+\mathcal{E}(\varepsilon C_{1,\varphi})]\sum_{(k,\alpha)\in B} C_{k-m,\alpha}.\varepsilon^{|\alpha|}(C_{1,\varphi})^{|\alpha|}T_\varepsilon{}^{m-k}<1 \tag{23}$$

If there exists constant $C>0$ which does not depend of ε such that

$$\forall (k,\alpha)\in B, \qquad T_\varepsilon{}^{m-k}\varepsilon^{|\alpha|}\leq \frac{1}{2C}$$

where $C = rC_f[K+\mathcal{E}(\varepsilon C_{1,\varphi})]\sum_{(k,\alpha)\in B} C_{k-m,\alpha}.\varepsilon^{|\alpha|}(C_{1,\varphi})^{|\alpha|}$.

So (23) is satisfied if

$$T_\varepsilon = \inf_{(k,\alpha)\in B}\left(\frac{1}{2C}\right)^{k-m}\left(\frac{1}{\varepsilon}\right)^{s}$$

◇

References

1. C. Wagschal, J. Math. Pure et Appl., **58**, 309-337 (1979).
2. D. Gourdin and M. Mechab, J. Math. Pures Appl., **75**, 569-593 (1996).
3. A. Arosio and S. Spagnolo, J. Math. Pures et Appl., **65**, 263-305 (1984).
4. J. M. Bony and P. Schapira, Invent. Math., **17**, 95-105 (1972).
5. Y. Choquet-Bruhat, C. R. Acad. Sci. Paris, **274**, Serie A, 843-845 (1972).
6. J. Leray and Y. Ohya, Colloque de Liege, CBRM, 105-144 (1964).
7. J. Leray and Y. Ohya, Math. Annalen, **170**, 167-205 (1967).
8. J. Leray and L. Waelbroeck, Colloque de Liege, CBRM, 145-152 (1964).

9th Int. Coll. on Differential Equations, pp. 185-190
D. Bainov (Ed)

Approximation of the McKendrick-Von Foester Equation for Population Growth

DAVID GREENHALGH
Department of Statistics and Modelling Science, University of Strathclyde, Livingstone Tower, 26 Richmond Street, Glasgow G1 1XH, U.K.

ISTVÁN GYÖRI AND ISTVÁN KOVÁCSVÖLGYI
Department of Mathematics and Computing, University of Vészprem, P.O. Box 158, 8201 Vészprem, Hungary

Abstract: The McKendrick-Von Foester equation for population growth is a key tool in mathematical population dynamics. However real populations do not exhibit the unbounded population growth predicted by this model and other factors such as the interaction of different generations of the population have the effect of limiting the ultimate population size. Boulanger [1] studied such a model with a small interaction term. He showed that the solution of the perturbed equation with interaction term uniformly approximates the solution of the McKendrick-Von Foester equation as the interaction term goes to zero on any finite time interval. However Boulanger showed this under quite complex and restrictive assumptions. In this paper we show how Boulanger's result can be proved under simpler and less restrictive conditions.
Key Words and Phrases: *Age-dependent population model, small perturbation parameters, asymptotic theory.*

1. Introduction

In this introduction we shall outline the background to the problem which we shall study. The **McKendrick-Von Foester** (McK-VF) equation for population growth is a key equation in mathematical population dynamics. Let $\rho_0(a,t)$ denote the density with respect to age of the number of individuals of age a at time t. This means that the total number of individuals between ages A_1 and A_2 ($A_1 < A_2$) at time t is given by

$$\int_{A_1}^{A_2} \rho_0(a,t)da.$$

Let $\mu_0(a)$ denote the age-dependent death rate at age a, $\phi(a)$ the density (with respect to age) of the initial age distribution and $b(a)$ the age-dependent birth rate (at age a). Then the McK-VF equation is

$$\frac{\partial \rho_0}{\partial a}(a,t) + \frac{\partial \rho_0}{\partial t}(a,t) = -\mu_0(a)\rho_0(a,t), \tag{1}$$

with boundary conditions

$$\rho_0(0,t) = \int_0^\infty b(a)\rho_0(a,t)da$$

and $\rho_0(a,0) = \phi(a)$.

Note that to the first order in Δa, $\rho_0(a,t)\Delta a$ is the number of individuals between ages a and $a+\Delta a$ at time t, and $\rho_0(a+\Delta a, t+\Delta a)\Delta a$ is the number of individuals between ages $a+\Delta a$ and $a+2\Delta a$ at time $t+\Delta a$, i.e. the same cohort Δa time units later. Thus

$$\rho_0(a+\Delta a, t+\Delta a)\Delta a = \rho_0(a,t)\Delta a - \mu_0(a)\rho_0(a,t)\Delta a^2 + o(\Delta a^2).$$

Using the Taylor series expansion of $\rho_0(a+\Delta a, t+\Delta a)$ the terms in Δa are identical. Equating the terms in Δa^2 we obtain the McK-VF equation. The first boundary condition equates two expressions for the birth rate at time t, the second says that $\phi_0(a)$ is the initial population density.

As ρ is a population density it is natural to look for solutions in the Banach space of Lebesgue integrable functions:

$$L^1[0,\infty) = \left\{ f : [0,\infty) \to \mathcal{R} \text{ such that } \int_0^\infty | f(a) | \, da \text{ exists} \right\},$$

with norm

$$\|\phi\| = \int_0^\infty | \phi(a) | \, da.$$

We assume that $b(a)$ is bounded, $\phi(a) \in L^1[0,\infty)$ and the consistency condition

$$\phi(0) = \int_0^\infty b(a)\phi(a)da \tag{2}$$

is satisfied. This consistency condition comes from equating the two expressions for $\rho_0(0,0)$ obtained from the boundary and initial conditions in the McK-VF equation (1). Then using the method of characteristics the solution to (1) is

$$\rho_0(a,t) = \begin{cases} \phi(a-t)\, e^{-\int_{a-t}^{a} \mu_0(s)ds}, & \text{if } a \geq t, \\ \\ B_0(t-a)\, e^{-\int_0^a \mu_0(s)ds}, & \text{if } t \geq a. \end{cases} \tag{3}$$

Here $B_0(t)$ is the unique positive solution of

$$B_0(t) = \int_0^t b(a)\, B_0(t-a)\, e^{-\int_0^a \mu_0(\xi)d\xi} + \int_t^\infty b(a)\, \phi(a-t)\, e^{-\int_{a-t}^{a} \mu_0(\xi)d\xi}. \tag{4}$$

By the contraction mapping theorem it is straightforward to show that (4) has a unique positive solution for $B_0(t)$.

A key parameter is the 'net reproduction number'

$$R = \int_0^\infty b(a) e^{-\int_0^a \mu_0(s)ds} da,$$

the expected number of offspring that a single individual produces during his or her whole lifetime. If $R = 1$ then the McKendrick-Von Foester equation has a stationary distribution which depends on the initial age distribution (Sharpe and Lotka [2], Samuelson [3], Webb [4]). If $R > 1$ then the McKendrick-Von Foester equation produces exponential growth.

However real populations do not exhibit unbounded exponential growth and often stabilising factors come into force. Such stabilising factors might be the interaction of the population and the environment or the interaction of individuals within a population. We shall show that if the interaction term is small and R is close to one then the solution to the McK-VF equation is close to that of the full model with interaction term over a finite interval. However as we have already noted, the asymptotic behaviour of the two solutions as $t \to \infty$ is very different. In the next section we shall outline in detail the problem which we shall study. Then we shall give and sketch the proof of our main result. A brief summary concludes the paper.

2. The Problem

Boulanger [1] investigated the following population dynamical model with age interaction:

$$\frac{\partial \rho_\epsilon}{\partial a} + \frac{\partial \rho_\epsilon}{\partial t} = -\rho_\epsilon\Big[\mu_\epsilon(a) + \int_0^\infty \beta_\epsilon(a,s)\rho_\epsilon(s,t)ds\Big], \tag{5i}$$

with boundary conditions

$$\rho_\epsilon(0,t) = \int_0^\infty b(a)\rho_\epsilon(a,t)da \tag{5ii}$$

and initial condition

$$\rho_\epsilon(a,0) = \phi(a). \tag{5iii}$$

Here $\mu_\epsilon(a) = \mu_0(a) - \epsilon$ where $\mu_0(a)$ is such that

$$R = \int_0^\infty b(a)e^{-\int_0^a \mu_0(\xi)d\xi}da = 1,$$

and $\mu_o(a) \geq \epsilon$ for all a. Also $\beta_\epsilon(a,s) = \epsilon\overline{\beta}(a,s)$ where $\overline{\beta}(a,s) : \mathcal{R}^+ \times \mathcal{R}^+ \to \mathcal{R}$.

Boulanger assumes:

(*i*) $\mu_\epsilon(a)$, $\beta_\epsilon(a,s)$ and $\phi(a)$ are non-negative and continuous;

(*ii*) $\exists\ 0 \leq a_1, a_2 < \infty$ such that $b(a) \equiv 0$ for $a \notin (a_1, a_2)$;

(*iii*) $lim_{a\to\infty} \int_0^a \mu_\epsilon(s)ds = \infty$;

(*iv*) $\exists\ A > 0$ such that $\phi(a)$ is decreasing in (A, ∞).

Boulanger asserts that Webb [4] shows that these conditions imply that equations (5) and condition (2) have a unique solution for $\rho_\epsilon(a,t)$. We disagree with this and claim that the following stronger conditions are necessary:

(*i*) $\mu_\epsilon(a)$, $b(a)$ and $\beta_\epsilon(a,s)$ are bounded, positive and continuous functions and $\phi \in L^1[0,\infty)$ is a continuous positive function;

(*ii*) $\mu_\epsilon(a)$ is also continuously differentiable with respect to a and its derivative is also bounded on $[0,\infty)$;

(*iii*) $\beta_\epsilon(a,s)$ is also continuously differentiable with respect to a and its derivative is also bounded on $(a,s) \in [0,\infty) \times [0,\infty)$.

Boulanger investigated the relationship between the solution of these equations and the McK-VF equation under the following additional condition:

Let $\nu_\epsilon(a,\xi)$ be the solution of the equation

$$\frac{\partial \nu_\epsilon}{\partial a} + \epsilon\frac{\partial \nu_\epsilon}{\partial t} = -\epsilon\nu_\epsilon\Big[-1 + p_0\int_0^\infty \overline{\beta}(a,s)\nu_\epsilon(s,\xi)\ell_0(s)ds\Big],$$

with boundary condition

$$\nu_\epsilon(0,\xi) = \int_0^\infty b(a)\ell_0(a)\nu_\epsilon(a,\xi)da$$

and initial condition

$$\nu_\epsilon(a,0) = 1.$$

Here p_0 and $\ell_0(a)$ are given by

$$p_0 = \frac{\left(\int_0^\infty b(s)\ell_0(s)\int_0^s \frac{\phi(u)}{\ell_0(u)}\,du\,ds\right)}{\int_0^\infty sb(s)\ell_0(s)ds} \qquad \text{and} \qquad \ell_0(a) = e^{-\int_0^a \mu_0(\xi)d\xi}.$$

Boulanger assumed $\nu_\epsilon(a,\xi)$ analytic in ϵ uniformly in a and ξ. This is a strong condition which is not easy to check. Under this assumption Boulanger showed that

$$\lim_{\epsilon\to 0}\sup_{\{a\geq 0,\ T\geq t\geq 0\}} \mid \rho_0(a,t) - \rho_\epsilon(a,t) \mid = 0,$$

i.e. the solution of the full equations (5) with interaction term approach the solutions of the McK-VF equation on any finite time interval as $\epsilon \to 0$. Our main goal in this paper is to prove a similar uniform approximating property under simpler and more general conditions.

3. Results

Our main result is:

Theorem 1. Suppose that $\rho_\epsilon(a,t)$ is the solution of (5) and also that
(i) $\mu_\epsilon : [0,\infty) \to [0,\infty)$ is continuous for each $\epsilon \geq 0$, $s_1(\epsilon) = \sup_{\{a\geq 0\}} \mid \mu_\epsilon(a) - \mu_0(a) \mid < \infty$ and $s_1(\epsilon) \to 0$ as $\epsilon \to 0$, (so that $\mu_\epsilon(a) \to \mu_0(a)$ uniformly as $\epsilon \to 0$);
(ii) $\beta_\epsilon : [0,\infty) \times [0,\infty) \to [0,\infty)$ is continuous and bounded for all $\epsilon > 0$ and $s_2(\epsilon) = \sup_{\{a\geq 0,\ s\geq 0\}} \mid \beta_\epsilon(a,s) \mid \to 0$ as $\epsilon \to 0$, (so that $\beta_\epsilon(a,s) \to 0$ uniformly as $\epsilon \to 0$);
(iii) $b : [0,\infty) \to [0,\infty)$ is continuous and bounded;
(iv) $\phi : [0,\infty) \to [0,\infty)$ is continuous and bounded and $\int_0^\infty \phi(a)da < \infty$.
Then for $T > 0$,

$$\sup_{\{a\geq 0,\ T\geq t\geq 0\}} \mid \rho_\epsilon(a,t) - \rho_0(a,t) \mid \to 0 \text{ as } \epsilon \to 0,$$

(so that $\rho_\epsilon(a,t) \to \rho_0(a,t)$ uniformly on $0 \leq a,\ 0 \leq t \leq T$).

This Theorem proves Boulanger's main result under less restrictive conditions.

Proof. We make the transformation

$$\tilde{\rho}_\epsilon(a,t) = \rho_\epsilon(a,t)e^{\int_0^a \mu_\epsilon(s)ds},$$

and write

$$\tilde{\beta}_\epsilon(a,s) = \beta_\epsilon(a,s)e^{-\int_0^s \mu_\epsilon(u)du},$$

$$\tilde{b}_\epsilon(a) = b(a)e^{-\int_0^a \mu_\epsilon(u)du},$$

and

$$\tilde{\phi}_\epsilon(a) = \phi(a)e^{\int_0^a \mu_\epsilon(u)du}.$$

Then equations (5) reduce to

$$\frac{\partial\tilde{\rho}_\epsilon}{\partial a} + \frac{\partial\tilde{\rho}_\epsilon}{\partial t} = -\tilde{\rho}_\epsilon(a,t)\int_0^\infty \tilde{\beta}_\epsilon(a,s)\tilde{\rho}_\epsilon(s,t)ds, \tag{6i}$$

with boundary conditions

$$\tilde{\rho}_\epsilon(0,t) = \int_0^\infty \tilde{b}_\epsilon(a)\tilde{\rho}_\epsilon(a,t)da \tag{6ii}$$

and initial condition

$$\tilde{\rho}_\epsilon(a,0) = \tilde{\phi}_\epsilon(a). \tag{6iii}$$

The consistency condition (2) becomes:

$$\tilde{\phi}_\epsilon(0) = \int_0^\infty \tilde{b}_\epsilon(a)\tilde{\phi}_\epsilon(a)da.$$

Using the method of characteristics we deduce that

$$\tilde{\rho}_\epsilon(t,a) = \begin{cases} \tilde{\phi}_\epsilon(a-t)\ e^{-\int_0^t\int_0^\infty \tilde{\beta}_\epsilon(a-t+\sigma,s)\tilde{\rho}_\epsilon(\sigma,s)dsd\sigma}, & \text{if } a \geq t, \\ B_\epsilon(t-a)\ e^{-\int_0^a\int_0^\infty \tilde{\beta}_\epsilon(\sigma,s)\tilde{\rho}_\epsilon(t-a+\sigma,s)dsd\sigma}, & \text{if } t \geq a \end{cases} \tag{7}$$

with boundary condition

$$B_\epsilon(t) = \int_0^t \tilde{b}_\epsilon(a)\ B_\epsilon(t-a)\ e^{-\int_0^a\int_0^\infty \tilde{\beta}_\epsilon(\sigma,s)\tilde{\rho}_\epsilon(t-a+\sigma,s)dsd\sigma}da$$
$$+ \int_t^\infty \tilde{b}_\epsilon(a)\ \tilde{\phi}_\epsilon(a-t)\ e^{-\int_0^t\int_0^\infty \tilde{\beta}_\epsilon(a-t+\sigma,s)\tilde{\rho}_\epsilon(\sigma,s)dsd\sigma}da.$$

For $\epsilon = 0$ these equations reduce to

$$\tilde{\rho}_0(a,t) = \begin{cases} \tilde{\phi}_0(a-t), & a \geq t, \\ B_0(t-a), & t \geq a, \end{cases}$$

and

$$B_0(t) = \int_0^t \tilde{b}_0(a)B_0(t-a)da + \int_t^\infty \tilde{b}_0(a)\tilde{\phi}_0(a-t)da.$$

Since the improper integrals in (7) are positive

$$\tilde{\rho}_\epsilon(t,a) \leq \begin{cases} \tilde{\phi}_\epsilon(a-t), & a \geq t, \\ B_\epsilon(t-a), & t \geq a. \end{cases}$$

The first step is to show that $B_\epsilon(t)$ is bounded above on $[0,T]$ independently of ϵ.

Lemma 1. Let $K_1 = \sup_{\{a\geq 0\}} b(a)$, $K_2 = K_1\int_0^\infty \phi(u)du$ and $M = K_2e^{K_1t}$. Then $B_\epsilon(t) \leq M$ on $[0,T]$, so $B_\epsilon(t)$ is bounded above on $[0,T]$ independently of ϵ.

Proof. See Greenhalgh, Györi and Kovácsvölgyi [5].

Next consider the integral equations (7). We shall show that the improper integrals in (7) are uniformly bounded in σ and that the bound tends to zero as ϵ does.

Lemma 2. (*i*) For $a \geq t \geq \sigma$ and $T \geq t$:

$$\int_0^\infty \tilde{\beta}_\epsilon(a-t+\sigma,s)\tilde{\rho}_\epsilon(\sigma,s)ds \leq s_2(\epsilon)\left[MT + \int_0^\infty \phi(s)ds\right].$$

(*ii*) For $T \geq t \geq a \geq \sigma$:

$$\int_0^\infty \tilde{\beta}_\epsilon(\sigma,s)\tilde{\rho}_\epsilon(t-a+\sigma,s)ds \leq s_2(\epsilon)\left[MT + \int_0^\infty \phi(s)ds\right].$$

Proof. See Greenhalgh, Györi and Kovácsvölgyi [5].

Thus both of the improper integrals in (7) are bounded above independently of σ, a and t by a bound that goes to zero as ϵ does.

Lemma 3. $\mid B_\epsilon(t) - B_0(t) \mid \leq M_{1,\epsilon}e^{K_1T}$ where $M_{1,\epsilon} \to 0$ as $\epsilon \to 0$.

Proof. See Greenhalgh, Györi and Kovácsvölgyi [5].

So $B_\epsilon(t) \to B_0(t)$ as $\epsilon \to 0$ uniformly on $[0,T]$. Putting these results together we deduce Theorem 1 (see Greenhalgh, Györi and Kovácsvölgyi [5] for details).

4. Summary and Conclusions

We have discussed an age-dependent population model previously analysed by Boulanger [1]. This model was a small perturbation of the linear McKendrick-Von Foester (McK-VF) equation for population growth. For the McK-VF equation a key parameter is the 'net reproduction number R', which is the total number of offspring produced by an individual during his or her whole lifetime. If $R = 1$ then the linear equation has a stationary distribution which depends on the initial age-distribution, whereas for $R > 1$ the linear model produces exponential growth. Thus a small perturbation in the parameters can produce a very different asymptotic behaviour for the linear model.

Boulanger investigated a related model where the death rate is increased by interaction between generations. He showed that the solution to a small perturbation of the McK-VF model converged uniformly to the solution of the linear model over a finite time range (even though the age range was infinite). However Boulanger proved this result under quite restrictive conditions.

We proved Boulanger's result without assuming his complex analyticity condition, and also generalised some of his other assumptions. Boulanger assumed that $\mu_\epsilon(a) = \mu_0(a) - \epsilon$, and $\lim_{a\to\infty} \int_0^a \mu_0(\xi)d\xi = \infty$, we merely assume that $\mu_\epsilon(a) \to \mu_0(a)$ uniformly as $\epsilon \to 0$. Boulanger assumed that $b(a)$ has finite support, we merely assume $b(a)$ bounded. He assumed that $\phi(a)$ is eventually monotone decreasing, we assume that $\phi(a)$ is bounded. Thus we have proved Boulanger's result under simpler assumptions. This work is discussed in further detail in Greenhalgh, Györi and Kovácsvölgyi [5].

Acknowledgements

We are grateful to the University of Strathclyde Staff Travel fund, the Royal Society, (United Kingdom) and the University of Vészprem, Hungary for financial support for this work.

References

1. E.N. Boulanger, J. Math. Biol., **32**, 521-533 (1994).
2. F.R. Sharpe and A.J. Lotka, Phil. Mag. J. Sci., **21**, 435-438 (1911).
3. P.A. Samuelson, Human Biol., **48**, 559-580 (1976).
4. G.F. Webb, Theory of Nonlinear Age-Dependent Population Dynamics, Marcel Dekker, New York (1985).
5. D. Greenhalgh, I. Györi and I. Kovácsvölgyi, Math. Comp. Modelling, (1998) (in press).

9th Int. Coll. on Differential Equations, pp. 191–195
D. Bainov (Ed)

Global Asymptotic Stability of FitzHugh–Nagumo System

MAKOTO HAYASHI
Department of Mathematics, College of Science and Technology, Nihon University, Funabashi, Chiba, 274-8501, Japan
E-mail: mhayashi@penta.ge.cst.nihon-u.ac.jp

Abstract: The result that the FitzHugh–Nagumo system has no homoclinic orbits is given. By virture of the result, the global asymptotic stability of an equilibrium point of the system is discussed.
Key Words and Phrases: *Liénard system, Homoclinic orbit, Global asymptotic stability.*
1991 Mathematics Subject Classification. 34C05, 34C25, 34C37.

1. Introduction

Our purpose in this paper is to consider on the existence of the homoclinic orbit of the FitzHugh–Nagumo system ([1] and [2]), which is obtained by simplifying some nerve model. The result for the homoclinic orbit of the system will be given in §2 and is proved. By virture of the result, the global asymptotic stability of an equilibrium point of the system will be stated in §3. Finally the phase portrait as an example illustrating our results will be given in §4.

To explicate the iron mechanism for the excitation and the conduction of nerve Hodgkin and Huxley ([3]) developed the system of four nonlinear ordinary differential equations as a model of nerve conduction in the squid giant axon (Loligo). Since the functions in the Hodgkin–Huxley system are so complicated, R. FitzHugh and J. Nagumo et al. simplified the system by introducing the two dimentional model

$$\begin{cases} \dot{w} = v - \frac{1}{3}w^3 + w + I \\ \dot{v} = \rho(a - w - bv), \end{cases} \tag{FHN}$$

where the dot ($\dot{}$) denotes differetiation and a, ρ, b are real constants such that

[C1] $\qquad a \in \mathbf{R}, \qquad \rho > 0, \qquad 0 < b < 1.$

The variables w corresponds to the potential difference through the axon membrane and v represents the potassim activation (sodium inactivation). The quantity I is the current through the membrane. The system (FHN) for special values of I has been investigated by using numerical methods and phase space analysis in [1] or [2].

The system (FHN) has a unique equilibrium point (x_I, y_I) for each $I \in \mathbf{R}$, where

$$x_I = \sqrt[3]{\{3(I + a/b) + \sqrt{9(I + a/b)^2 + 4(1/b - 1)^3}\}/2}$$

$$+\sqrt[3]{\{3(I+a/b)-\sqrt{9(I+a/b)^2+4(1/b-1)^3}\}/2}$$

and

$$y_I = (a - x_I)/b.$$

Instead of the parameter I we introduce a new parameter η. By the transformation $\eta = x_I$, $x = w - \eta$ and $y = v - a/b + \eta/b + \rho b(w - \eta)$, the system (FHN) is transformed to the following system:

$$\begin{cases} \dot{x} = y - \left\{\frac{1}{3}x^3 + \eta x^2 + (\eta^2 + \rho b - 1)x\right\} \\ \dot{y} = -\frac{\rho b}{3}\left\{x^3 + 3\eta x^2 + 3(\eta^2 + \frac{1}{b} - 1)x\right\} \end{cases} \tag{FNS}$$

The system (FNS) is called the FitzHugh nerve system and has a unique equilibruim point $E(0,0)$. Also we note that this is a system of the Liénard type.

2. Homoclinic Orbit

We say that, if the system (FNS) has the orbits whose α- and ω-limit sets are the equilibrium point, then they are homoclinic orbits of the system. The first result is following

Theorem 2.1 *The system* (FHN) *has no homoclinic orbits.*

We prove the Theorem for the system (FNS) is equivalent to the system (FHN). We can easily seen that if $\rho b \geq 1$, the equilibrium point E of the system (FHS) is a saddle point. Thus in the case the system has no homoclinic orbits. Hence it is quite reasonable to assume that $0 < \rho b < 1$. Let $\eta_0^2 = 1 - \rho b$. Then instead of [C1], we can assume the condition

[C2] $\qquad a \in \mathbf{R}, \qquad 0 < b < 1, \qquad 0 < \rho < 1/b.$

The theorem is proved by dividing into three cases (i) $\eta^2 > \eta_0^2$, (ii) $\eta^2 < \eta_0^2$ and (iii) $\eta^2 = \eta_0^2$. By considering the variational matrix about E, we see easily that,

(i) if $\eta^2 > \eta_0^2$, then E is a stable focus or node,

(ii) if $\eta^2 < \eta_0^2$, then E is a unstable focus or node.

Thus in the cases (i) and (ii) the system (FNS) has no homoclinic orbits.

We set

$$F(x) = \frac{1}{3}x^2(x + 3\eta)$$

and

$$g(x) = \frac{\rho b}{3}\{x^3 + 3\eta x^2 + 3(\eta^2 + \frac{1}{b} - 1)x\}.$$

To prove the case (iii) $\eta^2 = \eta_0^2$ we need the following lemma ([4]) for the system (FHS).

Lemma 2.1 *Suppose that there exists a* $\delta > 0$ *such that* $F(x) > 0$ *for* $0 < |x| < \delta$. *If*

$$\frac{1}{F(x)}\int_0^x \frac{g(\xi)}{F(\xi)}d\xi > \frac{1}{4} \quad \text{for} \quad 0 < x < \delta$$

are satisfied, then the system (FNS) *has no homoclinic orbits.*

Let

$$\delta = \frac{-3\eta + \sqrt{9\eta^2 + 12\gamma\rho b}}{2} > 0,$$

where γ is a constant such that

$$0 < \gamma < \min\left\{\frac{6(1-\rho b)}{\rho b},\ 3\right\}.$$

Then we have $0 < \delta < 3\eta$. Since $\eta^2 + 1/b - 1 > 0$, we get for $0 < x < \delta$

$$\frac{1}{F(x)}\int_0^x \frac{g(\xi)}{F(\xi)}d\xi > \frac{3\rho b}{x^2(x+3\eta)}\int_0^x d\xi = \frac{3\rho b}{x(x+3\eta)} > \frac{\gamma\rho b}{\delta(\delta+3\eta)} = \frac{1}{4}.$$

Thus from Lemma 2.1 the system (FNS) with $\eta^2 = \eta_0^2$ has no homoclinic orbits.

We have completed the proof of Theorem 2.1. □

Remark. The unique existence and non-existence of the closed orbit of the system (FNS) has been treated by [5], [6], [7], [8] and [9] etc. However, the result for a homoclinic orbit of the system was not given until now.

We denote

$$C^+ = \{(x,y)\ :\ x > 0 \text{ and } y = F(x)\} \quad \text{and} \quad C^- = \{(x,y)\ :\ x < 0 \text{ and } y = F(x)\}.$$

From the above theorem we have the following:

- The positive (resp., negative) semitrajectory of (FNS) passing through any point on the curve C^+ (resp., C^-) meets the negative y-axis.
- The negative (resp., positive) semitrajectory of (FNS) passing through any point on the curve C^+ (resp., C^-) meets the positive y-axis.

Therefore we conclude the following result.

Lemma 2.2 *All positive semitrajectories of* (FNS) *near the equilibrium point E keep on rotating around E.*

3. Global Asymptotic Stability

We say that the equilibrium point E is globally asymptotically stable if E is stable and if every orbit of the system (FNS) tends to E. By virture of Theorem 2.1, we get the following

Theorem 3.1 *The equilibrium point E* (0, 0) *of the system* (FNS) *is globally asymptotically stable if*

$$\eta^2 = \eta_0^2 \le \frac{1}{b} - 1 \tag{1}$$

or

$$\eta^2 \ge \eta_0^2 \ \text{and}\ \eta^4 - 4\eta^2\eta_0^2 + \eta_0^4 + 2(\frac{1}{2} - 1)\eta^2 - 4(\frac{1}{b} - 1)\eta_0^2 + 4(\frac{1}{b} - 1)^2 \ge 0 \tag{2}$$

or

$$2\{\eta_0^2 + (\frac{1}{b} - 1)\}^3 < \eta^2\{\eta^2 + 3(\frac{1}{b} - 1)\}^2. \tag{3}$$

We see from the uniqueness of orbits for the initial value problem and the Poincaré–Bendixson theorem that, if the system (FNS) satisfies the following conditions:

(iv) all orbits are bounded in the future,

(v) no closed orbits exist,

(vi) E is asymptotically stable,

then the unique equilibrium point E is globally asymptotically stable.

We shall check the above conditions to prove Theorem 3.1. Assume that the condition (1) or (2) or (3) holds.

(iv) From [10;Theorem 2.1] all orbits of the system (FNS) are bounded in the furture.

(v) From the results in [8] and [7], the system (FNS) has no closed orbits.

(vi) Suppose that E is not stable. Then from Lemma 2.2 every positive semitrajectories of (FNS) starting in the neighborhood of E keep on rotating around E and go away from E. Hence, by the fact (iv) and the Poincaré–Bendixson theorem, the system has a closed orbit. This is a contradiction to the fact (v).

Therefore the proof of Theorem 3.1 is completed. □

In [11] and [6], we gave the result that the system (FNS) has a unique closed orbit if $\eta^2 = \eta_0^2 > 1/b - 1$. Thus from Theorem 3.1 we have the following

Theorem 3.2 *The equilibrium point E of the system* (FNS) *is globally asymptotically stable if and only if* $\eta^2 = \eta_0^2 \leq 1/b - 1$.

4. A Numerical Example

We shall present the phase portrait as an example illustrating an application of Theorem 3.2. Let $b = 4/5$ and $\rho = 1$. Then the system satisfies the condition $\eta^2 = \eta_0^2 \leq 1/b - 1$. We see that the unique equilibrium point $(0,0)$ of the system is globally asymptotically stable as is shown in Figure below.

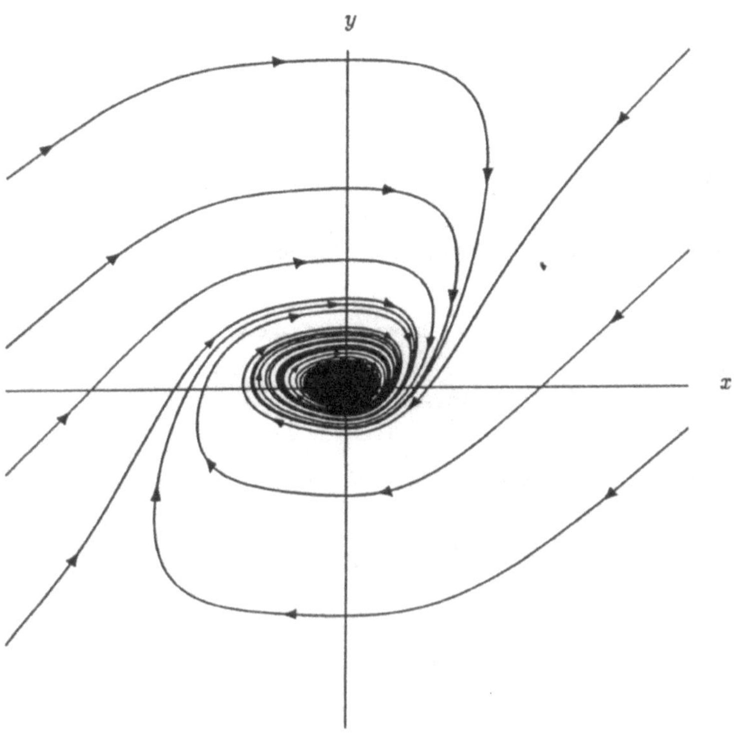

References

1. R. FitzHugh, Biophysical, **1**, 445–466 (1961).
2. J. Nagumo, S. Arimoto and S. Yoshizawa, Proc. IRE., **50**, 2061–2070 (1962).
3. A. L. Hodgkin and A. F. Huxley, J. Physiol., **117**, 500–544 (1952).
4. J. Sugie and T. Hara, Disc. Conti. Dynam. Syst., **2**, 237–254 (1996).
5. M. Hayashi, Quart. Appl. Math., (in press).
6. M. Hayashi and M. Tsukada, (submitted).
7. J. Sugie, Quart. Appl. Math., **49**, 543–554 (1991).
8. E. Kaumann and U. Staude, Lecture Notes in Math., **1017**, 313–321 (1983).
9. S. A. Treskov and E. P. Volokitin, Quart. Appl. Math., **54**, 601–607 (1996).
10. J. Graef, J. Diff. Eqs., **12**, 34–62 (1972).
11. M. Hayashi, (submitted).

9th Int. Coll. on Differential Equations, pp. 197–203
D. Bainov (Ed)

Oscillation Results for Semilinear Elliptic Equations

VELMER B. HEADLEY
Department of Mathematics, Brock University, St. Catharines, Ontario L2S 3A1, Canada

Abstract: In this paper we show how to extend, to the semilinear elliptic equation $(-1)^m \sum_{|\alpha|=|\beta|=m} D^\alpha[A_{\alpha,\beta}(x)D^\beta v]+c(x)\,|v|^{\gamma-1}\,v=0$ or its special case $(-\Delta)^m v+c(x)v\,|v|^{\gamma-1}=0$, known nonoscillation theorems for linear elliptic equations.

1. Preliminary Results

Let Ω be an unbounded open subset of $\mathbf{R}^n$, and let the semilinear elliptic differential operator L be defined as follows

$$Lu := \sum_{|\alpha|=0}^{m}\sum_{|\beta|=0}^{m}(-1)^{|\alpha|}D^\alpha\left[A_{\alpha,\beta}(x)D^\beta u\right]+V(x,u) \quad (x\in\Omega\subseteq\mathbf{R}^n), \tag{1.0.1}$$

where $V\in C(\Omega\times\mathbf{R};\mathbf{R})$, the multi-index notation is as in [1], and the coefficients $A_{\alpha\beta}(x)(|\alpha| \leq m,\ |\beta|\leq m,\ x\in\Omega)$ are real-valued, satisfy the symmetry conditions

$$A_{\alpha\beta}(x)=A_{\beta\alpha}(x)\,(|\alpha|\leq m,\ |\beta|\leq m,\ x\in\Omega), \tag{1.0.2}$$

and are sufficiently smooth on Ω.

The *ellipticity constant* $E_0(L;\Omega)$ for the differential operator L is defined as follows:

$$0<E_0(L;\Omega):=\inf\left\{\sum_{|\alpha|=|\beta|=m}A_{\alpha,\beta}(x)\xi^{\alpha+\beta} : x\in\Omega,\ \xi\in\mathbf{R}^n, |\xi|=1\right\}. \tag{1.0.3}$$

If k is any positive integer, and if G is any nonempty open subset of Ω, we define the norm $\|\cdot\|_{k,G}$ as in [1]; namely,

$$\|\varphi\|_{k,G} := \left(\sum_{|\alpha|\leq m}\int_G |D^\alpha\varphi|^2\right)^{1/2}. \tag{1.0.4}$$

We let $H^k(G)$ denote the Sobolev space $W^{k,2}(G)$ (see [2]); and we let $H_0^k(G)$ denote the completion of $C_0^\infty(G)$ with respect to the norm $\|\cdot\|_{k,G}$.

DEFINITION 1.1. If G is a bounded nonempty open subset of Ω that satisfies the hypotheses of [1, Lemma 9.1], and if the differential equation

$$Lu=0 \tag{1.1.1}$$

has a nontrivial solution u in $H_0^m(G) \cap C^{2m}(G)$ such that each $D^\alpha u\,(|\alpha| \leq m-1)$ is continuous in some neighborhood of each point of the boundary of G, then G is called a *nodal domain* for L. We will say that (1.1.1) is ***nodally oscillatory*** in Ω iff, for every $r > 0$, the region $\{x \in \Omega : |x| > r\}$ contains the closure of at least one nodal domain for L. We will say that (1.1.1) is ***nodally nonoscillatory*** in Ω iff (1.1.1) is not nodally oscillatory in Ω.

DEFINITION 1.2. Let $\mathbf{N}$ denote the set of all natural numbers. Following [3] and [4], we will say that a nontrivial, global, C^{2m} solution of (1.1.1) is *oscillatory* in Ω iff the (open) set $\{x \in \Omega : u(x) \neq 0\}$ is unbounded and is expressible as the union of a countable collection, $\{G_k : k \in \mathbf{N}\}$, of mutually disjoint, connected, bounded, open sets such that:

(i) $\forall k \in \mathbf{N}, \|u\|_{m,G_k} < \infty$;

(ii) given any $r > 0$, there exists at least one G_k contained in the region $\{x \in \Omega : |x| > r\}$.

We will say that a nontrivial, global, C^{2m} solution u of (1.1.1) is *nonoscillatory* in Ω iff u is not oscillatory in Ω.

REMARK 1.3. Note that if u is radially symmetric and is oscillatory in the sense of Definition 1.2, then each G_k in Definition 1.2 is a bounded, open, spherical shell such that u is identically zero on ∂G_k, the boundary of G_k.

Note also that if u is oscillatory in the sense of Definition 1.2 (but is not necessarily radially symmetric), then its zero-set, $\{x \in \Omega : u(x) = 0\}$, is unbounded, and therefore u is also oscillatory in the traditional sense.

DEFINITION 1.4. Let G be a nonempty open subset of Ω. Following [4], we will say that G has *bounded thickness* iff we can find a line Γ such that every line Γ' parallel to Γ has the property that every maximal connected subset of $\Gamma' \cap G$ has bounded diameter. The supremum of all such diameters is called the *thickness of G in the direction* Γ. Note that if G is bounded, then it has bounded thickness in every direction; the supremum of all such thicknesses will be called the *thickness* of G.

As examples, we note that the bounded spherical shell $\{x \in \Omega : r_1 < |x| < r_2\}$, where $0 < r_1 < r_2 < \infty$, has thickness $2\,(r_2^2 - r_1^2)^{1/2}$ in every direction, and that the unbounded cylindrical shell $\left\{(y_1, \ldots, y_n, y_{n+1}) \in \mathbf{R}^{n+1} : r_1 < \left(\sum_{k=1}^{n} y_k^2\right)^{1/2} < r_2\right\}$ has thickness $2\,(r_2^2 - r_1^2)^{1/2}$ in any direction orthogonal to the unit vector $(0, \ldots, 0, 1)$.

REMARK 1.5. Let

$$F_G[L;\varphi] := F[L;G;\varphi] := \sum_{|\alpha|=0}^{m} \sum_{|\beta|=0}^{m} \int_G A_{\alpha\beta}(x) D^\alpha \varphi D^\beta \varphi \, dx, \tag{1.5.1}$$

whenever the right-hand side of (1.5.1) exists. We state, as our first preliminary result, the following theorem that we obtained in an earlier paper [4].

THEOREM 1.6 [4, Theorem 2.8] Let $V(x,u) \equiv 0$. Suppose that the equation (1.1.1) has at least one nontrivial oscillatory solution in Ω, and that each member of the family $\{G_k : k \in \mathbf{N}\}$ in Definition 1.2 has thickness $\tau_k \in (0,\infty)$ and has a family of nonempty open subsets G'_{k,t_k} (having thickness t_k, with $0 < t_k \leq \tau_k$ and $G'_{k,\tau_k} := G_k$) which satisfy the following three hypotheses:

(H1) $\exists k_0 \in \mathbf{N}$ such that $\forall k \in \mathbf{N}$, if we replace G by G'_{k,t_k} in Definition 1.4, then the set $\Gamma' \cap G$ in Definition 1.4 has at most k_0 maximal connected subsets;

(H2) $\forall k \in \mathbf{N}$, the Lebesgue measure of G'_{k,t_k} depends continuously on t_k;

(H3) If $0 < t_{k_1} < t_{k_2}$, then $G'_{k_1,t_{k_1}} \neq G'_{k_2,t_{k_2}}$ and $G'_{k_1,t_{k_1}} \subset G'_{k_2,t_{k_2}}$.

Suppose also that for any bounded, open set $G \subset \Omega$, and for any φ in $C_0^\infty(G)$, we have

$$F_G[L;\varphi] \leq 0. \tag{1.6.1}$$

Then the equation (1.1.1) is nodally oscillatory.

REMARK 1.7. The hypothesis (1.6.1) is satisfied if a certain symmetric matrix S is negative semidefinite throughout Ω. The entries of the matrix S consist of the coefficients $A_{\alpha,\beta}(x)$ $(|\alpha| \leq m, |\beta| \leq m, x \in \Omega)$, arranged in an appropriate manner. For details, we refer the reader to [4, Remark 2.7].

2. Main Results

In this section, we will show how to extend, to the semilinear elliptic equations mentioned in the Abstract to this paper, known nonoscillation theorems for linear elliptic equations. Our proofs will make use of the following Sturmian theorem.

THEOREM 2.1. Define the differential operator M_1 as follows:

$$M_1 u := \sum_{|\alpha|,|\beta|=0}^{m} (-1)^{|\alpha|} D^\alpha[a_{\alpha\beta}(x) D^\beta u] \quad (x \in \Omega \subseteq \mathbf{R}^n), \tag{2.1.1}$$

where the coefficients $a_{\alpha,\beta}(x)$ $(|\alpha| \leq m, |\beta| \leq m, x \in \Omega)$ are real-valued and satisfy symmetry and regularity conditions analogous to those satisfied by the coefficients $A_{\alpha,\beta}(x)(|\alpha| \leq m, |\beta| \leq m, x \in \Omega)$. In particular, we suppose that each of the functions $a_{\alpha,\beta}(|\alpha| \leq m, |\beta| \leq m)$ satisfies the hypotheses of the global version of Gårding's inequality [1, Theorem 7.6], and belongs to $C^q(\Omega)$, where $q > (n/2) + 2m - 1 + |\alpha| + |\beta|$. Suppose that

$$\forall (x,u) \in \Omega \times \mathbf{R}, \qquad V(x,u) \equiv 0, \tag{2.1.2}$$

and that for every $\varphi \in C_0^\infty(\Omega)$,

$$\int_\Omega \varphi M_1 \varphi \leq \int_\Omega \varphi L \varphi. \tag{2.1.3}$$

[Note that if the coefficients of L dominate those of M_1, then (2.1.3) holds.] Suppose also that the equation (1.1.1) has a nontrivial oscillatory solution u, and that the family $\{G_k : k \in \mathbf{N}\}$ in Definition 1.2 has the property that for each positive integer k, the open set G_k has thickness $\tau_k \in (0,\infty)$ and has a family of nonempty open subsets G'_{k,t_k} (having thickness t_k, with $0 < t_k \leq \tau_k$ and $G'_{k,\tau_k} := G_k$), which satisfy the hypotheses (H1), (H2) and (H3) in the statement of Theorem 1.6. Then the equation

$$M_1 v = 0 \tag{2.1.4}$$

has at least one nontrivial oscillatory solution.

PROOF. It is known [5, p. 32, Theorem 1] that, for each positive integer k, the real number $\inf\left\{F[L;G_k;\varphi] : \varphi \in C_0^\infty(G_k), \|\varphi\|_{0,G_k} = 1\right\}$ is the smallest eigenvalue of the boundary-value problem

$$Lw = \lambda w \text{ in } G_k, \quad w \in H_0^m(G_k) \cap C^{2m}(G_k). \tag{2.1.5}$$

Since u is a nontrivial solution of (1.1.1), therefore zero is an eigenvalue of (2.1.5). It follows, from these observations and the hypothesis (2.1.3), that

$$\begin{aligned} &\inf\left\{F[M_1;G_k;\varphi] : \varphi \in C_0^\infty(G_k), \|\varphi\|_{0,G_k} = 1\right\} \\ \leq\ &\inf\left\{F[L;G_k;\varphi] : \varphi \in C_0^\infty(G_k), \|\varphi\|_{0,G_k} = 1\right\} \leq 0. \end{aligned} \tag{2.1.6}$$

If G is any nonempty open subset of Ω, let

$$\lambda_0(L;G) := \inf\left\{\|\varphi\|_{0,G}^{-2} F_G[L;\varphi] : 0 \not\equiv \varphi \in C_0^\infty(G)\right\}. \tag{2.1.7}$$

Using the statement (2.1.6) and the definition (2.1.7), together with the statement of [4,Lemma 2.4] and the proof of [4, Lemma 2.5], we see that there exist positive constants b_0, c_1 and a nonnegative constant c_3 such that

$$0 \geq \lambda_0(M_1; G_{k,t_k}) \geq (k_0 t_k)^{-2m} b_0^{-2} c_1 E_0(M_1;\Omega) - c_3. \tag{2.1.8}$$

The constants b_0 and c_1 depend on m and n; the constant c_3 depends on $m, n, E_0(M_1;\Omega), \sup\{|a_{\alpha\beta}(x)| : x \in \Omega, |\alpha| \leq m, |\beta| \leq m\}$ and the modulus of continuity for the principal coefficients $a_{\alpha\beta}(x)(|\alpha| = m, |\beta| = m, x \in \Omega)$. Using (2.1.8) and [4, Lemma 2.6], we see that, for each positive integer k,

$$\lim_{t_k \to 0+} \lambda_0(M_1; G'_{k,t_k}) = +\infty. \tag{2.1.9}$$

From (2.1.8), (2.1.9) and the fact that the function $t_k \longmapsto \lambda_0(M_1; G'_{k,t_k})$ is continuous and nondecreasing (by [4, Lemma 2.6]) we deduce that for each positive integer k there exist $\sigma_k(\in (0,t_k])$ and $G'_k := G'_{k,\sigma_k} \subseteq G'_{k,t_k}$ such that $\lambda_0(M_1; G'_k) = 0$. It follows from [4, Lemma 2.5] that, for each positive integer k, the equation $M_1[v_k] = 0$ has a nontrivial solution $v_k \left[\in H_0^m(G'_k) \cap C^{2m}(G'_k)\right]$. Let

$$w := \sum_{k \in N} v_k \chi_{G'_k} \text{ and } w_N := \sum_{k=1}^{N} v_k \chi_{G'_k} (N \in \mathbf{N}). \tag{2.1.10}$$

Since each v_k is in $H_0^m(G'_k) \cap C^{2m}(G'_k)$, it follows, from [1, Lemma 9.1], that

$$D^\alpha v_k = 0\,(|\alpha| \leq m-1) \text{ at each point of the boundary of } G'_k. \tag{2.1.11}$$

Furthermore, for each positive integer k, the set $\overline{G'_k}$ is compact, hence the continuity of the Euclidean norm $|\cdot| : \mathbf{R}^n \to [0,\infty)$ implies that there exist constants $r_k^* \geq 0, r_k^{**} \geq 0$ and points $y_k \in \overline{G'_k}$, $z_k \in \overline{G'_k}$ such that $r_k^* = |y_k| = \inf\left\{|x| : x \in \overline{G'_k}\right\}$ and $r^{**} = |z_k| = \sup\left\{|x| : x \in \overline{G'_k}\right\}$. [We note that since the set $\bigcup_{k \in \mathbf{N}} G_k$ is unbounded, therefore $\lim_{l \to \infty} r_l^* = \infty$ and $\lim_{l \to \infty} r_l^{**} = \infty$.]

Using the definition of M_1 given in (2.1.1), together with integrating by parts, the boundary conditions (2.1.11), an analog of the symmetry hypothesis (1.0.2) for M_1, and the differential equation $M_1[v_k] = 0$, we see that, $\forall N \in \mathbf{N}$ & $\forall \psi \in C_0^\infty(\Omega \cap \overline{B}(0; r_N^{**}))$,

$$\int_{\Omega \cap B(0;r_N^{**})} w_N M_1 \psi = \int_{\Omega \cap B(0;r_N^{**})} \sum_{k=1}^{N} v_k \chi_{G'_k} M_1 \psi = \sum_{k=1}^{N} \int_{G'_k} v_k M_1 \psi \tag{2.1.12}$$

$$= \sum_{k=1}^{N} \int_{G'_k} v_k \sum_{|\alpha|,|\beta| \leq m} (-1)^{|\alpha|} D^\alpha \left[a_{\alpha,\beta}(x) D^\beta \psi\right] dx$$

$$= \sum_{k=1}^{N} \int_{G'_k} \sum_{|\alpha|,|\beta| \leq m} (D^\alpha v_k)\, a_{\alpha\beta}(x) D^\beta \psi dx$$

$$= \sum_{k=1}^{N} \int_{G'_k} \sum_{|\alpha|,|\beta|\le m} (-1)^{|\beta|} \psi D^{\beta} \left(a_{\alpha\beta}(x) D^{\alpha} v_k\right) dx$$

$$= \sum_{k=1}^{N} \int_{G'_k} \sum_{|\beta|,|\alpha|\le m} (-1)^{|\alpha|} \psi D^{\alpha} \left(a_{\beta\alpha}(x) D^{\beta} v_k\right) dx$$

$$= \sum_{k=1}^{N} \int_{G'_k} \sum_{|\beta|,|\alpha|\le m} (-1)^{|\alpha|} \psi D^{\alpha} \left(a_{\alpha,\beta}(x) D^{\beta} v_k\right) dx = \sum_{k=1}^{N} \int_{G'_k} \psi M_1 v_k = 0.$$

Using a known regularity theorem [6, Theorem 5.II, p. 36], we now show that (2.1.12) implies the conclusion of Theorem 2.1, as follows. The definition of w_N given in (2.1.10) and the fact that the family $\{G'_k : k \in \mathbf{N}\}$ is disjoint imply that, for every multi-index α such that $\alpha \le 2m$, and for every $N \in \mathbf{N}$, we have

$$|D^{\alpha} w_N|^2 = \sum_{k=1}^{N} |D^{\alpha} v_k|^2 \chi_{G'_k}. \tag{2.1.13}$$

It is also clear that for every $N \in \mathbf{N}$,

$$w_N \in H^m\left(\Omega \cap B(0; r_N^{**})\right). \tag{2.1.14}$$

Using the hypothesis that each of the functions $a_{\alpha\beta}$ ($|\alpha| \le m, |\beta| \le m$) belongs to $C^q(\Omega)$, where $q > (n/2) + 2m - 1 + |\alpha| + |\beta|$, together with the statements (2.1.12) and (2.1.14) and the regularity theorem cited above, we see that, for every $N \in \mathbf{N}$, there exists a nontrivial function $p_N : \Omega \cap \overline{B}(0; r_N^{**}) \to \mathbf{R}$ such that $p_N = w_N$ a.e. in $\Omega \cap B(0; r_N^{**})$ [and, hence, $p_N = w$ a.e. in $\Omega \cap B(0; r_N^{**})$], $M_1 p_N = 0$ throughout $\Omega \cap B(0; r_N^{**})$, and, $\forall k \ge N, p_N \in H_0^m(G'_k) \cap C^{2m}\left(\Omega \cap B(0; r_k^{**})\right)$. It follows that there exists a nontrivial function $v : \Omega \to \mathbf{R}$ such that $v = w$ a.e. in Ω, the differential equation $M_1 v = 0$ holds throughout Ω, and, $\forall k \ge 1, v \in H_0^m(G'_k) \cap C^{2m}(\Omega)$. To show that v is oscillatory in Ω, we proceed as in [4, Proof of Theorem 2.1, Last 12 lines], replacing u in that proof with v. We omit the details.

REMARK 2.2. As our first illustration of the technique mentioned in the Abstract to this paper, we extend, to the semilinear elliptic equation

$$(-\Delta)^m v + c(x) v\, |v|^{\gamma-1} = 0 \qquad (x \in \Omega := \{x \in \mathbf{R}^n : |x| > r_1\}), \tag{2.2.1}$$

the nonoscillation portion of a result due to Müller-Pfeiffer [7, Theorem 1]. (Here, r_1 is a positive constant.)

THEOREM 2.3. Suppose that $0 < \gamma < 1$, and that m is even. Suppose also that either $n \ge 2m$ or $n < 2m$ and n is odd. Let

$$\prod_{k=0}^{m-1} \left[s + \left(\frac{n - 2m + 4k}{2} \right)^2 \right] = \sum_{j=0}^{m} q_j s^j. \tag{2.3.1}$$

Suppose also that there exists a constant $r_0 > r_1$ such that

$$c_-(x) < \left(q_0 + \sum_{j=1}^{m} \frac{[(2j-1)!!]^2 q_j}{4^j (\log |x|)^{2j}} \right) |x|^{-2m} \text{ whenever } |x| > r_0. \tag{2.3.2}$$

Then every bounded, nontrivial, global solution of (2.2.1) is nonoscillatory in the region $\Omega := \{x \in \mathbf{R}^n : |x| > r_1\}$.

PROOF. Let v_1 be a bounded, nontrivial, global solution of (2.2.1). Then

$$\exists K_2 > 0 \text{ such that } \forall x \in \Omega, |v_1(x)| \leq K_2. \tag{2.3.3}$$

Suppose (in order to obtain an eventual contradiction) that v_1 is oscillatory in Ω. Then Definition 1.2 implies that the (open) set $\Omega_1 := \{x \in \Omega : v_1(x) \neq 0\}$ is unbounded. Define the linear elliptic differential operators M_2 and M_3 as follows:

$$M_2 v := (-\Delta)^m v + (-1)^m c(x) K_2^{1-\gamma} |v_1(x)|^{\gamma-1} v \qquad (x \in \Omega_1); \tag{2.3.4}$$

$$M_3 v := (-\Delta)^m v + (-1)^{m+1} c_-(x) K_2^{1-\gamma} |v_1(x)|^{\gamma-1} v \qquad (x \in \Omega_1). \tag{2.3.5}$$

Since $0 < \gamma < 1$, the statement (2.3.3) implies that

$$\forall x \in \Omega_1, K_2^{1-\gamma} |v_1(x)|^{\gamma-1} \geq 1. \tag{2.3.6}$$

The statement (2.3.6) implies that

$$\forall x \in \Omega_1, -c_-(x) K_2^{1-\gamma} |v_1(x)|^{\gamma-1} \leq -c_-(x) \leq c(x). \tag{2.3.7}$$

Using the statement (2.3.7), together with the hypotheses that m is even and v_1 is oscillatory in Ω, we deduce from Theorem 2.1 that

$$\text{The equation} \quad M_3 v = 0 \quad \text{has at least one nontrivial oscillatory solution.} \tag{2.3.8}$$

Since m is even and $0 < \gamma < 1$, the statement (2.3.6) and the definition (2.3.5) imply that we can choose the positive constant K_2 so large that the symmetric matrix S mentioned in Remark 1.7 and corresponding to M_3 is negative semidefinite throughout Ω_1, hence we deduce from (2.3.8) and Theorem 1.6 that

$$\text{The equation} \quad M_3 v = 0 \quad \text{is nodally oscillatory.} \tag{2.3.9}$$

On the other hand, the hypothesis (2.3.2) and [7, Theorem 1] imply that the equation $M_3 v = 0$ is nodally nonoscillatory. This contradicts (2.3.9). The proof of Theorem 2.3 is now complete.

REMARK 2.4. As our second illustration of the technique mentioned in the Abstract to this paper, we extend, to the semilinear elliptic equation

$$M_4 v := (-1)^m \sum_{|\alpha|=|\beta|=m} D^\alpha \left[A_{\alpha,\beta}(x) D^\beta v\right] + c(x) |v|^{\gamma-1} v = 0, \tag{2.4.1}$$

a nonoscillation result due to Allegretto [8, Theorem 1].

THEOREM 2.5. Suppose that $0 < \gamma < 1$, that $0 \notin \overline{\Omega}$, that $n \neq 2$, and that m is even. Define the function $T : \mathbf{R} \rightarrow \mathbf{R}$ as follows

$$T(s) := (4 - s - n)^2 (n - s)^2 / 16 \tag{2.5.1}$$

$$+ \inf \left\{ k(k + n - 2) \left[k^2 + (n - 2)k + \frac{n^2 - 4n + 4s - s^2}{2}\right] : k \in \mathbf{N} \cup \{0\} \right\}.$$

Suppose also that

$$\limsup_{|x| \to \infty} |x|^{2m} c_-(x) < E_0(M_4; \Omega) \prod_{j=0}^{(m/2)-1} T(-4j). \tag{2.5.2}$$

Then every bounded, nontrivial, global solution of (2.4.1) is nonoscillatory in Ω.

PROOF. We proceed as in the proof of Theorem 2.3. Let v_1 be a bounded, nontrivial, global solution of (2.4.1). Then (2.3.3) holds. Suppose (in order to obtain an eventual contradition) that v_1 is oscillatory in Ω. Then Definition 1.2 implies that the (open) set $\Omega_1 := \{x \in \Omega : v_1(x) \neq 0\}$ is unbounded. Define the linear elliptic differential operators M_5 and M_6 as follows:

$$M_5 v := (-1)^m \sum_{|\alpha|=|\beta|=m} D^\alpha \left[A_{\alpha,\beta}(x) D^\beta v\right] + (-1)^m c(x) K_2^{1-\gamma} |v_1(x)|^{\gamma-1} v; \tag{2.5.3}$$

$$M_6 v := (-1)^m \sum_{|\alpha|=|\beta|=m} D^\alpha \left[A_{\alpha,\beta}(x) D^\beta v\right] + (-1)^{m+1} c_-(x) K_2^{1-\gamma} |v_1(x)|^{\gamma-1} v. \tag{2.5.4}$$

The statements (2.3.6) and (2.3.7) remain valid, and the statement (2.3.8) is replaced by the following statement:

$$\text{The equation} \quad M_6 v = 0 \quad \text{has at least one nontrivial oscillatory solution.} \tag{2.5.5}$$

Since m is even and $0 < \gamma < 1$, the statement (2.3.6) and the definition (2.5.4) imply that we can choose the positive constant K_2 so large that the symmetric matrix S mentioned in Remark 1.7 and corresponding to M_6 is negative semidefinite throughout Ω_1, hence we deduce from (2.5.5) and Theorem 1.6 that

$$\text{The equation} \quad M_6 v = 0 \quad \text{is nodally oscillatory.} \tag{2.5.6}$$

On the other hand, the hypothesis (2.5.2) and [8, Theorem 1] imply that the equation $M_6 v = 0$ is nodally nonoscillatory. This contradicts (2.5.6). The proof of Theorem 2.5 is now complete.

References

1. S. Agmon, Lectures on Elliptic Boundary Value Problems, Van Nostrand, Princeton (1965).
2. R. A. Adams, Sobolev Spaces, Academic Press, New York (1975).
3. V. B. Headley, J. Math. Anal. Appl., **143**, 379-393 (1989).
4. V. B. Headley, in: Proceedings of a Conference on Comparison Methods and Stability Theory, X. Z. Liu and D. Siegel (Eds.), Marcel Dekker, New York (1994) pp. 159-168.
5. S. G. Mikhlin, The Problem of the Minimum of a Quadratic Functional, State Publishing House for Technical Literature, Moscow (1952); English Translation, Holden-Day, New York (1965).
6. G. Fichera, Linear Elliptic Differential Systems and Eigenvalue Problems, Lecture Notes in Mathematics No. 8, Springer-Verlag, Berlin (1965).
7. E. Müller-Pfeiffer, Acta Math. Acad. Sci. Hungar., **38**, 139-150 (1981).
8. W. Allegretto, Pacific J. Math., **64**, 1-16 (1976).

9th Int. Coll. on Differential Equations, pp. 205–211
D. Bainov (Ed)

A Simple Model of the Ocean with Eastern Boundary Slope Current

P.F. HODNETT AND Y. YUAN
Department of Mathematics and Statistics, University of Limerick, Ireland
J. HUTHNANCE
Proudman Oceanography Laboratory, Birkenhead, England

Abstract: A model is developed to simulate how an eastern boundary slope current (known to exist at the eastern margin of oceans) interacts with the large scale ocean circulation. A simple, steady model of ocean circulation in the Northern Hemisphere (based on the thermocline equations) is used with an appropriate condition at the eastern boundary of the ocean to represent the influence of the slope current on the ocean circulation. The detailed dynamics of the slope current is not resolved here but the gross influence of the ocean circulation on the slope current is taken into account. The influence of the eastern slope current on the ocean circulation is simulated by assigning a non-zero value to the integrated mass transport at the eastern boundary of the ocean. The solution is for an equatorial region extending from the equator to 18.1°N, a subtropical region between 18.1°N and 53.5°N and a subpolar region between 53.5°N and 60°N. It transpires that the value of the coefficient of vertical thermal diffusion, k_z is critically important in determining **a)** a realistic distribution of temperature in the ocean and **b)** values of the mass transported across the eastern boundary of the ocean which accord with observations (i.e. of order one Sverdrup per 1000km). The value of k_z so determined for thermocline depth of 562m is $1.25\text{cm}^2/\text{sec}$. which confirms previous estimates.

1. Introduction

This paper attempts to address the question of how the large scale ocean circulation interacts with processes occurring at the eastern boundary of the ocean - in particular how the ocean circulation interacts with slope currents which are known to exist at the eastern margin of oceans (see [1]; [2]). This is done through developing a simple ocean model where the temperature is zonally uniform and on which an appropriate condition at the eastern boundary of the ocean represents the influence of the slope current on the ocean circulation. Since the detailed dynamics of the slope current is not considered therefore the detailed influence of the ocean circulation on the slope current is not described here. However the gross influence of the ocean circulation on the slope current is taken into account.

The model for the ocean circulation is based on the thermocline equations which provide an acceptable description of the dynamics in the interior of the ocean below the surface mixing (Ekman) layer and outside the ocean western boundary layer (ex. the Gulf Stream in the North Atlantic). Also assumed is that the temperature (or equivalently the density) does not depend on longitude. As noted by Welander [3] such regimes do exist in regions of the large oceans (for example the northern Pacific ocean). However, as noted by McCreary [4] longitudinal variation of temperature is important in an eastern thermal-diffusion boundary layer. Therefore, the neglect here of zonal variation in temperature assumes that such variations in the ocean interior (the domain of the model) are much less than across the eastern boundary domain (not described in detail here). In a recent paper Hodnett and Ansari [5] addressed some questions concerning how the large scale ocean circulation is affected by various eastern boundary shapes. As here, they assumed that the temperature is zonally uniform with the additional assumption that thermal diffusion is negligible by using what are called the ideal fluid thermocline equations. In order to obtain a realistic temperature field without thermal fronts it is necessary to include non-zero thermal diffusion. This necessitates numerically solving the energy equation for the temperature, T and also the resulting expression for the zonal velocity component, u. The Hodnett and Ansari [5] solution, with zero thermal diffusion, is analytical. Hodnett and Yuan [6] investigated the effect of including small thermal diffusion in the Hodnett and Ansari [5] model. They discovered that even small thermal diffusion gives rise to significantly different behaviour of the flow field in the eastern boundary region of the ocean. At the eastern boundary of the ocean both Hodnett and Ansari [5] and Hodnett and Yuan [6] take the normal (zonal) mass transport (integrated over the ocean depth) to be zero. With this eastern boundary condition there is no interaction between the ocean circulation and possible along-slope currents at the eastern boundary of the ocean. Here to simulate the influence of the eastern slope current on the ocean circulation the integrated mass transport at the eastern boundary of the ocean is assigned a non-zero value. Huthnance [7] estimates that the eastern slope current, primarily along-slope, exchanges cross-slope mass transports of the order of one Sverdrup per 1000km or $1\text{m}^2/\text{sec}$. Such a value for the transport of mass across the eastern boundary of the ocean is presumed here. It transpires that to determine both (a) a realistic temperature field (in relation to ocean observations) and (b) mass transports across the eastern boundary of the ocean of the order of $1\text{m}^2/\text{sec}$ then the magnitude of the vertical thermal diffusion coefficient must be carefully chosen.

2. Equations and Boundary Conditions

Following Hodnett and Ansari [5] the non-dimensional thermocline equations (with non-zero thermal diffusion and with temperature assumed independent of longitude) are:

$$\frac{\partial p}{\partial z} = T, \quad 2u\sin\phi = -\frac{\partial p}{\partial \phi}, \quad 2\nu\sin\phi = \frac{1}{\cos\phi}\frac{\partial p}{\partial \lambda}, \tag{2.1}$$

$$\nu = \tan\phi\frac{\partial w}{\partial z}, \tag{2.2}$$

$$w\frac{\partial T}{\partial z} + \nu\frac{\partial T}{\partial \phi} = k_z\frac{\partial^2 T}{\partial z^2}, \tag{2.3}$$

where z, ϕ, λ are in the upward, northward, eastward directions; w, ν, u are the corresponding velocity components; k_z is the vertical coefficient of thermal diffusion. Note that meridional thermal diffusion is neglected in equation (2.3). It was found here, through numerical experiment, that when k_ϕ (meridional thermal diffusion coefficient) is the same order as k_z (vertical thermal diffusion coefficient) then the influence of k_ϕ on the solution is negligible so that the solutions reported here have k_ϕ set at zero.

We take the same boundary conditions as in Hodnett and Ansari [5] which are $T = T_s(\phi)$ and $w = w_s(\phi)$ at the surface, $z = 0$ with $w = 0$ at $z = -\gamma$ (where γ is the nondimensional depth of a flat ocean: since nondimensimalisation is with respect to a vertical length scale of 562m therefore $\gamma = 1$ yields a thermocline depth of 562m while $\gamma = 2$ gives a depth of 1124m etc.). The expression for $w_s(\phi) = c\ (4.75 \cos 4.75\phi - \sin 4.75\phi \tan \phi)$ where c is a non-dimensional constant with value $c = 0.1272$ and this gives upwelling at the surface in the equatorial region between 0° and 18.1°N, downwelling in the subtropical region between 18.1°N and 53.5°N and upwelling in the subpolar region between 53.5°N and 60°N. The expression for $T(\phi)$ taken here is $T_s(\phi) = \alpha \cos(3\phi/2)$ where α is a nondimensional scale constant with value $\alpha = 4$ (this gives the temperature at the equator to be 20°C). The ocean is bounded by the latitude range $0° \leq \phi \leq 60°$N. The particular expression for $T_s(\phi)$ is chosen (a) to give surface temperature decreasing northward and (b) so that $T_s(\phi)$ is zero at $\phi = 60°$N to represent observations which show that the bottom water isotherm (i.e. $T = 0$) outcrops at the surface at approximately 60°N.

At the eastern boundary of the ocean we wish to simulate the presence of a northward flowing continental slope current which is fed by net inflow of the order of one Sverdrup per 1000km (= $1\text{m}^2/\text{sec.}$) from the southern section of the ocean and which is returned to the ocean by net outflow in the northern section. This is achieved by requiring that, at the eastern boundary of the ocean, the mass transport (integrated over the ocean depth) satisfies the condition

$$\int_{z=-\gamma}^{z=0} u dz = Af(\phi) \quad at \quad \lambda = \lambda_E, \tag{2.4}$$

where the line of longitude, $\lambda = \lambda_E$ represents the eastern boundary of the ocean and $f(\phi)$ is a given function which satisfies (in order to preserve mass continuity)

$$\int_{\phi=0}^{\phi=\pi/3} f(\phi) d\phi = 0, \tag{2.5}$$

where the nondimensional constant, A takes the value $A = 0.178$ (when the order of the mass transport is taken to be one Sverdrup per 1000km i.e. $1\text{m}^2/\text{sec}$). It is known that the along slope transports are not uniform (for example in the North Atlantic, transport appears to increase northwards off Portugal and again west of Britain - c.f. Huthnance [7]) and therefore the slope current accepts or yields up transport from or to the adjacent ocean interior nonuniformly with latitude. In the absence of definitive and continuous observational information which would identify the integrated mass transport across the slope (as a function of latitude) we here choose to use the simple linear function, $f(\phi) = 1 - 6\phi/\pi$. The width of the North Atlantic is represented by $\lambda_E = 5\pi/18$ and the North Pacific by $\lambda_E = 5\pi/9$.

We note that through the eastern boundary condition (2.4), the detailed dynamics of the ocean is influenced by the simulated presence of the continental slope current but the dynamics of the slope current is not resolved here. However, the gross properties of the slope current are represented by (2.4) i.e. the slope current transports mass in the northward direction, fed by net inflow of the order of 1 Sverdrup per 1000km from the southern section of the ocean ($0 \leq \phi \leq \pi/6$) which is returned to the ocean by net outflow in the northern section ($\pi/6 \leq \phi \leq \pi/3$). The slope current transport at latitude, ϕ, i.e.

$$\int_{\phi'=0}^{\phi'=\phi} Af(\phi')d\phi' = A\phi(1 - 3\phi/\pi), \tag{2.6}$$

is zero at $\phi = 0$ and $\phi = \pi/3$ and has a maximum value of $A\pi/12$ at $\phi = \pi/6$. When calculated in dimensional units this maximum value of the slope current transport is 5/3 Sverdrups (since $\pi/2$ radians represents 10^4km in distance along the slope).

3. Solution

When the temperature, T, is independent of longitude, λ then ν, w are both determined independently of T (i.e. are entirely wind forced) and are given by $\nu = w_s(\phi)\tan\phi/\gamma$ and $w = w_s(\phi)[1 + z/\lambda]$.

The assumption that the temperature field is zonally uniform has the following significant consequences. Since ν, w are then determined (independently of T) therefore the equation (2.3) for T becomes linear and hence facilitates analytical progress.

When the temperature, T has been determined then the zonal velocity component, u is given by

$$\begin{aligned} -2u\sin\phi &= \left[\int_{-\gamma}^{z} \frac{\partial T}{\partial \phi}dz' - \frac{1}{\gamma}\int_{-\gamma}^{0}\left(\int_{-\gamma}^{z}\frac{\partial T}{\partial \phi}dz'\right)dz + \frac{2(\lambda - \lambda_E)}{\gamma}\frac{d}{d\phi}[w_s(\phi)\sin^2\phi]\right] \\ &\quad - \frac{2A}{\gamma}f(\phi)\sin\phi. \end{aligned} \tag{3.1}$$

We note that the presence of the slope current does not effect ν, w and only effects the zonal velocity component, u through the term $Af(\phi)$ in expression (3.1) and since A/γ is small ($= 0.178$ when $\gamma = 1$; $= 0.089$ when $\gamma = 2$) the effect is significant only near where $u = 0$.

3.1. Solution for the Temperature

When the thermal diffusion coefficient k_z is non-zero equation (2.3) for the temperature, T must be solved numerically. On substitution of the known expressions for ν, w then equation (2.3) for T becomes

$$w_s(\phi)\left[(1+s)\frac{\partial T}{\partial s} + \tan\phi\frac{\partial T}{\partial \phi}\right] = \bar{k}_z\frac{\partial^2 T}{\partial s^2}, \tag{3.2}$$

where $s = z/\gamma$ and $\bar{k}_z = k_z/\gamma$. Equation (3.2) is parabolic in type and to ensure stability and convergence of the numerical solution the implicit Crank-Nicolson (see C.F. Gerald and P.O. Wheatley [8]) numerical scheme was used to solve the equation. Here we take the coefficient of vertical heat diffusion, $\bar{k}_z$, to be constant although McCreary [4] suggests that $\bar{k}_z$ varies inversely as the square of the buoyancy frequency. It is noted that the expression for $w_s(\phi)$ used here vanishes, at $\phi = 18.1°$N and $\phi = 53.5°$N, is positive in the equatorial region between 0° and 18.1°N, is negative in the subtropical region between 18.1°N and 53.5°N and is again positive in the subpolar region between 53.5°N and 60°N. This means that the left hand side of equation (3.2) vanishes at $\phi = 18.1°$N and $\phi = 53.5°N$ so that T is a linear function of s on these latitude lines. Also since $w_s(\phi)$ is positive when $0 \leq \phi \leq 18.1°N$, is negative when $18.1°N \leq \phi \leq 53.5°N$ and is positive when $53.5°N \leq \phi \leq 60°N$ then a separate numerical solution of equation (3.2) must be derived in each of these regions. The appropriate boundary conditions are

$$T = 4\cos(3\phi/2) \qquad at \quad s = 0, \tag{3.3}$$

$$T = 0 \qquad at \quad s = -1 \tag{3.4}$$

where (3.3) and (3.4) are applied in each separate latitudinal region together with

$$\frac{\partial T}{\partial \phi} = 0 \qquad at \quad \phi = 0 \quad \text{for} \quad 0 \leq \phi \leq 18.1°\text{N}, \tag{3.5}$$

where condition (3.5) represents the observed symmetry of the temperature field across the equator,

$$T = 4\cos(80.25°)[1 + s] \text{ at } \phi = 53.5°\text{N} \qquad \text{when } 18.1°\text{N} \leq \phi \leq 53.5°\text{N}, \tag{3.6}$$

$$\text{and when } 53.5°\text{N} \leq \phi \leq 60°\text{N}, \tag{3.7}$$

where conditions (3.6) and (3.7) both result from the fact that $\partial^2 T/\partial s^2 = 0$ on the latitude line $\phi = 53.5°N$. The problem is solved numerically for assigned values of the vertical thermal diffusion coefficient, $\bar{k}_z$ and Hodnett and Yuan [6] have reported the solution for the case when $\bar{k}_z$ is given the small value 0.01. Once T has been determined numerically, then u is calculated numerically through expression (3.1).

The temperature field reported by Hodnett and Yuan [6] for the case $\bar{k}_z = 0.01$ is not realistic near the latitudinal boundary between the equatorial and subtropical regions since the temperature gradients reported are much larger than those observed. Here the value of $\bar{k}_z$ was progressively increased until realistic looking temperature fields (in relation to observations) began to appear. Of course the related dimensional value of the vertical thermal diffusion coefficient should correspond to known estimates which Pedlosky [9] gives to be of order $1cm^2/sec.$ while Veronis [10] identifies a specific value of $1.3cm^2/sec.$

It is of interest to determine the mass transports across the eastern boundary from and into the ocean. If $s = -\bar{s}$ (where $\bar{s}$ varies with latitude ϕ) is the depth at which $u = 0$ at a particular latitude then the mass transport (across the eastern boundary) between the ocean bottom. (i.e. $s = -1$) and where $u = 0$ is given by

$$M_\ell = \gamma \int_{s=-1}^{s=-\bar{s}} u ds, \tag{3.8}$$

and the mass transport (at $\lambda = \lambda_E$) between $u = 0$ and the surface of the ocean is given by

$$M_u = \gamma \int_{s=-\bar{s}}^{s=0} u ds. \tag{3.9}$$

Huthnance [7] suggests these values should be of order $1m^2/sec$. When $\bar{k}_z = 0.25$ (dimensional value is $1.25cm^2/sec$ when $\gamma = 1$) the temperature field is shown in Fig. 1 and looks realistic in relation to observations. The values in ($m^2/sec.$) of the lower ocean mass transport, M_ℓ and the upper ocean mass transport, M_u are given in Table 1 (for $\gamma = 1$ when $A = 0.178$) and are seen to accord (at all latitudes) with what observations suggest and are of order 1m²/sec. Therefore the value of $\bar{k}_z$ (here a dimensional value of 1.25cm²/sec for the case $\gamma = 1$, i.e. thermocline depth of 562m) is critical in determining both a) a realistic temperature distribution and b) realistic values of cross-boundary mass transports (when the slope current transports approximately one Sverdrup per 1000km). These calculations confirm previous estimates (Veronis [10] and Pedlosky [9]) that the effective vertical thermal diffusion coefficient for ocean circulation as it effects eastern boundary processes is of order 1cm²/sec.

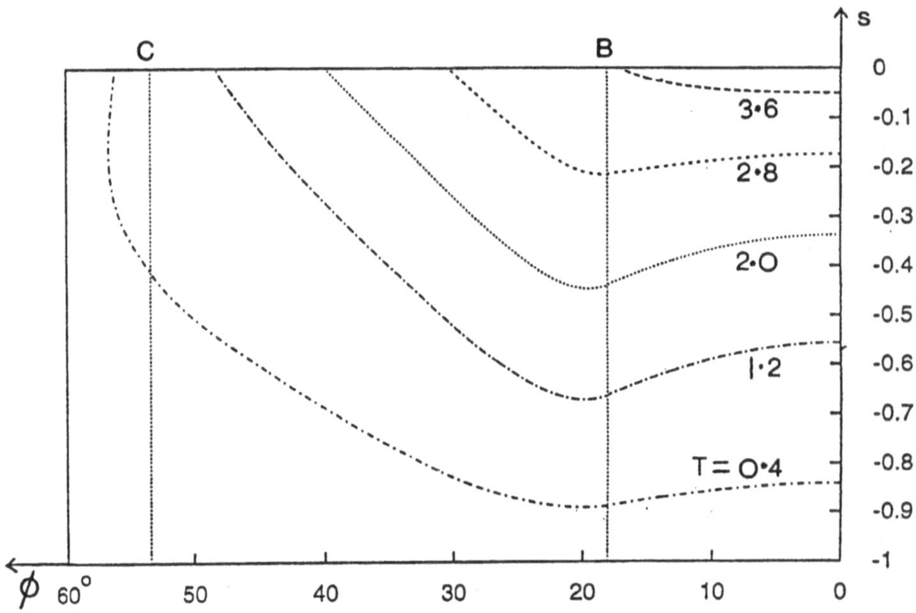

Figure 1: Temperature distribution ($\bar{k}_z = 0.25$)

Table 1: The typical values of M_ℓ and M_u (in m²/sec.) at the eastern boundary when $\bar{k}_z = 0.25$, $\gamma = 1$ and $A = 0.178$.

ϕ	0	10	18.1	30	40	50	60
M_u	−0.005	−1.906	−1.334	2.563	1.247	0.474	0.082
M_ℓ	0.995	2.565	1.731	−2.553	−1.590	−1.149	−1.083

4. Conclusions

When a zonally uniform ocean model connects to an eastern boundary current accepting or exporting transports of the order of one Sverdrup per 1000km it is critically important to carefully choose the value of the coefficient of vertical thermal diffusion, k_z in order to a) ensure a realistic distribution of temperature in the ocean and b) realistic values for the mass transported across the eastern boundary of the ocean. The value of k_z, so determined here, is $1.25cm^2/sec$ (for a thermocline depth of 562m) which confirms previous estimates of Veronis [10] and Pedlosky [9].

Acknowledgements

The work reported here received financial support through the CEC MAST II OMEX contract No. MAS2-CT93-0069.

References

1. J.M. Huthnance, Proc. Royal Soc. Edinburgh, **88B**, 83-101 (1986).
2. R.D. Pingree and B. LeCann, J. Marine Biol. Association of the U.K., **70**, 857-885 (1990).
3. P. Welander, J. Phys. Oceanography, **6**, 121-124 (1976).
4. J.P. McCreary, Phil. Trans. Royal Soc. London A, **302**, 385-413 (1981).
5. P.F. Hodnett and A.R. Ansari, Geophys. Astrophys. Fluid Dynamics, **78**, 73-93 (1994).
6. P.F. Hodnett and Y. Yuan, Proc. R. Ir. Acad., **97A**, 193-207 (1997).
7. J.M. Huthnance, Progress in Oceanography, **35**, 353-431 (1995).
8. C.F. Gerald and P.O. Wheatley, Applied Numerical Analysis, (3rd. edn), Reading, Mass. Addison-Wesley (1984).
9. J. Pedlosky, Geophysical Fluid Dynamics, Springer-Verlag, New York (1984).
10. G. Veronis, Deep-Sea Research, **16**, 301-323 (1969).

9th Int. Coll. on Differential Equations, pp. 213-220
D. Bainov (Ed)

New Algorithm of Curve Fitting for Non Decreasing Time Series Data

ICHIRO HOFUKU
Tokyo Metropolitan College of Technology, Higashi-Ooi 1-10-40, Shinagawa-ku, 140, Japan

KUNIO OSHIMA AND SHIGERU FURUICHI
Department of Electronics and Computer Science, Science University of Tokyo in Yamaguchi, Daigaku-dori 1-1-1, Onoda 756, Japan

MASANORI OHYA
Department of Information Science, Science University of Tokyo, Noda City, Chiba 278, Japan

Abstract: The purpose of this paper is to present a new algorithm of curve fitting for non decreasing discrete data. The function generated by this method is differentiable on the interval where the function is defined, moreover, the function is not only a set of piece wise smooth functions but consists of one function which can be express on the interval. We will discuss how to construct such a function along with the algorithm of curve fitting and to apply the non decreasing time series data.

1. Introduction

The purpose of this paper is to present a new algorithm of curve fitting for non decreasing data. The non decreasing data, we call, means the non decreasing time series data which is characteristically guarantee the convergence. There are many examples of non decreasing data. Figure 1 shows the one example which X-axis shows the date and Y-axis shows the cumulative number of bugs in software development corresponding to date. We call this non decreasing data as debugging data. In general, when we generate an approximated function for non decreasing data, it often happens that the accuracy of the approximated function is not quite satisfied as we expected. One of the typical cases is that the non decreasing data are increasing more rapidly than we expected (see Figure 1). In such a case, it is usually difficult to determine or choice a good approximated function in the sense that we find a best fitted function for the data by the use of known approximating methods.

A new algorithm which we will give is to generate a optimal function, named the locally modified function, which can be a best fitted one for any non decreasing data. The function generated by this algorithm is differentiable on the interval where the function is defined, and the function is not only a set of piecewise smooth functions but consists of one function which can be expressed on the interval where the function is defined. We will discuss how to construct such a function along with the algorithm of curve fitting and to apply it to non decreasing data.

2. Locally Modified Function

In this section, we will generate a based approximated function $f_1(t)$ for given debugging data $T_i(t_i, y_i)$ $(i = 1, 2, \cdots, 154)$, say data1, and generate the Locally Modified Function $Lf(t)$ to fit more in nearness on the interval $[t_i, t_j]$ which is given on my purpose. Figure 1 shows the data1 and one approximated function, which is called Gompertz function. As is shown in Figure 1, $f_1(t)$ is not so fitted as we expected that the convergent value of $f_1(t)$ is not to be so reliable than we expected.

2.1. Based Approximated Function

At first, we define the function $f_1(t)$ as follows;

Definition 2.1 *Let $t \geq 0$, and α, β, $\gamma > 0$, the function $f_1(t)$ is defined by*

$$f_1(t) = \gamma \exp\Big\{-\alpha \exp(-\beta t)\Big\}. \tag{2.1}$$

$f_1(t)$ is called a Gompertz function generated as one approximated function which is applied for debugging data in Software Reliability growth model. There are some other approximated functions in Software Reliability growth model by the name of Exponential Software Reliability growth model and S-shaped Software Reliability growth model and so on. We call these approximated functions as based approximated function. The advantage of generating the based approximated function is that it is possible to forecast the final cumulative number of bugs at the middle of debugging period. Figure 1 shows the data1 and $f_1(t)$, which has generated by considering on the interval where one software debugging is completed. As is shows in Figure 1, the $f_1(t)$ is not so fitted as we expected for data1.

The purpose of this paper is to modify the $f_1(t)$ to fit more in nearness of functional approximation for data1. At first, we will adopt the Gompertz function, we let $f_1(t)$, among the based approximated function and modify the $f_1(t)$ to fit more for data1. Of course, we may choose other approximated function in based approximated function if we want.

The outline of generating the more fitted approximated function is that we will change the derivative of $f_1(t)$ by multiplying some weight function by the right hand side of $\frac{df_1}{dt}$ depending on the location of data1 and $f_1(t)$ only on some interval $[t_i, t_j]$ $(i < j)$, and we have the modified function to fit more than $f_1(t)$ on the interval $[t_i, t_j]$ by solving the modified differential equation. We call this modified function as *locally modified function*, we let $Lf(t)$. The method of generating a weight function corresponding to the interval $[t_i, t_j]$ will show later. Now, we will give the method of generating $Lf(t)$.

For some (but any) interval $[t_i, t_j]$, the modification type of the $f_1(t)$ is as follows;

(2.1.a) For some interval $[t_i, t_j]$, we force to increase the $f_1(t)$.

(2.1.b) For some interval $[t_i, t_j]$, we force to decrease the $f_1(t)$, but the derivative of modified function won't be zero on such interval.

(2.1.c) For some interval $[t_i, t_j]$, we force to decrease the $f_1(t)$ so as to the derivative of modified function will almost zero.

2.2. *P-function*

In this section, we will give the method of generating the weight function corresponding to (2.1.a). (2.1.a) is increasing model in the sense that we will modify the $f_1(t)$ to force to increase.

At first, we differentiate the Eq(2.1) with respect to t, then we have,

$$\frac{df_1}{dt} = \alpha\beta\exp(-\beta t)f_1(t). \tag{2.2}$$

The method of generating the $Lf(t)$ is that we will modify the differential equation Eq(2.2) by multiplying some weight function by the right hand side of Eq(2.2) and solving with under the initial condition, $Lf(0) = \gamma\exp(-\alpha)$, we have $Lf(t)$ on the interval $[t_i, t_j]$. We call this weight function as *P-function*.

Now we will define the $P - function$ as follows;

Definition 2.2 *The function $P_{i,j}(b_{ij} : s)(t)$ is defined as follows;*

$$P_{i,j}(b_{ij} : s)(t) = 1 + s\exp\left\{b_{ij} \times \left(\frac{h_{ij}}{2}\right)^2\right\}\exp\left\{-b_{ij}\left(t - \frac{t_i + t_j}{2}\right)^2\right\}. \tag{2.3}$$

$$(b_{ij}, s > 0, \quad i \geq 0)$$

Parameter h_{ij} shows the $t_j - t_i$, and i, j means the interval where the $P - function$ effects. The method of finding the parameter b_{ij}, s will show later.

Modifying the Eq(2.2) by multiplying the function $P_{i,j}(b_{ij} : s)(t)$ by the right side of Eq(2.2), we have the new differential equation,

$$\frac{d(Lf_{ij}(b_{ij} : s)(t))}{dt} = \alpha\beta\exp(-\beta t)P_{i,j}(b_{ij} : s)(t)Lf_{ij}(b_{ij} : s)(t). \tag{2.4}$$

Solve the Eq(2.4) under the initial condition,

$$Lf_{ij}(b_{ij} : s)(0) = \gamma\exp(-\alpha),$$

we have $Lf_{ij}(b_{ij} : s)(t)$ which is characteristically more increasing than the $f_1(t)$ on the interval $[t_i, t_j]$.

[The Method of Finding the Parameter s]
The function $P_{i,j}(b_{ij} : s)(t)$ has following character,

$$P_{i,j}(b_{ij} : s)(t_i) = P_{ij}(b_{ij} : s)(t_j) = 1 + s. \tag{2.5}$$

Eq(2.5) means that the influence of the derivative of $Lf_{ij}(b_{ij} : s)(t)$ is less than s times as much as the derivative of all point except on the interval $[t_i, t_j]$. As we will want to generate the $Lf_{ij}(b_{ij} : s)(t)$ on condition that the influence of the derivative for other interval of $Lf_{ij}(b_{ij} : s)(t)$ is as less as possible, we let s be 10^{-2}. Figure 2 shows the sketch of the $P_{i,j}(b_{ij} : 10^{-2})(t)$.

[The Method of Finding the Parameter b_{ij}]
We will have the value of b_{ij} in Eq(2.3) such that the value of

$$|Lf_{ij}(b_{ij} : s)(t_j) - y_j| \tag{2.6}$$

is minimum. In case of modify the $f_1(t)$ on some intervals, we will generate the $Lf(T)$ one after another for corresponding interval where is separated by time series. So, if we will modify the $f_1(t)$ on the interval $[t_i, t_{i+1}]$, it has already modified on the interval $[t_{i-1}, t_i]$. So we will find the b_{ij} such that the value of Eq(2.6) is minimum.

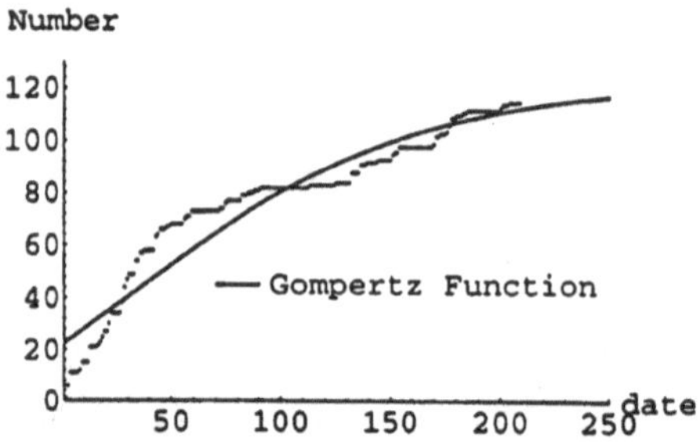

Figure 1: This graph shows the relation between the data1 and the $f_1(t)$.

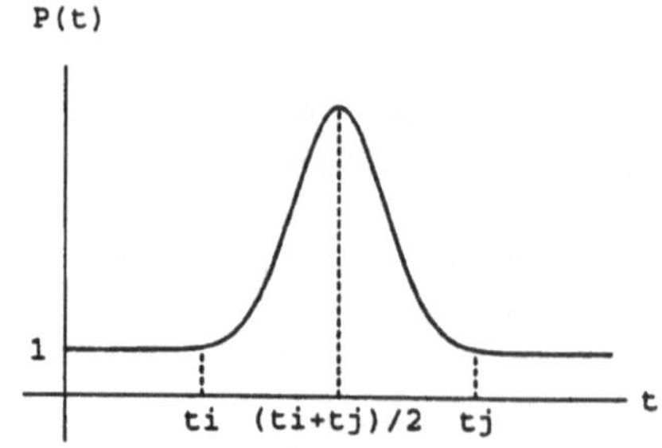

Figure 2: This graph shows the function $P_{i,j}(b_{ij},: 10^{-2})(t)$.

2.3. Q-function

We will give the method of generating the weight function for case (2.1.b) and (2.1.c). These cases are decreasing model in the sense that we force to decrease the $f_1(t)$ on the interval $[t_i, t_j]$ by multiplying some weight function. We call this weight function as *Q-function*. Elementary method of generating Q-function is similar in method to generate the P-function, but there is different in the point that it is often necessary for Q-function to transform it's weigh function corresponding to data's condition T_i, T_j. For some interval $[t_i, t_j]$, there are two case's location of two points T_i, T_j, which one is increasing and another is non decreasing (i.e. square). So we have to ready the two weight functions depending on the location of two points T_i, T_j. So, we will define the following functions;

Definition 2.3 *For some interval $[t_i, t_j]$, we are ready to two weight functions $Qmd-function$, $Qsd-function$ corresponding to the location of two points T_i, T_j as follows;*
(2.3.a) increasing $\rightarrow$ Qmd-function
(2.3.b) square $\rightarrow$ Qsd-fucntion

2.3.1. $Qmd-function$

Different from the P-function, there is the restriction on generating Qmd-function. At first, we define the $Qmd_{i,j}(b_{ij} : s)(t)$ as follows;

Definition 2.4 *The function $Qmd_{i,j}(b_{ij} : s)(t)$ is defined as follows;*

$$Qmd_{i,j}(d_{ij} : s)(t) = 1 - s\exp\left\{d_{ij} \times \left(\frac{h_{ij}}{2}\right)^2\right\}\exp\left\{-d_{ij}\left(t - \frac{t_i + t_j}{2}\right)^2\right\}. \qquad (d_{ij}, s > 0, \quad i \geq 0) \tag{2.7}$$

Modifying the Eq(2.2) by multiplying the $Qmd_{i,j}(d_{ij} : s)(t)$ by the right side of Eq(2.2), we have the new differential equation,

$$\frac{d(Lf_{ij}(d_{ij} : s)(t))}{dt} = \alpha\beta\exp(-\beta t)Qmd_{i,j}(d_{ij} : s)(t)Lf_{ij}(d_{ij} : s)(t). \tag{2.8}$$

Solve the Eq(2.8) with under the initial condition,

$$Lf_{ij}(d_{ij} : s)(0) = \gamma\exp(-\alpha).$$

We have $Lf_{ij}(d_{ij} : s)(t)$. $Lf_{ij}(d_{ij} : s)(t)$ is characteristically more decreasing than the $f_1(t)$ on the interval $[t_i, t_j]$. Now there is a problem occurred. If $Qmd_{i,j}(d_{ij} : s)(t) < 0$ at the peak point $t = (t_i + t_j)/2$, what has happen for $Lf_{ij}(d_{ij} : s)(t)$? The $Lf_{ij}(d_{ij} : s)(t)$ has the interval, where the derivative of $Lf_{ij}(d_{ij} : s)(t)$ is negative on the interval $[t_i, t_j]$. As the result of that, $Lf_{ij}(d_{ij} : s)(t)$ has, so called, the "Bump" on the interval $[t_i, t_j]$. So, we will add the condition not to occur the Bump for $Lf_{ij}(d_{ij} : s)(t)$ as follows;

$$Qmd_{i,j}(d_{ij} : s)\left(\frac{t_i + t_j}{2}\right) > 0. \tag{2.9}$$

As we let $s = 10^{-2}$, and from the inequality (2.9), the range of d_{ij} is as follows;

$$0 < d_{ij} < \frac{8}{{h_{ij}}^2} \ln 10. \tag{2.10}$$

Figure 3 shows the function $Qmd_{i,j}(d_{ij} : s)(t)$.

2.3.2. The improvement of $Qmd - function$

In applying the $Qmd - function$, there is another problem occurred.

Remark 2.5 *There is often the limit of it's application for the $Qmd - function$ on the interval $[t_i, t_j]$ because of the restriction of the parameter d_{ij}.*

As the restriction of parameter d_{ij}, it is often impossible to decrease the $f_1(t)$ fully depending on the location of two points T_i, T_j.

To avoid this case, we define the other function renamed Qmd-function as follows;

Definition 2.6 *We define the $Qmd_{i,j}(d_{ij} : s, k)(t)$ as follows;*

$$Qmd_{i,j}(d_{ij} : s, k)(t) = \sum_{m=1}^{2k-1}\left[1 - s\exp\left\{d_{ij} \times \left(\frac{1}{2} \times \frac{h_{ij}}{k}\right)^2\right\}\right.$$
$$\left.\exp\left\{-d_{ij}\left(t - \left(t_i + \frac{h_{ij}}{2k}m\right)\right)^2\right\}\right] - (2k-2). \tag{2.11}$$
$$(k = 1, 2, \cdots, \qquad d_{ij},\ s > 0)$$

The $Qmd_{i,j}(d_{ij} : s, k)(t)$ is equal to $Qmd_{i,j}(d_{ij} : s)(t)$ at $k = 1$ and has the minimum at $t = (t_i + t_j)/2$. As is same with the $Qmd_{i,j}(d_{ij} : s)$ we defined before, we will add the condition not to occur the Bump on the interval $[t_i, t_j]$ as follows;

$$Qmd_{i,j}(d_{ij} : s, k)\left(\frac{t_i + t_j}{2}\right) > 0. \tag{2.12}$$

Figure 4 shows the sketch of $Qmd_{i,j}(d_{ij} : s, k)(t)$ in case $h_{ij} = 1.0,\ k = 3,\ d_i = 160$. The range of d_{ij} for $k = 3$ is,

$$0 < d_{ij} < 165.059.$$

By comparing the two weight functions in Figure 3 and Figure 4, it is clear that the decreasing ratio of two weight functions is different.

[The Method of Finding the Parameter d_{ij}]

The method of finding d_{ij} is similar in method to find the b_{ij}. But $Qmd_{i,j}(d_{ij} : s, k)$ has the restrict to d_{ij} corresponding to each k. So, at first, we let $k = 1$ and find the d_{ij} so as to minimize the distance of modified function at $t = t_i$ and the y_i under the range of d_{ij} for $k = 1$. And then if the $Lf_{ij}(d_{ij} : s)(t)$ has Bump on the interval $[t_i, t_j]$, next, we let k be 2 and analyze similarly. And finally we have the d_{ij} which is satisfied with the range of d_{ij} corresponding to some k.

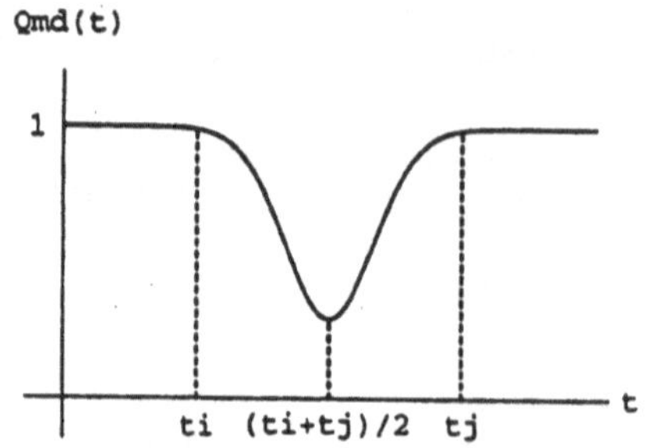

Figure 3: This graph shows the example for the function $Qmd_{i,j}(d_{ij} : 10^{-2})(t)$.

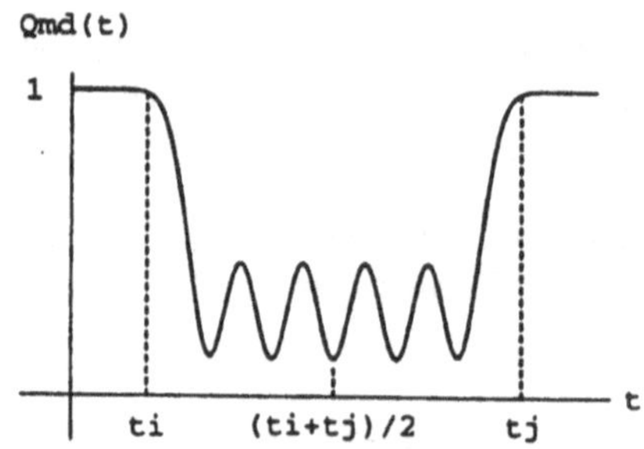

Figure 4: This graph shows the example for the function $Qmd_{i,j}(160 : 10^{-2}, 3)(t)$.

2.4. Qsd-function

We will give the method of generating the weight function for (2.1.c). (2.1.c) is decreasing model which the location of two points T_i, T_j is square. Here, in addition to the interval $[t_i, t_j]$, we consider other intervals $[t_{i-1}, t_i]$ and $[t_j, t_{j+1}]$. The method of modifying the $f_1(t)$ is that, at first, we will modify the $f_1(t)$ on the interval $[t_i, t_j]$ by use of some weight function so as to the derivative of $Lf_{ij}(t)$ is almost zero on it's interval and then modify the $f_1(t)$ on other intervals repeatedly by applying the P-function or Qmd-function depending on the location of data points corresponding to the interval.

At first of this method, we define the function $Qsd_{i,j}(m_i, n_i)(t)$ as follows;

Definition 2.7 *The function $Qsd_{i,j}(m_{ij}, n_{ij})(t)$ is defined as follows;*

$$Qsd_{i,j}(m_{ij}, n_{ij})(t) = \left[1 - \exp\left\{-m_{ij}\left(t - \frac{t_i + t_j}{2}\right)^2\right\}\right]^{n_{ij}}. \tag{2.13}$$

$$(\; m_{ij} > 0, \qquad n_{ij} \in N \;)$$

Now we add the condition for the $Qsd_{i,j}(m_{ij}, n_{ij})(t)$ as follows;

$$Qsd_{i,j}(m_{ij}, n_{ij})\left(t_i + \frac{h_{ij}}{10}\right) = Qsd_{i,j}(m_{ij}, n_{ij})\left(t_j - \frac{h_{ij}}{10}\right) \le \frac{1}{10} \tag{2.14}$$

$$Qsd_{i,j}(m_{ij}, n_{ij})(t_{i-1}) \ge \frac{99}{100}, \qquad Qsd_{i,j}(m_{ij}, n_{ij})(t_{j+1}) \ge \frac{99}{100} \tag{2.15}$$

and we let $h^* = min\{h_{(i-1)i}, h_{j(j+1)}\}$, Eq(2.15) follows

$$Qsd_{i,j}(m_{ij}, n_{ij})\left(\frac{t_i + t_j}{2} + \frac{h_{ij}}{2} + h^*\right) \ge \frac{99}{100}. \tag{2.16}$$

By the Eq(2.14) and the Eq(2.16),we have

$$\frac{\ln\left(\frac{1}{10}\right)}{\ln\left\{1 - \exp\left(-\frac{4m_{ij}h_{ij}^2}{25}\right)\right\}} \le n_{ij} \le \frac{\ln\left(\frac{99}{100}\right)}{\ln\left[1 - \exp\left\{-m_{ij}\left(h^* + \frac{h_{ij}}{2}\right)^2\right\}\right]}. \tag{2.17}$$

We will find the minimum n_{ij} satisfying the Eq(2.17), and adopt the m_{ij} corresponding to the n_{ij}. Figure 6 shows the graph of $h^* = 0.5$, $h_{ij} = 1.0$ and in this case, the value of n_{ij} and m_{ij} are

$$n_{ij} = 6, \qquad m_{ij} = 6.39275.$$

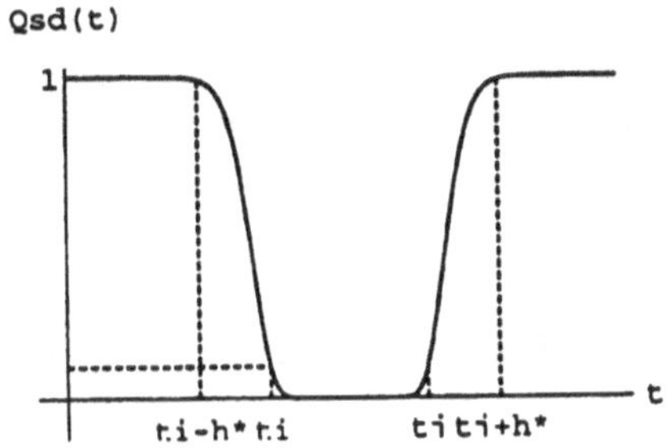

Figure 5: This graph shows the example for the function $Qsd_{ij}(m_{ij}, n_{ij})(t)$.

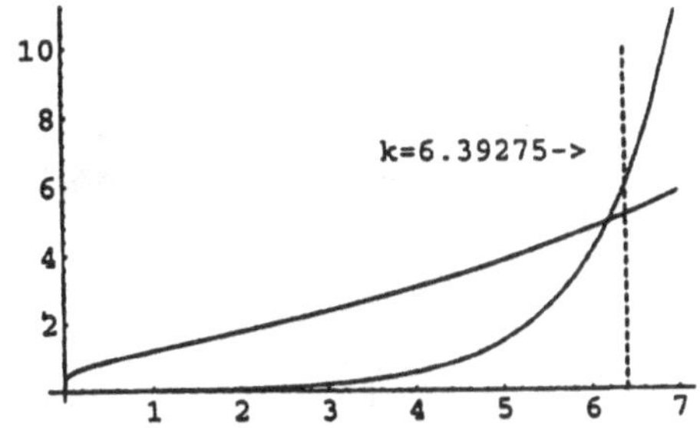

Figure 6: This graph shows the simulation to determine the value of n_{ij} and m_{ij} for the case of $h^* = 0.5$, $h_{ij} = 1.0$.

3. Analysis and Result

At the end of this paper, we will give an example of generating the $Lf(t)$ at the middle of debugging period. Figure 7 shows the relation between the data1 by considering the interval $(0, t_{154}]$ and $f_1(t)$. The parameter α, β and γ of $f_1(t)$ is,

$$\alpha = 0.483095, \qquad \beta = 0.483095, \qquad \gamma = 100.4 \ .$$

As is shown in Figure 7, it is not so fitted as we expected on the interval $(0, t_{154}]$. So we will modify $f_1(t)$ so as to fit more by use of the method we explained before. At first step of this method, we choose the T_1, T_2, T_3 and T_4 experimentially and each interval is separated, $[t_1, t_2]$, $[t_2, t_3]$ and $[t_3, t_4]$ by setting up the point T_1, T_2, T_3 and T_4. For each interval, we will modify $f_1(t)$ so as to fit more by the use of each weight function depending on the location of $f_1(t)$ and Point T_i on each interval. By harmonizing the each weight function for each interval, we have the weight function $w(t)$ as follows;

$$w(t) = P_{1,2}(0.00377734 : 10^{-2})(t) + Qsd_{2,3}(0.00263634, 3)(t) + P_{3,4}(0.028713)(t). \tag{3.1}$$

As multiplying $w(t)$ by the right hand side of $\frac{df_1}{dt}$, and solve, we have $Lf(t)$ is as follows;

$$\begin{aligned} Lf(t) = {} & 102.515 \exp\{\frac{-1.62108}{e^{0.0207944t}} \\ & + 0.0205483\sqrt{\pi} erf\{5.62223(1.67964 - 0.015818t)\} \\ & - 0.0760169\sqrt{\pi} erf\{6.8858(1.11283 - 0.0105454t)\} \\ & + 0.109731\sqrt{\pi} erf\{9.73799(0.546019 - 0.00527268t)\} \\ & + 0.0305635\sqrt{\pi} erf\{8.13536(-0.432486 + 0.00755468t)\} \\ & + 0.035461\sqrt{\pi} erf\{2.95074(-7.96142 + 0.057426t)\}, \end{aligned}$$

where,

$$erf(t) = \frac{2}{\sqrt{\pi}} \int_0^t e^{-u^2} du.$$

Figure 7 shows the relation between the data1 and locally modified function $Lf(t)$. Table 1 shows the square value of $f_1(t)$ and $Lf(t)$ for $(0, t_{154}]$ and convergent value of each function. It is clear that $Lf(t)$ is more fitted than $f_1(t)$. As the result of that we will adopt $Lf(t)$ as the more fitted approximated function than $f_1(t)$. And the forecast of cumulative number of bugs is 104.675 not the convergent value 100.4 of $f_1(t)$.

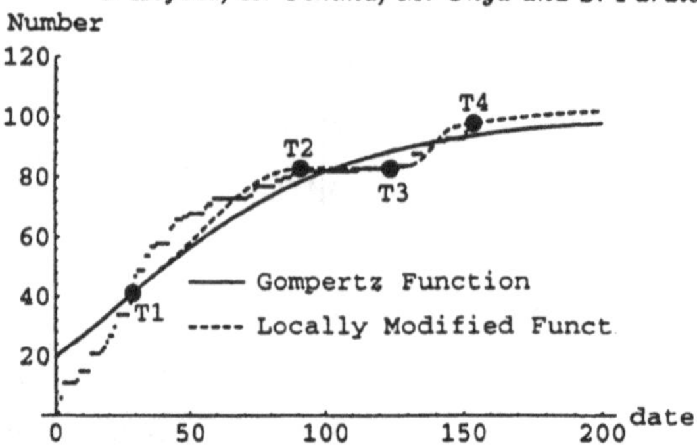

Figure 7: This Figure shows the relation between $f_1(t)$ and $Lf(t)$.

Table 1: This table shows the comparison of $f_1(t)$ and $Lf(t)$.

	square value	convergent value
$f_1(t)$	7315.8	100.4
$Lf(t)$	5613.66	104.675

References

1. K. Oshima, I. Hofuku and K. Horimoto, in: Selected Works of the 6-th Int. Coll. Diff. Eq., (1995) pp. 181-188.

2. K. Horimoto, I. Hofuku and K. Oshima, Nonlinear Analysis, Theory, Methods & Applications, **30**, 985-993 (1997).

3. K. Oshima, I. Hofuku and K. Horimoto, Nonlinear Analysis, Theory, Methods & Applications, **30**, 1007-1017 (1997).

9th Int. Coll. on Differential Equations, pp. 221-228
D. Bainov (Ed)

The Asymptotic Distribution for the Eigenfunctions of the Sturm–Liouville Equation

A. JODAYREE AKBARFAM AND E. POURREZA
Faculty of Mathematical Sciences, Tabriz University, Tabriz, Iran
E-Mail: tabriz_u@rose.ipm.ac.ir

Abstract: In this paper we study some results in connection with the eigenfunctions corresponding to the eigenvalues of the Sturm-Liouville of the form

$$y'' + (\lambda t - q(t))y = 0, \quad -1 \le t \le 1$$

subject to Dirichlet boundary condition

$$y(-1) = 0 = y(1).$$

From a general point of view, in the study of the asymptotic behavior of solutions of the (1), as $\lambda \to \infty, t = 0$ is called turning point because, depending on whether $t > 0$ or $t < 0$, an integral of (1) either oscillates or is monotonic for large values of λ. In [1], the higher-order eigenvalue asymptotics for Sturm-Liouville is found.

Inserting the asymptotics forms of the eigenvalues in asymptotic solutions,represents the major motivation of this paper.
Key Words and Phrases: *Turning points, Asymptotic solutions.*
1991 Mathematics Subject Classification. 34

1. Introduction

The equation

$$y'' + (\lambda t - q(t))y = 0, \quad -1 \le t \le 1 \tag{1}$$

with boundary condition

$$y(-1) = 0 = y(1)$$

has a double infinite sequence of real eigenvalues whose corresponding eigenfunctions have precisely n zeros in $(-1, 1)$, for all sufficiently large n [2]. The positive eigenvalues $\{u_n\}_{n=1}^{\infty}$ admit the asymptotic representation

$$\sqrt{u_n} = \frac{n\pi - \pi/4}{\int_0^1 \sqrt{t}dt} + \frac{1}{2n\pi}T_1 + O\left(\frac{1}{n^2}\right)$$

where q is m times continuously differentiable and

$$T_1 = \frac{5}{72\int_0^1 \sqrt{t}dt} + \frac{1}{2}\int_0^1 \frac{q(t)}{\sqrt{t}}dt.$$

The distribution of the negative eigenvalues $\{\lambda_n\}_{n=1}^{\infty}$ is of the form

$$\sqrt{-\lambda_n} = \frac{n\pi - \pi/4}{\int_{-1}^{0}\sqrt{-t}dt} + \frac{1}{2n\pi}T_2 + O\left(\frac{1}{n^2}\right)$$

where

$$T_2 = \frac{5}{72\int_{-1}^{0}\sqrt{-t}dt} + \frac{1}{2}\int_{-1}^{0}\frac{q(t)}{\sqrt{-t}}dt$$

for more detail see [1].

First, we find the asymptotic form of $U(t,\lambda)$, the solution of (1) with initial condition

$$U(-1,\lambda) = 0, \quad \frac{\partial U}{\partial t}(-1,\lambda) = 1. \tag{2}$$

The asymptotic nature of the approximation solution of the differential equation (1) has been investigated by many writers. Now we apply Olver's result in order to determine the asymptotic form of $U(t,\lambda)$. First we state a number of relevant properties of the functions Ai and Bi (the Airy functions). Then we also introduce certain auxiliary functions and constants associated with Ai and Bi which are used in Olver's paper [3].

2. Relevant Properties of Airy Functions

The Airy functions are the solutions of the differential equation

$$y'' = ty.$$

For real values of t the standard solutions are denoted by $Ai(t)$ and $Bi(t)$. They have the initial values

$$Ai(0) = \frac{1}{\sqrt{3}}Bi(0) = \frac{1}{3^{2/3}\Gamma(2/3)},$$

$$Ai'(0) = -\frac{1}{\sqrt{3}}Bi'(0) = -\frac{1}{3^{1/3}\Gamma(1/3)},$$

and satisfy the Wronskian relation

$$Ai(t)Bi'(t) - Ai'(t)Bi(t) = 1/\pi. \tag{3}$$

When t is positive $Ai(t), Bi(t), -Ai'(t)$ and $Bi'(t)$ are all positive and monotonic; when t is negative these functions are oscillatory, with diminishing period as $t \to -\infty$. Their precise asymptotic behavior is given in [4] by

$$Ai(t) = \frac{1}{2}\pi^{-1/2}t^{-1/4}e^{-(2/3)t^{3/2}}\{1 + O(t^{-3/2})\},$$

$$Ai'(t) = -\frac{1}{2}\pi^{-1/2}t^{1/4}e^{-(2/3)t^{3/2}}\{1 + O(t^{-3/2})\},$$

$$Bi(t) = \pi^{-1/2}t^{-1/4}e^{(2/3)t^{3/2}}\{1 + O(t^{-3/2})\},$$

$$Bi'(t) = \pi^{-1/2}t^{1/4}e^{(2/3)t^{3/2}}\{1 + O(t^{-3/2})\},$$

as $t \to +\infty$, and

$$Ai(t) = \pi^{-1/2}(-t)^{-1/4}\left\{\cos\left(\frac{2}{3}(-t)^{3/2} - \pi/4\right) + O(|t|^{-3/2})\right\}$$

$$Ai'(t) = \pi^{-1/2}(-t)^{1/4}\left\{\sin\left(\frac{2}{3}(-t)^{3/2} - \pi/4\right) + O(|t|^{-3/2})\right\}$$

$$Bi(t) = -\pi^{-1/2}(-t)^{-1/4}\left\{\sin\left(\frac{2}{3}(-t)^{3/2} - \pi/4\right) + O(|t|^{-3/2})\right\}$$

$$Bi'(t) = \pi^{-1/2}(-t)^{1/4}\left\{\cos\left(\frac{2}{3}(-t)^{3/2} - \pi/4\right) + O(|t|^{-3/2})\right\}$$

as $t \to -\infty$.

In order to have a simple way of estimating the magnitudes of the Airy functions, we define

$$\begin{aligned} E(t) &= e^{\frac{2}{3}t^{3/2}}, \quad (t > 0) \\ E(t) &= 1, (t \le 0); \quad E^{-1}(t) \equiv 1/E(t), \end{aligned} \tag{4}$$

and introduce four auxiliary functions $M(t), \chi(t), N(t)$ and $\psi(t)$, defined for all real values of t by the equations

$$\begin{aligned} Ai(t) &= E^{-1}(t)M(t)\sin\chi(t), \\ Bi(t) &= E(t)M(t)\cos\chi(t), \\ Ai'(t) &= E^{-1}(t)N(t)\sin\psi(t), \\ Bi'(t) &= E(t)N(t)\cos\psi(t). \end{aligned} \tag{5}$$

Thus

$$M^2(t) = E^2(t)Ai^2(t) + E^{-2}(t)Bi^2(t),$$
$$N^2(t) = E^2(t)Ai'^2(t) + E^{-2}(t)Bi'^2(t).$$

From the asymptotic forms of Ai and Bi we deduce that (see [3])

$$M(t) \sim \left(\frac{5}{4\pi}\right)^{1/2}\frac{1}{t^{1/4}}, \quad N(t) \sim \left(\frac{5}{4\pi}\right)^{1/2} t^{1/4}, \quad t \to +\infty,$$

$$M(t) \sim \frac{1}{\pi^{1/2}|t|^{1/4}}, \quad N(t) \sim \frac{|t|^{1/4}}{\pi^{1/2}}, \quad t \to -\infty. \tag{6}$$

The following constants occur in the subsequent analysis, (see [3])

$$\lambda_1 = max_{(-\infty,\infty)}\{\pi|t|^{1/2}M^2(t)\} = 1.430...,$$

$$\lambda_2 = max_{(-\infty,\infty)}\{\pi|t^{1/2}Bi(t)|E^{-1}(t)M(t)\} = 1.315... \tag{7}$$

Approximation theorem for the equation (1).

From [3], the differential equation (1) has solutions $W_1(t,\lambda), W_2(t,\lambda)$ given by

$$W_1(t,\lambda) = Ai(-\lambda^{1/3}t) + \epsilon_1$$

$$W_2(t,\lambda) = Bi(-\lambda^{1/3}t) + \epsilon_2$$

$$W_1'(t,\lambda) \equiv \frac{\partial W_1(t,\lambda)}{\partial t} = -\lambda^{1/3}\{Ai'(-\lambda^{1/3}t) + \eta_1\}$$

$$W_2'(t,\lambda) \equiv \frac{\partial W_2(t,\lambda)}{\partial t} = -\lambda^{1/3}\{Bi'(-\lambda^{1/3}t) + \eta_2\} \tag{8}$$

where from [3] error bounds ϵ_i, η_i, $i = 1, 2$ are of the form

$$\frac{E(-t\lambda^{1/3})}{M(-t\lambda^{1/3})}|\epsilon_1|, \frac{E(-t\lambda^{1/3})}{N(-t\lambda^{1/3})}|\eta_1| \leq \frac{1}{\lambda_1}\left\{e^{\frac{\lambda_1 F_1(-t)}{\sqrt{\lambda}}} - 1\right\},$$

$$\frac{E^{-1}(-t\lambda^{1/3})}{M(-t\lambda^{1/3})}|\epsilon_2|, \frac{E^{-1}(-t\lambda^{1/3})}{N(-t\lambda^{1/3})}|\eta_2| \leq \frac{\lambda_2}{\lambda_1}\left\{e^{\frac{\lambda_1 F_2(-t)}{\sqrt{\lambda}}} - 1\right\}. \tag{9}$$

In the above result, the functions F_1, F_2 are given by, for $|t| < 1$,

$$0 \leq F_1(t) = \int_t^1 \frac{q(v)}{v^{1/2}} dv$$

$$0 \leq F_2(t) = \int_{-1}^t \frac{q(v)}{v^{1/2}} dv$$

(note that these integrals are convergent) and λ_1, λ_2 are constants defined in (6) while $E(t), M(t)$ and $N(t)$ are auxiliary functions defined earlier.

If λ is large then the right side of (8) is clearly $O\left(\frac{1}{\sqrt{\lambda}}\right)$ uniformly with respect to t. Hence we have

$$|\epsilon_1| = E^{-1}(-t\lambda^{1/3})M(-t\lambda^{1/3})O\left(\frac{1}{\sqrt{\lambda}}\right),$$

$$|\eta_1| = E^{-1}(-t\lambda^{1/3})N(-t\lambda^{1/3})O\left(\frac{1}{\sqrt{\lambda}}\right),$$

$$|\epsilon_2| = E(-t\lambda^{1/3})M(-t\lambda^{1/3})O\left(\frac{1}{\sqrt{\lambda}}\right),$$

$$|\eta_2| = E(-t\lambda^{1/3})N(-t\lambda^{1/3})O\left(\frac{1}{\sqrt{\lambda}}\right). \tag{10}$$

Now, the solution of the equation (1) with initial condition (2) is of the form

$$U(t,\lambda) = \frac{W_1(-1,\lambda)W_2(t,\lambda) - W_1(t,\lambda)W_2(-1,\lambda)}{W_1(-1,\lambda)W_2'(-1,\lambda) - W_1'(-1,\lambda)W_2(-1,\lambda)}.$$

It is well known that the Wronskian of two linearly independent solutions of (1) is a nonzero constant, (for each λ), whence we can say that

$$U(t,\lambda) = \frac{W_1(-1,\lambda)W_2(t,\lambda) - W_1(t,\lambda)W_2(-1,\lambda)}{W_1(0,\lambda)W_2'(0,\lambda) - W_1'(0,\lambda)W_2(0,\lambda)}. \tag{11}$$

Now from (8), (9) and (3) we have

$$\begin{aligned} & W_1(0,\lambda)W_2'(0,\lambda) - W_1'(0,\lambda)W_2(0,\lambda) \\ &= -\lambda^{1/3}\{Ai(0) + \epsilon_1(0)\}\{Bi'(0) + \eta_2(0)\} + \lambda^{1/3}\{Ai'(0) + \eta_1(0)\}\{Bi(0) + \epsilon_2(0)\} \\ &= -\lambda^{1/3}\{Ai(0)Bi'(0) - Ai'(0)Bi(0)\} + \lambda^{1/3}O\left(\frac{1}{\sqrt{\lambda}}\right) \\ &= -\lambda^{1/3}/\pi + O\left(\frac{1}{\lambda^{1/6}}\right). \end{aligned} \tag{12}$$

Now let

$$p(t) = \int_{-1}^t \sqrt{-v}\,dv = (-2/3)(-t)^{3/2} + 2/3 \quad \text{for } t \leq 0.$$

Then by (4) and (6) for $-1 \le t < 0$ we have

$$E^{-1}(\lambda^{1/3})E(-t\lambda^{1/3}) = e^{-\sqrt{\lambda}p(t)},$$

$$E(\lambda^{1/3})E^{-1}(-t\lambda^{1/3}) = e^{\sqrt{\lambda}p(t)},$$

$$M(\lambda^{1/3})M(-t\lambda^{1/3}) \sim \frac{5}{4\pi(-t)^{1/4}\lambda^{1/6}} \quad \lambda \to \infty. \tag{13}$$

We need the following estimates for determining the asymptotic form of the solutions. From (5), (10) and (13) we have, noting that $\cos\chi(t) = O(1)$,

$$\epsilon_1(-1)Bi(-t\lambda^{1/3}) = E^{-1}(\lambda^{1/3})M(\lambda^{1/3})E(-t\lambda^{1/3})M(-t\lambda^{1/3})O\left(\frac{1}{\sqrt{\lambda}}\right) =$$

$$\begin{cases} e^{-p(t)\sqrt{\lambda}}\frac{1}{(-t)^{1/4}}O(\lambda^{-2/3}) & \text{if } -1 \le t < 0 \\ e^{-\frac{2}{3}\sqrt{\lambda}}O(\lambda^{-7/12}) & t = 0 \\ e^{-\frac{2}{3}\sqrt{\lambda}}\frac{1}{t^{1/4}}O(\lambda^{-2/3}) & \text{if } 0 < t \le 1 \end{cases}$$

$$\epsilon_2(t)Ai(\lambda^{1/3}) = E^{-1}(\lambda^{1/3})M(\lambda^{1/3})E(-t\lambda^{1/3})M(-t\lambda^{1/3})O\left(\frac{1}{\sqrt{\lambda}}\right) =$$

$$\begin{cases} e^{-p(t)\sqrt{\lambda}}\frac{1}{(-t)^{1/4}}O(\lambda^{-2/3}) & \text{if } -1 \le t < 0 \\ e^{-\frac{2}{3}\sqrt{\lambda}}O(\lambda^{-7/12}) & t = 0 \\ e^{-\frac{2}{3}\sqrt{\lambda}}\frac{1}{t^{1/4}}O(\lambda^{-2/3}) & \text{if } 0 < t \le 1 \end{cases}$$

$$\epsilon_1(t)Bi(\lambda^{1/3}) = E(\lambda^{1/3})M(\lambda^{1/3})E^{-1}(-t\lambda^{1/3})M(-t\lambda^{1/3})O\left(\frac{1}{\sqrt{\lambda}}\right) =$$

$$\begin{cases} e^{p(t)\sqrt{\lambda}}\frac{1}{(-t)^{1/4}}O(\lambda^{-2/3}) & \text{if } -1 \le t < 0 \\ e^{\frac{2}{3}\sqrt{\lambda}}O(\lambda^{-7/12}) & t = 0 \\ e^{\frac{2}{3}\sqrt{\lambda}}\frac{1}{t^{1/4}}O(\lambda^{-2/3}) & \text{if } 0 < t \le 1 \end{cases}$$

$$\epsilon_2(-1)Ai(-t\lambda^{1/3}) = E(\lambda^{1/3})M(\lambda^{1/3})E^{-1}(-t\lambda^{1/3})M(-t\lambda^{1/3})O\left(\frac{1}{\sqrt{\lambda}}\right) =$$

$$\begin{cases} e^{p(t)\sqrt{\lambda}}\frac{1}{(-t)^{1/4}}O(\lambda^{-2/3}) & \text{if } -1 \le t < 0 \\ e^{\frac{2}{3}\sqrt{\lambda}}O(\lambda^{-7/12}) & t = 0 \\ e^{\frac{2}{3}\sqrt{\lambda}}\frac{1}{t^{1/4}}O(\lambda^{-2/3}) & \text{if } 0 < t \le 1 \end{cases}$$

From the asymptotic behavior of Ai and Bi we also have:

$Ai(\lambda^{1/3})Bi(-t\lambda^{1/3}) =$

$$\begin{cases} \frac{1}{2\pi}(-t)^{-1/4}\lambda^{-1/6}e^{-p(t)\sqrt{\lambda}}\{1 + O(\lambda^{-1/2})\} & \text{if } -1 \le t < 0 \\ \frac{1}{2\sqrt{\pi}}\lambda^{-1/12}Bi(0)e^{-\frac{2}{3}\sqrt{\lambda}}\left\{1 + O\left(\frac{1}{\sqrt{\lambda}}\right)\right\} & t = 0 \\ -\frac{1}{2\pi}t^{-1/4}\lambda^{-1/6}e^{-\frac{2}{3}\sqrt{\lambda}}\left\{\sin\left(\frac{2}{3}t^{3/2}\sqrt{\lambda} - \pi/4\right) + O\left(\frac{1}{\sqrt{\lambda}}\right)\right\} & \text{if } 0 < t \le 1 \end{cases} \tag{14}$$

and

$Bi(\lambda^{1/3})Ai(-t\lambda^{1/3}) =$

$$\begin{cases} \frac{1}{2\pi}(-t)^{-1/4}\lambda^{-1/6}e^{p(t)\sqrt{\lambda}}\{1+O(\lambda^{-1/2})\} & \text{if } -1 \le t < 0 \\ \frac{1}{\sqrt{\pi}}\lambda^{-1/12}Ai(0)e^{\frac{2}{3}\sqrt{\lambda}}\left\{1+O\left(\frac{1}{\sqrt{\lambda}}\right)\right\} & t=0 \\ \frac{1}{\pi}t^{-1/4}\lambda^{-1/6}e^{\frac{2}{3}\sqrt{\lambda}}\left\{\cos\left(\frac{2}{3}t^{3/2}\sqrt{\lambda}-\pi/4\right)+O\left(\frac{1}{\sqrt{\lambda}}\right)\right\} & \text{if } 0 < t \le 1 \end{cases} \tag{15}$$

From above calculation we get

$$W_1(-1,\lambda)W_2(t,\lambda) - W_1(t,\lambda)W_2(-1,\lambda) =$$
$$Ai(\lambda^{1/3})Bi(-t\lambda^{1/3}) - Ai(-t\lambda^{1/3})Bi(\lambda^{1/3})\}+$$
$$\epsilon_1(-1)Bi(-t\lambda^{1/3})\epsilon_2(t)Ai(\lambda^{1/3}) - \epsilon_1(t)Bi(\lambda^{1/3})\epsilon_2(-1)Ai(-t\lambda^{1/3})+$$
$$\epsilon_2(-1)\epsilon_1(t)\epsilon_2(t)\epsilon_1(-1) =$$

$$\begin{cases} \frac{1}{2\pi}(-t)^{-1/4}\lambda^{-1/6}e^{-p(t)\sqrt{\lambda}}\{1+O(\lambda^{-1/2})\} - \frac{1}{2\pi}(-t)^{-1/4}\lambda^{-1/6}e^{p(t)\sqrt{\lambda}}\{1+O(\lambda^{-1/2})\}+ \\ e^{p(t)\sqrt{\lambda}}(-t)^{-1/4}O(\lambda^{-2/3}) + e^{-p(t)\sqrt{\lambda}}(-t)^{-1/4}O(\lambda^{-2/3}) \quad t\in(-1,0), \\ \frac{1}{2\sqrt{\pi}}\lambda^{-1/12}Bi(0)e^{-\frac{2}{3}\sqrt{\lambda}}\left\{1+O\left(\frac{1}{\sqrt{\lambda}}\right)\right\} - \frac{1}{\sqrt{\pi}}\lambda^{-1/12}Ai(0)e^{\frac{2}{3}\sqrt{\lambda}}\left\{1+O\left(\frac{1}{\sqrt{\lambda}}\right)\right\} \quad t=0, \\ -\frac{1}{2\pi}(t)^{-1/4}\lambda^{-1/6}e^{-\frac{2}{3}\sqrt{\lambda}}\left\{\sin\left(\frac{2}{3}t^{3/2}\sqrt{\lambda}-\pi/4\right)+O\left(\frac{1}{\sqrt{\lambda}}\right)\right\} \\ -\frac{1}{\pi}t^{-1/4}\lambda^{-1/6}e^{\frac{2}{3}\sqrt{\lambda}}\left\{\cos\left(\frac{2}{3}t^{3/2}\sqrt{\lambda}-\pi/4\right)+O\left(\frac{1}{\sqrt{\lambda}}\right)\right\} \\ +t^{-1/4}e^{\frac{2}{3}\sqrt{\lambda}}O(\lambda^{-2/3}) + t^{-1/4}e^{-\frac{2}{3}\sqrt{\lambda}}O(\lambda^{-2/3}) \quad t\in(0,1] \end{cases} \tag{16}$$

By using the asymptotic form for the Wronskian, we can find the leading term of $U(t,\lambda)$. For $-1 \le t < 0$, the above results and (11) show that

$$U(t,\lambda) = \frac{1}{2(-t)^{1/4}\sqrt{\lambda}}\left\{e^{\sqrt{\lambda}\int_{-1}^{t}\sqrt{-t}dt}\left(1+O\left(\frac{1}{\sqrt{\lambda}}\right)\right) - e^{-\sqrt{\lambda}\int_{-1}^{t}\sqrt{-t}dt}\left(1+O\left(\frac{1}{\sqrt{\lambda}}\right)\right)\right\}$$
$$= \frac{1}{(-t)^{1/4}\sqrt{\lambda}}\left(1+O\left(\frac{1}{\sqrt{\lambda}}\right)\right)\sinh(p(t)\sqrt{\lambda}) \quad \lambda\to\infty \tag{17}$$

where, as usual,

$$p(t) = \int_{-1}^{t}\sqrt{-t}dt.$$

Similarly by using (11), (14) and (15) with $t=0$ we have,

$$U(0,\lambda) = \left\{\frac{\pi^{1/2}Ai(0)}{\lambda^{5/12}}e^{\frac{2}{3}\sqrt{\lambda}} - \frac{\pi^{1/2}Bi(0)}{2\lambda^{5/12}}e^{-\frac{2}{3}\sqrt{\lambda}}\right\}\left\{1+O\left(\frac{1}{\sqrt{\lambda}}\right)\right\}$$
$$= \frac{\pi^{1/2}Ai(0)}{\lambda^{5/12}}\{e^{\frac{2}{3}\sqrt{\lambda}} - \frac{\sqrt{3}}{2}e^{-\frac{2}{3}\sqrt{\lambda}}\}\left\{1+O\left(\frac{1}{\sqrt{\lambda}}\right)\right\} \quad \lambda\to\infty \tag{18}$$

because $Ai(0) = \frac{Bi(0)}{\sqrt{3}}$. For $0 < t$, we have by (14–15),

$$U(t,\lambda) = \frac{1}{t^{1/4}\sqrt{\lambda}}\left\{\frac{e^{-\frac{2}{3}\sqrt{\lambda}}\sin\left(\frac{2}{3}t^{3/2}\sqrt{\lambda}-\pi/4\right)}{2} + e^{\frac{2}{3}\sqrt{\lambda}}\cos\left(\frac{2}{3}t^{3/2}\sqrt{\lambda}-\pi/4\right)\right\}$$
$$\times\left\{1+O\left(\frac{1}{\sqrt{\lambda}}\right)\right\}$$

$$= \frac{1}{t^{1/4}\sqrt{\lambda}}\left\{e^{\frac{2}{3}\sqrt{\lambda}}\cos\left(\frac{2}{3}t^{3/2}\sqrt{\lambda}-\pi/4\right)\right\}\left(1+O\left(\frac{1}{\sqrt{\lambda}}\right)\right)\Bigg\} \quad \lambda\to\infty. \tag{19}$$

Now we find $\frac{\partial U}{\partial t}(t,\lambda)$. Indeed we have

$$\frac{\partial U}{\partial t}(t,\lambda) = \frac{W_1(-1,\lambda)W_2'(t,\lambda)-W_1'(t,\lambda)W_2(-1,\lambda)}{W_1(0,\lambda)W_2'(0,\lambda)-W_1'(0,\lambda)W_2(0,\lambda)}. \tag{20}$$

From (8) we may write

$$W_1(-1,\lambda)W_2'(t,\lambda) = -\lambda^{1/3}\{Ai(\lambda^{1/3})+\epsilon_1(-1)\}\{Bi'(-\lambda^{1/3}t)+\eta_2(t)\}.$$

Now we calculate the error bounds in the last display. From (10), (6) and (13) we may write, for large λ,

$$\lambda^{1/3}Ai(\lambda^{1/3})\eta_2(t) = \lambda^{1/3}M(\lambda^{1/3})E^{-1}(\lambda^{1/3})N(-t\lambda^{1/3})E(-t\lambda^{1/3})O\left(\frac{1}{\sqrt{\lambda}}\right) =$$

$$\begin{cases} e^{-p(t)\sqrt{\lambda}}(-t)^{1/4}O(\lambda^{-1/6}) & \text{if } -1\le t<0\\ e^{-\frac{2}{3}\sqrt{\lambda}}O(\lambda^{-1/4}) & t=0\\ e^{-\frac{2}{3}\sqrt{\lambda}}t^{1/4}O(\lambda^{-1/6}) & \text{if } 0<t\le 1\end{cases}$$

$$\lambda^{1/3}\epsilon_1(-1)Bi'(-t\lambda^{1/3}) = \lambda^{1/3}E^{-1}(\lambda^{1/3})M(\lambda^{1/3})E(-t\lambda^{1/3})N(-t\lambda^{1/3})O\left(\frac{1}{\sqrt{\lambda}}\right) =$$

$$\begin{cases} e^{-p(t)\sqrt{\lambda}}(-t)^{1/4}O(\lambda^{-1/6}) & \text{if } -1\le t<0\\ e^{-\frac{2}{3}\sqrt{\lambda}}O(\lambda^{-1/4}) & t=0\\ e^{-\frac{2}{3}\sqrt{\lambda}}t^{1/4}O(\lambda^{-1/6}) & \text{if } 0<t\le 1\end{cases}$$

Similarly we have by (5), (4) and (16),

$$W_1'(t,\lambda)W_2(-1,\lambda) = -\lambda^{1/3}\{Ai'(-\lambda^{1/3}t)+\eta_1(t)\}\{Bi(\lambda^{1/3})+\epsilon_2(-1)\},$$

$$\lambda^{1/3}\eta_1(t)Bi(\lambda^{1/3}) = \lambda^{1/3}E^{-1}(-t\lambda^{1/3})M(\lambda^{1/3})E(\lambda^{1/3})N(-t\lambda^{1/3})O\left(\frac{1}{\sqrt{\lambda}}\right) =$$

$$\begin{cases} e^{p(t)\sqrt{\lambda}}(-t)^{1/4}O(\lambda^{-1/6}) & \text{if } -1\le t<0\\ e^{\frac{2}{3}\sqrt{\lambda}}O(\lambda^{-1/4}) & t=0\\ e^{\frac{2}{3}\sqrt{\lambda}}t^{1/4}O(\lambda^{-1/6}) & \text{if } 0<t\le 1,\end{cases}$$

$$\lambda^{1/3}\epsilon_2(-1)Ai'(-t\lambda^{1/3}) = \lambda^{1/3}E(\lambda^{1/3})M(\lambda^{1/3})E^{-1}(-t\lambda^{1/3})N(-t\lambda^{1/3})O\left(\frac{1}{\sqrt{\lambda}}\right) =$$

$$\begin{cases} e^{p(t)\sqrt{\lambda}}(-t)^{1/4}O(\lambda^{-1/6}) & \text{if } -1\le t<0\\ e^{\frac{2}{3}\sqrt{\lambda}}O(\lambda^{-1/4}) & t=0\\ e^{\frac{2}{3}\sqrt{\lambda}}t^{1/4}O(\lambda^{-1/6}) & \text{if } 0<t\le 1.\end{cases}$$

The leading terms in the display following (20) are found next

$-\lambda^{1/3}Ai(\lambda^{1/3})Bi'(-t\lambda^{1/3}) =$

$$\begin{cases} -\frac{1}{2\pi}(-t)^{1/4}\lambda^{1/3}e^{-p(t)\sqrt{\lambda}}\{1+O(\lambda^{-1/2})\} & \text{if } -1 \le t < 0 \\ -\frac{1}{2\sqrt{\pi}}\lambda^{1/4}Bi'(0)e^{-\frac{2}{3}\sqrt{\lambda}}\left\{1+O\left(\frac{1}{\sqrt{\lambda}}\right)\right\} & t = 0 \\ -\frac{1}{2\pi}t^{1/4}\lambda^{1/3}e^{-\frac{2}{3}\sqrt{\lambda}}\left\{\cos\left(\frac{2}{3}t^{3/2}\sqrt{\lambda}-\pi/4\right)+O\left(\frac{1}{\sqrt{\lambda}}\right)\right\} & \text{if } 0 < t \le 1 \end{cases} \tag{21}$$

$-\lambda^{1/3}Bi(\lambda^{1/3})Ai'(-t\lambda^{1/3}) =$

$$\begin{cases} \frac{1}{2\pi}(-t)^{1/4}\lambda^{1/3}e^{p(t)\sqrt{\lambda}}\{1+O(\lambda^{-1/2})\} & \text{if } -1 \le t < 0 \\ -\frac{1}{\sqrt{\pi}}\lambda^{1/4}Ai'(0)e^{\frac{2}{3}\sqrt{\lambda}}\left\{1+O\left(\frac{1}{\sqrt{\lambda}}\right)\right\} & t = 0 \\ -\frac{1}{\pi}t^{1/4}\lambda^{1/3}e^{\frac{2}{3}\sqrt{\lambda}}\left\{\sin\left(\frac{2}{3}t^{3/2}\sqrt{\lambda}-\pi/4\right)+O\left(\frac{1}{\sqrt{\lambda}}\right)\right\} & \text{if } 0 < t \le 1. \end{cases} \tag{22}$$

Combining the last six displays we get

$$\frac{\partial U}{\partial t}(t,\lambda) =$$

$$\begin{cases} (-t)^{1/4}\{1+O(\lambda^{-1/2})\}\cosh(p(t)\sqrt{\lambda}) & \text{if } -1 \le t < 0 \\ \pi^{1/2}\lambda^{-1/12}Bi'(0)\left\{\frac{1}{2}e^{-\frac{2}{3}\sqrt{\lambda}}+\frac{1}{\sqrt{3}}e^{\frac{2}{3}\sqrt{\lambda}}\right\}\left\{1+O\left(\frac{1}{\sqrt{\lambda}}\right)\right\} & t = 0 \\ t^{1/4}\left\{\frac{1}{2}e^{-\frac{2}{3}\sqrt{\lambda}}\cos\left(\frac{2}{3}t^{3/2}\sqrt{\lambda}-\pi/4\right)-e^{\frac{2}{3}\sqrt{\lambda}}\sin(\frac{2}{3}t^{3/2}\sqrt{\lambda}-\pi/4)\right\}\left\{1+O\left(\frac{1}{\sqrt{\lambda}}\right)\right\} & \\ & \text{if } 0 < t \le 1 \end{cases} \tag{23}$$

where the error terms are uniform in t if t is in some compact set which excludes $t = 0$.

Acknowledgements

This research is supported by a grant from Tabriz University.

References

1. A. Jodayree, AJSE, (in print).
2. F.V. Atkinson and A.B. Mingarelli, J. Reine Ang. Math., **375**, 380-393 (1987).
3. F.W.J. Olver, J. Soc. Indust. Appl. Math., **11**, 748-772 (1963).
4. G.N. Watson, Theory of Bessel Functions, Second edition, Cambridge University Press (1944).

9th Int. Coll. on Differential Equations, pp. 229-234
D. Bainov (Ed)

Existence of Group Invariant Solutions of a Certain Semilinear Elliptic Equation

RYUJI KAJIKIYA
Faculty of Engineering, Nagasaki Institute of Applied Science, 536 Aba-machi, Nagasaki, 851-0193, Japan
E-mail: kajikiya@bas.nias.ac.jp

Abstract: This paper studies solutions for the Emden-Fowler equation in the unit ball under the Dirichlet boundary condition together with the G invariance: $u(gx) = u(x)$ for $g \in G$ and $|x| < 1$. Here G is a closed subgroup of the orthogonal group. It is proved that there exist infinitely many solutions without radial symmetry if and only if G is not transitive on the unit sphere.
Key Words and Phrases: *Semilinear elliptic equation, group invariant solution.*

1. Introduction

We consider the existence of group invariant solutions of the Emden-Fowler equation,

$$(1.1) \qquad -\Delta u = |u|^{p-1}u, \qquad x \in B,$$

$$(1.2) \qquad u = 0, \qquad x \in \partial B,$$

$$(1.3) \qquad u(gx) = u(x), \qquad x \in B, \quad g \in G,$$

where B is a unit ball of $\mathbb{R}^n$, $n \geq 2$, $1 < p < (n+2)/(n-2)$ and G is a closed subgroup of the orthogonal group $O(n)$. Since each $g \in G$ is an orthogonal matrix, it holds that $|gx| = |x|$, and therefore $u(gx)$ is well-defined and (1.3) makes sense. We call a solution of (1.1)-(1.3) a G *invariant solution*. It is known that the problem (1.1)-(1.2) has infinitely many radially symmetric solutions. A radially symmetric solution becomes a G invariant solution for any G.

We consider the converse problem: Does there exist a solution which is G invariant and not radially symmetric? Since G is a closed subgroup of $O(n)$, it is a linear transformation group on $\mathbb{R}^n$ and on the unit sphere S^{n-1},

$$S^{n-1} = \{x \in \mathbb{R}^n : |x| = 1\}.$$

We say that G is *transitive* on S^{n-1} if and only if $\{gx : g \in G\} = S^{n-1}$ for any $x \in S^{n-1}$. Our answer to the problem is as follows.

Theorem 1. *The following two assertions are equivalent.*
(i) G is not transitive on S^{n-1}.

(ii) There exists a sequence $\{u_k\}_{k=1}^{\infty}$ *of solutions of (1.1)-(1.3) such that each* u_k *is* G *invariant, not radially symmetric and*

$$0 < \|u_1\|_{H_0^1(B)} < \|u_2\|_{H_0^1(B)} < \|u_3\|_{H_0^1(B)} < \cdots \nearrow \infty,$$

where $\|\cdot\|_{H_0^1(B)}$ *denotes the* $H_0^1(B)$ *Sobolev norm.*

Corollary 1. *If* $\dim G \leq n-2$, *then the assertion (ii) of Theorem 1 holds.*

Corollary 2. *If* G *is a finite subgroup of* $O(n)$, *then the assertion (ii) of Theorem 1 remains valid.*

By Cartan's theorem, a closed subgroup G of the Lie group $O(n)$ is a Lie subgroup. Therefore in Corollary 1, $\dim G$ denotes the dimension of G as a Lie group or a manifold.

From Theorem 1, we find a question: What kind of G is transitive? This problem has been solved by Montgomery, Samelson and Borel as below.

Theorem A. (Montgomery and Samelson [1], Borel [2]) *Let* $n \geq 2$ *and* G *be a connected closed subgroup of* $SO(n)$. *Then the following are equivalent.*

(i) G *is transitive on* S^{n-1}.

(ii) G *is* $O(n)$*-conjugate to one of the following subgroups :* $SO(n)$; $SU(m)$, $U(m)$ $(n = 2m)$; $Sp(m)$, $Sp(m)Sp(1)$, $Sp(m)U(1)$ $(n = 4m)$; $Spin(9)$ $(n = 16)$; $Spin(7)$ $(n = 8)$; G_2 $(n = 7)$.

If G is not necessarily connected, we have the next theorem from Theorem A.

Theorem B. *Let* $n \geq 2$ *and* G *be a closed subgroup of* $O(n)$. *Then the following are equivalent.*

(i) G *is transitive on* S^{n-1}.

(ii) The connected component of G *which has the unit matrix is* $O(n)$*-conjugate to one of the Lie groups listed in (ii) of Theorem A.*

2. Examples

In this section, we present some examples.

Example 2.1. Let $G = \{e, -e\}$, where e is the unit matrix. Then we have infinitely many solutions $\{u_k\}_{k=1}^{\infty}$ by Corollary 2 such that each u_k is even but not radially symmetric.

Example 2.2. Let

$$G = \{g \in O(n) : \text{each element of } g \text{ is equal to 1 or 0}\}.$$

Then there exist infinitely many solutions $\{u_k\}_{k=1}^{\infty}$ of (1.1)-(1.3) such that each u_k is not radially symmetric and

$$u_k(x_1, \cdots, x_n) = u_k(x_{\sigma(1)}, \cdots, x_{\sigma(n)}) \qquad \text{for any permutation } \sigma.$$

Example 2.3. Let $G = O(m) \times O(n-m)$. Then there exist infinitely many solutions $\{u_k\}_{k=1}^{\infty}$ such that $u_k = u_k(|x'|, |x''|)$ for $x' \in \mathbb{R}^m$, $x'' \in \mathbb{R}^{n-m}$ and $u_k \neq u_k(|x|)$.

3. Proof of Main Results

When G is transitive, a G invariant solution must be radially symmetric by the definition of transitivity. In this case, the assertion (ii) of Theorem 1 does not occur. Consequently, it holds that (ii) of Theorem 1 implies (i).

Let us prove that (i) implies (ii). Our approach is based on the variational method.

Definition 3.1. Define $I(\cdot)$ and $H_0^1(B,G)$ by

$$\begin{aligned} I(u) &\equiv \int_B \left(\frac{1}{2}|\nabla u|^2 - \frac{1}{p+1}|u|^{p+1}\right) dx, \\ H_0^1(B,G) &\equiv \{u \in H_0^1(B) : u(gx) = u(x) \text{ for } g \in G, x \in B\}. \end{aligned}$$

The functional $I(\cdot): H_0^1(B,G) \to \mathbb{R}$ is of class C^2 in the sense of Fréchet derivative. It is easy check that $I'(u) = 0$ if and only if u satisfies (1.1)-(1.3) in the distribution sense. Moreover, a weak solution becomes a classical solution by the elliptic regularity theory. Therefore, it is sufficient to find the critical points of $I(\cdot)$ without radial symmetry.

We first consider the G invariant eigenvalue problem:

$$\begin{cases} -\Delta u = \lambda u, & x \in B, \\ u = 0, & x \in \partial B, \\ u(gx) = u(x), & x \in B, \quad g \in G. \end{cases} \tag{3.1}$$

Let $\lambda_k = \lambda_k(G)$ denote the k-th eigenvalue, which is repeated a number of times equal to its multiplicity.

Lemma 3.1.([3]) *If G is not transitive on S^{n-1}, then there exist an integer m and $C > 0$ such that $0 \le m \le n-2$ and*

$$\lambda_k(G) \le C k^{2/(n-m)} \qquad \text{for all } k \in \mathbb{N}.$$

Using Lemma 3.1 with the symmetric mountain pass lemma by Ambrosetti and Rabinowitz [4], we obtain the G invariant critical values of $I(\cdot)$ in Lemma 3.2. To state the lemma, we introduce Krasnoselskii's genus.

Let K be a closed subset of $H_0^1(B,G)$ which satisfies $-u \in K$ for $u \in K$ and $0 \notin K$. Then the genus of K, which is denoted by $\gamma(K)$, is defined by the smallest integer k such that there exists an odd continuous mapping from K to $\mathbb{R}^k \setminus \{0\}$. If such a k does not exist, we define $\gamma(K) = \infty$.

Lemma 3.2. *There exists a sequence $\{\alpha_k\}_{k=1}^{\infty} \subset \mathbb{R}$ satisfying the following conditions.*

(i) $0 < \alpha_1 \le \alpha_2 \le \cdots$, $\lim_{k\to\infty} \alpha_k = \infty$.

(ii) Each α_k is a G invariant critical value of $I(\cdot)$. That is, there exists a $u_k \in H_0^1(\Omega, G)$ such that $I(u_k) = \alpha_k$ and $I'(u_k) = 0$.

(iii) If $\alpha_k = \alpha_{k+1} = \cdots = \alpha_{k+l} \equiv \alpha$, then the genus $\gamma(K_\alpha) \ge l+1$, where

$$K_\alpha = \{u \in H_0^1(\Omega, G) : I'(u) = 0 \text{ and } I(u) = \alpha\}.$$

(iv) There exists a constant $C > 0$ such that

$$\alpha_k \le C k^{2(p+1)/((n-m)(p-1))} \quad \text{for } k \in \mathbb{N},$$

where m is the integer appeared in Lemma 3.1.

Proof. The proof of Lemma 3.2 is due to the symmetric mountain pass lemma by Ambrosetti and Rabinowitz [4] except for (iv). Let $\lambda = \lambda_k(G)$ and $\phi_k(x)$ denote the k-th eigenvalue of (3.1) and the corresponding eigenfunction with $\|\phi_k\|_{L^2(B)} = 1$, respectively. Set

$$E_k = \operatorname{span}\{\phi_1, \cdots, \phi_k\} = \{\sum_{i=1}^{k} t_i\phi_i : t_i \in \mathbb{R}\}. \tag{3.2}$$

We denote the L^q- norm of u by $\|u\|_q$. The definition of $I(\cdot)$,

$$I(u) = \frac{1}{2}\|\nabla u\|_2^2 - \frac{1}{p+1}\|u\|_{p+1}^{p+1}, \tag{3.3}$$

implies that there exists a constant $R_k > 0$ such that

$$I(u) \leq 0 \qquad \text{for } \|u\|_{H_0^1} \geq R_k \text{ and } u \in E_k.$$

We may assume that $0 < R_1 < R_2 < \cdots \nearrow \infty$. Define α_k by

$$\alpha_k = \inf_{f\in\Gamma_k} \max_{u\in D_k} I(f(u)),$$

where

$$\begin{aligned} \Gamma_k &= \{f \in C(D_k, H_0^1(B,G)) : f \text{ is odd on } D_k \text{ and } f(u) = u \text{ on } \partial D_k\}, \\ D_k &= \{u \in E_k : \|u\|_{H_0^1} \leq R_k\}, \\ \partial D_k &= \{u \in E_k : \|u\|_{H_0^1} = R_k\}. \end{aligned}$$

The proof of (i), (ii) and (iii) are similar to [4]. (see also [5, pp 55-60].) Let us prove (iv). We choose $f = \mathit{identity} \in \Gamma_k$ to obtain

$$\alpha_k \leq \max_{u\in D_k} I(u) \leq \sup_{u\in E_k} I(u).$$

From definition (3.2) of E_k it follows that $\|\nabla u\|_2^2 \leq \lambda_k\|u\|_2^2$ for $u \in E_k$. Since $p > 1$, there exists a $C > 0$ such that $\|u\|_2 \leq C\|u\|_{p+1}$. Therefore we have the inequality $\|\nabla u\|_2^2 \leq C^2\lambda_k\|u\|_{p+1}^2$ for $u \in E_k$. Substituting this inequality into (3.3), we obtain

$$I(u) \leq \frac{1}{2}\|\nabla u\|_2^2 - C'\lambda_k^{-(p+1)/2}\|\nabla u\|_2^{p+1} \qquad \text{for } u \in E_k,$$

which implies $\sup_{u\in E_k} I(u) \leq C\lambda_k^{(p+1)/(p-1)}$. This inequality together with Lemma 3.1 shows (iv). The proof is complete.

In the next lemma, we investigate all of radially symmetric solutions for (1.1)-(1.2). For a radially symmetric solution $u = u(r)$, $r = |x|$, the problem (1.1)-(1.2) is rewritten as

$$u'' + \frac{n-1}{r}u' + |u|^{p-1}u = 0, \qquad 0 < r < 1, \tag{3.4}$$

$$u'(0) = u(1) = 0. \tag{3.5}$$

Lemma 3.3.([6], [7], [3]) *For each $k \geq 1$, there exists a unique solution u of (3.4)-(3.5) which has exactly k zeros in $[0,1]$ and $u(0) > 0$. Denote it by u_k. Then all solutions* *of (3.4)-(3.5) consist of* u_k, $-u_k$ $(k \in \mathbb{N})$ *and the zero solution. Set* $\beta_k = I(u_k)$. *Then we have:*

(i) $0 < \beta_1 < \beta_2 < \cdots \nearrow \infty$.

(ii) There exist constants $A > 0$ *and* $B > 0$ *such that*

$$Ak^{2(p+1)/(p-1)} \le \beta_k \le Bk^{2(p+1)/(p-1)} \quad \textit{for } k \ge 1.$$

Sketch of proof. Instead of the boundary value problem (3.4)-(3.5), we consider the initial value problem

$$w'' + \frac{n-1}{r}w' + |w|^{p-1}w = 0, \quad 0 < r < \infty, \tag{3.6}$$

$$w'(0) = 0, \quad w(0) = 1. \tag{3.7}$$

The proof of the lemma is based on the next two assertions.

Assertion 1. The initial value problem (3.6)-(3.7) has a unique global solution $w(r)$ on $[0, \infty)$ and moreover it is oscillatory (see [8, Corollary 6.7]), that is, $w(r)$ has infinitely many zeros, $t_1 < t_2 < t_3 < \cdots \nearrow \infty$.

Assertion 2. If $w(r)$ is a solution of (3.6), so is $\lambda^{2/(p-1)}w(\lambda r)$ for any $\lambda > 0$.

Observing the above two assertions, we set $u_k(r) \equiv t_k^{2/(p-1)}w(t_k r)$. It is a solution of the boundary value problem (3.4)-(3.5) with exactly k zeros in the interval $[0, 1]$ and satisfies $u_k(0) > 0$. The uniqueness of such a solution follows from Assertions 1 and 2. For the proof of (i) and (ii) of Lemma 3.3, see [3] or [6]. The proof is complete.

For the detailed proof of Lemmas 3.1, 3.2 and 3.3, refer to [3]. We now prove Theorem 1 and Corollaries 1 and 2 by combining Lemmas 3.2 and 3.3.

Proof of Theorem 1. As stated already, it is clear that (ii) implies (i). We show that (i) implies (ii). Let G be not transitive. In contradiction to (ii), suppose that the set of G invariant non-radial solutions is bounded in $H_0^1(B)$. Therefore there exists a $k_0 \ge 1$ such that any G invariant solution corresponding to α_k for $k \ge k_0$ must be radially symmetric. Since $\{\beta_k\}$ are all of radially symmetric critical values, it holds that $\{\alpha_k : k \ge k_0\} \subset \{\beta_k : k \ge 1\}$. If $\alpha_k = \alpha_{k+1}$ with some k, then $\gamma(K_\alpha) \ge 2$ by (iii) of Lemma 3.2. On the other hand, since $\alpha_k = \beta_j$ with a unique j, we find $K_{\alpha_k} = K_{\beta_j} = \{u_j, -u_j\}$, where u_j is a radially symmetric solution with exactly j zeros. By definition of genus, it holds that $\gamma(K_{\alpha_k}) = \gamma(K_{\beta_j}) = 1$ since K_{β_j} is a finite set. This is a contradiction. Therefore $\{\alpha_k\}$ must be strictly increasing. Then there exists a sequence $\{\mu(k)\}_{k=1}^{\infty}$ of integers such that $\alpha_k = \beta_{\mu(k)}$. Since $\{\mu(k)\}$ is strictly increasing, we see $k \le \mu(k)$, and so $\beta_k \le \beta_{\mu(k)} = \alpha_k$. Lemmas 3.2 and 3.3 show

$$Ak^{2(p+1)/(p-1)} \le Ck^{2(p+1)/((n-m)(p-1))} \quad \text{for all } k \ge k_0.$$

Since $m \le n - 2$ by Lemma 3.1, we obtain a contradiction by letting $k \to \infty$. This completes the proof of Theorem 1.

Proof of Corollaries 1 and 2. If $\dim G \le n - 2$, then G can not be transitive on S^{n-1}. Hence Theorem 1 shows Corollary 1.

Corollary 2 is clear because a finite group must not be transitive.

Acknowledgements

The author is supported in part by the Grant-in-Aid for Scientific Research (C) (No. 08640223), Ministry of Education, Science, Sports and Culture.

References

1. D. Montgomery and H. Samelson, Ann. of Math., **44**, 454-470 (1943).
2. A. Borel, C. R. Acad. Sci., Paris, **230**, 1378-1380 (1950).
3. R. Kajikiya, (preprint).
4. A. Ambrosetti and P. H. Rabinowitz, J. Funct. Anal., **14**, 349-381 (1973).
5. P. H. Rabinowitz, in: CBMS Regional Conf. Ser. in Math., **65**, Amer. Math. Soc., Providence (1986).
6. R. Kajikiya, Hiroshima Math. J., **20**, 259-276 (1990).
7. R. Kajikiya, Hiroshima Math. J., **21**, 111-161 (1991).
8. W.-M. Ni and R. D. Nussbaum, Comm. Pure Appl. Math., **38**, 67-108 (1985).

9th Int. Coll. on Differential Equations, pp. 235-240
D. Bainov (Ed)

Structured Programming Diagram PAD, PDE Solver, and Formation of Planetary System

Toshio KAWAI
Chitose Institute of Science and Technology, Chitose, Hokkaido, 066-8655, Japan
t-kawai@photon.chitose.ac.jp

Abstract: In view of a variety of topics dealt with in this colloquium, three topics are prepared. The first is the structured programming diagram PAD (category: software engineering). The second is a user-oriented general purpose PDE solver (category: numerical analysis of PDE). The last is the Saturn rings and our Planetary system formation (category: applications). Although these topics may seem to be separate, they are integrated parts in computational physics.

1. PAD

Programmers are often lost in the jungle of control flow like Fig. 1. The number of possible cases of the flow doubles at each branch. More complicated is the program when the control flow returns up the stream. In the common flow chart the complexity is hidden by a simple arrow, but actually a man orders the computer to do very complex processing which he himself does not intend. Man tends to overestimate his ability of logical thinking.

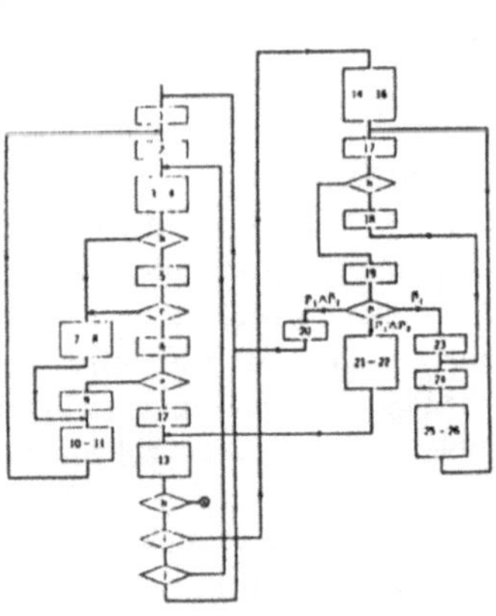

Figure 1: Flow chart

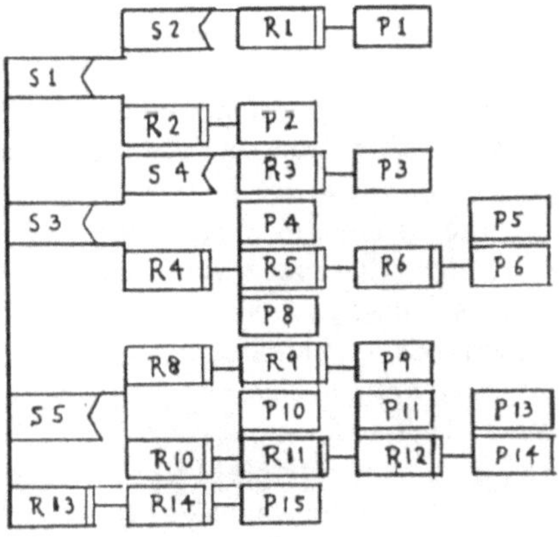

Figure 2: PAD

Structured programming by Dijkstra [1] is well-known. When we understand the spirit of the structured programming from another viewpoint, program can be expressed graphically by a tree, with leaves (unit of processing) on two kinds branches: selection and repetition. We can read the tree diagram from top to down and from left to right. One part can be connected to the whole at one point. I designed the shape of boxes and named the diagram as Problem Analysis Diagram, PAD. Indeed the jungle changes to a tree (Fig. 2) [2].

It propagated from person to person, and in 20 years most programs in Japan are produced by using PAD. I cannot express its effectiveness quantitatively, but it is contributing the productivity and reliability of Japanese software.

If a computational physicist aims at discoveries as experimental physicists do, he will meet unexpected output. How can he prove that his result is not due to bugs? Writing a correct program is an essential issue, and it is achieved by design, not by debugging.

2. PDE Solver–Distran–

As we all know very seriously through painful experiences, many kinds of numerical anomaly can arise in the solution of partial differential equations. For example, consider a steady solution of an ordinary differential equation

$$\dot{x} = -ax + b. \tag{1}$$

Of course the solution is b/a, simply obtained by one division. But how about a case when a is negative or zero? Thus algebraic solution $A^{-1}B$ may not express the physical reality.

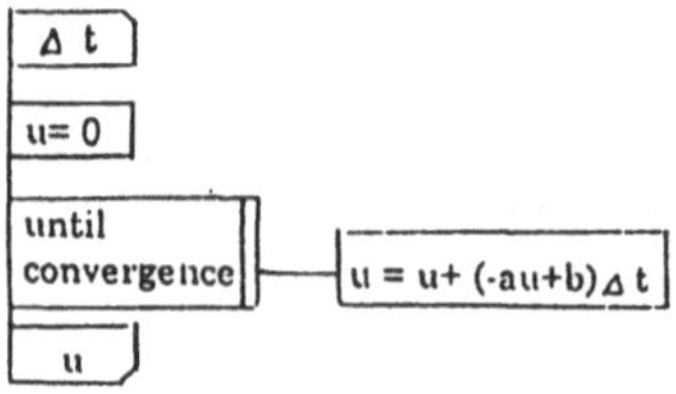

Figure 3: Nature algorithm in PAD

Problems originating from physics are safely simulated with the algorithm of nature: evolution with interaction. The algorithm of nature is expressed in Fig.3 in PAD. The only source of numerical accidents if the Δt (and Δx, Δy, Δz); the finite size of discretization. Nature algorithm is ideal because we need not worry about all kinds of numerical accidents.

Its demerit so far was the long computing time. Huge or cheap personal computers are now solving this difficulty. Once we make up our mind to adopt nature algorithm, then programs can be generated automatically by machine: the idea of a program translator for distributed systems (distran). We have only to write problems in mathematical equations. We are now free from bugs.

Fig. 4 is an example of a problem specification for distran (a 2-D version on Personal Computer). The output like Fig. 5 (actually color) is obtained in a minute to solve a problem having 50×50 mesh points. With a massively parallel processor (MPP) of Tera FLOPS (10^{12} floating point per second), usual 3-D problems with $1000 \times 1000 \times 1000$ meshes will be solved in an hour.

MPP is an indispensable tool for computational physics. The author has been participating in the development project of an MPP named PACS [3]. Development of

comment	Convection-diffusion
problem name	example 60
space dimension	2
vector variables	v=v2
equations	$v_x = 1.0$
	$v_y = 1.0$
	s=exp(-r/2.0)
	r=x*x+y*y
	dt(u)=-(v_x*dx(u))-(v_y*dy(u))+dxx(u)+dyy(u)+s
boundary conditions	u=0.0 on r0
	u=free on r1
initial condition	u=0.0
region	$-5.0 \leq X \leq 5.0$
	$-5.0 \leq Y \leq 5.0$
time	$0.0 \leq T \leq 8.0$

Figure 4: Specifying a problem in distran

MPP throughout the world is generally slow because of a kind of common sense that programming for MPP is very difficult. But I insist that MPP should be used without programming. This can be the final solution of the difficulty. MPP makes automatic programming possible, and the automation helps MPP to be welcome for wider user community. The two tools should go hands in hands, and tools like distran is considered to be very essential.

3. Saturn Rings

Naive questions in nature which have been left unsolved can now be explored with computers. The first target is the Saturn rings. Trillion of ice particles orbiting Saturn collide each other. I myself tried to simulate tens of particles without success. One morning a hint hit me that the collisions should be inelastic, not elastic. Then rings appeared really on the screen. Indeed, elasticity is an idealized notion, but it is not real.

Once it was seen on the display, the theory followed readily. We give a proof for the simplest case that all particles are spheres with common mass, orbits are on $x - y$ plane and velocities are transformed according to a special collision law:

$$\boldsymbol{v}_i' - \boldsymbol{v}_j' = -e(\boldsymbol{v}_i - \boldsymbol{v}_j), \tag{2}$$

$$\boldsymbol{v}_i' + \boldsymbol{v}_j' = \boldsymbol{v}_i + \boldsymbol{v}_j, \tag{3}$$

or

$$\boldsymbol{v}_i' = \alpha\boldsymbol{v}_i + \beta\boldsymbol{v}_j, \tag{4}$$

$$\boldsymbol{v}_j' = \beta\boldsymbol{v}_i + \alpha\boldsymbol{v}_j. \tag{5}$$

Total kinetic energy K increases by

$$\Delta K = -\frac{1}{2}(1 - e^2)(\boldsymbol{v}_i - \boldsymbol{v}_j)^2 \leq 0. \tag{6}$$

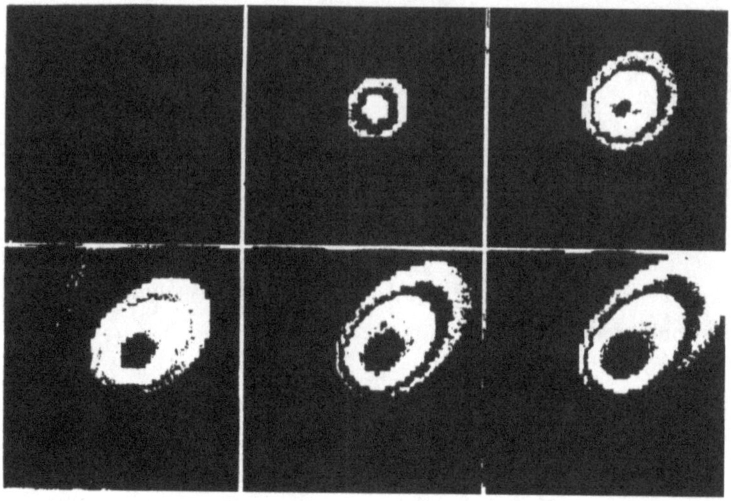

Figure 5: Solution of the problem specified in Fig. 4

If we further assume that specific angular momentum $\boldsymbol{r} \times \boldsymbol{v}$ is initially common for all particles, then any pair of particles collides each other forever (a property of elliptic Kepler orbit). From Cauchy's convergence theorem,

$$\Delta K \to 0, \quad |\boldsymbol{v}_i - \boldsymbol{v}_j| \to 0. \tag{7}$$

In physical terms, all orbits cross with all other orbits forever. The two velocities at the crossing points tend to be equal, which means convergence of orbits to one common orbit (a sharp ring). It is actually demonstrated in Fig. 6. All assumptions above are for the sake of simple proof, and easily removed [4].

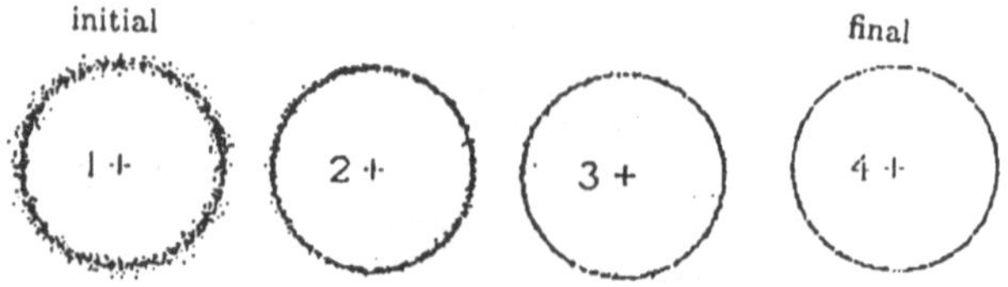

Figure 6: Formation of a sharp ring by inelastic collisions

Sharp rings are discovered to be a logical consequence of inelastic collision. Astronomers have been caught by a prejudice that collisions cannot sharpen a ring, but will rather blur it instead. They justified the prejudice by an energetical argument which turns out to be incomplete.

Rings are common when particles in central field collide inelastically. We can demonstrate generation of rings easily in personal computers like Fig. 7. Here inelastically colliding free particles are confined in a perfectly reflecting sphere. This phenomenon can also be proved by a simple logic [4]. Rings are now found by Hubble telescope often in the universe. I would like to propose to have a ring around the earth if it is useful for our life.

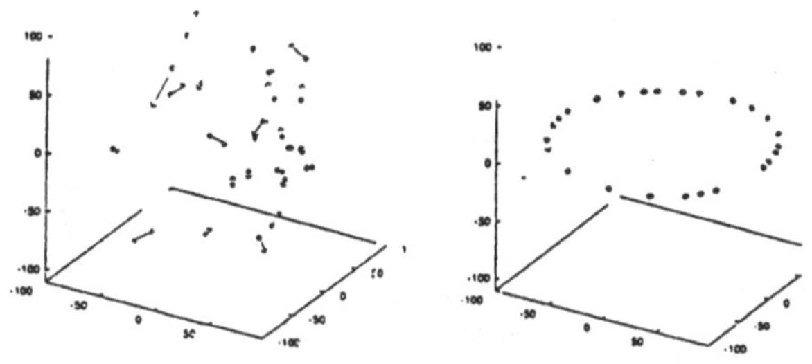

Figure 7: Inelastically colliding free particles in an elastic circle

4. Planetary System

How was the earth born? Everybody agrees that the earth is the result of coalescence of many particles upon collisions. Mercury, Venus, ... Pluto are the result of fusion upon collision of millions of planetesimals around the Sun. This problem can be simulated by integrating millions of equations of motion for hundred millions of years. A particle is attracted by the Sun plus other millions of particles. This is an "astronomical" computational problem. The special purpose computer for such problems, GRAPE, gravity pipe), was constructed in Japan [5], and it aims to attain 1000 Tera FLOPS. Still we have so far succeeded simulating 1000 particles for 1000 years, without any sign of planet formation.

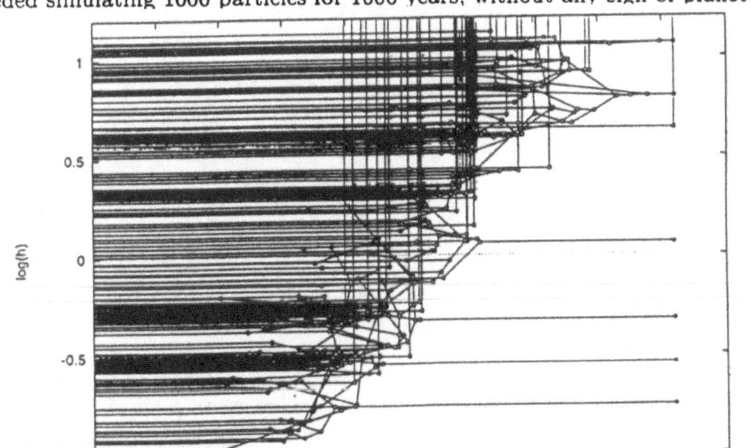

Figure 8: Evolution of our planetary system. Coagulation of particles by collision vs time (year). Vertical axis corresponds to the logarithm of distance from the Sun.

I formulated the dynamics of orbital elements, which are constant of motion if the mutual gravity is neglected. The dynamics of the slowly varying orbital elements was rigorously expressed, and solved by an event-driven way. The details are described in reference [4]. The formulation revealed that each particle has two big events through a life: one collision and one scattering (equivalent to many small scatterings). Thus for a system of 1000 initial particles, about 2000 events occur until the end of evolution. We have only to compute 2000 events to trace one history of planetary system in several seconds of computing time with a workstation.

A typical result is shown in Fig. 8. Distance of particles are shown in the range from Mercury to Pluto in logarithmic scale. We see the location of seven final planets in this example. Statistics of 1000 such histories show that the location (distance from the Sun) of the final planets obey the geometric series with the common ratio 1.85.

Titius and Bode found the well-known empirical correlation as shown in Fig. 9. We see in this figure that the real distribution can be seen as a sample of the general law of the geometric series. Our model has six parameters: the number of initial particles, total mass of the planets, and so on. For a large range of the six parameters the result was the same: a geometric law with the common ratio of 1.85. This is a magic number at present. Recently, several planetary systems are being observed. According to the present simulation, I would dare to predict that they are arranged just like our planetary system with common ratio 1.85.

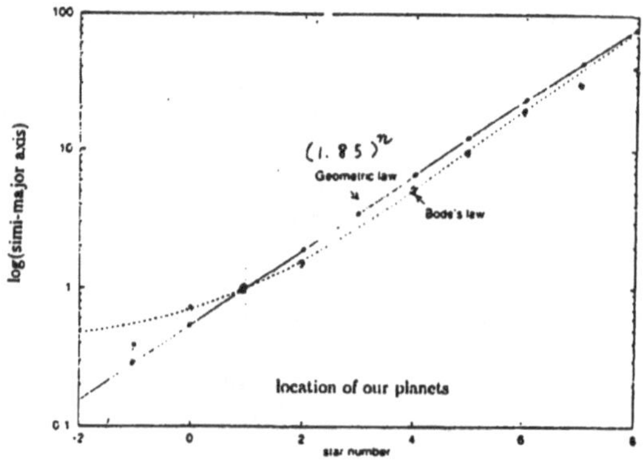

Figure 9: Configuration of final planets

5. Conclusion

Computer is a giant. Although we are short, we can extend our horizon if we are on his shoulder. I myself could look into Nature, which was never possible without computers. Computational physics have prospective future, and so general software tools (PAD and distran) are the important, as well as specific applications and hardware like PACS and GRAPE. These topics are closely related inside. We need certain range of interest even as a specialist. I hope the prosperity of this colloquium which covers a wide variety of fields, in spite of the difficulty in the modern age of accelerated specialization.

References

1. O. Dahl, E.W. Dijkstra and C.A.R. Hoare, Structured Programming, Academic Press (1972).
2. Y. Futamura et al., Japan Ann. Rev. in Electronics, Computers and Telecommunications, **12**, 97–115 (1984).
3. Y. Iwasaki et al., Nuclear Physics B, **17**, 259–262 (1990).
4. T. Kawai, Int. J. Mod. Phys. B, **12**, 309–360 (1998).
5. J. Makino and M. Taiji, Scientific Simulation with Special-Purpose Computers, John Wiley and Sons (1998).

9th Int. Coll. on Differential Equations, pp. 241–244
D. Bainov (Ed)

Coefficients of an Asymptotic Expansion of logDet for Elliptic Operators with Parameter

YOONWEON LEE
Department of Mathematics, Inha University, Inchon, 402-751, Korea

Abstract: For classical elliptic pseudodifferential operators $A(\lambda)$ of order $m > 0$ with parameter λ of weight $\chi > 0$, it is known that $logDet_\theta A(\lambda)$ admits an asymptotic expansion as $\lambda \to +\infty$. In this paper we show, with some assumptions, that the coefficient of $\lambda^{-\frac{n}{\chi}}$ can be expressed by the value of a zeta function at 0 for some elliptic ψDO on $M \times S^1 \times \cdots \times S^1$ multiplied by $\frac{m}{c_{n-1}}$.
Key Words and Phrases: *Determinant of elliptic operators, asymptotic series, zeta function.*
1991 Mathematics Subject Classification. 58G15, 58G26.

1. Introduction

The regularized determinant of an elliptic operator on a closed oriented Riemannian manifold is a spectral invariant, which is important in topology, geometry and mathematical physics. It is defined by eigenvalues of an elliptic operator of Fredholm index 0, specially a self-adjoint operator.

In early 1970's, Ray and Singer defined an analytic torsion as an analytic counterpart of Reidemeister torsion by using the determinant of Laplacian acting on each p-form.

It is well-known that on a closed oriented Riemannian manifold the spectrum of an elliptic operator A with an Agmon angle θ is discrete and consisting of eigenvalues. Then, we can define a zeta function for A by $\zeta_{A,\theta}(s) = \sum_{\lambda_i} \lambda_i^{-s}$, where the sum runs over all nonzero eigenvalues of A. It is well defined for $Re(s) > \frac{ord(A)}{dim M}$. It was shown by Seeley ([3]) that $\zeta_{A,\theta}(s)$ can be continued meromorphically so that $\zeta_{A,\theta}(s)$ is regular at 0. Then the regularized determinant $Det_\theta(A)$ is defined by $e^{-\zeta'_{A,\theta}(0)}$.

It is a kind of generalization of the usual determinant but it is not easy to calculate. It cannot be computed by an intergal of any density on the base manifold. With this reason we consider a one parameter family of elliptic operators $A(\lambda)$ and an asymptotic series of $logDet_\theta(A(\lambda))$ as $\lambda \to +\infty$.

2. Asymptotic Series of *logDet*

Let M be a compact oriented Riemannian manifold of dimension d and let $E \xrightarrow{\pi} M$ be a vector bundle of rank k. Let $A(\lambda) : C^\infty(E) \longrightarrow C^\infty(E)$ be a classical elliptic pseudodifferential operator of order $m > 0$ with parameter λ of weight $\chi > 0$, where λ is a nonnegative real number.That is,the symbol of $A(\lambda)$ has an asymptotic expansion as in (3.1).

We assume that there is an angle θ such that the principal symbol $a_m(x, \xi, \lambda)$ does not have any eigenvalues on the ray $\{z \in \mathbf{C} | z = \rho e^{i\theta}, \rho \geq 0\}$ for $|\xi| + |\lambda|^{\frac{1}{\chi}} \neq 0$ and that

$A(\lambda)$ does not have any eigenvalues in a sector $L_{[\theta-\epsilon,\theta+\epsilon]} = \{z \in \mathbf{C} | \theta - \epsilon \le \arg z \le \theta + \epsilon\}$ for some small $\epsilon > 0$. We call this θ an Agmon angle. In fact from the compactness of the set $\{(x,\xi,\lambda) | x \in M, |\xi|^2 + |\lambda|^{\frac{2}{\chi}} = 1\}$,we know that $a_m(x,\xi,\lambda)$ does not have any eigenvalues in a sector $L_{[\theta-\delta,\theta+\delta]}$ for sufficiently small $\delta > 0$.

It is shown in [2] that as $\lambda \longrightarrow +\infty$, $logDet_\theta A(\lambda)$ admits an asymptotic expansion of the following form:

$$logDet_\theta A(\lambda) \sim \pi_{-d}\lambda^{\frac{d}{\chi}} + \pi_{-d+1}\lambda^{\frac{d-1}{\chi}} + \cdots + \pi_0 + \pi_1\lambda^{-\frac{1}{\chi}} \cdots + \pi_n\lambda^{-\frac{n}{\chi}} + \cdots + \sum_{j=0}^{d} q_j \lambda^{\frac{j}{\chi}} \log\lambda$$

as $\lambda \to +\infty$. Here each coefficient π_i and q_j is computable in terms of the asymptotic symbol of $A(\lambda)$.

Let $\sum_{j=0}^{\infty} a_{m-j}(x,\xi,\lambda)$ be an asymptotic symbol of $A(\lambda)$ for some local coordinate U. We also assume that for each j, the function $\tilde{a}_{m-j} : U \times \mathbf{R}^d \times \mathbf{R} \times \cdots \mathbf{R} \longrightarrow \{k \times k \text{ matrices}\}$ defined by $\tilde{a}_{m-j}(x,\xi,\lambda_1,\cdots,\lambda_n) = a_{m-j}(x,\xi,\sqrt{\lambda_1^2 + \cdots + \lambda_n^2}^{\chi})$ is smooth. Then from $A(\lambda)$ we can construct a classical elliptic pseudodifferential operator P_n with the same Agmon angle θ on $M \times S^1 \times \cdots \times S^1$, where P_n is uniquely determined up to smoothing operators. Let $\zeta_{P_n}(s)$ be the zeta function constructed from the eigenvalues of P_n.

The purpose of this note is to show that $\pi_n = \frac{m}{c_{n-1}}\zeta_{P_n}(0)$, where $c_{n-1} = \frac{2\sqrt{\pi}^n}{\Gamma(\frac{n}{2})}$, the volume of standard S^{n-1}.

3. Construction of P_n

Suppose that in a local coordinate system U, the asymptotic symbol of $A(\lambda)$ is $\sigma(A(\lambda)) \sim \sum_{j=0}^{\infty} a_{m-j}(x,\xi,\lambda)$. Then

$$a_{m-j}(x,\xi,\lambda) : U \times \mathbf{R}^d \times [0,\infty) \longrightarrow \{k \times k \quad \text{matrices}\} \tag{3.1}$$

with

$$a_{m-j}(x,t\xi,t^\chi\lambda) = t^{m-j}a_{m-j}(x,\xi,\lambda) \quad \text{for} \quad t > 0.$$

Now for each j, we extend a_{m-j} to

$$\tilde{a}_{m-j} : U \times S^1 \times \cdots \times S^1 \times \mathbf{R}^d \times \mathbf{R} \times \cdots \times \mathbf{R} \longrightarrow \{k \times k \text{ matrices}\} \quad \text{by}$$

$$(x,t_1,\cdots,t_n,\xi,\lambda_1,\cdots,\lambda_n) \mapsto a_{m-j}(x,\xi,\sqrt{\lambda_1^2 + \cdots + \lambda_n^2}^{\chi}).$$

Then $\tilde{a}_{m-j}$ is smooth by the assumption of the smoothness of $a_{m-j}(x,\xi,\sqrt{\lambda_1^2 + \cdots + \lambda_n^2}^{\chi})$. Note that $\tilde{a}_{m-j}(x,t_1,\cdots,t_n,\xi,\lambda_1,\cdots,\lambda_n)$ is an homogeneous function of degree $m-j$ with respect to $\xi, \lambda = (\lambda_1,\cdots,\lambda_n)$.

Consider a diagram

$$\begin{array}{ccc} p^*E & \longrightarrow & E \\ \big\downarrow & & \big\downarrow \pi \\ M \times S^1 \times \cdots \times S^1 & \xrightarrow{\ p\ } & M \end{array}$$

where p is the natural projection.

Choose an elliptic ψDO $\tilde{A}_n : C^\infty(p^*E) \longrightarrow C^\infty(p^*E)$ whose asymptotic symbol is $\sigma(\tilde{A}_n) \sim \sum_{j=0}^{\infty} \tilde{a}_{m-j}(x,t_1,\cdots,t_n,\xi,\lambda_1,\cdots,\lambda_n)$ in a local coordinate $U \times S^1 \times \cdots \times S^1$.

Since the principal symbol $a_m(x,\xi,\sqrt{\lambda_1^2+\cdots+\lambda_n^2}^{\chi})$ of $\tilde{A}_n$ does not have any eigenvalues in the sector $L_{[\theta-\delta,\theta+\delta]}$ for $|\xi|+\sqrt{\lambda_1^2+\cdots+\lambda_n^2}^{\frac{1}{\chi}}\neq 0$, there exists $R>0$ such that Spectrum$(\tilde{A}_n)\cap\{z||z|\geq R,\ \theta-\delta\leq\arg z\leq\theta+\delta\}$ is empty (see [1] for details). Since the spectrum of $\tilde{A}_n$ is discrete, there are only finitely many eigenvalues of $\tilde{A}_n$ in $\{z||z|< R,\ \theta-\delta\leq\arg z\leq\theta+\delta\}$. Note that each eigenspace corresponding to an eigenvalue is a finite-dimensional vector space by the ellipticity of $\tilde{A}_n$ of order $m>0$. Let Q_n be the span of {eigenvectors of $\tilde{A}_n$ whose eigenvalues are $\rho e^{i\theta}$ for some ρ, $0\leq\rho\leq R$}. Then Q_n is a finite-dimensional vector space. Define $\phi_n: C^\infty(p^*E)\to C^\infty(p^*E)$ as the natural projection onto Q_n.

Define $P_n=\tilde{A}_n-Re^{i\theta}\phi_n$. Then P_n is injective with an Agmon angle θ and the asymptotic symbol of P_n is exactly the same as the asymptotic symbol of $\tilde{A}_n$. Define $\zeta_{P_n}(s)=\sum_{\lambda_i}\lambda_i^{-s}$, where λ_i's are all the eigenvalues of P_n. Then by [1] (also see [3]), $\zeta_{P_n}(s)$ is regular at 0 with

$$\zeta_{P_n}(0)=-\frac{e^{i\theta}}{m(2\pi)^{d+n}}\int_{M\times S^1\cdots\times S^1}d\ vol(x,t_1,\cdots,t_n)$$

$$\int_{|\xi|^2+\lambda_1^2+\cdots+\lambda_n^2=1}d(\xi,\lambda_1,\cdots,\lambda_n)'\int_0^\infty \mathrm{tr}\ \tilde{r}_{-m-d-n}(x,t_1,\cdots,\xi,\lambda_1,\cdots,\lambda_n,e^{i\theta}\mu)d\mu,$$

where $\tilde{r}_{-m-d-n}$ is the homogeneous part of degree $-m-d-n$ in the asymptotic symbol of the resolvent $(P_n-\mu I)^{-1}$. Note that $\tilde{r}_{-m-d-n}(x,t_1,\cdots,t_n,\xi,\lambda_1,\cdots,\lambda_n,e^{i\theta}\mu)$ does not depend on $t_1,\cdots,t_n$ in this case.

Recall that $\log Det_\theta A(\lambda)\sim\pi_{-d}\lambda^{\frac{d}{\chi}}+\cdots+\pi_0+\pi_1\lambda^{-\frac{1}{\chi}}+\cdots+\pi_n\lambda^{-\frac{n}{\chi}}+\cdots+\sum_{j=0}^d q_j\lambda^{\frac{j}{\chi}}\log\lambda$ as $\lambda\to+\infty$. Then our goal is to prove the following theorem.

Theorem. $\pi_1=\frac{m}{c_{n-1}}\zeta_{P_n}(0)$ for each $n\geq 1$.

Remark When $n=1$, $\pi_1=\frac{m}{2}\zeta_{P_1}(0)$, which was shown in [4]. This theorem is an extension of the result in [4].

4. Proof of the Theorem

From the Appendix of [2] we can derive the local formula

$$\pi_n=\frac{-e^{i\theta}}{(2\pi)^d}\int_M\int_{\mathbf{R}^d}\int_0^\infty \mathrm{tr}r_{-m-d-n}(x,\xi,1,e^{i\theta}\mu)d\mu d\xi d\ vol(x),$$

where r_{-m-d-n} is the homogeneous part of degree $-m-d-n$ in the asymptotic symbol of the resolvent $(A(\lambda)-\mu I)^{-1}$.

Note the relation

$$\tilde{r}_{-m-d-n}(x,t_1,\cdots,t_n,\xi,\lambda_1,\cdots,\lambda_n,e^{i\theta}\mu)=r_{-m-d-n}(x,\xi,\sqrt{\lambda_1^2+\cdots+\lambda_n^2}^{\chi},e^{i\theta}\mu).$$

Then

$$\zeta_{P_n}(0)=-\frac{2e^{i\theta}\cdot(2\pi)^n}{m(2\pi)^{d+n}}\int_M d\ vol(x)\int_{\substack{|\xi|^2+\lambda_1^2+\cdots+\lambda_{n-1}^2=1\\ \lambda_n>0}}d(\xi,\lambda_1,\cdots,\lambda_n)'$$

$$\int_0^\infty \mathrm{tr}r_{-m-d-n}(x,\xi,\sqrt{\lambda_1+\cdots+\lambda_n^2}^{\chi},e^{i\theta}\mu)d\mu$$

since the integrand is even in λ_n.

Set

$$\text{(I)}=\int_{\substack{|\xi|^2+\lambda_1^2+\cdots+\lambda_{n-1}=1\\ \lambda_n>0}}d(\xi,\lambda_1,\cdots,\lambda_n)'\int_0^\infty \mathrm{tr}r_{-m-d-n}(x,\xi,\sqrt{\lambda_1^2+\cdots+\lambda_n^2}^{\chi},e^{i\theta}\mu)d\mu.$$

Then using the projection from the upper hemisphere to
$\{(\xi, \lambda_1, \cdots, \lambda_{n-1}) \in \mathbf{R}^d \times \mathbf{R}^{n-1} | |\xi|^2 + \lambda_1^2 + \cdots + \lambda_{n-1}^2 < 1\}$,we obtain

$$(\mathrm{I}) = \int_{|\xi|^2+\lambda_1^2+\cdots+\lambda_{n-1}^2<1} \int_0^\infty tr r_{-m-d-n}(x, \xi, (\sqrt{1-|\xi|^2})^\times, e^{i\theta}\mu) \times$$

$$(1 - |\xi|^2 - \lambda_1^2 - \cdots - \lambda_{n-1}^2)^{-\frac{1}{2}} d\mu d\lambda_1 \cdots d\lambda_{n-1} d\xi$$

$$= \int_{|\xi|<1} \int_{\lambda_1^2+\cdots+\lambda_{n-1}^2<1-|\xi|^2} (1 - |\xi|^2 - \lambda_1^2 - \cdots - \lambda_{n-1}^2)^{-\frac{1}{2}} d\lambda_1 \cdots d\lambda_{n-1}$$

$$\times \int_0^\infty tr r_{-m-d-n}(x, \xi, (\sqrt{1-|\xi|^2})^\times, e^{i\theta}\mu) d\mu d\xi$$

$$= \frac{c_{n-1}}{2} \int_{|\xi|<1} \int_0^\infty (1-|\xi|^2)^{\frac{n-2}{2}} tr r_{-m-d-n}(x, \xi, (\sqrt{1-|\xi|^2})^\times, e^{i\theta}\mu) d\mu d\xi$$

$$= \frac{c_{n-1}}{2} \int_{|\xi|<1} \int_0^\infty (1-|\xi|^2)^{-\frac{m+d+2}{2}} \times$$

$$tr r_{-m-d-n}\left(x, \frac{\xi_1}{\sqrt{1-|\xi|^2}}, \cdots, \frac{\xi_d}{\sqrt{1-|\xi|^2}}, 1, e^{i\theta} \frac{\mu}{\sqrt{1-|\xi|^2}^m}\right) d\mu d\xi,$$

since the weight of μ is m. Here $c_{n-1} = \frac{2\sqrt{\pi}^n}{\Gamma(\frac{n}{2})}$, the volume of standard S^{n-1}.

Consider a map $\Phi : \mathbf{R}^d \times (0, \infty) \longrightarrow \{\xi \in \mathbf{R}^d | |\xi| < 1\} \times (0, \infty)$ defined by

$$(\eta_1, \cdots, \eta_d, \nu) \mapsto \left(\frac{\eta_1}{\sqrt{1+|\eta|^2}}, \cdots, \frac{\eta_d}{\sqrt{1+|\eta|^2}}, \frac{\nu}{\sqrt{1+|\eta|^2}^m}\right) = (\xi_1, \cdots, \xi_d, \mu).$$

Then Φ is a diffeomorphism with $\det(J(\Phi)) = (1+|\eta|^2)^{-\frac{m+d+2}{2}} = (1-|\xi|^2)^{\frac{m+d+2}{2}}$. Hence we can get

$$(\mathrm{I}) = \frac{c_{n-1}}{2} \int_{\mathbf{R}^d} \int_0^\infty \operatorname{tr} r_{-m-d-n}(x, \eta, 1, e^{i\theta}\nu) d\nu d\eta.$$

■

Acknowledgements

The research of the author is supported in part by BSRI-97-1436 and Inha University Research Fund.

References

1. R. Seeley, Proc. Symp. Pure Math., **10**, 288-307 (1967).
2. D. Burghelea, L. Friedlander and T. Kappeler, Journal of Funct. Anal., **107**, 34-65 (1992).
3. M. Wodzicki, Invent. Math., **66**, 115-135 (1982).
4. Y. Lee, Proc. Amer. Math. Soc., **124**, 3885-3888 (1996).
5. M. A. Shubin, Pseudodifferential Operators and Spectral Theory, Springer-Verlag, Berlin/New York (1985).

9th Int. Coll. on Differential Equations, pp. 245-250
D. Bainov (Ed)

Accurate Splatting of Octrees

J-Y LIU AND I. GARGANTINI
Department of Computer Science, The University of Western Ontario, London, Ontario, Canada N6A 5B7
e-mail: irene@csd.uwo.ca

Abstract: Using the special properties of the linear octrees, lists of displayable voxels are created. A list contains all the voxels which, aligned according to the chosen viewing direction, project onto the same pixel. These lists can be used to define accurate footprints in order to display a *3D-Scene* using transparency, reflexivity or thickness. Four sets of data, obtained with this method, are visualized: they refer to acquisitions made via Magnetic Resonance, Computer Tomography and X-ray crystallography.
Key Words and Phrases: *Volume visualization, footprints, splatting of octree nodes.*
1991 Mathematics Subject Classification. 68U05.

1. Generalities

The visualization of octrees appeared already in the eighties [1, 2, 3] – the most efficient approach being the exploitation of the sorting of the octrees' locational codes to lign up the nodes along the viewing direction. The present paper determines accurate footprints for the projection of octrees' nodes, while introducing, with respect to previous work, three innovations: modularity, delay of classification, choice among binary, semitransparent and thickness rendering. In 2D, the basic idea is illustrated by the example shown in part (a) of Figure 1, where the screen is reduced to a single line, fragmented into consecutive bins ordered from left to right. The object is represented by the shaded regions, labelled $10d, 13d, 22d, 3dd$. An immediate, but not accurate, way would be to fill the bins with the object's nodes which project onto the very same spot, that is: $bin(0) = bin(1) = bin(2) = ... = bin(5) = \{22d\}; bin(12) = \{3dd\}; bin(14) = \{10d, 3dd\}; bin(15) = \{10d, 13d, 3dd\}; bin(19) = \{13d\}$.

In 3D, a given *3D-Scene* is referenced to a World Coordinate (WC) system and bounded by a $(2^n \times 2^n \times 2^n)$-raster $n \geq 1$, with the origin located at the north, front, west corner of the raster. The Viewing-Reference Coordinate (VRC) system has its origin identical with that of the WC. The projection plane is located perpendicularly to the viewing direction.

The mapping between the two systems is: $P^{(v)} = M_{v \leftarrow w} P^{(w)} = R_x(\phi) \cdot R_y(-\theta) \cdot P^{(w)}$, where $P^{(v)}$ is a point in VRC, and $P^{(w)}$ is its image in WC; ϕ is the rotation angle around the X-axis and $(-\theta)$ is the clockwise rotation around the Y-axis. The projection of a vector $\vec{v_1}$ on vector $\vec{v_2}$ is given by: $l = |v_1| \cos\theta = |v_1| \frac{v_{1x}v_{2x}+v_{1y}v_{2y}}{|v_1||v_2|} = \frac{v_{1x}v_{2x}+v_{1y}v_{2y}}{|v_2|}$.

Viewing priorities. Consider again Figure 1 (2D). When the viewer is located along the NW direction, the first block visible to him is the NW one; once this is removed, the NE and the SW become visible; once also these are removed, the SE block appears. If

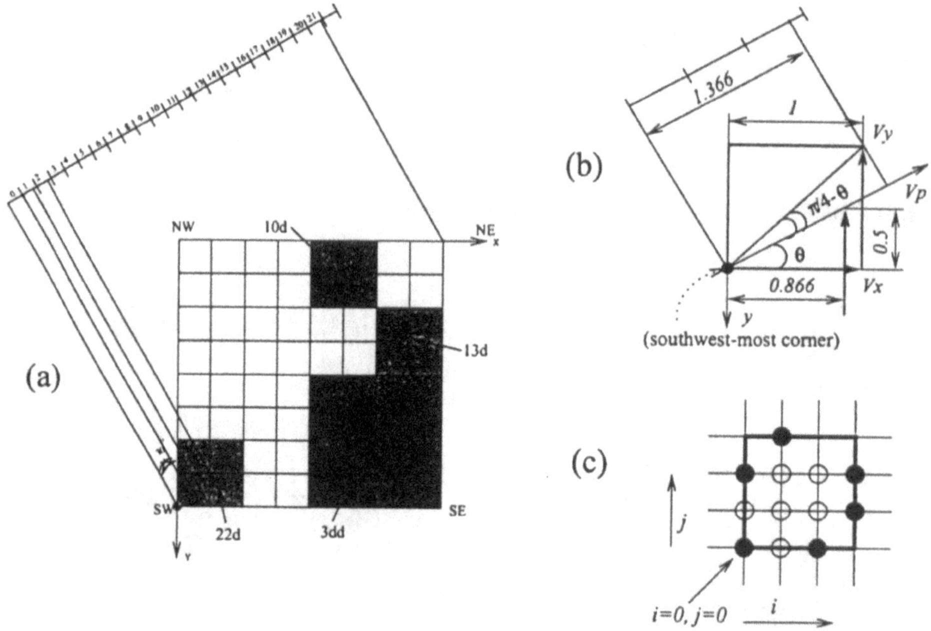

Figure 1: (a) Projecting nodes into bins; (b) Projecting the most SW element; (c) Template of a voxel in 3D

the blocks are organized in the NW → NE → SW → SE sequence (or NW → SW → NE → SE) we don't have to resort directly to the viewer's system to know which part of the data to examine first. We can follow the viewing priority by recursively processing the octree nodes in the order NW, NE, SW and SE (labelled 0,1,2 and 3 respectively).

x	y	z	Priority Order	x	y	z	Priority Order
> 0	> 0	> 0	0,1,2,4,3,5,6,7	< 0	< 0	< 0	7,6,5,3,4,2,1,0
< 0	> 0	> 0	1,0,3,5,2,4,7,6	> 0	< 0	< 0	6,7,4,2,5,3,0,1
> 0	< 0	> 0	2,3,0,6,1,7,4,5	< 0	> 0	< 0	5,4,7,1,6,0,3,2
< 0	< 0	> 0	3,2,1,7,0,6,5,4	> 0	> 0	< 0	4,5,6,0,7,1,2,3

Table 1: Visiting the octal nodes according to the viewing direction

As a 2D raster can have $2^n \times 2^n$ elements, pixels which belong to the same quadrant can be grouped together - and processed all at the same time, as it will be seen later. Look at the $(2^3 \times 2^3)$-raster, always in Figure 1. If four pixels are grouped together, as in the SW quadrant, there is no need to specify the last digit; we can use a 'don't care symbol', d: that particular block is identified by '22d', and the biggest SE block becomes '3*dd*'. Note that the numeric encoding of a block incorporates, implicitly, its location with respect to the origin (set at the corner of the NW quadrant); furthermore, if these codes are related to the recursive subdivision of the raster by quadrants, they also identify the pointer to the elements of the successive subdivision.

In 3D the priority views can be expressed as per Table 1. Initially, the given *3D-*

Scene is represented as a set of locational (octal) codes, each set sorted according to one of the eight principal directions. Only four sorts (including the initial one) are required since the others can be obtained by reversing the order of the elements (see Table 1).

2. Creating Footprints

The shape of a projected node depends on the position of the screen. We call the projection of the simple element (in 2D or 3D) a *template*; the projection of a node consisting of more than one element, a *footprint*. A footprint is built by means of templates.

The 2D case. As commonly done, we set the projection line normal to the viewing direction anywhere in front of the raster; its length should insure that all elements of the raster can be projected onto it. Its origin is set coincidental with the projection of the SW-most corner of the raster (see (b) of Figure 1). This very same corner is the reference for evaluating all templates. We now position the bins to cover all possible projections, each bin's length equalling half of a voxel's edge. Then, denoting θ the angle formed by the projection line with the X-axis, the number of bins is given by $0.5 \times$ *numberbins* $= \lceil 2^n\sqrt{2}\cos(\pi/4 - \theta)\rceil$. In the previous example ((b) of Figure 1), θ is thirty degrees and *numberbins* $= 22$.

Template of a $(2^0 \times 2^0)$-node. This is the projection of the SW-most element. The corresponding bins will have only one element corresponding to the diagonal projected onto the projection line. This diagonal can be expressed by means of a vector obtained by moving, from the SW corner of the raster (see Figure 1) one unit in the positive X-direction and one unit in the negative Y-direction, thus obtaining vector $\vec{V} = \{V_x, V_y\} = \{1, -1\}$. Similarly, the projection line is represented by vector $\vec{V_p}$. For a screen line with a thirty degree slope, as in part (b) of Figure 1, the projection line is represented by $\vec{V_p} = \{V_{px}, V_{py}\} = \{0.866, -.5\}$, so that the length of the projection of $\vec{V}$ on vector $\vec{V_p}$ is 1.366, thus roughly occupying three bins. All other voxels which do not coincide with the SW-most element of the raster are dealt with using suitable offsets.

Footprint of a $(2^1 \times 2^1)$-node. This consists of four voxels with codes '0', '1', '2' and '3'. We know that voxel labeled '0' occupies three bins (using the previous template); we now have to determine *from which bin* to start. The most SW corner of node '0' of a (2×2)-node can be expressed by vector $V_x = 0, V_y = -1$ (referenced, as before, to the SW corner of the raster itself). This vector gives, on the projection line, $l = 0.5$, thus establishing that the set of bins for '0' starts at $bin(1)$. This, in turns, means that $bin(1), bin(2), bin(3)$ will contain the projection of voxel '0' of a (2×2)-node. Similarly, voxel with code '1' belongs to $bin(3)$, $bin(4)$ and $bin(5)$; voxel with code '2' belongs to $bin(0)$, $bin(1)$ and $bin(2)$; voxel with code '3' belongs to $bin(2)$, $bin(3)$ and $bin(4)$. We have thus determined the projection of a (2×2)-node by filling the bins as follows: $bin(0) = \{2\}; bin(1) = \{0, 2\}; bin(2) = \{0, 2, 3\}; bin(3) = \{0, 1, 3\}; bin(4) = \{1, 3\}; bin(5) = \{1\}$.

Footprints of a $(2^2 \times 2^2)$-node. Since a $2^2 \times 2^2$-node consists of four 2×2-nodes, we can use the footprint of a 2×2-node assuming that the $2^2 \times 2^2$-node is located at the reference position. Its four children have codes '0d', '1d', '2d' and '3d'. The offset of the projection of node with code '0d' can be calculated by evaluating the projection of the vector $\vec{V}$ from the reference position to the node's SW-most corner. We get $\vec{V} = \{0, -2\}$ and $l = 1$, meaning that the projection of the node with code '0d' starts at bin(2). We proceed similarly for the other three nodes, thus obtaining eleven bins: $bin(0) = \{22\}; bin(1) = \{20, 22\}; bin(2) = \{02, 20, 22, 23\}; bin(3) = \{00, 02, 20, 21, 23, 32\}; bin(4) = \{00, 02, 03, 21, 23, 30, 32\}; bin(5) = \{00, 01, 03, 12, 21, 30, 32, 33\}$, et cetera.

The 3D case. As before, we consider footprints of different groups of nodes. In this case a node is defined in World Coordinates (WC), then mapped to the Viewing-

Reference Coordinate (VRC) system and, finally, positioned onto the screen. Since we define a template (or footprint) and then translate the latter according to the original position of a node, we also need a local coordinate system (identified by i and j) which links the projection of a node to the projection of the front-NW corner of a node. We denote this corner *ref-corner* and its projection *ref-point*.

Template of a voxel-node. Note that the projection of a voxel is a convex polygon with four or six vertices, depending on the viewing direction. This projection is obtained by evaluating the projections of the voxel's corners, determining their convex hull and their rectangular extent. The latter is used to determine the dimensions of the screen, and, therefore, the number of bins required to display the entire *3D-Scene*. The template is obtained by filling the convex hull. Part (c) of Figure 1 illustrates the template of one isolated voxel viewed from the direction $\vec{V} = \{V_x, V_y, V_z\} = \{1, 2, 3\}$. The filled circles represent the polygon's vertices, which are the projections of the corners of the voxel. The unfilled circles are determined by scan conversion. Filled and unfilled circles represent the projected pixels forming the template. In the example *ref-corner*=(0,0,0) in WC and *ref-point* is given by $i = j = 1$.

Footprint of a $(2^1 \times 2^1 \times 2^1)$-node. Following the previous convention, this node is assumed to be located at the origin of the raster with its extent calculated from the projections of the node's corners. We map the template of each voxel onto the extent, processing each voxel according to the viewing priorities expressed by Table 1. Let's choose again $\vec{V} = \{V_x, V_y, V_z\} = \{1, 2, 3\}$, so that we can use the templates generated in the previous section. We get $\min x = -2, \min y = -2$, and $\max x = 4, \max y = 4$ thus obtaining a 7-pixels-wide and 7-pixels-high rectangular extent. The raster origin, $(X, Y, Z) = (0, 0, 0)$, projects onto $i = 2$ and $j = 2$. We now process each voxel of the node according to above-mentioned priority. We position the template so that its *ref-point* coincides with the actual projection of the *ref-corner* on hand (this varies as the voxel changes from '0' to '7'). As an example, take voxel with code '7' at $(X, Y, Z) = (1, 1, 1)$ in WC, having coordinates $(X', Y') = (1, 0)$ in VCR and values $i = 1, j = 0$ on the extent. This voxel's template starts at $i = 2$ and $j = 1$. A node with higher aggregation of voxels can be processed in a similar manner, using the footprints of the node's children.

3. The Algorithm

So far we have created footprints for each kind of nodes, each footprint being determined by the resolution of the initial *3D-Scene* and the angle determined by the viewing direction with respect to the raster. Since we consider parallel projections, the viewer is set to the infinity. The user controls the refinement of the footprints by choosing the size of the screen's pixels (set, in (2), to 0.5). The algorithm implementing the display focuses on creating lists of *all* displayable voxels, each indexed by the pixel onto which the voxel projects. To each voxel is associated an opacity function or gray level. In our splatting, the list of displayable voxels is used in three ways: (i) to provide a raw sketch of the object; (ii) to simulate the reflection of light on the object's surface (binary classification); (iii) to simulate the penetration of light into the object (semi-transparency). In the first case we simply count the number of voxels indexed by the screen's pixel and paint that pixel with an intensity proportional to that number over a scale of gray levels in the range [0,255]. Figure 2d is obtained in this way. In the second case we only consider the front nodes and project those using the corresponding footprints. In the third case we use all the nodes in each bin coupled with a transparency formula based on both Levoy's and Udupa's renderings [4, 5, 6]. The corresponding function 'display [j] [i]' can be found in *'/csd/ftp/pub/csd-technical-reports/522'*.

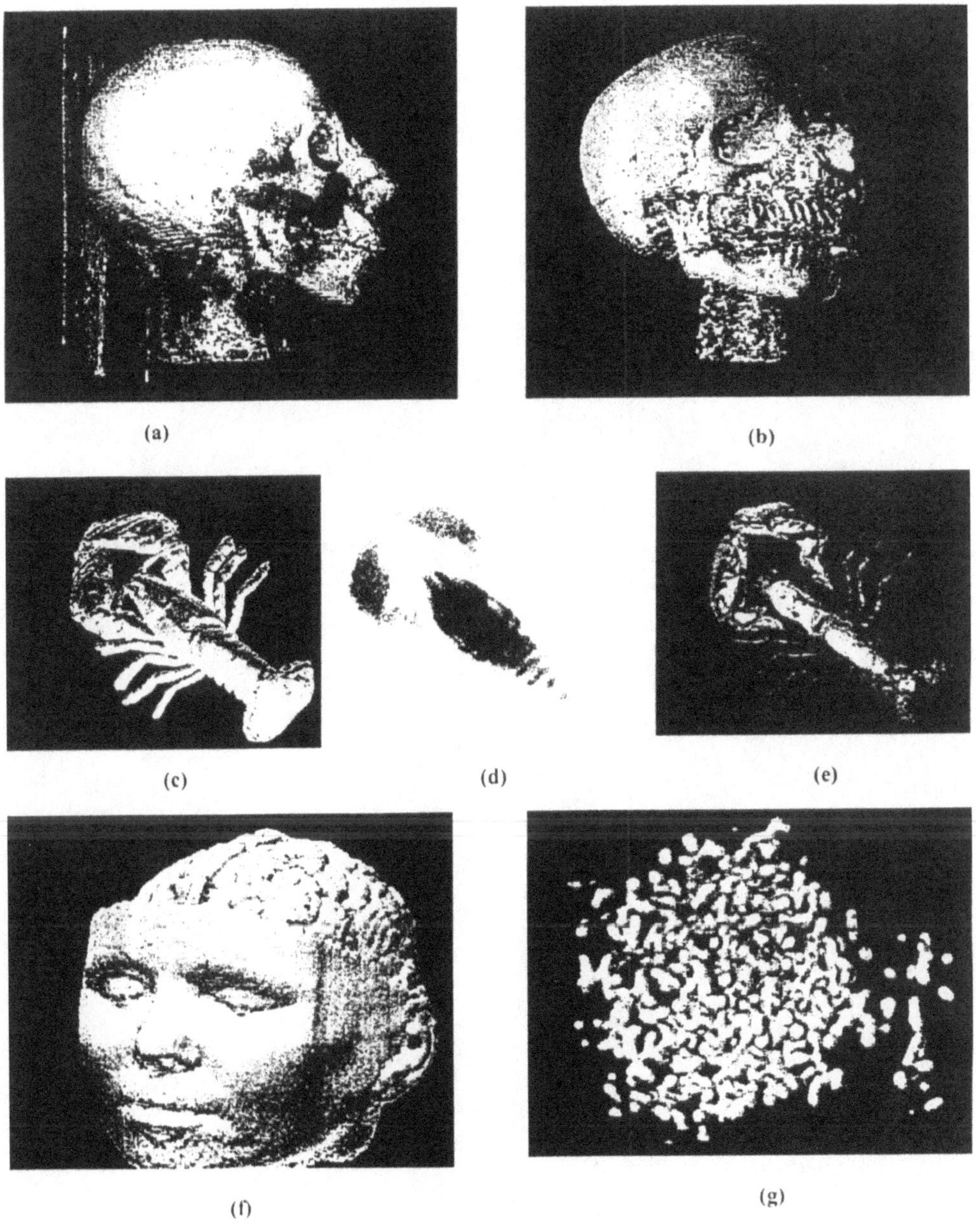

(a) (b) (c) (d) (e) (f) (g)

Figure 2: Splatting as applied to acquired data

9th Int. Coll. on Differential Equations, pp. 251-256
D. Bainov (Ed)

Singular Linear ODEs

GRZEGORZ ŁYSIK
Institute of Mathematics PAN, 00-950 Warszawa P.O.B. 137, Poland

Key Words and Phrases: *Singular differential equations, Laurent series, Mellin transformation, functional equations.*
1991 Mathematics Subject Classification. 34A20, 34A30, 44A15.

Let

(1) $$P = P\left(x, \frac{d}{dx}\right) = \sum_{i=0}^{n} a^i(x) \frac{d^i}{dx^i}$$

be a linear ordinary differential operator with meromorphic coefficients at zero

(2) $$a^i(x) = \begin{cases} \sum_{j=j_0}^{\infty} a^i_j x^j & \text{for } 0 < |x| < r \quad \text{if } i = 0, \dots, n-1, \\ x^n & \text{if } i = n. \end{cases}$$

The *Newton diagram* for P is defined by

$$\mathcal{N}_P = \{(i,j) \in \mathbb{N}_0 \times \mathbb{Z} : a^i_{i+j} \neq 0\};$$

the *Katz invariant* by

(3) $$\kappa_P = \min\{\kappa \in \mathbb{R} : \text{ there are no points of } \mathcal{N}_P \text{ below } \{(i,j) : j = \kappa(i-n)\}\};$$

and the *characteristic polynomial* W is the polynomial W_0 in the expression

(4) $$x^{-z} P(x^z) = \sum_{j=-k}^{\infty} W_j(z) x^z.$$

The point zero is a

1. regular one if $\kappa \leq -1$;
2. regular singular if $-1 < \kappa \leq 0$;
3. irregular singular if $\kappa > 0$.

In the first case the solutions of $Pu = 0$ are analytic functions. In the second one we have

Theorem 1 (Fuchs). *Let zero be a regular singular point of P and f analytic at zero. Then any solution of $Pu = f$ is given by a convergent series of the type*

$$\sum_{\mu=0}^{m}\sum_{p=0}^{q_\mu}\sum_{\gamma=0}^{\infty} C_{\mu,p,\gamma} ln^p x \cdot x^{\rho_\mu+\gamma},$$

where $\rho_0 = 0, \rho_1, ..., \rho_m$ are the roots of the characteristic polynomial, $q_\mu \in \mathbb{N}_0$ for $\mu = 0, 1, ..., m$.

The above theorem is no longer true in the case of irregular singular point. However there exist formal power series solutions, they need not to be convergent.

Example 1 (Euler). The equation $x^2u' = u - x$ has a formal divergent power series solution $\hat{u}(x) = \sum_{\gamma=1}^{\infty}(\gamma-1)!x^\gamma$.

During the last few years a special procedure called *multisummability* was developed to deal with divergent power series solutions of differential equations (Ecalle [1], Braaksma [2], Malgrange [3],...). Applying this procedure to the Euler equation compute the Borel transform of $\hat{u}$

$$B\hat{u}(\zeta) := \sum_{\gamma=1}^{\infty}\frac{(\gamma-1)!}{\Gamma(\gamma)}\zeta^{\gamma-1} = \frac{1}{1-\zeta} \quad \text{for } \zeta \neq 1.$$

Since the function $g(\zeta) = 1/(1-\zeta)$ is of exponential growth along any ray d starting from zero and omitting the point 1 we can compute the *Laplace transform* of g along d

$$\mathcal{L}_d g(x) = \int_d \frac{1}{1-\zeta} exp\Big(-\frac{\zeta}{x}\Big)d\zeta, \; x \in S(d,\pi) := \{x : \; |arg\, x - arg\, d| < \pi/2\}.$$

Gluing together $\mathcal{L}_d g$ over d with $0 < arg\, d < 2\pi$ we get a function u defined in $S = S(-\pi/2 < arg\, x < 5\pi/2)$. Then u is a solution of the Euler equation and $\hat{u}$ is the asymptotic expansion of u in S.

On the other hand, since for the Euler equation $\mathcal{N}_P$ lies below the characteristic level it is natural to look for its solutions in the form

$$u(x) = \sum_{\gamma=-\infty}^{0} a_\gamma x^\gamma + ln x \sum_{\gamma=-\infty}^{0} b_\gamma x^\gamma.$$

Comparing both sides of $x^2u' = u - x$ we get $b_\gamma = (-1)^{\gamma-1}/(-\gamma)!$ for $\gamma \in -\mathbb{N}_0$, $a_0 = C$, $a_\gamma = d_\gamma + (-1)^\gamma C/(-\gamma)!$ with $d_\gamma = (-1)^\gamma \sum_{i=1}^{-\gamma}\frac{(-1)^i}{i \cdot i!(-\gamma-i)!}$ for $\gamma \in -\mathbb{N}$, (so $|d_\gamma| \leq 1/(-\gamma-1)!$). Thus,

$$u(x) = Ce^{-1/x} - ln x \cdot e^{-1/x} + \sum_{\gamma=-\infty}^{-1} d_\gamma x^\gamma$$

and the series converges for $x \in \widetilde{\mathbb{C}\setminus\{0\}}$.

It appears ([4], Ch. IV, Th. 10.1) that the fundamental system of solutions of the homogeneous equation $Pu = 0$ consists of Laurent type series. However the coefficients arising in this kind of representation are very difficult to find as shows

Example 2. Looking for a solution of the equation $x^2u' = (1 + x^2)u$ in the form of Laurent series $u = \sum_{\gamma=-\infty}^{\infty} a_\gamma x^\gamma$ we find

$$(\gamma-1)a_{\gamma-1} = a_\gamma + a_{\gamma-2} \quad \text{for } \gamma \in \mathbb{Z}.$$

So we can fix arbitrarily $a_0 = C$. Putting $a_1 = a$ we find that, in general, both series $\sum_{\gamma=0}^{\infty} a_\gamma x^\gamma$ and $\sum_{\gamma=-\infty}^{-1} a_\gamma x^\gamma$ are divergent. However they will converge for a very special choice of $a/C = J_1(2)/J_2(2)$ (J_n – the Bessel function of order n), since the general solution is given by

$$u(x) = Ce^{x-1/x} = C \sum_{\gamma=-\infty}^{\infty} \Big(\sum_{i=\max(0,-\gamma)}^{\infty} \frac{(-1)^i}{i!(i+\gamma)!} \Big) x^{\gamma}.$$

After this introduction let us present a method of computing Laurent type series solutions to a class of singular equations. First we introduce

Definition. By a *Laurent type series of Gevrey order* $\kappa > 0$ and radious of convergence $r > 0$ we mean a series

$$u(x) = \sum_{\mu=1}^{m} \sum_{p=0}^{q} \sum_{\gamma=-\infty}^{\infty} C_{\mu,p,\gamma} ln^p x \cdot x^{\rho_\mu+\gamma} \tag{6}$$

convergent for $0 < |x| < r$, where $m \in \mathbb{N}, q \in \mathbb{N}_0$, $\rho_1, ..., \rho_m \in \mathbb{C}$ with $\rho_\nu \neq \rho_\mu \bmod 1$ for $1 \le \nu < \mu \le m$ and with some $C < \infty$

$$|C_{\mu,p,\gamma}| \le C \frac{C^{|\gamma|}}{(|\gamma|!)^{1/\kappa}} \quad \text{for } \gamma \in -\mathbb{N}.$$

Theorem 1 ([5]). *Let P be a linear differential operator (1) with meromorphic coefficients (2) at zero. Assume one of the following cases*

Case 1°. P is of order 1;

Case 2°. $\mathcal{N}_P \subset \mathbb{N}_0 \times \mathbb{N}_0$;

Case 3°. $\mathcal{N}_P \subset \mathbb{N}_0 \times (-\mathbb{N}_0)$;

Case 4°. P is a composition of operators satisfying any of the Cases 1° – 3°.

If f is a Laurent type series of Gevrey order κ then every solution of $Pu = f$ is a Laurent type series of Gevrey order $\max(\kappa, \kappa_P)$, where κ_P is the Katz invariant of P.

Sketch of the proof. First of all we note that Case 4° follows by the other cases. Also Case 1° is trivial, since we have an integral formula for solutions. In the Cases 2° and 3° let us consider a Cauchy problem

$$\begin{cases} P(x, \frac{d}{dx})u = f, \\ u(t) = u_0, ..., u^{(n-1)}(t) = u_{n-1}, \end{cases} \tag{7}$$

with $0 < t < r$. We can assume that $t = 1$. We shall apply to (7) the Mellin transformation defined for functions of the form (6) by

$$\mathcal{M}^{\pm}u(z) = \int_{\gamma^{\pm}} u(x) x^{-z-1} dx$$

where $\gamma^{\pm} = \{x \in \tilde{B}(r) : \mp \arg x \ge 0, |x| = 1\}$. Then

$$\mathcal{M}u(z) = \sum_{\mu=1}^{m} \sum_{p=0}^{q} \sum_{\gamma=-\infty}^{\infty} C_{\mu,p,\gamma} \frac{-p!}{(z-\rho_\mu-\gamma)^{p+1}} \quad \text{for } z \notin \{\rho_1, ..., \rho_m\} + \mathbb{Z};$$

$$S := b(\mathcal{M}u) = \sum_{\mu=1}^{m} \sum_{p=0}^{q} \sum_{\gamma=-\infty}^{\infty} C_{\mu,p,\gamma} (-1)^p \delta^{(p)}_{(\rho_\mu+\gamma)} \in L'_{(-\infty, lnr)}(\mathbf{L}) \text{ with } \mathbf{L} = \bigcup_{\mu=1}^{m} (\rho_\mu + \mathbb{R})$$

and

$$u(x) = \frac{1}{2\pi i} S[x^{\cdot}] \quad \text{for } x \in \tilde{B}(r).$$

By the operational properties of the Mellin transformation (7) is transformed into a functional equation for the function $G = \mathcal{M}u$

$$(8) \qquad W(z)G(z) + \sum_{\substack{j=-k \\ j\neq 0}}^{\infty} W_j(z-j)G(z-j) = H(z).$$

In the Cases 2° and 3°, (8) can be solved by the method of successive approximations. For instance in Case 2° the approximations of the solution G are given by

$$\widetilde{G}_\nu(z) := G_\nu(z) - G_{\nu-1}(z) = (-1)^\nu \sum_{l=\nu}^{\infty} A(\nu, l; z)H(z-l), \ \nu \in \mathbf{N}$$

where for $l \in \mathbf{N}$ and $\nu \in \mathbf{N}$ (with W_j defined by (4))

$$A(\nu, l; z) = \sum_{\substack{i_1+\ldots+i_\nu=l \\ 1\leq i_\mu, \mu=1,\ldots,\nu}} \frac{W_{i_1}(z-i_1)W_{i_2}(z-i_1-i_2)\cdot\ldots\cdot W_{i_\nu}(z-i_1-\ldots-i_\nu)}{W(z)W(z-i_1)W(z-i_1-i_2)\cdot\ldots\cdot W(z-i_1-\ldots-i_\nu)}.$$

Next we prove the convergence of $\sum \widetilde{G}_\nu = G$ and compute the boundary value $b(G)$ of G, which is equal to the sum of derivatives if δ at $\rho_\mu + \mathbf{Z}$. Then $u(x) = \frac{1}{2\pi i} b(G)[x^{\cdot}]$ is a Laurent type series solution of (7).

Example 3. Consider the problem

$$P\Big(x, \frac{d}{dx}\Big)u = x^2\frac{d^2u}{dx^2} - \Big(x + \frac{1}{x}\Big)u = 0, \ u(1) = 0, \ u'(1) = -1.$$

Then $k = 1, \kappa_P = 1/2, W(z) = z(z-1), W_{-1} = W_1 = -1$. Since $\kappa_P \notin \mathbf{N}_0$, P is not a composition of operators of order 1. So P does not satisfy assumptions of Theorem 1. Taking the Mellin transform we arrive at the equation

$$W(z)G(z) - G(z+1) - G(z-1) = 1.$$

The approximations $\widetilde{G}_\nu$ are given by $\widetilde{G}_\nu(z) = \sum_{l=0}^{\nu} A(\nu, -\nu+2l; z)$ with

$$A(\nu, -\nu+2l; z) = \sum_{\substack{i_1+\ldots+i_\nu=l \\ i_\mu\in\{0,1\}, \mu=1,\ldots,\nu}} \frac{1}{W(z)W(z+1-2i_1)\cdot\ldots\cdot W(z+\nu-2i_1-\ldots-2i_\nu)},$$

$l \in \{0, 1, \ldots, \nu\}$. Since $\widetilde{G}_\nu$ has a pole at zero of order $1+\nu$ (for $l = \lfloor \nu/2 \rfloor$) the sum $\sum \widetilde{G}_\nu$ does not converge to any function $G \in \mathcal{O}(\mathbf{C}\setminus\mathbf{Z})$ having polynomial growth near $\mathbf{Z}$.

In the following we shall assume that $\kappa_P > 0$. Put $\widetilde{x} = x^{\kappa_P}$. Then the coefficients of a new operator $\widetilde{P}$ are analytic functions in $\widetilde{x}^{1/\kappa_P}$ and the Katz invariant of $\widetilde{P}$ is 1. So we can assume that $\kappa_P = 1$ and consequently the operator P can be written in the form

$$P\Big(x, \frac{d}{dx}\Big) = Q\Big(x^2\frac{d}{dx}\Big) + \sum_{i=0}^{n-1} g^i(x)\Big(x^2\frac{d}{dx}\Big)^i$$

where Q is a polynomial of degree n and $g^i(x) = \sum_{j=1}^{\infty} g^i_j x^j$, $i = 0, \ldots, n-1$ are analytic in $B(r)$. Assume $r > 1$ and consider the Cauchy problem (7) with $t = 1$.

Changing the independent variable $s(x) = exp\{1 - 1/x\}$ we get

$$(9) \qquad \begin{cases} Q\big(s\frac{d}{ds}\big)w + \sum_{i=0}^{n-1} g^i\big(\frac{1}{-\ln s/e}\big)\big(s\frac{d}{ds}\big)^i w = h, \\ w(1) = w_0, \ldots, w^{(n-1)}(1) = w_{n-1}, \end{cases}$$

where $h(s) = f\left(\frac{1}{-lns/e}\right), w(s) = u\left(\frac{1}{-lns/e}\right)$, $w_0 = u_0, w_1 = u_1, w_2 = u_1 + u_2, \ldots$ and so on. Since $\lim_{s\to 0} g^i\left(\frac{1}{-lns/e}\right) = 0$ for $i = 0, \ldots, n-1$ we have obtained an equation with regular singular point at zero but with coefficients being generalized analytic functions of the form

$$g^i\left(\frac{1}{-lns/e}\right) = \int_{l_\theta} \psi^i(\alpha) s^\alpha e^{-\alpha} d\alpha \quad \text{for } Re\left(e^{i\theta}(lns-1)\right) < -1/r,$$

$$\text{where } l_\theta = [0, e^{i\infty\theta}) \text{ and } \psi^i(\alpha) = \sum_{j=1}^{\infty} g^i_j \frac{\alpha^{j-1}}{(j-1)!}, \; i = 0, \ldots, n-1$$

is an entire function satisfying $|\psi^i(\alpha)| \le C_{r'} exp\{|\alpha|/r'\}$ *for any* $r' < r$.

The equations of this type were studied by Bogdan Ziemian in [6]. Under suitable conditions he proved the existance of generalized analytic solutions with positive radii of convergence. But his proof does not give that the convergence radii of solutions are > 1.

To solve (9) we apply the Mellin transformation. Putting $G(z) = \mathcal{M}w(z)$ we get a convolution equation

$$Q(z)G(z) + \int_{l_\theta} A^0(\alpha, z) G(z-\alpha) e^{-\alpha} d\alpha = \Phi(z) \tag{10}$$

where

$$A^0(\alpha, z) = \sum_{i=0}^{n-1} \psi_i(\alpha)(z-\alpha)^i, \; \Phi(z) = \mathcal{M}h(z) + W(z),$$

(W is a polynomial of degree $\le n-1$ dependent on $u_0, \ldots, u_{n-1}$).

We solve (10) by the method of successive aproximations according to the scheme

$$G^0_\theta(z) = \frac{\Phi(z)}{Q(z)}$$

$$G^{\nu+1}_\theta(z) = \frac{1}{Q(z)}\left\{\Phi(z) - \int_{l_\theta} A^0(\alpha, z) G^\nu_\theta(z-\alpha) e^{-\alpha} d\alpha\right\}.$$

Putting $\tilde{G}^{\nu+1}_\theta = G^{\nu+1}_\theta - G^\nu_\theta$ we find for $z \notin \bigcup_{\mu=0}^{m}(\rho_\mu + \bar{l}_\theta)$

$$\tilde{G}^{\nu+1}_\theta(z) = \frac{(-1)^{\nu+1}}{Q(z)} \int_{l_\theta} A^\nu_\theta(\gamma, z) \frac{\Phi(z-\gamma)}{Q(z-\gamma)} e^{-\gamma} d\gamma$$

where for $\gamma \in l_\theta, \nu \in \mathbb{N}$

$$A^\nu_\theta(\gamma, z) = \int \frac{A^0(\alpha_1, z)}{Q(z-\alpha_1)} \cdot \frac{A^0(\alpha_2, z-\alpha_1)}{Q(z-\alpha_1-\alpha_2)} \cdot \ldots \cdot \frac{A^0(\alpha_\nu, z-\alpha_1-\ldots-\alpha_{\nu-1})}{Q(z-\alpha_1-\ldots-\alpha_\nu)}$$
$$\times A^0(\gamma-\alpha_1-\ldots-\alpha_\nu, z-\alpha_1-\ldots-\alpha_\nu) d\alpha$$

with the integration over $T^\nu_\theta(\gamma) = \{\alpha \in (l_\theta)^\nu : \alpha_1 + \ldots + \alpha_\nu < \gamma\}, \gamma \in l_\theta$.

For the convenience in the following we assume that $f = h = 0$. Let $\rho_1, \ldots, \rho_m$ be the roots of Q with multiplicity $k_1, \ldots, k_m$. Fix θ so that $(\rho_\nu + l_\theta) \cap (\rho_\mu + l_\theta) = \emptyset$ for $1 \le \nu < \mu \le m$. Then

$$G_\theta \stackrel{\text{def}}{=} \sum_{\nu=1}^{\infty} \tilde{G}^\nu_\theta \in \mathcal{O}\left(\mathbb{C} \setminus \bigcup_{\mu=1}^{m}(\rho_\mu + \bar{l}_\theta)\right)$$

satisfies near $\rho_\mu + \bar{l}_\theta$

$$|G_\theta(z)| \le \begin{cases} C\frac{1}{|z-\rho_\mu|^L} exp\{|z|/r'\} & \text{if } k_\mu = 1, \\ C exp\Big\{\frac{L}{|z-\rho_\mu|^{k_\mu - 1}} + \frac{|z|}{r'}\Big\} & \text{if } k_\mu > 1. \end{cases}$$

Next changing θ it can be shown that G_θ extends to a function $G \in \mathcal{O}\big(\widetilde{\mathbb{C}\setminus\{\rho_1,\dots,\rho_m\}}\big)$ satisfying exponential estimations and solving (10). So, the boundary value $b(G) = \sum_{\mu=1}^m T_\mu$ where

$$T_\mu \text{ is a Laplace} \begin{cases} \text{distribution on } \bar{l_\theta} & \text{if } k_\mu = 1, \\ \text{ultradistribution on } \bar{l_\theta} \text{ of the class } (p!)^{k_\mu/(k_\mu - 1)} & \text{if } k_\mu > 1 \end{cases}$$

and $T_{\mu|l_\theta}$ extends holomorphically to a function of exponential growth at ∞.

Finally the solution of (9) is given by

$$w(s) = \frac{1}{2\pi i} b(G)[s^{\cdot}]$$

and recalling that $s = exp\{1 - 1/x\}$ we get the solution of (7).

Example 1 (continued). Let us solve the Cauchy problem for the Euler equation $x^2u' - u = -x$, $u(1) = u_0$. Putting $s = exp\{1 - 1/x\}$, $w(s) = u\big(\frac{1}{-lns/e}\big)$ we get $sw' - w = \frac{1}{lns/e}$, $w(1) = u_0$. Applying the Mellin transformation we obtain $(z-1)G(z) = -u_0 + \int_{l_\theta} \frac{e^{-\alpha}}{z-\alpha} d\alpha$. Next we compute the boundary value $b(G) = 2\pi i\Big((u_0 + A)\delta_{(1)} + \int_{l_\theta} ln(\alpha - 1)\frac{d}{d\alpha}\big(e^{-\alpha}\delta_{(\alpha)}\big)d\alpha\Big)$ where $A = \big(\gamma - \sum_{j=1}^{\infty}\frac{1}{j!j}\big)/e$. So the solutions are given by

$$w(s) = (u_0 + A)s + lns/e \cdot \int_{l_\theta} ln(\alpha - 1)s^\alpha e^{-\alpha} d\alpha$$

and

$$u(x) = \Big[u_0 e - \sum_{j=1}^{\infty}\frac{1}{j!j} + \sum_{j=1}^{\infty}\frac{1}{j!j}\frac{1}{x^j} - lnx\Big]e^{-1/x}.$$

References

1. J. Ecalle, Introduction à L'accélération et à ses Applications, Traveaux en Cours, Hermann (1993).
2. B. Braaksma, J. Diff. Equations, **92**, 45–75 (1991).
3. B. Malgrange, Exposition. Math., **13**, 163–222 (1995).
4. P. Hartman, Ordinary Differential Equations, John Wiley & Sons, New York (1964).
5. G. Łysik, Demonstratio Math., (in press).
6. B. Ziemian, Dissertationes Math., **354**, Warszawa (1996).

9th Int. Coll. on Differential Equations, pp. 257-260
D. Bainov (Ed)

Fine Regularity for Nonlinear Nonvariational Parabolic Systems

MARIO MARINO AND ANTONINO MAUGERI
Dipartimento di Matematica, Università di Catania, Viale A. Doria 6, 95125 Catania, Italy
JOACHIM NAUMANN
Institut für Angewandte Mathematik, Humboldt-Universität zu Berlin, Unter den Linden 6, D-10099 Berlin, Germany

Abstract: We prove the Hölder continuity of the strong solutions u to nonlinear nonvariational basic parabolic systems for $n \leq 6$, of the gradient Du for $n \leq 4$, of the hessian matrix $H(u)$ for $n = 2$.

Key Words and Phrases: *Nonlinear parabolic systems, Basic systems, Caccioppoli-type inequality, $\mathcal{L}^{2,\lambda}$-theory.*

1. Introduction

Let Ω be a bounded open set in $\mathbb{R}^n, n \geq 2$, of class C^2, with generic point $x = (x_1, x_2, \dots, x_n)$. If T is a positive real number, we denote by Q the cylinder $\Omega \times (-T, 0)$ and by X the point (x,t) of $\mathbb{R}^n_x \times \mathbb{R}_t$. If $u(X)$ is a vector $Q \to \mathbb{R}^N, N$ integer ≥ 1, we set

$$D_i u = \frac{\partial u}{\partial x_i},\ Du = (D_1 u, D_2 u, \dots, D_n u),\ H(u) = \{D_{ij} u\} = \{D_i D_j u\},\ i,j = 1,2,\dots,n;$$

obviously Du and $H(u)$ are elements of $\mathbb{R}^{nN}$ and $\mathbb{R}^{n^2 N}$, respectively.

Setting

$$W^p(Q,\mathbb{R}^N) = \left\{ v : v \in L^p(-T,0,H^{2,p}(\Omega,\mathbb{R}^N)), \frac{\partial v}{\partial t} \in L^p(Q,\mathbb{R}^N) \right\},$$

$$W^p_0(Q,\mathbb{R}^N) = \left\{ v \in W^p(Q,\mathbb{R}^N) : v \in L^p(-T,0,H^{1,p}_0(\Omega,\mathbb{R}^N)), v(x,-T) = 0 \right\},$$

where $p \in [1,+\infty[$ and $H^{2,p}(\Omega,\mathbb{R}^N)$, $H^{1,p}_0(\Omega,\mathbb{R}^N)$ are the usual Sobolev spaces, let $u \in W^2(Q,\mathbb{R}^N)$ be a solution of the nonlinear nonvariational "basic" parabolic system

$$a\Big(H(u)\Big) - \frac{\partial u}{\partial t} = 0 \text{ in } Q. \tag{1.1}$$

We are concerned with regularity properties of the solution u to system (1.1).

This study is preliminary to the regularity of solutions $u \in W^2(Q,\mathbb{R}^N)$ of the complete system

$$a\Big(X,u,Du,H(u)\Big) - \frac{\partial u}{\partial t} = f(X,u,Du).$$

In fact the operator $a\Big(H(\cdot)\Big) - \dfrac{\partial}{\partial t}$ plays in *the theory of nonlinear system* the same role of the heat operator $\Delta - \dfrac{\partial}{\partial t}$ in *the linear one.*

Then it is important to know what are the principal properties of the operator

$$a\Big(H(\cdot)\Big) - \frac{\partial}{\partial t},$$

when we work in the framework of nonlinear discontinuous operators. A good assumption that allows to the operators to remain discontinuous and ensures the solvability of the boundary problems is the following condition (A), introduced by S. Campanato:

(A) *There exist three positive constants α, γ and δ with $\gamma + \delta < 1$ such that, $\forall u \in \mathbb{R}^N$, $\forall p \in \mathbb{R}^{nN}$, $\forall \tau, \eta \in \mathbb{R}^{n^2N}$ and for a.e. $X \in Q$ we have:*

$$\Big\| \sum_{i=1}^{n} \tau_{ii} - \alpha\Big[a\Big(X, u, p, \tau + \eta\Big) - a\Big(X, u, p, \eta\Big)\Big]\Big\|^2 \le \gamma \|\tau\|^2 + \delta \Big\| \sum_{i=1}^{n} \tau_{ii} \Big\|^2$$

(see [1], [2] and [3] for more details).

Under the assumptions (A) and $a(0) = 0$, S. Campanato (see [2]) and M. Marino - A. Maugeri (see [4] and [5]) obtained, with different techniques, the same result: if $u \in W^2(Q, \mathbb{R}^N)$ is a solution of the "basic" parabolic system

$$a\Big(H(u)\Big) - \frac{\partial u}{\partial t} = 0 \text{ in } Q,$$

then

Du is Hölder-continuous in Q, if $n = 2$;
u is Hölder-continuous in Q, if $n \le 4$.

This result says that a solution of the nonlinear system $a\Big(H(u)\Big) - \dfrac{\partial u}{\partial t} = 0$ is regular in Q for *small values of* n, whereas for *large values of* n it is partially regular.

Now we are able to give an answer to the open problem if the solution $u \in W^2(Q, \mathbb{R}^N)$ to the system (1.1) is Hölder-continuous in Q also for $n \ge 5$. We shall prove that

— *if $n \le 6$, u is Hölder-continuous in Q;*
— *if $n \le 4$, Du is Hölder-continuous in Q;*
— *if $n = 2$, $H(u)$ and $\dfrac{\partial u}{\partial t}$ are Hölder-continuous in Q.*

2. Preliminary Results

To prove our results we need some preliminary theorems, which are of independent interest.

Let $u \in W^2(Q, \mathbb{R}^N)$ be a solution of the "basic" parabolic system

$$a\Big(H(u)\Big) - \frac{\partial u}{\partial t} = 0 \text{ in } Q, \tag{2.1}$$

where $a(\xi)$ is a vector of $\mathbb{R}^N$, continuous onto $\mathbb{R}^{n^2N}$, satisfying the conditions (A) and $a(0) = 0$.

The following statements hold.

Theorem 2.1. *$Du \in W^2_{\text{loc}}(Q, \mathbb{R}^{nN})$ and, $\forall Q(2\sigma) = Q(X^0, 2\sigma) \subset Q$, the following inequalities hold*

$$\text{(2.2)} \quad \int_{Q(\sigma)} \Big(\|H(Du)\|^2 + \|\frac{\partial (Du)}{\partial t}\|^2\Big)\, dX \le c\sigma^{-2}\Big\{\sigma^{-2} \int_{Q(2\sigma)} \|D(u - P_{Q(2\sigma)})\|^2\, dX +$$

$$+\int_{Q(2\sigma)}\|H(u-P_{Q(2\sigma)})\|^2\,dX\Big\}\le c\int_{Q(2\sigma)}\|\frac{\partial}{\partial t}(u-P_{Q(2\sigma)})\|^2\,dX+$$
$$+c\sigma^{-2}\int_{Q(2\sigma)}\|H(u-P_{Q(2\sigma)})\|^2\,dX,$$

where $P_{Q(2\sigma)}$ is the vector-polinomial in x of degree ≤ 2, in t of degree ≤ 1 such that:

$$\int_{Q(2\sigma)}D^\alpha(u-P_{Q(2\sigma)})\,dX=0,\quad \forall\alpha:|\alpha|\le 2, \tag{2.3}$$

$$\int_{Q(2\sigma)}\frac{\partial}{\partial t}(u-P_{Q(2\sigma)})\,dX=0\;,\;\int_{Q(2\sigma)}D\frac{\partial}{\partial t}(u-P_{Q(2\sigma)})dX=0.$$

Theorem 2.2. *There exists a number $\bar q>2$ such that, $\forall q\in(2,\bar q)$, it results*

$$Du\in W^q_{\rm loc}(Q,\mathbb{R}^{nN})$$

and, $\forall Q(2\sigma)\subset\subset Q$, with $\sigma\in(0,1)$, the following estimate holds

$$\left[\fint_{Q(\sigma)}\left(\|H(Du)\|^q+\|\frac{\partial(Du)}{\partial t}\|^q\right)dX\right]^{\frac1q}\le c\left[\fint_{Q(2\sigma)}\left(\|H(Du)\|^2+\|\frac{\partial(Du)}{\partial t}\|^2\right)dX\right]^{\frac12}. \tag{2.4}$$

Theorem 2.3. $\frac{\partial u}{\partial t}\in W^2_{\rm loc}(Q,\mathbb{R}^N)$ *and, $\forall Q(2\sigma)\subset Q$, the following inequalities hold*

$$\int_{Q(\sigma)}\left(\left\|H\left(\frac{\partial u}{\partial t}\right)\right\|^2+\left\|\frac{\partial^2u}{\partial t^2}\right\|^2\right)dX\le c\sigma^{-2}\Big\{\sigma^{-2}\int_{Q(\frac32\sigma)}\left\|\frac{\partial}{\partial t}(u-P_{Q(\frac32\sigma)})\right\|^2dX+ \tag{2.5}$$
$$+\int_{Q(\frac32\sigma)}\left\|D\frac{\partial}{\partial t}(u-P_{Q(\frac32\sigma)})\right\|^2dX\Big\}\le c\sigma^{-4}\Big\{\int_{Q(2\sigma)}\left\|\frac{\partial}{\partial t}(u-P_{Q(2\sigma)})\right\|^2dX+$$
$$+\int_{Q(2\sigma)}\left\|H(u-P_{Q(2\sigma)})\right\|^2dX\Big\}.$$

Theorem 2.4. *There exists a number $\bar q>2$ such that, $\forall q\in(2,\bar q)$, it results*

$$\frac{\partial u}{\partial t}\in W^q_{\rm loc}(Q,\mathbb{R}^N)$$

and, $\forall Q(2\sigma)\subset\subset Q$, with $\sigma\in(0,1)$, the following estimate holds

$$\left[\fint_{Q(\sigma)}\left(\|H(\frac{\partial u}{\partial t})\|^q+\|\frac{\partial^2u}{\partial t^2}\|^q\right)dX\right]^{\frac1q}\le c\left[\fint_{Q(2\sigma)}\left(\|H(\frac{\partial u}{\partial t})\|^2+\|\frac{\partial^2u}{\partial t^2}\|^2\right)dX\right]^{\frac12}. \tag{2.6}$$

3. Main Results

The results of Hölder regularity of u, Du and $H(u)$ follow from the following estimates:
Theorem 3.1. *There exists a number $q'>2$ such that, $\forall Q(\sigma)\subset Q$, with $\sigma\le1$, $\forall\tau\in(0,1)$, it results if $n>2$*

$$\int_{Q(\tau\sigma)}\left\|H(u)-\big(H(u)\big)_{Q(\tau\sigma)}\right\|^2dX+\int_{Q(\tau\sigma)}\left\|\frac{\partial u}{\partial t}-\left(\frac{\partial u}{\partial t}\right)_{Q(\tau\sigma)}\right\|^2dX\le \tag{3.1}$$
$$\le c\tau^{4+n(1-\frac{2}{q'})}\Big\{\int_{Q(\sigma)}\left\|H(u)-\big(H(u)\big)_{Q(\sigma)}\right\|^2dX+$$

$$+\int_{Q(\sigma)}\left\|\frac{\partial u}{\partial t}-\left(\frac{\partial u}{\partial t}\right)_{Q(\sigma)}\right\|^2 dX\Big\}+c(\tau\sigma)^{4+n(1-\frac{2}{q'})}\left[\int_{Q(\sigma)}\left(\left\|\frac{\partial H(u)}{\partial t}\right\|^{q'}+\left\|\frac{\partial^2 u}{\partial t^2}\right\|^{q'}\right)dX\right]^{\frac{2}{q'}};$$

if $n=2$

$$\int_{Q(\tau\sigma)}\left\|H(u)-\big(H(u)\big)_{Q(\tau\sigma)}\right\|^2 dX+\int_{Q(\tau\sigma)}\left\|\frac{\partial u}{\partial t}-\left(\frac{\partial u}{\partial t}\right)_{Q(\tau\sigma)}\right\|^2 dX\le \tag{3.2}$$

$$\le c\tau^{4+2(1-\frac{2}{q'})}\int_{Q(\sigma)}\left\|H(u)-\big(H(u)\big)_{Q(\sigma)}\right\|^2 dX+$$

$$+c(\tau\sigma)^{4+2(1-\frac{2}{q'})}\Big\{\Big[\int_{Q(\sigma)}\left(\left\|\frac{\partial H(u)}{\partial t}\right\|^{q'}+\left\|\frac{\partial^2 u}{\partial t^2}\right\|^{q'}\right)dX\Big]^{\frac{2}{q'}}+$$

$$+\left(\int_{t^0-\sigma^2}^{t^0}\left\|\frac{\partial u(x_0,t)}{\partial t}\right\|^{q'}dt\right)^{\frac{2}{q'}}+\int_{t^0-\sigma^2}^{t^0}\left\|\frac{\partial Du(x,t)}{\partial t}\right\|^2_{H^{1,q'}(B(\sigma))}dt\Big\}.$$

Now let us remark that from (3.2) it immediately follows

Theorem 3.2. *If* $n=2$, *then* $H(u), \dfrac{\partial u}{\partial t}\in C^{0,1-\frac{2}{q'}}(Q)$.

Moreover, taking into account the following Poincaré-type inequality

$$\int_{Q(\tau\sigma)}\left\|D(u-P_{Q(\tau\sigma)})\right\|^2 dX\le c\tau^2\sigma^2\int_{Q(\tau\sigma)}\left(\left\|H(u-P_{Q(\tau\sigma)})\right\|^2+\left\|\frac{\partial}{\partial t}(u-P_{Q(\tau\sigma)})\right\|^2\right)dX,$$

from (3.1) and (3.2) it follows

Theorem 3.3. *Du is Hölder-continuous in* Q, *if* $n\le 4$.

Finally, from the estimate

$$\int_{Q(\tau\sigma)}\left\|u-P_{Q(\tau\sigma)}\right\|^2 dX\le c\tau^4\sigma^4\int_{Q(\tau\sigma)}\left(\left\|H(u-P_{Q(\tau\sigma)})\right\|^2+\left\|\frac{\partial}{\partial t}(u-P_{Q(\tau\sigma)})\right\|^2\right)dX,$$

it follows

Theorem 3.4. *u is Hölder-continuous in* Q, *if* $n\le 6$.

References

1. S. Campanato, Rend. Sem. Mat. Fis. Milano, **60**, 113–131 (1990).
2. S. Campanato, Atti Accad. Naz. Lincei Rend. Cl. Sci. Fis. Mat. Natur. (9), **2**, 129–136 (1991).
3. S. Campanato, Quaderno I, Dottorato di Ricerca in Matematica, Catania (1991).
4. M. Marino and A. Maugeri, Le Matematiche, **51**, 349–368 (1996).
5. M. Marino and A. Maugeri, in: Proc. Eighth Int. Coll. Diff. Equations, D. Bainov (Ed.), VSP, Utrecht (1998), pp. 293–300.
6. P. Cannarsa, Boll. U.M.I. (5), **18-C**, 291–315 (1981).
7. M. Marino and A. Maugeri, Rend. Mat. (7), **13**, 499–527 (1993).

9th Int. Coll. on Differential Equations, pp. 261-266
D. Bainov (Ed)

Overlapping Multi-Subdomain Asynchronous Fixed Point Methods for Elliptic Boundary Value Problems

J.- C. MIELLOU AND M. LAARAJ
UMR CNRS de Mathématiques 6623, Université de Franche-Comte, Besançon, France
miellou@comte-univ.fcomte.fr, laaraj@vega.univ-fcomte.fr

M. J. GANDER
Ecole Polytechniques, Laboratoire de Mathématiques Appliquées, Palaiseau, France
mgander@cmapx.polytechnique.fr

1. Introduction

We present and analyze asynchronous iterations for the solution of second order elliptic partial differential equations based on an overlapping domain decomposition. Asynchronous iterations constitute a general framework for fixed point methods, where the considered fixed point mapping is defined on a product space. Our general formulation includes standard relaxation algorithms such as Jacobi and Gauss-Seidel and their block versions. Indeed the block versions can be related to the additive and multiplicative Schwarz method respectively. In addition to the standard situations, asynchronous iterations can describe more involved synchronous or asynchronous parallel algorithms. For recent developments for the multi-subdomain multiplicative Schwarz (Gauss-Seidel) method we refer to [1] and [2]. For truly asynchronous methods for overlapping subdomain decompositions we refer to [3] for first partial results.

One important feature of the results presented here is the use of weighted L^∞ norms, which allows us to obtain a stronger convergence property than the usual L^∞ one, even for the standard algorithms (see [4] V.2, p. 294), provided that the problem and its decomposition are sufficiently regular (for a particular case with L shaped domain and less regularity see [5]).

Our results are particularly well suited for pseudo stationary problems arising from implicit or semi-implicit schemes for evolution equations. For the repetitive solution of the large scale problems which are usually well conditioned, the asynchronous fixed point method can be competitive to other methods. We refer to [6] for a future paper in this direction.

After the introduction, we present in the second section the problem formulation, introduce the notation used in the sequel and state our basic assumptions. In the third section we define the linear mapping $\mathcal{T}$ which defines the substructured solution process. Then we define the linear fixed point mapping T which is the composition of $\mathcal{T}$ with a suitable restriction operator R. We prove that T is a linear mapping in a suitable function space context. We also study the contraction property of T. We finally introduce an affine mapping whose linear part is T and whose fixed point is the solution of the elliptic partial differential equation. We state in the closing proposition the convergence of asynchronous iterations applied to the approximation of this affine fixed point mapping.

2. Problem Description, Notation and Assumptions

Let Ω be an open bounded domain of $\mathbb{R}^n$ with boundary $\partial\Omega$. We consider the Dirichlet problem

$$Au = f \text{ in } \Omega, \quad u = g \text{ on } \partial\Omega \tag{2.1}$$

with given $f \in L^p(\Omega)$ and $g \in L^q(\partial\Omega)$ and we assume that (2.1) has a unique solution in a suitable function space.

To formulate our algorithm, we need an overlapping decomposition of Ω with certain overlap properties. We construct such a decomposition by first decomposing the domain Ω into m non-overlapping open subdomains Ω'_i,

$$\Omega'_i \cap \Omega'_j = \emptyset,\ i \neq j, \quad \bigcup_{i=1}^{m} \overline{\Omega'}_i = \overline{\Omega}. \tag{2.2}$$

The boundary $\partial\Omega'_i$ of Ω'_i consists in general of a part inside Ω, which we call $\Gamma'_i := \partial\Omega'_i \cap \Omega$ and a part coinciding with the boundary of Ω, which we denote by $\gamma'_i := \partial\Omega'_i \cap \partial\Omega$. We now construct from this non-overlapping decomposition of Ω the desired overlapping decomposition which will be used by our algorithm. We enlarge Ω'_i to obtain the overlapping subdomains Ω_i, $\Omega'_i \subset \Omega_i \subset \Omega$, and we define the corresponding boundaries of the overlapping subdomains by

$$\Gamma_i = \partial\Omega_i \cap \Omega, \quad \gamma_i = \partial\Omega_i \cap \partial\Omega. \tag{2.3}$$

We require strict overlap, which means that the closures $\overline{\Omega'}_i$ and $\overline{\Gamma}_i$ satisfy

$$\overline{\Omega'}_i \cap \overline{\Gamma}_i = \emptyset, \quad i = 1, \dots, m. \tag{2.4}$$

The linear part of the affine mapping we need in order to introduce our algorithm is based on the homogeneous subproblems

$$Av_i = 0 \text{ in } \Omega_i, \quad v_i = 0 \text{ on } \gamma_i, \quad v_i = \varphi_i \text{ on } \Gamma_i, \ \varphi_i \in L^p(\Gamma_i) \tag{2.5}$$

and we need the following regularity of subdomain solutions

$$v_i \in \begin{cases} C^\infty(\overline{\Omega'}_i) & \text{if } \gamma'_i = \emptyset \\ C^1(\overline{\Omega'}_i) & \text{otherwise.} \end{cases} \tag{2.6}$$

Assumption 2.1 *The elliptic operator A satisfies both the standard maximum principle with respect to Ω_i and $\partial\Omega_i$ and the Hopf maximum principle with respect to γ'_i (compare [7]).*

For the exchange of information between subdomains, we will also employ the index notation

$$\Gamma_{i,j} = \Gamma_i \cap \Omega'_j, \ j \in J(i) \tag{2.7}$$

where the index set J is defined by

$$J(i) = \left\{ j : \Gamma_i \cap \Omega'_j \neq \emptyset, j \neq i \right\}. \tag{2.8}$$

Assumption 2.2 *There exists an indefinitely often continuously differentiable positive Greens function associated with A on each subdomain,*

$$G_i(x,y) > 0, \quad x,y \in \Omega_i, \; x \neq y.$$

To the Greens function corresponds a Poisson kernel $\frac{\partial G_i}{\partial \nu_i}(x,y)$ where $\frac{\partial}{\partial \nu_i}$ denotes the unit outward normal derivative with respect to $\partial\Omega_i$ (compare [8], p. 342 (consequence 5.1)).

Remark 2.3 *Weaker regularity assumptions are be possible for the subdomains Ω_i on the basis of the exterior ball property (cf [9], p. 16) and Hölder regularity of the boundaries.*

Assumption 2.4 *The Poisson Kernel $\frac{\partial G_i}{\partial \nu_i}(x,y)$ is positive and continuously differentiable on $\overline{\Omega}_i \times \partial\Omega_i$, $x \neq y$.*

Assumption 2.5 *There exists a positive eigenfunction $e \in C(\overline{\Omega})$ associated with the smallest eigenvalue λ of A ,*

$$Ae = \lambda e, \; \lambda \in \mathbb{R}^+. \tag{2.9}$$

Remark 2.6 *The required regularity (2.6) can be obtained on the basis of Assumptions (2.2) and (2.4).*

3. Linear Mappings Associated with Overlapping Subdomains

3.1. The linear mapping $\mathcal{T}$

We denote all the interior boundaries by $\Gamma = \cup_{i=1}^{m}\Gamma_i$ and we consider the space $L^1(\Gamma) := \prod_{i=1}^{m}\prod_{j\in J(i)} L^1(\Gamma_{i,j})$. We will furthermore need the function space $\mathcal{C} := \prod_{i=1}^{m} C(\overline{\Omega'}_i)$. Note that $\mathcal{C} \neq C(\overline{\Omega})$. We now define the linear mapping

$$\mathcal{T} : L^1(\Gamma) \longrightarrow \mathcal{C}, \quad \mathcal{T} : w \longmapsto v'.$$

For a given function $w \in L^1(\Gamma)$ we first write component-wise

$$w = \{\dots, w_i, \dots\}_{i=1,\dots,m}, \quad w_i = \{\dots, w_{i,j}, \dots\}_{j\in J(i)} \in L^1(\Gamma_i)$$

and then we compute using w the solutions v_i of the subproblems

$$Av_i = 0 \text{ in } \Omega_i, \quad v_i = w_{i,j} \text{ on } \Gamma_{i,j}, \; j \in J(i), \quad v_i = 0 \text{ on } \gamma_i. \tag{3.1}$$

Now we take the restriction $v'_i = v_i|_{\overline{\Omega'}_i}$ and we define the linear operator $\mathcal{T}_i$ by $\mathcal{T}_i(w) := v'_i$. Finally we set

$$\mathcal{T}(w) := v' = \{\dots, \mathcal{T}_i(w), \dots\} \tag{3.2}$$

thus defining the linear mapping $\mathcal{T}$. On using the Greens function and Poisson kernel, we can express v'_i and hence $\mathcal{T}_i$ and $\mathcal{T}$ in analytic form,

$$v'_i(x) = \sum_{j\in J(i)} \int_{\overline{\Gamma}_{i,j}} \frac{\partial G_i}{\partial \nu_j}(x,y) w_{i,j}(y)dy = \int_{\overline{\Gamma}_i} \frac{\partial G_i}{\partial \nu_j}(x,y) w_i(y)dy \quad \forall\, x \in \overline{\Omega'}_i. \tag{3.3}$$

Proposition 3.1 *$\mathcal{T} \in \mathcal{L}(L^1(\Gamma), \mathcal{C})$ is a linear isotone mapping with respect to the natural order.*

3.2. The Linear Mapping T

We first define the space $C_e(\overline{\Omega'}_j)$ containing all the elements of $C(\overline{\Omega'}_j)$ which can be endowed with the norm $|w_j|_{e,\infty} = \max_{x\in\overline{\Omega'}_j} \frac{|w_j(x)|}{e(x)}$ where $e(x)$ denotes the eigenfunction in Assumption 2.5. We define furthermore $C_e = \prod_{j=1}^m C_e(\overline{\Omega'}_j)$ endowed with the norm $|w|_{e,\infty} = \max_{j=1,\dots,m} |w_j|_{e,\infty}$ where $w = \{\dots, w_j, \dots\}$.

Then we define the mapping

$$T : C \longrightarrow C_e, \quad T : v' \longmapsto \tilde{v}'$$

on using the mapping $\mathcal{T}$. We first define for the components $v'_j \in C(\overline{\Omega'}_j)$ of $v' \in C$ the restriction operator $R_{i,j}$, $j \in J(i)$, by

$$w_{i,j} = R_{i,j}(v'_j) = v'_j|_{\Gamma_{i,j}}.$$

Denoting by $w_i = \{\dots, w_{i,j}, \dots\}_{j\in J(i)}$ we define the restriction operator R_i by

$$R_i(v') = w_i = \left\{\dots, R_{i,j}(v'_j), \dots\right\}_{j\in J(i)}$$

and finally collecting components we get $w = \{\dots, w_i, \dots\}$ and we can define the restriction operator R from C to $\prod_{i=1}^m \prod_{j\in J(i)} C(\overline{\Gamma}_{i,j})$ $(\subset L^1(\Gamma))$ by

$$R(v') := w = \{\dots, R_i(v'), \dots\}.$$

Now we can define the mapping T_i as the composition of $\mathcal{T}_i$ and R, $T_i(v') := v'_i = \mathcal{T}_i \circ R(v')$ and collecting the v'_i we find the mapping T to be

$$T(v') := \tilde{v}' = \mathcal{T} \circ R(v'). \tag{3.4}$$

Proposition 3.2 *$T \in \mathcal{L}(C, C_e)$ is a linear isotone mapping with respect to the natural order on C and C_e. The norm of T is given by*

$$||T||_{\mathcal{L}(C,C_e)} = \max_{i=1\dots m} \max_{x\in\overline{\Omega'}_i} \frac{(T\mathbf{1})_x}{e(x)} = \tilde{\mu}$$

where $\mathbf{1}$ is the function $\mathbf{1}(x) = 1 \ \forall\, x \in \overline{\Omega}$.

Remark 3.3 *The operator $\mathcal{T}$ is also in $L(\prod_{i=1}^m \prod_{j\in J(i)} C(\overline{\Gamma}_{i,j}), C_e)$, and $\mathcal{T}$ is also an isotone linear mapping with respect to the natural order on $\prod_{i=1}^m \prod_{j\in J(i)} C(\overline{\Gamma}_{i,j})$ and C_e.*

Remark 3.4 *Propositions 3.1 and 3.2 do not assert that the linear mappings $\mathcal{T}$ or T are contractions.*

3.3. Contraction Property of T

On using Assumption (2.5) we consider the subproblems

$$Av_i = 0 \text{ in } \Omega_i, \quad v_i = e \text{ on } \partial\Omega_i \tag{3.5}$$

and also the restriction

$$v'_i = v_i|_{\overline{\Omega'}_i}. \tag{3.6}$$

We define furthermore $e_i := e|_{\overline{\Omega'}_i}$.

Lemma 3.5 $\frac{v_i'(x)}{e_i(x)}$ *is defined and continuous on* $\overline{\Omega'}_i$, *and* $\max_{x\in\overline{\Omega'}_i} \frac{v_i'(x)}{e_i(x)} \le \mu_i < 1$.

Proposition 3.6 $T \in \mathcal{L}(C, C_e)$ *is a contraction with contraction constant* $\mu = \max_{i=1,\dots,m} \mu_i$.

Remark 3.7 *The proof of Proposition 3.6 shows that* μ *is an upper bound on* $\|T\|_{\mathcal{L}(C,C_e)}$ *and the bound is reached if* $w(x) = e(x)$ *and is thus sharp. One furthermore finds that* μ *is decreasing when the size of overlap increases and* μ *is decreasing when* ε *decreases in the case* $A = -\varepsilon\Delta + I$. *We refer to [6] for numerical experiments and comparison of algorithms.*

3.4. The Fixed Point Mapping

Let us consider an element w such that $w_i - u_i \in C(\overline{\Omega'}_i)$. This implies, with a trivial extension that

$$\delta w = w - u \in C. \tag{3.7}$$

Let us first solve the subproblems

$$Au_i = f \text{ in } \overline{\Omega}_i, \quad u_i = 0 \text{ on } \Gamma_i, \ u_i = g \text{ on } \gamma_i, \quad i = 1, \dots, m. \tag{3.8}$$

Restricting u_i to Ω_i',

$$u_i' = u_i|_{\overline{\Omega'}_i} \tag{3.9}$$

we consider the new subproblems

$$Av_i = f \text{ in } \overline{\Omega}_i, \quad v_i = w \text{ on } \Gamma_i, \quad v_i = g \text{ on } \gamma_i, \quad i = 1, \dots, m \tag{3.10}$$

and we get the restricted values

$$v_i' = v_i|_{\overline{\Omega'}_i} \tag{3.11}$$

so that the fixed point mapping is given by

$$v_i' = T_i(w) + u_i'. \tag{3.12}$$

Proposition 3.8 *If for* w *satisfying (3.7), we choose* $u^0 = T(w) + u'$ *where* $u' = \{\dots, u_i', \dots\}$, u_i' *defined by (3.8), (3.9) then the asynchronous iterations initialized by* u^0, *applied to the affine fixed point mapping* $F(w) = T(w) + u'$ *give rise to a sequence of iterates which converges, with respect to the uniform weighted norm* $|\ |_{e,\infty}$ *towards the solution* u^* *of problem (2.1).*

The proof is based on the use of El Tarazi's theorem [10] and on Proposition 3.6. Compare also [3] for previous partial results and numerical experiments.

References

1. L. Badea, Numer. Math., **55**, 61-81 (1989).
2. L. Badea, SIAM J. Numer. Anal., **28**, 179-204 (1991).
3. J.C. Miellou, L. Giraud, A. Laouar and P. Spiteri, in: Progress in Partial Differential Equations: the Metz Surveys, Papers from the 'Metz days' conferences, held in Metz, France, during 1989-1990, M. Chipot et al. (Ed.), Harlow: Longman Scientific Technical, (ISBN 0-582-07024-4), Pitman Res. Notes Math. Ser. 249 (1991) pp. 166-183.
4. R. Courant and D. Hilbert, Methods of Mathematical Physics, Volume II, Interscience Publishers, a division of John Wiley & Sons, New York (1962).
5. M. Laaraj and J.C. Miellou, Etude Numérique de la Méthode Alternée de Schwarz pour des Problèmes Stationnaires de Convection Diffusion 27 Congres d'Analyse Numérique, SUPER-BESSE 29 Mai-3 Juin, (1995).
6. L. Giraud, R. Guivarch, P. Spiteri and J.C. Miellou, (in preparation).
7. M.H. Protter and H.F. Weinberger, Maximum Principles in Differential Equations, Prentice-Hall, New Jersey (1967).
8. J. Nečas, Les Méthodes Directes en Théorie des Equations Elliptiques, Masson, Prague (1967).
9. C.V. Pao, Nonlinear Parabolic and Elliptic Equations, Plenum Press, New York (1992).
10. M.N. El Tarazi, Numer. Math., **39**, 325-340 (1982).

9th Int. Coll. on Differential Equations, pp. 267-272
D. Bainov (Ed)

Numerical Analysis of Natural Convection of a Viscoelastic Fluid, Using a Spectral Finite Difference Scheme

YOSHIHIRO MOCHIMARU
Department of International Development Engineering, Tokyo Institute of Technology, Tokyo 152-8552, Japan

Abstract: A newly established spectral finite difference scheme is applied to a steady two-dimensional natural convection problem of a viscoelastic fluid enclosed in a non-circular cavity heated from below. Conformal mapping of a physical plane to a circular region is adopted to get a steady-state heat & fluid flow field for various Grashof numbers and for various Weisenberg numbers.
Key Words and Phrases: *Natural Convection, Spectral Difference Scheme, Viscoelastic Fluid.*

1. Introduction

Steady-state natural convection of a viscoelastic fluid enclosed in a vertical non-circular two-dimensional region is numerically analyzed, based on a system of governing partial differential equations. Emphasis is placed on the effectiveness of a spectral finite difference scheme [1, 2, 3, 4] developed especially for analyzing partial differential equations with boundary and/or initial conditions. Another feature of the current analysis is given in a view point of the ability of the scheme to get a steady-state solution even for relatively highly viscoelastic fluid flow. With traditional numerical methods, steady-state solutions available for viscoelastic fluid flow are limited for a moderate range of a Weisenberg number (i.e. a ratio of a characteristic elastic force to a viscous force.) In the current analysis, assumed is the following: The wall geometry is symmetric with respect to a vertical axis and concave along the boundary. Also to reproduce phenomena with stronger circulation (non-linear deviation from a pure conduction), it is assumed that the upper part of the boundary wall is kept at a cooler constant temperature, and that the lower part of the boundary wall is kept at a higher constant temperature.

2. Analysis

Let (x, y) be a Cartesian coordinate system such that the x-coordinate is horizontal, and that the y-axis coincides with the symmetrical axis of the domain, the positive y-direction being upward. Hereafter all the length are assumed to be made dimensionless with respect to a suitable length a. Introduced is the following conformal mapping from (x, y) plane to (α, β) plane such that

$$(1) \qquad \exp(\alpha + i\beta) = \frac{(1 + i\eta)}{(1 - i\eta)}, \qquad \eta = f(x + iy),$$

where f is an analytic function. In case of a situation where the boundary is given by $\mathrm{Im}(f) = 0$ and along the boundary $\mathrm{Real}(f)$ is a monotonusly increasing function from $-\infty$ to $+\infty$ with increasing β, the boundary is mapped into $\alpha = 0$. That is, in the ξ-plane $[\xi \equiv \exp(\alpha + i\beta)]$ the boundary constitutes a circle of a unit radius. Without loss of generality, the function f is assumed to have a character that inverse convertion from $\alpha + i\beta$ to $x + iy$ is unique, and that the region inside the domain is expressed by $-\infty < \alpha \leq 0$. As a concrete example, the following is considered:

(2)
$$f \equiv -icz^2 + z + \frac{1}{z},$$
$$z \equiv x + iy, 0 < c < \frac{2}{3\sqrt{3}}.$$

Eqs.(1) and (2) can be inverted to give

(3)
$$z = -i\left\{\frac{1}{3c} + \sqrt{-\frac{A}{3c}}\left(F - \frac{1}{F}\right)\right\},$$

where

(4)
$$A \equiv \frac{1}{3c} - \frac{\xi - 1}{\xi + 1},$$

(5)
$$B \equiv 1 - \frac{2}{27c^2} + \frac{1}{3c}\frac{\xi - 1}{\xi + 1},$$

(6)
$$F \equiv \sqrt[3]{\sqrt{1 - \frac{27cB^2}{4A^3}} + \frac{3B}{A}\sqrt{-\frac{3c}{4A}}}.$$

In Eq.(3) $\sqrt{\ }$ means a principal value, whereas in Eq.(6) $\sqrt[3]{\ }$ and $\sqrt{\ }$ mean principal values for almost all cases. (Exceptional cases occur if the argument of the second square root becomes real and negative. In this case, the sign of the second square root has to be changed.)

3. Basic Equations

Just simply to express the steady-state viscoelastic behaviour, as a constitutive equation the so-called second order fluid model is introduced: $\tau^{ij} = \mu d^{ij} - \lambda(\delta/\delta t)d^{ij}$, where τ^{ij} : extra stress tensor, d^{ij} : rate of deformation tenosr, $\delta/\delta t$: convective time derivative, μ : viscosity, λ : fluid material constant (≥ 0). In a two-dimensional laminar natural convection, the equation of (dimensionless) vorticity transport, the relationship between vorticity and a stream function, and the (dimensionless) energy equation (neglecting dissipation terms) are

(7)
$$\frac{\partial}{\partial t}\left[\frac{K}{r^2}\left\{(1-\gamma)\varsigma + \gamma\bar{\zeta}\right\}\right] + \frac{1}{r}\frac{\partial(\bar{\zeta}, \psi)}{\partial(\alpha, \beta)} = \frac{1}{\sqrt{Gr}}\left(\frac{\partial^2}{\partial r^2} + \frac{1}{r}\frac{\partial}{\partial r} + \frac{1}{r^2}\frac{\partial^2}{\partial \beta^2}\right)\varsigma + \frac{1}{r}\frac{\partial(T, y)}{\partial(r, \beta)},$$

(8)
$$\frac{K}{r^2}\varsigma + \left(\frac{\partial^2}{\partial r^2} + \frac{1}{r}\frac{\partial}{\partial r} + \frac{1}{r^2}\frac{\partial^2}{\partial \beta^2}\right)\psi = 0,$$

$$\frac{K}{r^2}\frac{\partial T}{\partial t}+\frac{1}{r}\frac{\partial(T,\psi)}{\partial(r,\beta)}=\frac{1}{Pr\sqrt{Gr}}\left(\frac{\partial^2}{\partial r^2}+\frac{1}{r}\frac{\partial}{\partial r}+\frac{1}{r^2}\frac{\partial^2}{\partial\beta^2}\right)T \tag{9}$$

respectively, where

$$\bar{\zeta}=\zeta+\epsilon\frac{r^2}{K}\left(\frac{\partial^2}{\partial r^2}+\frac{1}{r}\frac{\partial}{\partial r}+\frac{1}{r^2}\frac{\partial^2}{\partial\beta^2}\right)\zeta, \tag{10}$$

$$K\equiv\frac{\partial(x,y)}{\partial(\alpha,\beta)}=\left|\frac{dz}{d(\alpha+i\beta)}\right|^2, \tag{11}$$

$$r\equiv\exp\alpha, \tag{12}$$

$$\epsilon\equiv\lambda/(\rho a^2), \tag{13}$$

Gr : Grashof number based on a length a and the temperature difference T_l-T_u, Pr : Prandtl number, t : dimensionless time based on a/U, $T\equiv(T'-T_u)/(T_l-T_u)$, T' : local temperature, T_l : constant lower wall temperature, T_u : constant upper wall temperature, $U\equiv(\nu/a)\sqrt{Gr}$, γ : parameter introduced for getting higher numerical stability ($\gamma\le 0$; although originally γ should be equal to unity, the value of γ is independent of the steady-state solution if it is attained.) ζ : dimensionless vorticity based on U/a, ν : kinematic viscosity, ψ : dimensionless stream function based on Ua. In case of $\epsilon=0$, it stands for an incompressible Newtonian fluid.

4. Boundary Conditions

For dynamic conditions, no slip condition at the wall applies, that is

$$\psi(r=1,\beta,t)=0, \tag{14}$$

$$\frac{\partial}{\partial r}\psi(r=1,\beta,t)=0. \tag{15}$$

For a thermal boundary condition, due to a constant wall temperature a Dirichlet condition applies:

$$T(\alpha=0,\beta_0<|\beta|\le\pi,t)=0, \tag{16}$$

$$T(\alpha=0,0\le|\beta|<\beta_0,t)=1, \tag{17}$$

$$\beta_0=2\tan^{-1}\left\{-cr_0^2\sin 2\phi+\left(r_0+\frac{1}{r_0}\right)\sin\phi\right\}, \tag{18}$$

$$r_0\equiv\sqrt{2\cos^2\phi+\cos^2 2\phi}, \tag{19}$$

$$\phi=\cos^{-1}\left\{\left(\sqrt{\sqrt[3]{4/c^2}+6}-\sqrt{\sqrt[3]{4/c^2}-2}\right)\Big/4\right\}. \tag{20}$$

[Half of the cavity width, x_0, is given by $x_0=r_0\sin\phi$.]

5. Spectral Formulation of the Solution

Throughout the present analysis, steady-state solution is sought, using a time marching way. That is, the said problem is treated as an initial/boundary value problem. Since the configuration of the boundary, dynamic boundary conditions, and thermal boundary conditions are symmetric with respect to the y-axis, a symmetric flow and thermal pattern is possible if a symmetrical initial flow and thermal condition is given. (In the current analysis at an initial state ($t = 0$) fluid is assumed to be at rest with a uniform temperature $T = T_l$.) Thus, among many possibilities the following Fourier expansion is introduced:

(21) $$\zeta = \sum_{n=1}^{\infty} \zeta_n(r,t) \sin n\beta,$$

(22) $$\psi = \sum_{n=1}^{\infty} \psi_n(r,t) \sin n\beta,$$

(23) $$T = \sum_{n=0}^{\infty} T_n(r,t) \cos n\beta.$$

Then the boundary condition (14) reduces to

(24) $$\psi_n(1,t) = 0, \qquad (n \geq 1),$$

and the boundary condition (15) becomes

(25) $$\frac{\partial}{\partial r}\psi_n(1,t) = 0, \qquad (n \geq 1).$$

Direct discretization of Eq.(15) or Eq.(25) is known to produce numerical instability a great deal, so that with the aid of Eq.(8), introduced is the following:

(26) $$\frac{1}{h^2}\psi_n(1-h,t) + \frac{1}{\pi}\int_0^{\pi} \sin n\beta (K\zeta)_{r=1-h} d\beta = 0,$$

where $1 - h$ is the r-coordinate value at a grid point adjacent to the boundary. Thermal boundary conditions (16) and (17) reduce to

(27) $$T_0(1,t) = \frac{\beta_0}{\pi},$$

(28) $$T_n(1,t) = \frac{2\sin n\beta_0}{n\pi}, \qquad (n \geq 1).$$

6. Initial Conditions

As stated in the previous section, initial conditions are expressed as

(29) $$\psi(r,\beta,t=0) = \zeta(r,\beta,t=0) = 0,$$

(30) $$T(|r| < 1,\beta,t=0) = 0,$$

$$T(r=1,\beta_0<|\beta|\leq\pi,t=0)=0, \tag{31}$$

$$T(r=1,|\beta|<\beta_0,t=0)=1, \tag{32}$$

which become

$$\psi_n(r,t=0)=\zeta_n(r,t=0)=0 \qquad (n\geq 1), \tag{33}$$

$$T_n(|r|<1,t=0)=0 \qquad (n\geq 0). \tag{34}$$

Eqs.(27) and (28) apply to $T_0(1,t=0)$ and $T_n(1,t=0),(n\geq 1)$ respectively.

7. Auxiliary Conditions at $r=0(\alpha=-\infty)$

The condition that scalor variables and their gradients are continuous at $r=0$ gives

$$\frac{\partial}{\partial r}T_0(0,t)=0, \tag{35}$$

$$T_n(0,t)=0, \qquad (n\geq 1), \tag{36}$$

$$\psi_n(0,t)=\zeta_n(0,t)=0, \qquad (n\geq 1). \tag{37}$$

8. Numerical Integration Scheme

Truncating the series (21) – (23) up to a suitable order, and decomposing the basic governing eqs.(7),(8), and (9) into corresponding Fourier components gives rise to a system of simultaneous non-linear partial differential equations. For a component decomposition, addition formulae of trigonometric functions etc. can be utilized. The resulting system of differential equations can be discretized using a finite difference approximation in time and space, which can be integrated semi-implicitly in a time-marching scheme with time to get a steady-state solution.

9. Numerical Results

Mean Nusselt number along a heated part Nu_m (based on a reference length a) is given by

$$Nu_m=\frac{1}{L_0}\int_0^{\beta_0}\frac{\partial}{\partial\alpha}T(\alpha=0,\beta,t)d\beta=\frac{1}{L_0}\left\{\beta_0\frac{\partial}{\partial r}T_0(1,t)+\sum_{n=1}^{\infty}\frac{\sin n\beta_0}{n}\frac{\partial}{\partial r}T_n(1,t)\right\}, \tag{38}$$

where L_0 is one half of the circumference of the heated part and is given by

$$L_0\equiv\int_0^{\beta_0}\sqrt{K(\alpha=0)}d\beta. \tag{39}$$

Figures 1 and 2 show steady-state streamlines and isotherms respectively for $c=0.2$ at $Gr=10^3, Pr=7, \epsilon=0.1, [\sqrt{Gr}\epsilon$ (Weisenberg number) $=3.16$]. Figures 3 and 4 show steady-state streamlines and isotherms respectively for $c=0.2$ at $Gr=10^3, Pr=7, \epsilon=2.0, [\sqrt{Gr}\epsilon=63.2]$. Figures 5 and 6 show steady-state streamlines and isotherms respectively for $c=0.2$ at $Gr=2\times10^3, Pr=7, \epsilon=2, [\sqrt{Gr}\epsilon=89.4]$.

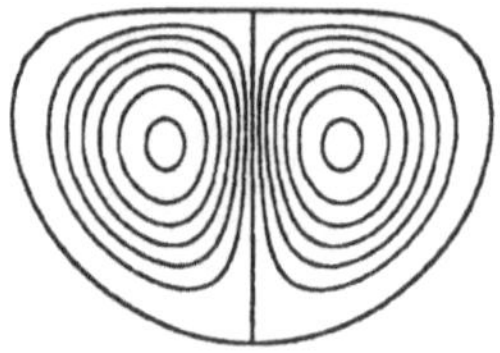

Fig.1 Steady-state streamlines for $c = 0.2$ at $Gr = 10^3$, $Pr = 7$, $\epsilon = 0.1$, $\delta\psi$(difference of streamlines) $= 0.005$.

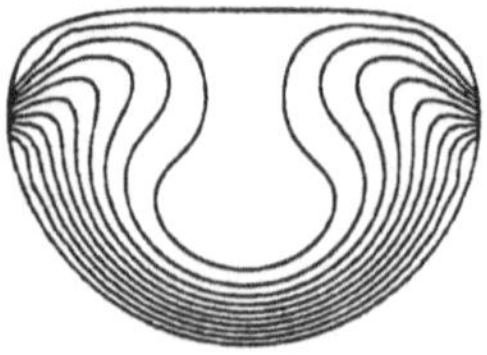

Fig.2 Steady-state isotherms for $c = 0.2$ at $Gr = 10^3$, $Pr = 7$, $\epsilon = 0.1$, δT (difference of isotherms) $= 0.1$, $Nu_m = 3.47$.

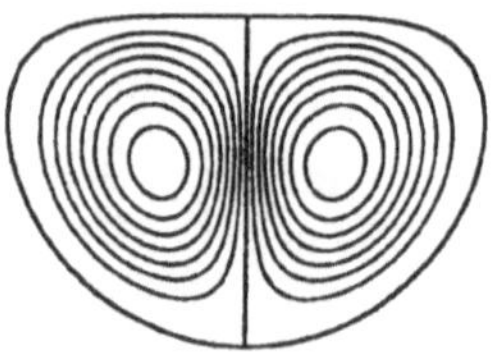

Fig.3 Steady-state streamlines for $c = 0.2$ at $Gr = 10^3$, $Pr = 7$, $\epsilon = 2.0$, $\delta\psi = 0.001$.

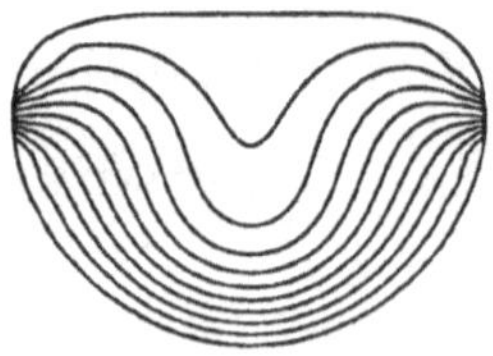

Fig.4 Steady-state isotherms for $c = 0.2$ at $Gr = 10^3$, $Pr = 7$, $\epsilon = 2.0$, $\delta T = 0.1$, $Nu_m = 2.23$.

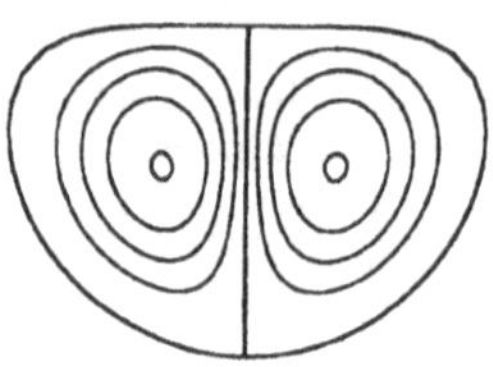

Fig.5 Steady-state streamlines for $c = 0.2$ at $Gr = 2 \times 10^3$, $Pr = 7$, $\epsilon = 2.0$, $\delta\psi = 0.0025$.

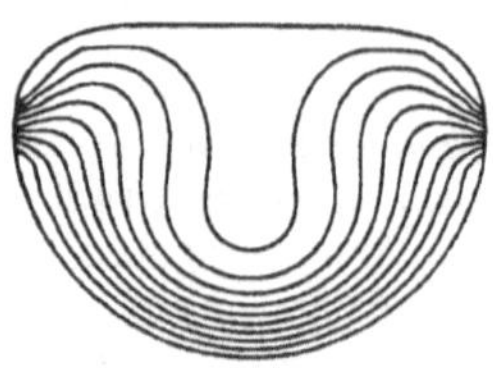

Fig.6 Steady-state isotherms for $c = 0.2$ at $Gr = 2 \times 10^3$, $Pr = 7$, $\epsilon = 2.0$, $\delta T = 0.1$, $Nu_m = 2.68$.

10. Conclusion

The proposed current spectral finite difference scheme is found to be also effective for analyzing natural convection of a relatively highly viscoelastic fluid.

References

1. Y. Mochimaru, in: Proc. 26th Ann. Iranian Math. Conf., **2**, 245–254 (1995).
2. Y. Mochimaru, in: 7th Int. Coll. on Differential Equations, D. Bainov (Ed.), VSP, Utrecht (1996) pp. 243–252.
3. Y. Mochimaru, in: 8th Int. Coll. on Differential Equations, D. Bainov (Ed.), VSP, Utrecht (1997) pp. 311–318.
4. Y. Mochimaru, Advances in Computational Heat Transfer, 411–417 (1997).

9th Int. Coll. on Differential Equations, pp. 273-279
D. Bainov (Ed)

Two-Step Variable Metric Conjugate Gradient Algorithms

ISSAM A.R. MOGHRABI
Natural Science Division, Lebanese American University, P.O.Box 13-5053, Beirut, Lebanon

Abstract: This paper considers the behavior of a particular class of conjugate gradient (CG) algorithms with a two-step variable metric memory, of which the best known is the original member introduced in 1977 by Perry. Part I is concerned with direct modifications of Perry's method [1]; part II addresses the improvement of the more recent developments due to Shanno, as inspired by the work done in [1]. This class has the "intermediate" storage property that it replaces the (n by n) matrix approximating G^{-1}, the inverse Hessian of the general function $f(x)$, by four (n by 1) vectors, thus requiring much less computer space than quasi-Newton (QN) methods for n much greater than 4, although it requires twice that of the standard CG-methods. However, taking account of the relative rates of convergence this class is the most efficient at "intermediate" n. In this paper three new modifications to Perry's algorithm are introduced, both with and without exact line-searches, which still generate conjugate directions. Finally a numerical study is reported in which the performance of the new algorithms is compared to that of Perry's algorithm and this gives encouraging results over the twenty-seven well-known test functions studied.

1. Introduction

It is well-known that the QN-methods, and especially the BFGS [2,3,4,5] method, exhibit more rapid convergence than CG-methods when minimizing a general nonlinear continuous and differentiable function $f(x)$, $x \in \mathbf{R}^n$ [6]. However, when n is large, QN-methods become impractical (or inefficient) due to the fact they require the storage and updating of the n by n matrix approximating G^{-1}, i.e., storage of at least (n by n)/2, allowing for symmetry. At the same time CG-methods [7] generally require the storage of only two (n by 1) vectors containing past information on the function and its derivatives. For this reason CG-methods tend to replace the QN- methods whenever the storage space is relatively scarce, and this must correspond to some dimensionality n.

The problem under consideration is that defined in the first paragraph of this section. CG- methods are iterative, and we shall assume that the function value $f(x_k)$ and its gradient vector $g(x_k)$ are known at iteration k. At each iteration a new point x_{k+1} is defined by taking a step in the direction d_k :

$$x_{k+1} = x_k + \lambda_k d_k \tag{1}$$

and a new direction d_{k+1} is defined by

$$d_{k+1} = -g_{k+1} + \beta_k d_k \tag{2}$$

where λ_k is a positive scalar, β_k is a scalar which satisfies the conjugacy condition:

$$d_k^T G d_i = 0 \qquad \text{for all} \quad i < k. \tag{3}$$

For exact line-searches (ELS) the scalar λ is determined by minimizing the one-dimensional function $f(x_k + \lambda d_k)$.

2. Perry's Two-Step CG-Method

Perry developed [8] an interim procedure of this type which generates mutually conjugate directions using ELS. He constructed his method by taking a linear combination of d_k and $-g_{k+1}$ defined over a different metric represented by the n by n positive definite matrix H so that:

$$d_{k+1} = -H_k g_{k+1} + \beta_k d_k \tag{4}$$

where β_k can be constructed to satisfy

$$y_k^T d_{k+1} = 0, \tag{5}$$

where

$$y_k = g_{k+1} - g_k. \tag{6}$$

Equation (5) is the conjugate-direction requirement so that, using (4),

$$\beta_k = (y_k^T H_k g_{k+1}) / y_k^T d_k. \tag{7}$$

For simplicity, in the remainder of this paper subscripts (k) have been omitted and subscripts $(k+1)$ replaced by asterisks.

Perry in his method made the following assumptions:

$$\text{(i)} \quad d = -Hg, \tag{8}$$

$$\text{(ii)} \quad Hy = d + Hg^*, \tag{9}$$

$$\text{(iii)} \quad d^* = -Hg^* + \gamma Hy \tag{10}$$

where

$$\gamma = (y^T H g^*)/(y^T H y) \tag{11}$$

and he replaced the full memory BFGS matrix updating formula:

$$H^* = H - \frac{\nu y^T H}{\nu^T y} - \frac{H y \nu^T}{\nu^T y} + \left(1 + \frac{y^T H y}{\nu^T y}\right) \frac{\nu \nu^T}{\nu^T y} \tag{12}$$

where

$$\nu = x^* - x \tag{13}$$

by partial memory, which is represented by the linear combination of the two vectors $\mathbf{c} = \mathbf{Hy}$ and $\mathbf{d}$ rather than the n by n matrix $\mathbf{H}$.

Algorithm (I) (Perry 1977):

For a given initial point x :

Step 1: Set $d = -g$.

Step 2: Find a scalar $\lambda > 0$ which minimizes $f(x + \lambda d)$ and set $x^* = x + \lambda d$.

Step 3: Check for convergence; if not go to step 4.

Step 4: Define a new vector c by

$$c = \begin{cases} y, & \text{for } \quad k = 1, \\ y^* - \dfrac{\nu^T y^*}{\nu^T y} y + \left[\left(1 + \dfrac{y^T y}{\nu^T y}\right) \dfrac{\nu^T y^*}{\nu^T y} - \dfrac{y^T y^*}{\nu^T y}\right] \nu & \text{for } \quad k \geq 2. \end{cases} \tag{14}$$

Step 5: Define the new search direction d^* by

$$d^* = \left(-1 + \frac{d^T g^*}{d^T y}\right) c - \left[-\frac{c^T g^*}{d^T y} + \frac{c^T y . d^T g^*}{(d^T y)^2} + \frac{\lambda d^T g^*}{d^T y} - 1\right] d. \tag{15}$$

3. New Algorithm (II)

In this section we present the first new algorithm, which is a modification of Perry's algorithm (I) and is based on ELS.

Consider first the following new self-scaling BFGS update:

$$H^* = H - \frac{H y y^T H}{y^T H y} + w w^T + \sigma \frac{\nu \nu^T}{\nu^T y} \tag{16}$$

where

$$\sigma = (y^T H y)/(\nu^T y) \tag{17}$$

and

$$w = (y^T H y)^{1/2} (\nu / \nu^T y - H y / y^T H y). \tag{18}$$

It is clear that for different variable-metric methods it is possible to generate corresponding two-step variable-storage CG-methods by using exactly the same analysis as in Perry's paper.

Hence, using (16) we define the new search direction d^* by

$$d^* = -H^* g^* = \left[H - \frac{\nu y^T H}{\nu^T y} - \frac{H y \nu^T}{\nu^T y} + \left(\sigma + \frac{y^T H y}{\nu^T y}\right) \frac{\nu \nu^T}{\nu^T y}\right] g^*. \tag{19}$$

Let us define the new vector **c** by

$$c = H y. \tag{20}$$

Substituting (20) in (19) yields

$$d^* = -H g^* + \frac{c^T g^*}{\nu^T y} \nu + \frac{\nu^T g^*}{\nu^T y} c - 2 \frac{y^T c}{\nu^T y} . \frac{\nu^T g^*}{\nu^T y} \nu. \tag{21}$$

But $H g = -d$ and $H y = c$ gives

$$H g^* = c - d \tag{22}$$

so that (21) becomes

$$d^* = \left(-1 + \frac{d^T g^*}{d^T y}\right) c - \left[-\frac{c^T g^*}{d^T y} + 2 \frac{y^T c . d^T g^*}{(d^T y)^2} - 1\right] d. \tag{23}$$

The new search direction $\mathbf{d}^*$ in (23) using ELS is identical to (15) in the case of a quadratic objective function and then both reduce to the standard Hestenes and Stiefel CG-method [7].

To derive the vector c we know that for the first iteration $H_1 = I$ and hence $c_1 = H_1 y_1 = y_1$. For $k \geq 2$ we can repeat the new updating formula in (16) by

$$H^* = I - (\nu y^T)/(\nu^T y) - (y\nu^T)/(\nu^T y) + 2(y^T y)/(\nu^T y).(\nu\nu^T)/(\nu^T y) \tag{24}$$

and hence

$$c = H^* y^* = \left[y^* - \frac{\nu^T y^*}{\nu^T y} y - \frac{y^T y^*}{\nu^T y}\nu + 2\frac{y^T y.\nu^T y^*}{(\nu^T y)^2}\nu \right] \tag{25}$$

as required.

4. The Effect of Using Inexact Linear Searches on the New Direction $\mathbf{d}^*$

In this section we reconsider algorithms I and II, adapted to employ inexact line searches (ILS). Brodlie in [9] expressed the Huang two-parameter family of correction formulae [10]

$$H^* = H + \rho(\nu\nu^T)/(\nu^T y) - (Hyy^T H)/(y^T Hy) + \varphi w w^T \tag{26}$$

where w is defined as in (18) and ρ, φ are scalar parameters which satisfy the conditions

$$\rho \neq 0, \qquad \varphi \neq -(\nu^T y)^2/[(\nu^T H^{-1}\nu)(y^T Hy) - (\nu^T y)] \tag{27}$$

as

$$\text{(a)} \quad Hy = Hg^* - Hg = Hg^* + \nu/\lambda, \tag{28}$$

$$\text{(b)} \quad y^T Hy/\nu^T y - 1/\lambda = y^T Hg^*/\nu^T y, \tag{29}$$

$$\text{(c)} \quad y^T Hg^*/\lambda y^T Hy = (y^T Hg^*/\nu^T y)(1 - y^T Hg^*/y^T Hy). \tag{30}$$

In our case using the same known equations (see [9]) the new search direction $\mathbf{d}^*$ can be expressed as:

$$d^* = -\rho(\nu^T g^*/\nu^T y)\nu - \mu(I - \nu y^T/\nu^T y)Hg^* \tag{31}$$

where

$$\mu = 1 - y^T Hg^*/y^T Hy - \varphi(\nu^T g^*/\nu^T y - y^T Hg^*/y^T Hy). \tag{32}$$

Equation (31) generates a family of QN-updates by considering different parameters ρ and φ. The new self-scaling (BFGS) update (16) can be considered as a special case of (31) by setting $\varphi = 1$ and $\rho = \sigma$ while the standard BFGS corresponds to $\rho = 1$ and $\varphi = 1$.

Algorithm (III):

We now consider generating the BFGS update i.e. $\rho = 1$ and $\varphi = 1$: the new search direction can be determined by:

$$d^* = -(\nu^T g^*/\nu^T y)\nu - (1 - d^T g^*/d^T y)[Hg^* - (c^T g^*/\nu^T y)\nu]. \tag{33}$$

Let $\alpha = (1 - d^T g^*/d^T y)$, equation (33) then becomes

$$d^* = -\alpha c + [\alpha(1 + c^T g^*/d^T y) - \lambda d^T g^*/d^T y]d \tag{34}$$

using (22), and where the vector c is the same as defined in (14).

Algorithm (IV):

Again, we consider generating the new self-scaled BFGS update, i.e., $\rho = \sigma$ and $\varphi = 1$. Now d^* can be expressed as:

$$d^* = -\alpha c + [\alpha(1 + c^T g^*/d^T y) - c^T y.d^T g^*/(d^T y)^2]d \tag{35}$$

using (20) and (22) and where the vector c is the same as defined in (25).

We emphasize that all four algorithms are restarted with $-g$ whenever $k = n$ or the Powell restarting criterion [11] is satisfied, namely:

$$|g^T g^*| > 0.2 g^{*T} g^* \tag{36}$$

in order to standardize the performance comparisons.

5. Numerical Comparison

This section reports some numerical tests for twenty four standard test functions carried out on pentium 100 Mhz, using Turbo C and double precision programs. The test function can be found in the standard literature. Perry only published results for six test functions and only up to dimensionality ten. For consistency of testing, the Perry algorithm I employs a cubic linear search (rather than his quadratic version) and the same convergence test as algorithms II, III, and IV namely:

$$f(x_{i+1}) \leq f(x_i) + 10^{-4} s_i^T g(x_i); \tag{37}$$

$$s_i^T g(x_{i+1}) > 0.9 s_i^T g(x_i). \tag{38}$$

The results obtained for Perry's algorithm are consistent with those he obtained at low dimensionality. The comparative performances of these four algorithms are simply evaluated by considering the total number of iterations and also the total number of function/gradient evaluations required to reduce the function value to within 1×10^{-10} of its minimum value. Numerical results for the twenty seven test functions are recorded in table 1. The table also presents the percentage performance of algorithms II, III, IV relative to Perry's algorithm I as the norm.

The totals and overall percentages conceal some local variation in performance, and weights the comparison towards high dimensionality; but in less than a quarter of the cases does algorithm I beat any of the new algorithms (generally better in both number of iterations and of function evaluations), but the advantage is seldom better than marginal.

All three new algorithms perform similarly to Perry's algorithm on low dimensionality problems. However, all the three new algorithms perform better than the original Perry's algorithm for the moderate and large dimensionality test functions. In particular, comparing algorithm III with algorithm I there is an overall gain of about 15 % in number of iterations and about 25 % in number of function evaluations. Hence, in conclusion, the new two-step variable storage conjugate gradient algorithms become competitive with the best CG and QN methods over a slightly greater range in n than Perry's algorithm and may be preferred on storage considerations.

Table 1

Test functions	n	Perry's (I) NOI (NOF)	Algr. (II) NOI (NOF)	Algr. (III) NOI (NOF)	Algr. (IV) NOI (NOF)
ROSEN	2	36 (96)	36 (96)	33 (83)	33 (83)
CUBIC	2	18 (52)	18 (54)	18 (51)	18 (51)
BEALE	2	10 (26)	10 (26)	10 (26)	10 (26)
BOX	2	8 (54)	8 (54)	8 (54)	8 (54)
FREUD	2	10 (26)	10 (26)	10 (26)	14 (40)
BIGGS	3	17 (59)	17 (61)	17 (59)	17 (61)
RECIPE	3	5 (16)	5 (16)	5 (16)	5 (16)
HELICAL	3	39 (166)	33 (144)	35 (73)	41 (92)
POWELL	4	63 (218)	34 (106)	65 (231)	57 (177)
WOOD	4	26 (60)	33 (73)	22 (52)	27 (61)
OREN	10	16 (93)	17 (109)	20 (128)	17 (107)
DIXON	10	22 (47)	21 (45)	25 (52)	24 (50)
EX-ROSEN	20	32 (90)	32 (86)	29 (79)	29 (74)
EX-POWELL	20	64 (232)	63 (259)	35 (104)	64 (212)
NON-DIGN.	20	29 (82)	29 (80)	24 (66)	23 (59)
SUM-QUAR.	25	10 (81)	10 (82)	9 (60)	9 (60)
SHALLOW	40	8 (21)	8 (21)	10 (25)	10 (25)
EX-ROSEN	60	32 (90)	32 (86)	29 (79)	29 (74)
EX-POWELL	60	70 (270)	66 (277)	35 (104)	64 (212)
EX-WOOD	60	26 (60)	26 (61)	23 (54)	27 (61)
EX-WOOD	80	26 (60)	26 (61)	23 (54)	27 (61)
NON-DIGN.	90	29 (81)	29 (79)	29 (78)	23 (57)
EX-ROSEN	100	32 (90)	32 (86)	29 (79)	29 (74)
EX-POWELL	100	70 (270)	66 (277)	35 (104)	67 (229)
TOTAL	**NOI (NOF)**	**688 (2340)**	**661 (2265)**	**578 (1737)**	**672 (2016)**

Performance of new 2-step CG-algorithms in relation to Perry's 2-step CG-algorithm (I)

	Perry's (I)	Algor. (II)	Algor. (III)	Algor. (IV)
NOI	100	96.0	84.0	97.7
NOF	100	96.7	74.2	86.1

All the algorithms terminate when $\|g^*\|_2 \leq 10^{-10}$.

References

1. A.Y. Ylias and D. Hutchinson, Two-Step VM CG Modification to Perry's Algorithm, Report 89.24, Leeds University (1989).

2. C.G. Broyden, Journal of Institute Mathematics and its Applications, **6**, 76-90 (1970).

3. R. Fletcher, Computer Journal, **13**, 317-322 (1970).

4. D. Goldfard, Mathematics of Computation, **24**, 23-26 (1970).

5. D.F. Shanno, Mathematics of Computation, **24**, 647-657 (1970).

6. G.P. McCormick and K. Ritter, Mathematical Programming, **3**, 101-111 (1972).

7. M.R. Hestenes and E. Stiefel, Journal of Research of the National Bureau of Standards, **49**, 409-436 (1952).

8. A. Perry, A Class of Conjugate Gradient Algorithms with a Two-Step Variable Metric Memory, Discussion Paper 269, Center of Mathematical Studies in Economics and Management Science, Northwestern University (1977).

9. K.W. Brodlie, Mathematical Programming, **12**, 344-355 (1977).

10. H.Y. Huang, Journal of Optimization Theory and Application, **5**, 405-423 (1970).

11. M.J.D. Powell, Mathematical Programming, **12**, 241-254 (1977).

9th Int. Coll. on Differential Equations, pp. 281–288
D. Bainov (Ed)

Nonsmooth Variational Methods and Applications to Discontinuous Boundary Value Problems

DUMITRU MOTREANU
Department of Mathematics, University of Iaşi, RO-6600 Iaşi, Romania
PANAGIS D. PANAGIOTOPOULOS
Department of Civil Engineering, Aristotle University, GR-54006, Thessaloniki, Greece; Faculty of Mathematics and Physics, RWTH, D-52062 Aachen, Germany

Abstract: The aim of our study is to develop a critical point theory for the functionals of the form

$$I = \Phi + \Psi$$

where Φ is locally Lipschitz and Ψ is proper, convex and lower semicontinuous. This approach extends and unifies the works of K. C. Chang [1] and A. Szulkin [2]. The applications deal with nonlinear elliptic boundary value problems with discontinuities.
Key Words and Phrases: *Critical point, weak solution.*

Introduction

The concrete problems in unilateral Mechanics, which involve both monotone and nonmonotone boundary conditions, deal with nonsmooth and nonconvex energy functionals expressed as the sum of a locally Lipschitz function $\Phi : X \to \mathbb{R}$ and a proper, convex and lower semicontinuous function $\Psi : X \to \mathbb{R} \cup \{+\infty\}$. Various examples in Mechanics leading to such problems are described in the book of Panagiotopoulos [3] and the one of Naniewicz and Panagiotopoulos [4].

If $\Phi \in C^1(X;\mathbb{R})$, this model represents an efficient critical point theory in the sense of Szulkin [2]. If $\Psi = 0$, it reduces to the one studied by Chang [1]. The theory of Chang has now been used in several directions by Goeleven, Motreanu and Panagiotopoulos [5] to study the qualitative properties of the solutions of hemivariational inequalities arising in unilateral Mechanics. If $\Phi \in C^1(X;\mathbb{R})$ and $\Psi = 0$ the problem reduces to the classical critical point theory for Φ (see Ambrosetti and Rabinowitz [6]).

The aim of this paper is to consider the general problem formulated above. It combines the approach of Chang with the one of Szulkin in a nontrivial way. Indeed the two mentioned approaches are completely different and absolutely not "additive". We derive a general minimax principle with linking which is then used to generalize the well-known Mountain-Pass Theorem. Using the resulting theory, we study a nonsmooth unilateral elliptic problem.

1. The General Setting

The aim of this Section is to set up the framework of our critical point theory. Namely, we deal with lower semicontinuous functionals (in short, l.s.c.) $I : X \to (-\infty, +\infty]$ on a real Banach space X satisfying the following structure hypothesis

(H) $I = \Phi + \Psi$, with $\Phi : X \to \mathbb{R}$ locally Lipschitz and $\Psi : X \to (-\infty, +\infty]$ proper (i.e., $\not\equiv +\infty$), convex and l.s.c..

Assumption (H) extends the corresponding statement of Szulkin [2] where Φ is supposed to be continuously differentiable as well as the theory of Chang [1] which is obtained for $\Psi = 0$.

In order to develop a critical point theory for functionals of type (H), we need to introduce the basic concepts of critical point and Palais-Smale condition.

Definition 1.1 An element $u \in X$ is called a critical point of functional I in (H) if

$$\Phi^0(u; v-u) + \Psi(v) - \Psi(u) \geq 0, \quad \forall\, v \in X.$$

Definition 1.2 The functional $I : X \to (-\infty, +\infty]$ in (H) is said to satisfy the Palais - Smale condition at the level $c \in \mathbb{R}$ (in short, $(\mathrm{PS})_c$) if every sequence $\{u_n\} \subset X$ satisfying $I(u_n) \to c$ and

$$\Phi^0(u_n; v-u_n) + \Psi(v) - \Psi(u_n) \geq -\varepsilon_n \|v - u_n\|, \quad \forall\, v \in X,$$

for a sequence $\{\varepsilon_n\} \subset \mathbb{R}^+$ with $\varepsilon_n \to 0$, contains a convergent subsequence. If $(\mathrm{PS})_c$ is verified for all $c \in \mathbb{R}$, I is said to satisfy the Palais - Smale condition (in short, (PS)).

The notation Φ^0 stands for the generalized directional derivative of Clarke [7], namely

$$\Phi^0(u; v) = \limsup_{w \to u, t \downarrow 0} \frac{1}{t}(\Phi(w + tv) - \Phi(w))$$

for all $u, v \in X$.

For the functional $I : X \to (-\infty, +\infty]$ in (H) we denote by $K(I)$ the set of critical points of I in the sense of Definition 1.1. Given any number $c \in \mathbb{R}$ let us set

$$K_c(I) := K(I) \cap I^{-1}(c).$$

For a latter use we put

$$I_a := \{v \in X : I(v) \leq a\}, \quad \forall\, a \in \mathbb{R}.$$

Following Szulkin [2] the result below points out some relevant aspects of Definition 1.2.

Proposition 1.1 (i) Any limit point u of a (Palais - Smale) sequence $\{u_n\}$ entering Definition 1.2 belongs to $K_c(I)$.

(ii) The inequality in Definition 1.2 is equivalent to

$$\Phi^0(u_n; v-u_n) + \Psi(v) - \Psi(u_n) \geq \langle z_n, v - u_n \rangle, \quad \forall\, v \in X,$$

for some sequence $\{z_n\} \subset X^*$ with $z_n \to 0$.

Proof. (i) The assertion follows from the upper semicontinuity of Φ^0 (see Clarke [7]).

(ii) The argument relies on Lemma 1.3 of Szulkin [2].

2. A Deformation Result

The aim of this Section is to extend Szulkin's deformation result ([2], Proposition 2.3) to the functionals of type (H). The proof of our deformation result is based on the next two preliminary properties.

Lemma 2.1 Let the functional $I : X \to (-\infty, +\infty]$ satisfy the conditions (H) and $(PS)_c$ for some $c \in \mathbb{R}$ and let U be a neighborhood of $K_c(I)$. Then, for each $\bar{\varepsilon} > 0$, there is an $\varepsilon \in (0, \bar{\varepsilon})$ such that, corresponding to every $u_0 \in I^{-1}([c-\varepsilon, c+\varepsilon])\backslash U$, there exists an $v_0 \in X$ satisfying

$$\Phi^0(u_0; v_0 - u_0) + \Psi(v_0) - \Psi(u_0) < -3\varepsilon\|v_0 - u_0\| .$$

Proof. One argues by contradiction on the basis of condition $(PS)_c$.

Lemma 2.2 Under the assumptions of Lemma 2.1 let an $\varepsilon \in (0, \bar{\varepsilon})$ be fixed as there stated. Then for each $u_0 \in I_{c+\varepsilon}\backslash U$, there exist $v_0 \in X$ and a neighborhood U_0 of u_0 in X such that Φ is Lipschitz continuous on U_0 of Lipschitz constant $K > 0$ and the relations below hold:

$$\Phi^0(u; v_0 - u) + \Psi(v_0) - \Psi(w) \le K(\|u - v_0\| + \|w - v_0\|), \ \forall\, u, w \in U_0,$$

and

$$\Phi^0(u; v_0 - u) + \Psi(v_0) - \Psi(w) \le -3\varepsilon\|v_0 - w\|, \quad \forall\, u, w \in U_0 \ \text{with} \ I(w) \ge c - \varepsilon.$$

If $u_o \in K(I)$ one can take $v_0 = u_0$. Otherwise the data v_0, U_0 and a number $\delta_0 > 0$ can be chosen so that $v_0 \notin \overline{U}_0$ and

$$\Phi^0(u; v_0 - u) + \Psi(v_0) - \Psi(w) \le -\delta_0\|v_0 - w\|, \quad \forall\, u, w \in U_0.$$

Proof. One discusses separately the cases $u_0 \in K(I)$ and $u_0 \notin K(I)$ making use essentially of Proposition 1.3 and Lemma 2.1.

Now we are in a position to formulate our deformation result for nonsmooth functionals of form (H).

Theorem 2.1 Let the functional $I : X \to (-\infty, +\infty]$ and the number $c \in \mathbb{R}$ be such that conditions (H) and $(PS)_c$ are satisfied. Let U be a neighborhood of $K_c(I)$ and let $\bar{\varepsilon} > 0$ be a fixed number. Then there exists $\varepsilon \in (0, \bar{\varepsilon})$ such that for every compact subset $A \subset X\backslash U$ with

$$c \le \sup_A I \le c + \varepsilon$$

there exist a closed neighborhood W of A in X and a deformation $\alpha : W \times [0, \bar{s}] \to X$, $\bar{s} > 0$, i.e. a continuous map with $\alpha(\cdot, 0) = id_W$, satisfying

$$\|u - \alpha(u, s)\| \le s, \quad \forall\, u \in W, \quad \forall\, s \in [0, \bar{s}]; \tag{1}$$

$$I(\alpha(u, s)) - I(u) \le Ms, \quad \forall\, u \in W, \quad \forall\, s \in [0, \bar{s}], \tag{2}$$

with a constant $M > 0$ independent of u and s;

$$I(\alpha(u, s)) - I(u) \le -2\varepsilon s, \tag{3}$$

for all $u \in W$ with $I(u) \ge c - \varepsilon$, $\forall\, s \in [0, \bar{s}]$;

$$\sup_{u \in A} I(\alpha(u, s)) - \sup_{u \in A} I(u) \le -2\varepsilon s, \quad \forall\, s \in [0, \bar{s}]. \tag{4}$$

Furthermore, if W_0 is a closed subset of X with $W_0 \cap K(I) = \emptyset$, W and α can be chosen so that

$$I(\alpha(u,s)) - I(u) \le 0, \quad \forall\, u \in W \cap W_0, \quad \forall\, s \in [0, \bar{s}]. \tag{5}$$

Proof. Consider an $\varepsilon \in (0, \bar{\varepsilon})$ determined by Lemma 2.2. Then, by Lemma 2.2, for each point $u_0 \in A$ there exists an open neighborhood U_0 of u_0 in X and some $v_0 \in X$ verifying the properties (1)-(5). In addition, we can suppose

$$u_0 \in K(I) \Longrightarrow U_0 \cap W_0 = \emptyset.$$

Take also another neighborhood U_0' of u_0 with $\overline{U}_0' \subset U_0$.

The compactness of A insures that there exists finite open coverings $\{U_i\}_{1 \le i \le m}$ and $\{U_i'\}_{1 \le i \le m}$ of A in X, where every pair of sets U_i, U_i' corresponds to a point $u_i \in A$ in the same way as does U_0, U_0' for u_0 above and $\overline{U}_i' \subset U_i$. Let v_i denote the point corresponding to u_i as v_0 was obtained by means of u_0. Taking, if necessary, a finer covering of A in X we may assume

$$u_i \in K(I) \Longrightarrow \text{dist}\,\{u_i, U_j\} > 0, \quad \forall\, j \ne i\,.$$

We introduce

$$V = \cup_{i=1}^{m} U_i$$

and, for each $i = 1, ..., m$, $\sigma_i : V \to I\!R$ by

$$\sigma_i = \rho_i / \sum_{j=1}^{m} \rho_j\,,$$

where $\rho_i : X \to I\!R$ is given by

$$\rho_i(x) = \text{dist}\,\{x, X \backslash U_i\}, \quad \forall\, x \in X.$$

Now, we define the required deformation $\alpha : V \times [0, +\infty) \to X$ by the following explicit formula. If $u \in U_{i_0} \backslash \cup_{j \ne i_0} U_j$ with $u_{i_0} \in A \cap K(I)$ we put

$$\alpha(u,s) := \begin{cases} u + s(u_{i_0} - u)/\|u_{i_0} - u\|, & 0 \le s < \|u_{i_0} - u\| \\ u_{i_0}, & s \ge \|u_{i_0} - u\| \end{cases}$$

and for all other $u \in V$

$$\alpha(u,s) = u + s \sum_i \sigma_i(u)(v_i - u)/\|v_i - u\|\,.$$

One can check that there exists a number $\bar{s} > 0$ such that $\alpha : V \times [0, \bar{s}] \to X$ is well defined and continuous. Then one verifies that the relations (1)-(5) hold.

For nonsmooth problems with symmetries it is very useful to have versions of Theorem 2.1 in the case of symmetric functionals satisfying assumption (H). The corollary below deals with such a case, namely nonsmooth even functionals.

Corollary 2.1 Assume in additon to the requirements of Theorem 2.1 that the functionals Φ and Ψ are even, i.e., $\Phi(u) = \Phi(-u)$ and $\Psi(u) = \Psi(-u)$ for all $u \in X$. If the set A is symmetric with respect to the origin, i.e., $A = -A$, then W can be chosen to be symmetric with respect to the origin and the deformation $\alpha : W \times [0, \bar{s}] \to X$ can be constructed so that $\alpha(\cdot, s) : W \to X$ is odd for each $s \in [0, \bar{s}]$, i.e., $\alpha(-u, s) = -\alpha(u, s)$, $\forall\, u \in W$.

Proof. Since I is even, its critical points arise in symmetric pairs $(u, -u)$. Thus, in view of the symmetry of A, the elements U_i and U_i' of the open coverings considered in the proof of Theorem 2.1 can be taken pairwise symmetric with respect to the origin. Then, starting with a deformation α as given by Theorem 2.1, we define $\beta : W \times [0, \bar{s}] \to X$ as

$$\beta(u, s) = \frac{1}{2}(\alpha(u, s) - \alpha(-u, s)), \quad \forall\ (u, s) \in W \times [0, \bar{s}]\ .$$

Then one can follow an analogous reasoning to the one developed in the proof of Theorem 2.1 to complete the proof.

Remark 2.2 Following the pattern of Corollary 2.1, under appropriate conditions, we can construct deformations $\alpha(u, s)$ with properties (1)-(5) such that each mapping $\alpha(\cdot, s)$ is equivariant with respect to the action of a topological group G on X, i.e. $\alpha(gu, s) = g\alpha(u, s)$, $\forall u \in X$, $\forall g \in G$. The situation treated in Corollary 2.1 corresponds to the group $G = \{id_X, -id_X\}$.

3. Nonsmooth Minimax Principles

The deformation result in Theorem 2.1 is now applied to provide a general minimax principle for the nonsmooth functionals verifying hypothesis (H). We recall the definition of linking.

Definition 3.1 Let S be a nonempty subset of the Banach space X and let Q be a compact topological submanifold of X with nonempty boundary ∂Q (in the sense of manifolds with boundary). We say that S and Q link if the next properties hold

$$S \cap \partial Q = \emptyset$$

and

$$f(Q) \cap S \neq \emptyset$$

whenever $f \in \Gamma$, where

$$\Gamma := \{f \in C(Q, X) :\ f|_{\partial Q} = id_{\partial Q}\}\ .$$

The theorem below is our main result of this Section. It extends and unifies various nonsmooth (and a fortiori smooth) minimax principles.

Theorem 3.1 Let the functional $I : X \to (-\infty, +\infty]$ on the Banach space X satisfy assumptions (H) and (PS). Let S and Q link in the sense of Definition 3.1. Assume further that

$$\sup_Q I \in I\!R, \quad b := \inf_S I \in I\!R, \quad a := \sup_{\partial Q} I < b\ .$$

Then the number

$$c := \inf_{f \in \Gamma} \sup_{x \in Q} I(f(x)),$$

where

$$\Gamma = \{f \in C(Q, X) :\ f|_{\partial Q} = id_{\partial Q}\}\ ,$$

is a critical value of I with

$$c \geq b\ .$$

This means in particular that $K_c(I)$ is nonempty.

Proof. According to the linking property of Definition 3.1 we know that $f(Q) \cap S \neq \emptyset$ for all $f \in \Gamma$. Hence $c \geq b$.

We have to show that c is a critical value of I. Arguing by contradiction we assume that $K_c(I) = \emptyset$. Then we can take $U = \emptyset$ in Theorem 2.1. Applying Theorem 2.1 with $\bar{\varepsilon} = c - a$, we obtain an $\varepsilon \in (0, \bar{\varepsilon})$.

The class Γ does not generally satisfy the property $\alpha(f(\cdot), s) \in \Gamma$ whenever $f \in \Gamma$, where α is the deformation in Theorem 2.1. Therefore it is necessary to enlarge Γ to fulfil the foregoing requirement. Let us set

$$\Gamma_1 := \{f \in C(Q, X) : \ f|_{\partial Q} \text{ and } id_{\partial Q} \text{ are homotopic as}$$

$$\text{maps from } \partial Q \text{ into } I_{c-\frac{\varepsilon}{4}} \text{ and } f(\partial Q) \subset I_{c-\frac{\varepsilon}{2}}\}.$$

Corresponding to the family Γ_1 we introduce the minimax value

$$c_1 = \inf_{f \in \Gamma_1} \sup_{x \in Q} I(f(x)) \ . \tag{6}$$

One verifies that

$$c = c_1 \ . \tag{7}$$

The set Γ_1 is a closed subset of the Banach space $C(Q, X)$ with respect to the usual uniform norm

$$\|f\| = \sup_{x \in Q} \|f(x)\|, \quad \forall \ f \in C(Q, X).$$

Let us introduce the functional $\Pi : C(Q, X) \to (-\infty, +\infty]$ by

$$\Pi(f) \doteq \sup_{x \in Q} I(f(x)), \quad \forall \ f \in C(Q, X) \ .$$

It is straightforward to see that Π is lower semicontinuous.

We apply to $\Pi : \Gamma_1 \to (-\infty, +\infty]$ Ekeland's variational principle (see Ekeland [8]) which, taking into account relations (6), (7), gives rise to some $f \in \Gamma_1$ satisfying

$$c \leq \Pi(f) \leq c + \varepsilon$$

and

$$\Pi(g) - \Pi(f) \geq -\varepsilon\|g - f\|, \quad \forall \ g \in \Gamma_1 \ .$$

Now we invoke Theorem 2.1 that provides the deformation $\alpha : W \times [0, \bar{s}] \to X$ corresponding to the compact set $A = f(Q)$. Now we can show that for $\bar{s} > 0$ small enough we have that

$$\alpha(f(\cdot), s) \in \Gamma_1, \quad \forall \ s \in [0, \bar{s}].$$

Then we arrived at a contradiction which proves that our initial assumption that c is not a critical value of I is false. This completes the proof of Theorem 3.1.

Corresponding to different situations of linking in Definition 3.1, Theorem 3.1 implies useful minimax results for functionals satisfying hypothesis (H). Here we point out only the appropriate version of Mountain Pass Theorem.

Corollary 3.1. Assume that the functional $I : X \to (-\infty, +\infty]$ on the Banach space X satisfies (H), (PS), $I(0) = 0$,

(i) there exist constants $\alpha > 0$ and $\rho > 0$ such that $I(u) \geq \alpha$ for all $\|u\| = \rho$;

(ii) there exists $e \in X$ with $\|e\| > \rho$ and $I(e) \leq 0$.

Then the number

$$c = \inf_{f \in \Gamma} \sup_{t \in [0,1]} I(f(t)),$$

where

$$\Gamma = \{f \in C([0,1], X) : f(0) = 0, f(1) = e\},$$

is a critical value of I with $c \geq \alpha$.

Proof. It is sufficient to apply Theorem 3.1 for $S = \{u \in X : \|u\| = \rho\}$ and $Q = [0; e] = \{te : t \in [0,1]\}$.

Remark 3.3 The results in this Section extend the minimax principles in Szulkin [2] to the case where the functional Φ of (H) is assumed to be locally Lipschitz instead continuously differentiable. We stress that we did not impose additional assumptions in comparison with those in Szulkin [2]. Another feature of our approach is that Theorem 3.1, on the basis of linking property of Definition 3.1, unifies various specific minimax results. If we take $\Psi = 0$, then we find the minimax results of Chang [1] for locally Lipschitz functionals (see Theorems 3.3, 3.4 in Chang [1]), again presented in a unifying manner by means of the linking concepts in Definition 3.1.

4. A Variational-Hemivariational Inequality

This Section is devoted to an application of our nonsmooth minimax principles.

Let Ω be a bounded domain in $\mathbb{R}^N$ with a C^1 boundary $\partial\Omega$. Consider in the Sobolev space $H_0^1(\Omega)$ the cone of nonnegative functions, i.e.,

$$K = \{u \in H_0^1(\Omega) : u(x) \geq 0 \text{ for a.e. } x \in \Omega\}.$$

Denote by $\Psi : H_0^1(\Omega) \to (-\infty, +\infty]$ the indicator function of K.

There are given $g \in L^2(\Omega)$ satisfying

$$g < 0 \quad \text{a.e. in } \Omega$$

and a (Carathéodory) function $j : \Omega \times \mathbb{R} \to \mathbb{R}$ such that $j(\cdot, y) : \Omega \to \mathbb{R}$ is measurable, $\forall\, y \in \mathbb{R}$, and $j(x, \cdot) : \mathbb{R} \to \mathbb{R}$ is locally Lipschitz, $\forall x \in \Omega$.

The function $j : \Omega \times \mathbb{R} \to \mathbb{R}$ is required to fulfil the conditions below

(j_1) $|\xi| \leq c(1 + |y|^{p-1})$, $\forall\, x \in \Omega$, $y \in \mathbb{R}$, $\xi \in \partial j(x, y)$, with constants $c > 0$ and $1 \leq p < \frac{2N}{N-2}$ if $N \geq 3$;

(j_2) $\liminf_{y \to 0} y^{-2} j(x, y) \geq 0$ uniformly with respect to $x \in \Omega$ and $j(x, 0) = 0$;

(j_3) $\mu^{-1} j^0(x, y; y) \leq j(x, y)$ for all $x \in \Omega$ and $y \in \mathbb{R}$, $y \geq 0$, where μ is a constant with $\mu > 2$;

(j_4) there is $e \in H_0^1(\Omega) \setminus \{0\}$ such that

$$\int_\Omega j(x, e(x))dx < 0.$$

Denote by λ_1 the first eigenvalue of $-\Delta$ on $H_0^1(\Omega)$.

Theorem 4.1 Under the assumptions (j_1)-(j_4) for $j : \Omega \times \mathbb{R} \to \mathbb{R}$, for every $\lambda < \lambda_1$ the variational-hemivariational inequality: find $u \in K$ such that

$$\int_\Omega \nabla u \cdot (\nabla v - \nabla u)dx + \int_\Omega j^0(x, u(x); v(x) - u(x))dx \geq$$

$$\geq \lambda \int_\Omega u(v - u)dx + \int_\Omega g(v - u)dx, \quad \forall\, v \in K,$$

has a nontrivial solution.

Proof. One applies Corollary 3.1 for $I = \Phi + \Psi : H_0^1(\Omega) \to (-\infty, +\infty]$, with $\Phi : H_0^1(\Omega) \to I\!R$ expressed by

$$\Phi(u) = \frac{1}{2}\|u\|_{H_0^1}^2 + \int_\Omega j(x, u(x))dx - \frac{\lambda}{2}\|u\|_{L^2}^2 - \int_\Omega gudx, \quad \forall\, u \in H_0^1(\Omega).$$

Remark 4.1 Two adhesively connected von Kármán plates subjected to elastoplastic boundary conditions, or to unilateral contact boundary conditions, or to friction boundary conditions lead to variational - hemivariational inequalities. The corresponding buckling problem leads to an eigenvalue problem of the type studied here.

Acknowledgement

The paper has been completed after the death of Professor P. D. Panagiotopoulos.

References

1. K. C. Chang, J. Math. Anal. Appl., **80**, 102–129 (1981).
2. A. Szulkin, Ann. Inst. Henri Poincaré, Analyse Non Linéaire, **3**, 77–109 (1986).
3. P. D. Panagiotopoulos, Hemivariational Inequalities. Applications in Mechanics and Engineering, Springer-Verlag, Berlin (1993).
4. Z. Naniewicz and P. D. Panagiotopoulos, Mathematical Theory of Hemivariational Inequalities and Applications, Marcel Dekker, Inc., New York (1995).
5. D. Goeleven, D. Motreanu and P.D.Panagiotopoulos, Nonlinear Anal., TMA, **20**, 9–16 (1997).
6. A. Ambrosetti and P. H. Rabinowitz, J. Functional Anal., **14**, 349–381 (1973).
7. F. H. Clarke, Optimization and Nonsmooth Analysis, John Wiley and Sons, New York (1983).
8. I. Ekeland, Bull. (New Series) Amer. Math., **1**, 443–474 (1979).

9th Int. Coll. on Differential Equations, pp. 289–292
D. Bainov (Ed)

Mean Value Properties for the Wave Equation

MASAHARU NISHIO
Department of Mathematics, Osaka City University, Osaka 558-8585, Japan

TAKAYUKI SUGIMOTO AND NORIAKI SUZUKI
Graduate School of Mathematics, Nagoya University, Nagoya 464-8602, Japan

Abstract: We prove a mean value property for solutions of the n-dimensional wave equation. Two converses of this result are also discussed.

Key Words and Phrases: *mean value property, wave equation.*

1. Introduction and Results

It is well known that every solution u of the one dimensional wave equation is represented as

$$u(x,t) = f(x+t) + g(x-t) \tag{1}$$

for some C^2-functions f and g. Observing this fact, we found the following mean value property: If a C^2-function $u(x,t)$ satisfies $u_{xx} = u_{tt}$ on a neighborhood of the rectangle $\{(x,t) \in \mathbf{R}^2; |x|+|t| \leq 1\}$, then

$$\int_{-1}^{1} u(x,0)\,dx = \iint_{|x|+|t|<1} u(x,t)\,dxdt. \tag{2}$$

The purpose of this note is to generalize (2) to the case of higher dimensions. We have the following theorem.

Theorem 1. *Let $n \geq 1$ and let $D = \{(X,t) \in \mathbf{R}^{n+1}; |X|+|t| < 1\}$, which we call a double characteristic cone. If a C^2-function u satisfies the n-dimensional wave equation*

$$\Delta u(X,t) = \frac{\partial^2}{\partial t^2} u(X,t) \tag{3}$$

on a neighborhood of $\overline{D}$, the closure of D, then we have

$$\frac{\omega_n}{2^n} \int_B u(X,0)(1-|X|^2)^{\frac{n-1}{2}}\,dX = \omega_{n-1} \iint_D u(X,t)\{(1-|X|)^2 - t^2\}^{\frac{n-1}{2}}\,dXdt, \tag{4}$$

where $B = \{X \in \mathbf{R}^n; |X| < 1\}$ denotes the unit ball and $\omega_n = \frac{2\pi^{n/2}}{n\Gamma(n/2)}$ is its volume ($\omega_0 := 1$).

Two converses of the above result are established; the equality (4) characterizes the solution of the wave equation and the double characteristic cone D. These are analogues of harmonic functions on a ball (cf. [1], see also [2]). To be exact, we prove the following.

Theorem 2. *Let Ω be a domain in $\mathbf{R}^{n+1}$. A C^2-function u satisfies the wave equation (3) on Ω if and only if u satisfies*

$$\frac{\omega_n r}{2^n}\int_{B(X_0,r)} u(X,t_0)(r^2-|X-X_0|^2)^{\frac{n-1}{2}}\,dX \tag{5}$$

$$= \omega_{n-1}\int\int_{D(X_0,t_0,r)} u(X,t)\{(r-|X-X_0|)^2-(t-t_0)^2\}^{\frac{n-1}{2}}\,dX\,dt$$

for all $(X_0,t_0)\in\Omega$ and $r>0$ such that $D(X_0,t_0,r)\subset\Omega$, where $D(X_0,t_0,r):=\{(X,t)\in\mathbf{R}^{n+1};|X-X_0|+|t-t_0|<r\}$ and $B(X_0,r):=\{X\in\mathbf{R}^n;|X-X_0|<r\}$.

Theorem 3. *Let Ω be a domain in $\mathbf{R}^{n+1}$ such that $\{(X,0)\in\mathbf{R}^{n+1};|X|<1\}\subset\Omega$. Assume that the interior of the closure of Ω coincides with Ω. If every solution u of the n-dimensional wave equation on $\overline{\Omega}$ satisfies*

$$\frac{\omega_n}{2^n}\int_B u(X,0)(1-|X|^2)^{\frac{n-1}{2}}\,dX=\omega_{n-1}\int\int_\Omega u(X,t)|(1-|X|)^2-t^2|^{\frac{n-1}{2}}\,dX\,dt, \tag{6}$$

then Ω is a double characteristic cone $\{(X,t)\in\mathbf{R}^n;|X|+|t|<1\}$.

The proof of Theorem 1 is based on the Ásgeirsson mean value theorem for the ultrahyperbolic equation. We shall recall it in section 2. The proofs of Theorems are given in section 3.

2. Ásgeirsson's Mean Value Theorem

We denote a point $(X,T)\in\mathbf{R}^{2n}$ as $X=(x_1,\cdots,x_n)\in\mathbf{R}^n$ and $T=(t_1,\cdots,t_n)\in\mathbf{R}^n$. We consider a solution U of the ultrahyperbolic equation

$$\Delta_X U(X,T)=\Delta_T U(X,T) \tag{7}$$

where $\Delta_X=\sum_{j=1}^n\partial^2/\partial x_j^2$ and $\Delta_T=\sum_{j=1}^n\partial^2/\partial t_j^2$.

In 1937, L. Ásgeirsson [3] proved the following mean value theorem.

Theorem A(cf. [4, p.747]). *Let $U(X,T)$ be a solution of the ultrahyperbolic equation* (7) *in a domain $\tilde\Omega\subset\mathbf{R}^{2n}$. If $\{(X,T)\in\mathbf{R}^{2n};|X-X_0|+|T-T_0|<r\}\subset\tilde\Omega$, then*

$$\int_{|X-X_0|<r}U(X,T_0)\,dX=\int_{|T-T_0|<r}U(X_0,T)\,dT. \tag{8}$$

By using Theorem A, we can obtain

Proposition 1. *Let $\tilde D=\{(X,T)\in\mathbf{R}^{2n};|X|+|T|<1\}$. If $U(X,T)$ satisfies* (7) *in a neighborhood of $\overline{\tilde D}$, then*

$$\frac{\omega_n}{2^n}\int_B u(X,0)(1-|X|^2)^{\frac{n-1}{2}}\,dX=\int\int_{\tilde D}u(X,T)\,dX\,dT. \tag{9}$$

Proof. Let $Y\in\mathbf{R}^n$ with $|Y|<1$. From (8),

$$\int_{|X-Y|<1-|Y|}U(X,0)\,dX=\int_{|T|<1-|Y|}U(Y,T)\,dT. \tag{10}$$

Integrating both side of (10) with respect to Y over $\{|Y|<1\}$, we obtain

$$\int_{|Y|<1}dY\int_{|X-Y|<1-|Y|}U(X,0)\,dX=\int_{|Y|<1}dY\int_{|T|<1-|Y|}U(Y,T)\,dT. \tag{11}$$

Of course, we may write the right hand of (11) as $\iint_{\tilde{D}} u(X,T)\,dX\,dT$. From the Fubini theorem, the left hand side of (11) is equal to

$$(12) \qquad \int_{|Y|+|X-Y|<1} dY \int_{|X|<1} U(X,0)\,dX.$$

Since $\{Y \in \mathbf{R}^n; |Y| + |X-Y| < 1\}$ is the ellipsoid with one major axis $1/2$ and other minor axies $\sqrt{1-|X|^2}/2$, so its volume is equal to $2^{-n}\omega_n(1-|X|^2)^{(n-1)/2}$. Hence (9) follows from (11) and (12). This completes the proof.

3. Proof of Theorems

Proof of Theorem 1. This follows from Proposition 1 easily. In fact, let $u(X,t)$ be a solution of the wave equation (3) on some neighborhood of $\overline{D}$. Then the function $U(X,T)$ defined by

$$U(X,T) = u(X,t_1), \quad \text{where } T = (t_1,\cdots,t_n)$$

satisfies (7) on a neighborhood of $\overline{\tilde{D}}$. Therefore by (9)

$$\begin{aligned}
\frac{\omega_n}{2^n}\int_B u(X,0)(1-|X|^2)^{\frac{n-1}{2}}\,dX &= \frac{\omega_n}{2^n}\int_{|X|<1} U(X,0)(1-|X|^2)^{\frac{n-1}{2}}\,dX \\
&= \iint_{|X|+|T|<1} U(X,T)\,dX\,dT \\
&= \int_{t_2^2+\cdots+t_n^2<(1-|X|)^2-t^2} dt_2\cdots dt_n \iint_{|X|+|t|<1} u(X,t)\,dX\,dt \\
&= \omega_{n-1}\iint_D u(X,t)\{(1-|X|)^2-t^2\}^{\frac{n-1}{2}}\,dX\,dt.
\end{aligned}$$

This completes the proof.

Proof of Theorem 2. The only if part follows from (4) easily by the aide of translation and dilation. To show if part, let u be a C^2-function on Ω satisfying (5). We shall show $\Delta u(X_0,t_0) = \partial^2 u/\partial t^2(X_0,t_0)$. We may assume $(X_0,t_0) = (0,0)$. Using Taylor formula,

$$u(X,t) = a + \sum_{j=1}^n a_j x_j + bt + \frac{1}{2}\{\sum_{i,j=1}^n a_{ij}x_ix_j + 2\sum_j b_j x_j t + ct^2\} + o((|X|+|t|)^2)$$

where

$$a := u(0,0),\ a_j := \frac{\partial u}{\partial x_j}(0,0),\ b := \frac{\partial u}{\partial t}(0,0)$$

$$a_{ij} := \frac{\partial^2 u}{\partial x_i \partial x_j}(0,0),\ b_j := \frac{\partial^2 u}{\partial x_j \partial t}(0,0),\ c := \frac{\partial^2 u}{\partial t^2}(0,0)$$

we obtain from (5) that

$$(13) \qquad \frac{\omega_n r}{2^n}\int_{B(0,r)}\{a + \frac{1}{2}\sum_{j=1}^n a_{jj}x_j^2\}(r^2-|X|^2)^{\frac{n-1}{2}}\,dX$$

$$= \omega_{n-1}\iint_{D(0,0,r)}\{a + \frac{1}{2}\{\sum_{j=1}^n a_{jj}x_j^2 + ct^2\}\}\{(r-|X|)^2-t^2\}^{\frac{n-1}{2}}\,dX\,dt + o(r^{2n+2}).$$

On the other hand, since the function $v(X,t) := a + \sum_{j=1}^n a_{jj}(x_j^2+t^2)/2$ satisfies the wave equation, (5) again gives

$$(14) \qquad \frac{\omega_n r}{2^n}\int_{B(0,r)}\{a + \frac{1}{2}\sum_{j=1}^n a_{jj}x_j^2\}(r^2-|X|^2)^{\frac{n-1}{2}}\,dX$$

$$= \omega_{n-1} \int\int_{D(0,0,r)} \{a + \frac{1}{2}\{\sum_{j=1}^{n} a_{jj}(x_j^2 + t^2)\}\}\{(r - |X|)^2 - t^2\}^{\frac{n-1}{2}}\, dX\, dt.$$

From (13) and (14), it follows

$$(c - \sum_{j=1}^{n} a_{jj}) r^{2n+2} \int\int_{|Y|+|s|<1} s^2\{(1 - |Y|)^2 - s^2\}^{\frac{(n-1)}{2}}\, dY\, ds = o(r^{2n+2}).$$

This shows $\sum_j^n a_{jj} = c$, which implies $\Delta u(0,0) = \partial^2 u/\partial t^2(0,0)$, as desired. This completes the proof.

Proof of Theorem 3. Let D be a double characteristic cone as in Theorem 1 and Ω be a domain in $\mathbf{R}^{n+1}$ with $B \times \{0\} \subset \Omega$ and satisfying (6). To show $\Omega \subset D$, we contrary suppose $\Omega \setminus \overline{D} \neq \emptyset$. Then there are point (X_1, t_1) in Ω and positive number η such that $|X_1| \neq 0, t_1 \neq 0$ and

$$D_\eta := \{(X,t) \in \mathbf{R}^{n+1}; |X - X_1| + |t - t_1| < \eta\} \subset \Omega \setminus \overline{D}.$$

Since $|X_1| + |t_1| > 1$, we may also assume

$$\frac{X_1 \cdot X}{|X_1|} + \frac{t_1 t}{|t_1|} > 1 \quad \text{for } \forall (X,t) \in D_\eta \tag{15}$$

where $X_1 \cdot X$ denotes the inner product.

Now take a C^2-function f on $\mathbf{R}$ such that $f(x) > 0$ if $x > 1$ and $f(x) = 0$ if $x \leq 1$ and define the function v by

$$v(X,t) = f(\frac{X_1 \cdot X}{|X_1|} + \frac{t_1 t}{|t_1|}).$$

Then clearly v satisfies the wave equation on $\mathbf{R}^{n+1}$, $v \geq 0$ and

$$v(X,0) = 0 \quad \text{on } B, \quad v(X,t) > 0 \quad \text{on } D_\eta. \tag{16}$$

Hence by (6) and (16) we have

$$\begin{aligned} 0 &= \frac{\omega_n}{2^n} \int_B v(X,0)(1 - |X|^2)^{\frac{n-1}{2}}\, dX = \omega_{n-1} \int\int_\Omega v(X,t)|(1 - |X|)^2 - t^2|^{\frac{n-1}{2}}\, dX\, dt \\ &\geq \omega_{n-1} \int\int_{D_\eta} v(X,t)|(1 - |X|)^2 - t^2|^{\frac{n-1}{2}}\, dX\, dt > 0. \end{aligned}$$

This contradiction leads to $\Omega \setminus \overline{D} = \emptyset$, and hence $\Omega \subset D$. Moreover, since $u \equiv 1$ is a solution of the wave equation, (4) and (6) give

$$\int\int_\Omega |(1 - |X|)^2 - t^2|^{\frac{n-1}{2}}\, dX\, dt = \int\int_D |(1 - |X|)^2 - t^2|^{\frac{n-1}{2}}\, dX\, dt.$$

This imples $D \setminus \overline{\Omega} = \emptyset$. Thus $\Omega = D$ follows from the assumption that Ω is equal to the interior of $\overline{\Omega}$. This completes the proof.

References

1. I. Netuka and J. Veselý, in: Classical and Modern Potential Theory and Applications, K. GowriSankran et al. (Eds.) Kluwer Academic Publishers (1994) pp. 359–398.

2. S. C. Çhu, Amer. Math. Monthly, **74**, 711–713 (1967).

3. L. Ásgeirsson, Math. Ann., **113**, 321–346 (1937).

4. R. Courant and D. Hilbert, Methods of Mathematical Physics, vol. II, Interscience Pub., New York (1966).

9th Int. Coll. on Differential Equations, pp. 293-298
D. Bainov (Ed)

A Note on the Solutions of $r''(x)r(x) = g(x)r'^2(x)$

EDWARD OMEY
EHSAL , Stormstraat 2, 1000 Brussels, Belgium,
e-mail: omedward@prof.ehsal.be

Abstract. In this paper we study the asymptotic behaviour of nonnegative solutions of the second order differential equation $r''(x)r(x) = g(x)r'^2(x)$.

In section 2 we first consider the case where $g(x) \to c$ where $-\infty \le c \le +\infty$ and obtain some first order results. In section 3 we take into account the rate at which $g(x)$ converges to c and we obtain several second-order results.

1. INTRODUCTION

A positive and measurable function $f: \mathbb{R}^+ \to \mathbb{R}^+$ is called regularly varying with (real) index α, if $f(x)$ satisfies the following asymptotic relation: as $x \to \infty$,

$$\frac{f(tx)}{f(x)} \to t^\alpha, \quad \text{for all} \quad t > 0. \tag{1.1}$$

Notation: $f \in RV(\alpha)$.

Functions satisfying (1.1) have proved to be useful in a variety of asymptotic problems concerning the limit behaviour of functions and processes. Basic properties can be found in [1] and [2].

If the index $\alpha > -1$ and if f is integrable on finite intervals one can prove that (1.1) implies

$$\frac{\int_0^x f(t)dt}{xf(x)} \to \frac{1}{\alpha+1} \tag{1.2}$$

and

$$\frac{\int_0^x \int_0^t f(s)dsdt}{x\int_0^x f(s)ds} \to \frac{1}{\alpha+2}. \tag{1.3}$$

By setting $r(x) = \int_0^x \int_0^t f(s)dsdt$ and combining (1.1), (1.2) and (1.3) we find that $r(x)$ satisfies the following differential equation:

$$r(x)r''(x) = g(x)r'^2(x). \tag{1.4}$$

In (1.4), $g(x)$ is a function satisfying $g(x) \to \frac{\alpha+1}{\alpha+2}$.

In this note we start from (1.4) and study the asymptotic behaviour of nonnegative solutions of (1.4) under various conditions on $g(x)$.

Equation (1.4) appears among others in the context of extreme value theory [3]. In that case $r''(x)$ denotes the tail of a probability distribution and (1.4) then characterizes the domain of attraction of the three classical max-stable distributions. Equation (1.4) also appears in the context of positive solutions of nonlinear elliptic equations. In [4], the author takes $r(x) = \int_x^\infty \sqrt{f(t)}dt$ and uses the class of functions $f(x)$ for which (1.4) holds with $g(x) \to L$, a constant.

In what follows we assume that all appearing functions are measurable; limit expressions of the form "$\to$" or "$\sim$" are expressions with respect to a variable tending to infinity. The quantity x° denotes a sufficiently large value of x.

In order to formulate our results we shall need several classes of functions related to RV.

Definitions. [1, 5, 6]

(i) A positive function $a(x): \mathbb{R} \to \mathbb{R}^+$ is called Beurling-varying (=: BV) if it satisfies

(1.5) $$\frac{a(x)}{x} \to 0, \qquad \text{and}$$

(1.6) $$\frac{a(x+ya(x))}{a(x)} \to 1, \quad \text{for all real } y.$$

It is called self-neglecting (=: SN) if (1.5) holds and if (1.6) holds locally uniformly (l.u.) in y.

(ii) Let $a(x) \in SN$ and c a real number. A positive function $f(x): \mathbb{R}^+ \to \mathbb{R}^+$ is in the class $\Gamma(c, a(x))$ if

$$\frac{f(x+ya(x))}{f(x)} \to \exp(cy), \qquad \text{for all real } y.$$

The function $f(x)$ is in the class $\Gamma(a(x))$ if $c = 1$.

(iii) Let $a(x) \in RV(0)$. A positive function $f(x): \mathbb{R}^+ \to \mathbb{R}^+$ is in the class $PI(a(x))$ if for all positive y,

$$\frac{(f(xy) - f(x))}{a(x)} \to \log(y).$$

All these classes have been useful in a variety of problems in analysis and probability theory, cf. [1, 2, 6, 7, 8] and the references given there.

2. BEHAVIOUR OF $r(x)$ IF $g(x) \to c$

In this section we study the asymptotic behaviour of nonnegative solutions of

(2.1) $$r(x)r''(x) = g(x)r'^2(x)$$

where

(2.2) $$g(x) \to c, \qquad -\infty \le c \le +\infty.$$

Our first result is the following

Proposition 2.1. *Suppose that $r(x)$ is a nonnegative solution of (2.1) and suppose that (2.2) holds.*

(i) If $c < 1$ or $c > 1$, then $r(x) \in RV(\beta)$ where $\beta = 1/(1-c)$ (here $\beta = 0$ if $c = +\infty$ or $-\infty$).

(ii) If $c = 1$, then $r(x) \in \Gamma(c, a(x))$ where $a(x) = |r(x)/r'(x)| \in SN$ and $c = +1$ or -1 depending on the sign of $r'(x)$.

Proof. For further use we define $f(x) := r(x)/r'(x)$. Using (2.1) it follows immediately that

$$f'(x) = 1 - g(x). \tag{2.3}$$

Using (2.2) we obtain $f'(x) \to 1 - c$ and consequently also that $f(x)/x \to 1 - c$. If c is different from 1, the latter relation shows that $xr'(x)/r(x) \to \beta$ and result (i) follows.

If $c = 1$, then $f'(x) \to 0$ and $f(x)/x \to 0$. This implies that $x + yf(x) \to +\infty$ for all real y.

Now, let $e > 0$ and choose x° so that $|f'(x)| \le e$, $x \ge x^\circ$. For $x \ge x^\circ$ and real y we have

$$f(x + yf(x)) - f(x) = \int_x^{x+yf(x)} f'(u)du.$$

Hence, $|f(x + yf(x)) - f(x)| \le e|yf(x)|$. This implies that

$$\limsup \left| \frac{f(x + yf(x))}{f(x)} - 1 \right| \le e|y|.$$

Since e was arbitrary, it follows that $a(x) := |f(x)| \in SN$.

To finish the proof, note that (2.1), $r(x) \ge 0$ and $c = 1$ imply that $r''(x) > 0$, $x \ge x^\circ$ and hence that $r'(x)$ is nondecreasing. If $r'(x) > 0$, we integrate the relation $r'(x)/r(x) = 1/f(x)$ to obtain

$$r(x) = r(x^\circ) \exp \int_{x^\circ}^{x} \frac{1}{f(u)} du. \tag{2.4}$$

It follows from (2.4) that for $x \ge x^\circ$ and $y \ge 0$,

$$\frac{r(x + yf(x))}{r(x)} = \exp \int_0^y \frac{f(x)}{f(x + vf(x))} dv.$$

Using $f(x) \in SN$, it follows that $r(x + yf(x))/r(x) \to \exp(y)$ which is the defining property of the class Γ.

The case $y < 0$ and $r'(x) < 0$ can be treated in a similar way.

This proves the result. □

Corollary 2.2. *Under the conditions of Proposition 2.1:*

(i) If $c < 0$, then $r'(x) \in RV(\beta - 1)$ and $-r''(x) \in RV(\beta - 2)$.

(ii) If $c = 0$, then $r'(x) \in RV(0)$.

(iii) If $0 < c < 1$, then $r'(x) \in RV(\beta - 1)$ and $r''(x) \in RV(\beta - 2)$.

(iv) If $c > 1$, then $-r'(x) \in RV(\beta - 1)$ and $r''(x) \in RV(\beta - 2)$.

Proof. In Proposition 2.1 we showed that $xr'(x)/r(x) \to \beta$. If $c < 1$, it follows that $r'(x) > 0$, $x \geq x^\circ$, and using (2.1) we obtain

$$\frac{xr''(x)}{r'(x)} \to \beta c. \tag{2.5}$$

From (2.5), cases (i) and (iii) follow.

If $c > 1$, then we have $r'(x) < 0$, $x \geq x^\circ$, and using (2.1) we obtain

$$\frac{-xr''(x)}{r'(x)} \to \beta c. \tag{2.6}$$

From (2.6), case (iv) follows. Finally, if $c = 0$, (2.3) shows that $f(x)/x \to 1$ so that $r'(x) > 0$, $x \geq x^\circ$. Now (ii) follows from (2.4). □

3. SECOND-ORDER BEHAVIOUR

We can obtain more refined results if we assume more about the rate at which $g(x)$ converges to c. We consider several cases.

Proposition 3.1. (the case $c = +\infty$) *Assume that $g(x) \to +\infty$ and that $g(x) \in RV(\alpha)$, $\alpha \geq 0$. Then $r(x) \in RV(0)$, $-r'(x) \in RV(-\alpha-1)$ and $r''(x) \in RV(-\alpha-2)$.*

Moreover,

(i) if $\alpha > 0$, then $r(x) \to a$ where $0 < a < \infty$ and

$$g(x)(a - r(x)) \to \frac{a(\alpha+1)}{\alpha},$$

(ii) if $\alpha = 0$, then $1/r(x) \in PI(a(x))$ where

$$a(x) = \frac{1}{r(x)g(x)}.$$

Proof. Using (2.3) and $g(x) \to +\infty$ we have $f'(x) \sim -g(x)$. Since $g(x) \in RV(\alpha)$ this implies that $f(x) \sim -xg(x)/(\alpha+1)$ and consequently that

$$\frac{r'(x)}{r(x)} \sim -(\alpha+1)/xg(x). \tag{3.1}$$

Since $r'(x) < 0$ for $x \geq x^\circ$, we know that $r(x) \to a$ for some a, $0 \leq a < \infty$.

(i) If $\alpha > 0$, we shall show that $a > 0$. To do this, by integrating relation (3.1) we obtain that

$$\log(r(x^\circ)) - \log(r(x)) \leq K \int_{x^\circ}^{x} \frac{1}{ug(u)} du$$

for some constant $K > 0$. Since $\alpha > 0$, the integral converges as $x \to +\infty$ and hence

$$\limsup \log(r(x^\circ)) - \log(r(x)) < +\infty.$$

It follows that $r(x) \to a$ where $0 < a < +\infty$. Using this in relation (3.1), we obtain $-r'(x) \sim a(\alpha+1)/xg(x) \in RV(-\alpha-1)$ and result (i) follows.

Using (2.1), we also have $r''(x) \sim a(\alpha+1)^2/x^2g(x) \in RV(-\alpha-2)$.

(ii) Next we consider $\alpha = 0$. It follows from (3.1) that

$$\left(\frac{1}{r(x)}\right)' \sim \frac{1}{xg(x)r(x)} \in RV(-1).$$

Hence $1/r(x) \in PI(1/(g(x)r(x)))$. Using (3.1) once more we also have $-r'(x) \in RV(-1)$ and from (2.1) we have $r''(x) \in RV(-2)$.

This proves the result. □

In our next result we consider the case $c = 1$. For convenience, we only prove a result for the case where $r'(x) > 0$. A similar result holds when $r'(x) < 0$.

Proposition 3.2. (the case $c = 1$ and $r'(x) > 0$) *Assume that $g(x) \to 1$ and that $h(x) := 1 - g(x) \in RV(\alpha)$, $\alpha \leq 0$. Then $r(x) \in \Gamma(f(x))$ with remainder term as follows:*

$$\frac{r(x + yf(x))}{r(x)} - e^y \sim -e^y h(x)\frac{y^2}{2}.$$

Proof. Using (2.3) we have $f'(x) = h(x)$ and from (2.4) it follows that

$$\log\left(\frac{r(x + yf(x))}{r(x)}\right) = \int_0^y \frac{f(x)}{f(x + uf(x))}du.$$

From this it follows that

$$\log\left(\frac{r(x + yf(x))}{r(x)}\right) - y = -\int_0^y \int_0^u f'(x + wf(x))dw \frac{f(x)}{f(x + uf(x))}du.$$

Since $f'(x) = h(x) \in RV(\alpha)$, $\alpha \leq 0$, and since $f(x) \in SN$, it follows that $f(x + wf(x)) \sim f(x)$ l.u. in w and that $f'(x + wf(x)) \sim h(x)$ l.u. in w. Hence we obtain

$$\log\left(\frac{r(x + yf(x))}{r(x)}\right) - y \sim -h(x)\frac{y^2}{2}.$$

The result follows. □

Remark. Under the conditions of Proposition 3.2, one can often conclude that not only $r(x) \in \Gamma$ with remainder term, but also that $\log(r(x))$ is regularly varying. If $h(x) \in RV(\alpha)$ we can distinguish several cases. Since $f'(x) = h(x)$ we have:

(a) if $\alpha = 0$ we have $f(x) \sim xh(x) \in RV(1)$ and hence

$$\log(r(x)) \in PI(1/h(x));$$

(b) if $-1 < \alpha < 0$ we have $f(x) \sim xh(x)/(\alpha + 1)$ and hence

$$\log(r(x)) \sim \frac{\alpha + 1}{\alpha h(x)};$$

(c) if $\alpha < -1$, $f(x) \to A$ and $A - f(x) \sim xh(x)/(-\alpha - 1)$. Using $a = 1/A$ it follows that $r'(x)/r(x) - a \sim a^2 xh(x)/(-\alpha - 1)$.

If $\alpha > -2$, we obtain $\log(r(x)) = ax + O(1)x^2h(x)$;
if $\alpha < -2$, we obtain $\log(r(x)) = ax + b + O(1)x^2h(x)$;
if $\alpha = -2$, we obtain $\log(r(x)) - ax \in PI(a^2x^2h(x))$.

In our final result we consider the case $c = 0$.

Proposition 3.3. (the case $c = 0$) *Assume that $g(x) \to 0$ and that $g(x) \in RV(\alpha)$, $\alpha \leq 0$. Then $r(x) \in RV(1)$ and $r''(x) \in RV(\alpha - 1)$. Moreover,*

(i) if $\alpha < 0$, then $r'(x) \to a$, $0 < a < \infty$ and

$$a - r'(x) \sim \frac{ag(x)}{-\alpha},$$

(ii) if $\alpha = 0$, then $r(x)/x \in PI(a(x))$ and $r'(x) \in PI(a(x))$ where $a(x) = r(x)g(x)/x$. Also $r'(x) - r(x)/x \sim a(x)$.

Proof. If $c = 0$, we proved that $r'(x) \in RV(0)$, $r(x) \in RV(1)$ and $f(x)/x \to 1$. Using this and (2.1) we obtain that

$$r''(x) \sim g(x)r(x)/x^2 \in RV(\alpha - 1). \tag{3.2}$$

Since $r''(x) > 0$ and $r'(x) > 0$, $x \geq x^\circ$, we see that $r'(x) \to a$ for some a where $0 < a \leq +\infty$. If $\alpha < 0$, it follows from (3.2) that $a < +\infty$ and consequently that

$$r''(x) \sim a\frac{g(x)}{x} \in RV(\alpha - 1) \tag{3.3}$$

and

$$a - r'(x) \sim a\frac{g(x)}{-\alpha} \in RV(\alpha). \tag{3.4}$$

If $\alpha = 0$, it follows from (3.2) that $r''(x) \in RV(-1)$ and this implies that $r'(x) \in PI(a(x))$ with $a(x) = r(x)g(x)/x$.

Now consider $w(x) := xr'(x) - r(x)$.

Using (2.1) and $xr'(x)/x \to 1$ we see that

$$w'(x) = \frac{xg(x)r'^2(x)}{r(x)} \sim a(x) \in RV(0).$$

Hence $w(x) \sim xa(x)$ and consequently $r'(x) - r(x)/x \sim a(x)$.

This shows that also $r(x)/x \in PI(a(x))$. □

Remark. By integrating (3.4) once again, we can obtain expressions of the form $r(x) - ax = b + O(1)xg(x)$ $(\alpha < -1)$, $r(x) - ax = O(1)xg(x)$ $(\alpha > 1)$, or $r(x) - ax \in PI(xg(x))$ $(\alpha = -1)$.

References

1. N.H. Bingham, C.M. Goldie and J.L. Teugels, Regular Variation, Cambridge University Press (1987).
2. J.L. Geluk and L. de Haan, Regular Variation, Extensions and Tauberian Theorems, CWI tract 40, M. C. Amsterdam (1987).
3. L. de Haan, On Regular Variation and its Applications to the Weak Convergence of Sample Extremes, CWI tracts, M. C. Amsterdam (1970).
4. S.D. Taliaferro, Radial Symmetry of Positive Solutions of Nonlinear Elliptic Equations, preprint, Dept. of Math., Texas A&M University, May 1997.
5. S. Bloom, Proc. Amer. Math. Soc., **54**, 243–250 (1976).
6. G.E. Peterson, J. London Math. Soc., **5**, 182–190 (1972).
7. M. Hacik and E. Omey, Publ. Inst. Math. N. S. (Béograd), **49**, 189–200 (1991).
8. E. Omey, in: Proc. 7th Int. Coll. on Differential Equations, D. Bainov, (Ed.), VSP, Utrecht (1997) 295–303.

9th Int. Coll. on Differential Equations, pp. 299–306
D. Bainov (Ed)

Properties of General Elementary Matrices in Mathematical Science

KUNIO OSHIMA
Science University of Tokyo in Yamaguchi, 1-1-1, Daigakudori, Onoda, Ymaguchi, Japan 759
ICHIRO HOFUKU
Tokyo Metropolitan College of Technology, 1-10-40, Higashiooi, Shinagawa-ku, Tokyo, Japan 140
SIGERU FURUICHI
Science University of Tokyo in Yamaguchi, 1-1-1, Daigakudori, Onoda, Ymaguchi, Japan 759

Abstract: The purpose of the paper is to introduce a unified treatment of General Elementary Matrices (abbreviate GEM) and to give an opportunity applying to mathematical science. We will first give a definition of GEM and then study those properties according to the classical structure of matrices. We next will define GEM as a linear transformation which maps a vector a to a vector b, and give several properties of the linear transformation.
Key Words and Phrases: *matrix, matrix structure*

1. Introduction

Let C be the complex field, $M_n(C)$ be the set of n-square complex matrices and C^n be the set of $n \times 1$ complex vectors.

Definition 1.1 *For non-zero scalars α and β in C and non-zero vectors x and y in C^n, a general elementary matrix (abbreviate GEM) is defined by*

$$E = \alpha I - \beta x y^H \qquad (1.1)$$

where I is the identity matrix.

The superscript H on the vector y indicates the conjugate transpose of y and if the superscript is T then this means that T is the transpose of y.

In this paper we first list several well-known results which we make use later. Then we will give algebraic structures, determinant, singularity (non-singularity), the minimal polynomial, eigenvectors, eigenvalue, and diagonalizabillity of GEM. Finally, we discuss some comments on GEM, particularly, in connection with a special linear transformation.

2. Previous Results

The following results and definition can be found in Householder [1], however we will list them for the sake of completeness.

Theorem 2.1 *A matrix B in $M_n(C)$ is of rank 1 if and only if $B = xy^H$ where x and y are vectors in C^n.*

Definition 2.2 *A matrix of the form $E(x, y; \beta) = I - \beta xy^H$ will be called an elmentary matrix [1].*

The set of all elementary matrices contains important non-singular ones which are frequently used in applied linear algebra. Some of these are the followings:

Elementary triangular matrix;

$$E(x_i, e_i; 1) \tag{2.1}$$

where $e_j^T x_i = 0$ whenever $j \leq i$.

Elementary permutation matrix;

$$E(e_i - e_j, e_i - e_j; 1). \tag{2.2}$$

Elementary Hermitian matrix;

$$E(x, x; 2), \ x^T x = 1. \tag{2.3}$$

Matrices (2.1) and (2.2) are used in the LU decomposition, and matrix (2.3) is used in the forming Hessenburg matrix for a given matrix A in $M_n(C)$. For further properties and applications of those matrices, see [1].

3. General Properties of GEMs

We will give the formal definition of a general elementary matrix and it's fundamental properties.

Definition 3.1 *For non-zero complex numbers α and β, and non-zero complex vectors x and y, a matrix of the form*

$$E = \alpha I - \beta xy^H \tag{3.1}$$

will be called a general elementary matrix, and we define the set

$$\mathcal{E} = \{\alpha I - \beta xy^H \epsilon \ M_n(C)\}. \tag{3.2}$$

Now for fixed vectors (but arbitrary vectors) u and v, we define the

$$\mathcal{RE} = \{\alpha I - \beta uv^H \cup I \cup \Theta\} \tag{3.3}$$

where I is the identity matrix and Θ is the zero matrix in $M_n(C)$.

Theorem 3.2 *The set $\mathcal{RE}$ forms a commutative ring with unity.*

Proof. Let $E_1 = \alpha_1 I - \beta_1 uv^H$ and $e_2 = \alpha_2 I - \beta_2 uv^H$, then

$$E_1 + E_2 = (\alpha_1 + \alpha_2)I - (\beta_1 + \beta_2)uv^H.$$

Since the set $\mathcal{RE}$ contains Θ and I, it is clear from matrix operations that the set $\mathcal{RE}$ satisfies all axioms to become a commutative ring with unity except the multiplication of two matrices are closed and commutative. If we see them, then the distributive properties follows immediately. Let $E_1 = \alpha_1 I - \beta_1 uv^H$ and $E_2 = \alpha_2 I - \beta_2 uv^H$, then the direct computation shows that

$$E_1 E_2 = \alpha_1 \alpha_2 I - \alpha_2 \beta_1 + \alpha_2 \beta_1 - \beta_1 \beta_2 (v^H u) uv^H = E_2 E_1.$$

Hence it forms a commutative ring with unity. □

As a result of Theorem 3.2, it follows by induction on n that

$$(\alpha I - \beta uv^H)^n = \alpha^n I - \{\sum_{r=1}^{n} (-1)^{r+1} {}_nC_r \alpha^{n-r} \beta^r (v^H u)^{r-1}\} uv^H. \tag{3.4}$$

Now if we take $|\alpha| < 1$ and $|\beta| < 1$, then for sufficiently small $|(v^H u)|$,

$$\lim_{n \to \infty} (\alpha I - \beta uv^H)^n = \Theta \tag{3.5}$$

and if we take $\alpha = 1$ and $|\beta| < 1$, then for sufficiently small $|(v^H u)|$,

$$\lim_{n \to \infty} (\alpha I - \beta uv^H) = I. \tag{3.6}$$

Therefore the set $\mathcal{E}$ essentially contains the identity and zero.

The following properties will be the characterizations of GEM. We will make use of these to apply to mathematical science.

Theorem 3.3 *The determinant of GEM is*

$$det(E) = det(\alpha I - \beta xy^H) = \alpha^n - \alpha^{n-1} \beta y^H x \tag{3.7}$$

where n is the order of E.

Proof. Using the Schur's complement [2],

$$det = \begin{vmatrix} A & B \\ -C & D \end{vmatrix} = det(A) det(D + CA^{-1}B). \tag{3.8}$$

Set $A = \alpha I$, $B = -\beta x$, $C = y^H$ and $D = I$. It follows that

$$det(E) = \alpha^n - \alpha^{n-1} \beta y^H x.$$

□

The next three corollaries follow directly:

Corollary 3.4 *Let E be GEM then E is singular if and only if*

$$y^H x = \frac{\alpha}{\beta}. \tag{3.9}$$

If E^{-1} exists then

$$E^{-1} = \frac{1}{\alpha}(I - \{\frac{\beta}{(y^H x - \alpha)}\} xy^H). \tag{3.10}$$

Corollary 3.5 *The minimal polynomial of GEM (the monic annihilating polynomial of least degree) is the degree two and the coefficients of the polynomial are* $-2\alpha+\beta y^H x$ *and* $\alpha^2-\alpha\beta y^H x$*, i.e.,*

$$m_E(\lambda)=\lambda^2+(-2\alpha+\beta y^H x)\lambda+\alpha^2-\alpha\beta y^H x. \tag{3.11}$$

Proof. Consider $det(\lambda I-E)=(\lambda-\alpha)^{n-1}(\lambda-\alpha+\beta y^H x)$, hence the minimal polynomial $m_E(\lambda)$ is $(\lambda-\alpha)(\lambda-\alpha+\beta y^H x)$. □

Notice that if E is not diagonalizable (we will see in next section) then the minimal polynomial is $(\lambda-\alpha)^2$.

We have investigated that the determinant of GEM is

$$det(E)=\alpha^n-\alpha^{n-1}\beta y^H x$$

in Theorem 3.3, and we apply this result to find exact eigenvalues and eigenvectors of GEM. We will start with the following two lemmas:

Lemma 3.6 *The eigenvalues of E are* α *and* $\alpha-\beta y^H x$ *with the multiplicities being* $n-1$ *and 1, respectively.*

Proof. By setting $det(\lambda I-E)$, we have that

$$det(\lambda I-E)=(\lambda-\alpha)^{n-1}(\lambda-\alpha+\beta y^H x). \tag{3.12}$$

□

Lemma 3.7 *The nonzero vector* x *in the component of* E *is the eigenvector corresponding to the eigenvalue of* $\alpha-\beta y^H x$.

Proof. Let x be the vector in E, then consider the following equation;

$$Ex=(\alpha I-\beta xy^H)x=\alpha x-\beta(y^H x)x=\{\alpha-\beta(y^H x)\}x, \tag{3.13}$$

which shows that x is the eigenvector of E and it's corresponding eigenvalue is $\{\alpha-\beta(y^H x)\}$. □

Combining Lemma 3.6 and Lemma 3.7, one will have the following characterization of eingenvectors and eigenvalues of GEM.

Theorem 3.8 *Let E be GEM then:*

(a) *the vector* x *is the eigenvector of GEM and it's corresponding eigenvalue is* $\alpha-\beta y^H x$ *if and only if* x *is the eigenvector of* βxy^H *and it's corresponding eigenvalue is* $y^H x$.

(b) *a nonzero vector* u *is an eigenvector of GEM and it's corresponding eigenvalue is* α *if and only if* u *is an eigenvector of* βxy^H *and it's corresponding eigenvalue is* 0.

Proof.

(a) It follows from Lemma 3.7 that

$$(\alpha I-\beta xy^H)x=(\alpha-\beta y^H x)x$$

and

$$(xy^H)x=(y^H x)x.$$

(b) For a nonzero vector u, we have that

$$(\alpha - \beta xy^H)u = \alpha u$$

and

$$\beta xy^H u = 0$$

which is

$$\beta xy^H u = 0u.$$

□

Note that eigenvalues of GEM are all α if and only if $y^H x = 0$. It is also clear that the matrix xy^H is the zero-matrix if and only if either vector x or y is 0. For such a case, GEM is just a scalar matrix αI, hence GEM is diagonal.

Corollary 3.9 *Suppose that the vectors x and y are non-zero then GEM is diagonalizable if and only if $y^H x \neq 0$.*

Proof. If E is diagonalizable then there exists a non-singular matrix S such that

$$SES^{-1} = S(\alpha I - \beta xy^H)S^{-1} = diag(\alpha - \beta y^H x, \alpha, \alpha, \ldots, \alpha), \tag{3.14}$$

hence

$$SES^{-1} = \alpha I - \beta Sxy^H S^{-1} = \alpha I - \beta y^H x E_{ii} \tag{3.15}$$

where E_{ii} is the matrix whose (i,i) component is 1, zero otherwise.

Conversely, if $y^H x \neq 0$ then it follows from Theorem 3.8 that x is the eigenvector such that

$$\beta(xy^H)x = \beta(y^H x)x.$$

Now we can choose a set of linearly independent vectors such that

$$\{x, u_2, u_3, \ldots, u_n\}$$

and then form the matrix

$$S = (x, u_2, u_3, \ldots, u_n).$$

Now S is clearly non-singular and we can form

$$S\beta xy^H S^{-1} = diag(y^H x, 0, 0, \ldots, 0).$$

It follows from $\alpha I = \alpha SS^{-1}$ that E is diagonalizable.

□

We will give an example which GEM is not diagonalizable:

$$E = \alpha I - \beta e_1 e_2^T = \begin{pmatrix} \alpha & \beta \\ 0 & \alpha \end{pmatrix}. \tag{3.16}$$

4. Hermitian GEMs

In this section, we will investigate Hermitian GEM and others.

Lemma 4.1 *GEM is normal if and only if x and y are linearly dependent.*

Proof. Let E be a normal GEM, then $\bar{\alpha} - \bar{\beta}y^H x$ is an eigenvalue with it's eigenvector x by Theorem 3.8. Since E is normal, $\bar{\alpha} - \bar{\beta}y^H x$ must be an eigenvalue of E^H with it's eigenvector x. Since

$$E = \alpha I - \beta x y^H, \ E^H = \bar{\alpha} I - \bar{\beta} y x^H, \tag{4.1}$$

it follows that

$$(xy^H)x = (yx^H)x$$

and

$$x(y^H x) = y(x^H x).$$

Since E is normal, E is diagonalizable. It follows from Corollary 3.9 that $y^H x$ is not equal to zero. Hence

$$x = \left(\frac{x^H x}{y^H x}\right) y. \tag{4.2}$$

The converse part is rather clear. □

Theorem 4.2 *GEM is Hermitian if and only if it satisfies the following two conditions;*

(1) α and β are real numbers

(2) x and y are linearly dependent.

Proof. If (1) and (2) are satisfied then $E = E^H$ by Theorem 4.1. We may write

$$E = \alpha I - \beta x x^H. \tag{4.3}$$

Conversely, if E is Hermitian then it's eigenvalues are all real. Hence α is real and β must be real by $x^H x > 0$. □

Corollary 4.3 *Hermitian GEM is positive definite (or positive semidefinite) if and only if $\alpha > \beta x^H x > 0$ (or $\alpha \geq \beta x^H x > 0$).*

Proof. Let $E = \alpha I - \beta x x^H$ where α and β are real, then all eigenvalues of E are positive (or non-negative) if and only if $\alpha > \beta x^H x > 0$ (or $\alpha \geq \beta x^H x > 0$) by Theorem 4.2. □

Corollary 4.4 *Hermitian GEM is a projector if and only if $\alpha = 1$ and $\beta = \frac{1}{x^H x}$.*

Proof. Let $E = \alpha I - \beta x x^H$ where α and β are real, then we need to derive that $E = E^2$.

$$E^2 = \alpha^2 I - (2\alpha\beta - \beta^2 x^H x) x x^H = \alpha I - \beta x x^H = E. \tag{4.4}$$

Hence $\alpha = 1$ and $\beta = \frac{1}{x^H x}$. □

Corollary 4.5 *Hermitian GEM projector is positive semidefinite and Hermitian GEM is unitary if and only if $\alpha = 1$ and $\beta = \frac{2}{x^H x}$.*

Proof. The first is a direct result of Corollary 4.3 and 4.4. Let E be unitary then $\alpha = 1$, and hence $2\alpha\beta - \beta^2 x^H x = 0$ implies that $\beta = \frac{2}{x^H x}$. □

5. GEM as a Linear Transformation

In this section, we will discuss uses of GEM as a linear transformation, and derive verious kinds properties of GEM as linear transformations.

We first observe GEM as a linear transformation.

Theorem 5.1 *Let a and b be vectors such that $y^H a \neq 0$, then there exists GEM such that $Ea = b$. In this case,*

$$E = \alpha I - \frac{1}{y^H a}(\alpha a - b)y^H. \tag{5.1}$$

Proof. The equation $(\alpha I - \beta x y^H)a = b$ implies that if $y^H a \neq 0$ then

$$x = \frac{\alpha a - b}{\beta(y^H a)}.$$

Hence, letting $\beta = 1$, it follows that

$$E = \alpha I - \frac{1}{y^H a}(\alpha a - b)y^H.$$

□

We will consider the special cases of GEM as a linear transformation, and show each of the property.

Lemma 5.2 *Let a and b be vectors such that $y^H a \neq 0$, then the inverse of GEM such that $E^{-1}b = a$, if exists, will be the following,*

$$E^{-1} = \alpha I - \frac{1}{y^H b}(\alpha b - a)y^H, \tag{5.2}$$

where $y^H b \neq 0$.

Proof. The direct calculation shows that

$$(\alpha I - \frac{1}{y^H b}(\alpha b - a)y^H)b = (\alpha b - \frac{1}{y^H b}(\alpha b - a)(y^H)b = \alpha b - (\alpha b - a) = a. \tag{5.3}$$

□

In order to apply GEM to mathematical science, we will consider the special cases of GEM as a linear transformation, and show each of the property, mainly, an eigenvalue and corresponding its eigenvector.

Corollary 5.3 *The following statements are true:*

$$\text{The eigenequation of } E_a = I - \frac{1}{a^H a}(a - b)a^H \text{ is } E_a\,(a - b) = \frac{a^T b}{a^T a}(a - b). \tag{5.4}$$

$$\text{The eigenequation of } E_a^{-1} = I - \frac{1}{a^H b}(b - a)a^H \text{ is } E_a^{-1}\,(b - a) = \frac{a^T a}{a^T b}(b - a). \tag{5.5}$$

The eigenequation of $E_b = I - \frac{1}{b^H a}(a-b)b^H$ *is* $E_b\,(a - b) = \frac{b^T b}{b^T a}(a - b)$. (5.6)

The eigenequation of $E_b^{-1} = I - \frac{1}{b^H b}(b-a)b^H$ *is* $E_b^{-1}\,(b - a) = \frac{b^T a}{b^T b}(b - a)$. (5.7)

It is expected that the properties that we have discussed can be applied to mathematical science, especially, to the feilds of numerical linear algebra and mathematical biology [3],[4] and [5].

References

1. A. Householder, The Theory of Matrices in Numerical Analysis, Dover Publications Inc., New York (1964).

2. P. Lancaster and M. Tismenetsky, The Theory of Matrices Second Edition with Applications, Academic Press, New York (1985).

3. S. Horiguchi and W.L. Miranker, Linear Algebra Appl., **115**, 641-658 (1989).

4. K. Oshima and S. Okamoto, Bulletin of the Faculty of Science, Ibaraki University, Series A. Mathematics, **22**, 9-13 (1990).

5. G. Sierksma, Linear Algebra Appl., **26**, 175-201 (1979).

9th Int. Coll. on Differential Equations, pp. 307-310
D. Bainov (Ed)

Oblique Derivative Problem in Morrey Spaces

DIAN K. PALAGACHEV
Politecnico di Bari, Department of Mathematics, Via E. Orabona 4, 70125, Bari, Italy
e-mail: dian@pascal.dm.uniba.it
MARIA ALESSANDRA RAGUSA
Department of Mathematics, Università di Catania, Viale A. Doria 6, 95125, Catania, Italy
e-mail: maragusa@dipmat.unict.it

Abstract: We present some recent results on strong solvability and global regularity in Morrey spaces $W^{2,p,\lambda}(\Omega)$ for the regular oblique derivative problem for second order uniformly elliptic operators with principal coefficients belonging to the Sarason class VMO of functions with vanishing mean oscillation.
Key Words and Phrases: *Uniformly elliptic operator, oblique derivative problem, VMO.*
1991 Mathematics Subject Classification. 35J25, 35B45, 35R05.

1. Introduction and Main Results

This note is devoted to the study of global regularity in Morrey spaces of the strong solutions to the regular oblique derivative problem for uniformly elliptic operators with discontinuous coefficients. Precisely, let Ω be a bounded domain in $\mathbb{R}^n$, $n \geq 3$, with $C^{1,1}$-smooth boundary $\partial\Omega$. Consider the unit vector field $\ell(x) : \partial\Omega \to \mathbb{R}^n$ and the first-order boundary operator $\mathcal{B} \equiv \partial/\partial\ell(x) + \sigma(x)$, $x \in \partial\Omega$. For $x \in \Omega$, we consider the second order uniformly elliptic operator $\mathcal{L} \equiv a^{ij}(x)D_{ij}$. It is assumed that a^{ij}'s belong to the Sarason class of functions with vanishing mean oscillation VMO (cf. [1]). Concerning the boundary operator $\mathcal{B}$, we suppose that it is *regular*, i.e., the vector field $\ell(x)$ is *never* tangential to the boundary of Ω.

Our interest is devoted to the study of global regularity and strong solvability in Morrey spaces for the next oblique derivative problem

$$(1)\qquad \begin{cases} \mathcal{L}u \equiv a^{ij}(x)D_{ij}u = f(x) & \text{for almost all } x \in \Omega, \\ \mathcal{B}u \equiv \partial u/\partial\ell(x) + \sigma(x)u = \varphi(x) & \text{in the trace sense on } \partial\Omega. \end{cases}$$

The interests to boundary value problems for elliptic operators with VMO principal coefficients increased significantly in the last ten years (cf. [2] for an excellent presentation of the current state of the theory). The Dirichlet problem for such kind of operators has been well studied both in the linear ([3]) and in the quasilinear ([4]) cases. In the papers [5] and [6], the regular oblique derivative problem for elliptic operators with VMO principal coefficients has been studied in the framework of Sobolev scale $W^{2,p}(\Omega)$, $1 < p < \infty$, for linear and quasilinear operators, respectively. The results of [5] have been refined to the case of elliptic operators with lower order terms ([7]). Recently, the $W^{2,p}$-theory developed in [7] has been applied also to the study of *non-regular* oblique derivative problem in Sobolev spaces (see [8]).

In the present note we announce global regularity result in Morrey spaces for the strong solutions of (1). Precisely, we prove that $f \in L^{p,\lambda}(\Omega)$ and $\varphi \in W^{(p,\lambda)}(\partial\Omega)$, $1 < p < \infty$, $0 < \lambda < n$, imply that the second derivatives $D_{ij}u$ of any strong solution to (1) belong to the Morrey space $L^{p,\lambda}(\Omega)$. By virtue of the standard technique, this regularizing property ensures also strong solvability in $W^{2,p,\lambda}(\Omega)$ of (1) for any $f \in L^{p,\lambda}(\Omega)$ and $\varphi \in W^{(p,\lambda)}(\partial\Omega)$. Finally, the imbedding relations between Morrey's and Hölder's spaces yield Hölder continuity of the gradient Du of the strong solutions to (1) for suitable values of p and λ.

To list the assumptions on the data of the problem (1), let us start with some useful definitions. The classical Sobolev space of functions having weak derivatives up to order k which belong to $L^p(\Omega)$ will be denoted by $W^{k,p}(\Omega)$. Let $p \in (1,\infty)$ and $\lambda \in (0,n)$. The function $f \in L^1_{\text{loc}}(\Omega)$ belongs to the the Morrey space $L^{p,\lambda}(\Omega)$ if

$$\|f\|_{L^{p,\lambda}(\Omega)} \equiv \left(\sup_{\substack{r>0\\ x\in\Omega}} \frac{1}{r^\lambda} \int_{B_r(x)\cap\Omega} |f(y)|^p dy \right)^{1/p} < \infty,$$

where $B_r(x)$ stands for the n-dimensional ball of radius r and centered at the point x. The subspace of $W^{k,p}(\Omega)$ consisting of functions with k-th order derivatives lying in $L^{p,\lambda}(\Omega)$ will be denoted by $W^{k,p,\lambda}(\Omega)$. A natural norm in that space is given by

$$\|u\|_{W^{k,p,\lambda}(\Omega)} = \|u\|_{L^{p,\lambda}(\Omega)} + \|D^k u\|_{L^{p,\lambda}(\Omega)}.$$

Further, to interpret the boundary condition in (1) in the trace sense on $\partial\Omega$, we need also the space of functions defined on $\partial\Omega$ which are traces of functions lying in $W^{1,p,\lambda}(\Omega)$. That class is well studied by Campanato in [9]. Thus, set $W^{(p,\lambda)}(\partial\Omega)$ for the Banach space formed by functions φ defined on $\partial\Omega$ and having the finite norm

$$\|\varphi\|_{W^{(p,\lambda)}(\partial\Omega)} = \left\{ \sup_{\substack{r>0\\ x'\in\partial\Omega}} \frac{1}{r^{\bar\lambda}} \int_{B_r(x')\cap\partial\Omega} |\varphi(x')|^p d\sigma_{x'} \right.$$

$$\left. + \sup_{\substack{r>0\\ x',\bar x'\in\partial\Omega}} \frac{1}{r^\lambda} \int_{B_r(x')\cap\partial\Omega} \int_{B_r(\bar x')\cap\partial\Omega} \frac{|\varphi(x')-\varphi(\bar x')|^p}{|x'-\bar x'|^{p+n-2}} d\sigma_{x'} d\sigma_{\bar x'} \right\}^{1/p},$$

with $\bar\lambda = \max\{\lambda-1, 0\}$.

To formulate the regularity assumptions on the coefficients of $\mathcal{L}$, we need also to recall the definitions of the John–Nirenberg space ([10]) of functions with bounded mean oscillation (BMO) and the Sarason class VMO of the functions with vanishing mean oscillation ([1]). A locally integrable function $f(x)$ is said to belong to BMO if $\|f\|_* \equiv \sup_{B\subset\mathbb{R}^n} \frac{1}{|B|}\int_B |f(x)-f_B|dx < \infty$. Here f_B stands for the integral average $\frac{1}{|B|}\int_B f(x)dx$ of the function $f(x)$ over the set B, and B ranges in the class of balls of $\mathbb{R}^n$. For $f(x) \in BMO$, set $\theta(r) = \sup_{\substack{\rho\le r\\ x\in\mathbb{R}^n}} \frac{1}{|B_\rho|}\int_{B_\rho} |f(x)-f_{B_\rho}|dx$. We say that $f(x) \in VMO$ if $\lim_{r\to 0^+}\theta(r) = 0$.

We will suppose that $\mathcal{L}$ is a uniformly elliptic operator with VMO coefficients:

$$\left\{ \begin{array}{l} \exists\, \mu>0:\ \mu|\xi|^2 \le a^{ij}(x)\xi_i\xi_j \le \mu^{-1}|\xi|^2 \quad \forall\xi\in\mathbb{R}^n, \text{ for a.a. } x\in\Omega, \\ a_{ij}(x) \in VMO(\Omega). \end{array} \right. \tag{2}$$

As it concerns the boundary operator $\mathcal{B}$, we suppose

$$\ell_i(x),\ \sigma(x) \in C^{0,1}(\partial\Omega), \quad \ell(x)\cdot\nu(x) > 0, \quad \sigma(x) < 0 \quad \forall\, x \in \partial\Omega, \tag{3}$$

with $\nu(x)$ being the unit *inward* normal to $\partial\Omega$. The simple geometric meaning of (3) is that the field $\ell(x)$ is nowhere tangential to $\partial\Omega$.

The main result of the note is the next

Theorem 1. *Suppose* (2), (3), $p \in (1,\infty)$ *and* $\lambda \in (0,n)$. *Let* $u \in W^{2,p}(\Omega)$ *be a solution of the problem* (1) *with* $f \in L^{p,\lambda}(\Omega)$ *and* $\varphi \in W^{(p,\lambda)}(\partial\Omega)$.

Then $D_{ij}u \in L^{p,\lambda}(\Omega)$ *and there is an absolute constant* C *such that*

$$\|u\|_{W^{2,p,\lambda}(\Omega)} \le C\left(\|u\|_{L^{p,\lambda}(\Omega)} + \|f\|_{L^{p,\lambda}(\Omega)} + \|\varphi\|_{W^{(p,\lambda)}(\partial\Omega)}\right). \tag{4}$$

An immediate consequence of Theorem 1 is the well-posedness of the oblique derivative problem (1) in the Morrey space $W^{2,p,\lambda}(\Omega)$. In fact, the Aleksandrov–Bakelman–Pucci maximum principle ([11, Theorem 2.6.2]) and the inverse mapping theorem imply

Corollary 1. *Let* (2) *and* (3) *be fulfilled,* $p \in (1,\infty)$ *and* $\lambda \in (0,n)$. *Then for every* $f \in L^{p,\lambda}(\Omega)$ *and* $\varphi \in W^{(p,\lambda)}(\partial\Omega)$ *the problem* (1) *is uniquely solvable in the class* $W^{2,p,\lambda}(\Omega)$.

Combining Theorem 1 with the known properties of Morrey spaces for suitable p and λ (cf. [9]), we obtain the next global Hölder regularity for the gradient Du of the strong solutions to (1).

Corollary 2. *Let* $u \in W^{2,p}(\Omega)$ *be a strong solution of* (1) *with* $f \in L^{p,\lambda}(\Omega)$ *and* $\varphi \in W^{(p,\lambda)}(\partial\Omega)$ *and suppose* $n-p<\lambda<n$. *Then* $Du \in C^{\alpha}(\overline{\Omega})$ *with* $\alpha = 1-(n-\lambda)/p$.

2. Sketch of the Proof of Theorem 1

Let $\tilde\varphi$ be a trace on $\{x_n=0\}$ of a $W^{1,p,\lambda}(\{x_n>0\})$-function, $x=(x_1,\dots,x_{n-1},x_n)=(x',x_n)$. It follows from [9, Theorem I.1] that $\tilde\varphi \in W^{(p,\lambda)}(\{x_n=0\})$. Now, take a function $\eta(y') \in C_0^2(\mathbb{R}^{n-1})$ such that $\int_{\mathbb{R}^{n-1}}\eta(y')dy' = 1$. Fix an arbitrary $R>0$ and denote $B_R^+ = B_R(x_0)\cap\{x_n=0\}$, $x_0=(x_0',0)$. Without loss of generality we may take $\Gamma_R = B_R(x_0)\cap\{x_n=0\}$ above. Later, having $\tilde\varphi \in W^{(p,\lambda)}(\Gamma_R)$ we suppose that $\tilde\varphi$ is extended to all $\mathbb{R}^{n-1}$ as a function with a compact support, preserving its $W^{(p,\lambda)}$-norm.

Supposing that the boundary $\partial\Omega$ is locally flatten near the point x_0 such that $\Omega\subset\{x_n>0\}$, we recall that the *regular obliqueness condition* (3) ensures $\ell_n(x_0)\neq 0$. Consider now the function

$$\phi(x) = \phi(x',x_n) = \frac{x_n}{(2-n)\ell_n(x_0)}\int_{\mathbb{R}^{n-1}}\tilde\varphi(x'-x_ny')\eta(y')dy'. \tag{5}$$

Essential step in the proof of Theorem 1 is ensured by the next

Lemma 1. *The function* $\phi(x)$ *belongs to* $W^{2,p,\lambda}(B_R^+)$ *and satisfies*

$$\phi(x',0)=0, \quad \frac{\partial\phi}{\partial x_n}(x',0) = \frac{\tilde\varphi(x')}{\ell_n(x_0)} \qquad \text{for} \quad x'\in\Gamma_R. \tag{6}$$

Moreover,

$$\|\phi\|_{W^{2,p,\lambda}(B_R^+)} = \|\phi\|_{L^{p,\lambda}(B_R^+)} + \|D^2\phi\|_{L^{p,\lambda}(B_R^+)} \le C(R)\|\tilde\varphi\|_{W^{p,\lambda}(\Gamma_R)} \tag{7}$$

with $\lim_{R\to 0^+} C(R) = 0$.

The proof of Lemma 1 consists of a very careful analysis of formula (5) and successive estimations of the norms $\|\phi\|_{L^{p,\lambda}(B_R^+)}$ and $\|D^2\phi\|_{L^{p,\lambda}(B_R^+)}$. It should be pointed that the existence of the second derivatives $D^2\phi \in L^p$ follows from the uniform boundedness in L^p of the differential quotients for the first derivatives $D\phi$.

Once having Lemma 1, the problem (1) reduces locally near any point $x_0 \in \partial\Omega$ to a homogeneous boundary value problem. The proof of Theorem 1 is then completed almost in the same manner as done in [5] and [12]. Precisely, an explicit representation formula is employed for the second derivatives of the solution to the homogeneous problem (1). That formula is written in terms of singular integrals with Calderón–Zygmund kernels. A crucial role in deriving the estimate (4) is played by the VMO-assumption on the coefficients a_{ij} and by the bound (7).

References

1. D. Sarason, Trans. Amer. Math. Soc., **207**, 391–405 (1975).
2. F. Chiarenza, in: Nonlinear Analysis, Function Spaces and Applications, Vol. 5, Prague (1994).
3. F. Chiarenza, M. Frasca and P. Longo, Trans. Amer. Math. Soc., **336**, 841-853 (1993).
4. D.K. Palagachev, Trans. Amer. Math. Soc., **347**, 2481–2493 (1995).
5. G. Di Fazio and D.K. Palagachev, Comment. Math. Univ. Carolinae, **37**, 537-556 (1996).
6. G. Di Fazio and D.K. Palagachev, Bull. Austral. Math. Soc., **53**, 501–513 (1996).
7. A. Maugeri and D.K. Palagachev, Forum Math., **10**, 393–405 (1998).
8. A. Maugeri, D.K. Palagachev and C. Vitanza, C. R. Acad. Sci. Paris, Sér. I, **327**, 53–58 (1998).
9. S. Campanato, Ann. Scuola Norm. Superiore di Pisa, **15**, 263–281 (1961).
10. F. John and L. Nirenberg, Commun. Pure Appl. Math., **14**, 415-426 (1961).
11. N.S. Trudinger, Nonlinear Second Order Elliptic Equations, Lecture Notes of Math. Inst. of Nankai Univ., Tianjin, China (1986).
12. G. Di Fazio and M.A. Ragusa, J. Funct. Anal., **112**, 241-256 (1993).

9th Int. Coll. on Differential Equations, pp. 311-318
D. Bainov (Ed)

Applying Probabilistic Neural Networks to the Multifont Recognition Problem

MARCIN PAPRZYCKI AND SEAN BOWERS
Department of Computer Science and Statistics, University of Southern Mississippi, Hattiesburg, MS 39406-5106, USA
{m.paprzycki, sean.bowers}@usm.edu

Abstract: Neural networks are considered one of the best tools for the pattern recognition problem. In 1990 D. Specht introduced a new class of Probabilistic Neural Networks which were initially unnoticed in the computational practice due primarily to their large computer memory requirement. In this note we present and discuss results of experiments assessing the usability of Probabilistic Neural Networks to the multifont recognition problem.
Key Words and Phrases: *probabilistic neural networks, pattern recognition, multifont character recognition problem, performance.*

1. Introduction

Since their re-introduction into the computer world (1989, [1]) Neural Networks (NNs) have been successfully applied to a number of pattern recognition problems (see [2] and the references collected there). In 1989-90 D. Specht introduced a new class of NNs based on a different activation function modeling Bayesian strategy for estimation of non-parametric estimators for probability density functions [3, 4, 5]. Initially, Probabilistic Neural Networks (PNNs) did not became popular as they require a very large amount of memory (the size of the network is directly proportional to the size of the training set) which was not easily available in computers used in the early 1990's. It is only now, when the computer memory became very cheap and easily available, that we observe a resurgence of interest in the PNNs. Major vendors of NN software either already have a PNN module in their NN packages (Ward Systems in NeuroShell 2.0 [6] and Trajan Software in Trajan 3.0 [7]) or plan to add such a module shortly (Mathworks in Matlab Neural Networks Package [8]). This led us to investigate the potential of PNNs applied to the pattern recognition. In earlier work we have reported the initial results of a series of experiments that investigated the performance of PNNs applied to the recognition of computer generated characters. We also compared the performance of PNNs with that of Ward-nets (a sophisticated version of backpropagation networks) [9]. The results indicated that PNNs have a definite potential of becoming an effective tool for solving the printed character recognition problem and that their performance was quite favorable in comparison with Ward-nets. At the same time, the discussions during and after the conference presentation led us to believe that a much broader investigation would be necessary to properly assess the potential of PNNs.

The aim of this note is to report on such an attempt. We have applied the NeuroShell implementation of PNNs to the multifont pattern recognition problem. The problem studied is a slight simplification of a general problem of recognition of computer-printed characters (as only the 26 upper case Latin letters are used). It is assumed that a computer generated printed character (belonging to one of the existing computer font-sets) is scanned and digitized and presented to a computer to be recognized. For simplicity, it is assumed that one letter is presented at a time, each image presented does represent a letter (even though some of the letters may be slightly fuzzy, e.g. due to the printing process) and letters may belong to different fonts. We studied the performance of PNNs applied to various aspects of this problem.

The remaining part of the paper is organized as follows. In Section 2 we discuss how the data was generated. Section 3 describes the experiments and discusses their results. Finally, the proposed directions of future research are presented.

2. Data Generation

In the initial paper [9] we presented results based on 5 fonts represented as 10×10 bit-maps (shifted and fuzzy data were also used in the experiments). Here, we expanded the data set used for experimentation. Using a Delphi program, provided by Daniel Hellsson of Unseen Technology, we have generated a set of bit-maps of size 14×14 for the capital letters from the following fonts (which are standard fonts available on almost any PC running Windows 95):

Arial	Desdemona	School
Bookman	Haetten	Script
Brush	Impact	Tahoma
Colonna	Kino	TNR
Corsiva	Monocorsiva	Verdana
Courier	New Century	

In all cases letters were centered in the bit-map and 0's filled the "borders" of the map (so the effective letter image was at most 12×12).

We have also generated (as separate sets) fuzzy representations of each letter of each font. We have used the following algorithm to generate them: for each bit of the bit-map representation we have generated a random number and this bit was flipped (from 0 to 1 or from 1 to 0) with probability K/100 (where K denotes the level of "fuzzification"). Thus, e.g. for the data of type fuzzy 1 each bit of the map was flipped with probability 1/100 while for data type fuzzy 5 each bit was flipped with probability 5/100.

This data has been preprocessed to match the requirements of the NeuroShell environment (using software developed specifically for this project). In the PNN setup we used ($14*14 = 196$) input nodes and 26 output nodes (corresponding to the 26 letters of the alphabet). The number of nodes in the hidden layer varied depending on the size of the input data (for PNNs there is one node in the hidden layer for each element in the input data set).

3. Experimental Results

In the first series of experiments we established the similarity between various fonts as "perceived" by the PNN used as a pattern classifier. We have trained the PNN with a

Table 1: Similarities between fonts (sorted by success).

	Verdana	New Century	Tahoma	Bookman	Colonna	School	Corsiva	Desdem	Arial	Haetten	TNR	Courier	Impact	Kino	Brush	Monocorsiva	Script	
Verdana	X	19	16	11	5	7	5	7	10	6	8	7	6	6	3	1	3	120
New Century	20	X	10	10	6	13	5	8	8	4	6	7	4	3	3	2	2	111
Tahoma	17	9	X	10	5	4	4	7	12	7	4	4	5	3	3	1	3	98
Bookman	13	9	12	X	8	6	2	5	9	6	6	4	2	4	3	1	0	90
Arial	9	7	14	8	11	2	2	4	X	4	11	3	5	3	1	0	3	87
Courier	7	9	7	8	11	8	5	6	2	6	2	X	3	6	2	1	1	84
Haetten	6	7	4	10	4	3	4	6	5	X	4	6	11	6	2	1	0	79
Colonna	4	5	6	8	X	3	4	2	5	4	8	6	4	2	8	6	2	77
Impact	7	5	5	6	5	3	2	6	4	14	6	1	X	6	4	1	1	76
TNR	6	7	6	8	8	4	0	4	10	5	X	1	6	2	4	2	0	73
School	6	9	1	5	5	X	11	5	2	3	3	3	3	1	6	5	4	72
Desdem	10	7	7	7	5	4	2	X	4	3	4	4	3	5	1	1	1	68
Kino	5	4	5	3	1	2	3	6	2	4	2	6	4	X	2	3	1	53
Brush	5	1	5	6	5	3	3	2	2	3	4	1	2	3	X	3	1	49
Monocorsiva	0	3	0	0	5	4	19	3	0	1	0	2	1	2	5	X	2	47
Corsiva	3	3	2	1	3	4	X	3	1	0	1	3	0	3	2	11	1	41
Script	1	2	3	0	1	1	2	0	0	0	1	2	1	3	5	2	X	24
	119	106	103	101	88	71	73	74	76	70	70	60	60	58	54	41	25	

given font (26 elements in the training set and thus 26 nodes in the hidden layer). We then presented to it all remaining fonts and scored how many letters have been recognized correctly. Table 1 summarizes our findings and contains the best results obtained for various values of the smoothing factor. (The smoothing factor parameter is applied to the trained network and represents the degree of expected generalization when the new data is being presented to it). In our experiments we have found that the best results have been obtained for a relatively wide interval of values of the smoothing factor; ranging from approximately 0.9 to 2.5. For the values of the smoothing factor outside of this interval the input data was either not recognized at all or the number of incorrect answers was higher. In the Table, for each font, we also represent the total number of correct answers obtained when the font was used to train the network (row-total), and when the font was recognized by networks trained with other fonts (column- total).

It can be observed that some fonts are relatively good as trainers. In particular, PNNs trained with Verdana and New Century recognize substantially more letters of other fonts (in average 7.5 and 6.93 letters per font respectively) than networks trained with the remaining fonts. Verdana and New Century also seem rather easy to recognize by other fonts as they have the highest scores in columns. At the same time Script and Corsiva are extremely bad choices as training sets (they lead to the recognition of about 1.5 and 2.56 letters per font) while Script and Monocorsiva are the most difficult to recognize.

Table 2: Best and worst trainers and their fuzzy representations as training sets.

	V NC	V1 NC1	V12 NC12	C S	C1 S1	C12 S12
Arial	7	6	6	0	1	2
Bookman	9	10	10	0	0	1
Brush	3	4	5	7	7	6
Colonna	5	4	5	2	4	3
Corsiva	9	7	7	x	x	x
Courier	7	7	7	4	3	3
Dresden	7	7	8	2	3	2
Haetten	4	5	5	0	1	2
Impact	6	6	6	0	0	0
Kino	4	4	4	3	3	3
Monocorsiva	1	2	2	10	9	10
New Century	x	x	x	4	4	4
School	16	17	17	4	4	4
Script	3	3	3	x	x	x
Tahoma	15	16	16	4	3	5
TNR	6	5	6	2	2	2
Verdana	x	x	x	2	2	3
TOTAL	102	103	107	44	46	50

In the second series of experiments we have decided to see if by simply combining the best and the worst trainers and adding some fuzzy data to the training set we can generate additional diversity to improve the recognition of the remaining fonts. In Table 2 we represent the results of this experiment. Here V stands for Verdana, NC for New

Century, S for Script, C for Corsiva, 1 for a level 1 fuzzy data corresponding to a given font and 2 for a level 2 fuzzy data corresponding to this font. Thus V12 NC12 means that Verdana and New Century and their level 1 and 2 fuzzy representations have been combined and used as the training set and applied to the remaining fonts. As previously, the best results are presented and the totals describing the training power of a network is also provided. Finally, x means that the data is not provided since a given font was a part of the training set and thus the font was recognized correctly.

As can be observed, for the best trainers (V and NC), instead of an improvement, a performance decrease occurred. The average number of letters recognized per font dropped from 7.5 and 6.93 for Verdana and New Century to 6.8 letters for the combined set. However, if the fact that Verdana and New Century recognize each other very well (Verdana recognizes 19 letters of New Century while New Century recognizes 20 letters of Verdana) is taken into account then the assessment of the situation differs. After excluding New Century from the first row in Table 1 the performance of Verdana trained network drops to 6.73 letters per font. Similarly, after excluding Verdana from the second row of Table 1 the performance of New Century trained network drops to 6.06 letters per font. Thus the performance of combined Verdana and New Century improves over the performance of each one of them separately. For the worst trainers, the performance also improves from 1.44 and 2.56 letters for Script and Corsiva to 2.93 letters for the combined data. Here, the effect of Corsiva's presence in Scripts results in Table 1 (and vice versa) is minimal and will be neglected.

Table 3: Effects of adding fuzzy data of various levels to the Verdana training set.

	V	V + 1	V + 2	V + 3	V + 4	V + 5
Arial	10	8	8	7	10	10
Bookman	11	11	10	11	10	11
Brush	3	3	3	5	4	5
Colonna	5	5	5	3	7	6
Corsiva	5	3	4	3	3	3
Courier	7	7	6	7	7	7
Dresden	7	8	7	7	6	8
Haetten	6	7	6	6	5	4
Impact	6	6	6	6	6	6
Kino	6	3	3	3	3	4
Monocorsiva	1	1	1	3	1	1
New Century	19	10	10	10	10	11
School	7	7	9	9	7	8
Script	3	3	3	3	3	3
Tahoma	16	16	16	16	16	16
TNR	8	7	7	6	8	8
TOTAL	120	105	104	105	106	111

It can also be seen that the addition of fuzzy data seems to improve the overall network quality. In the case of the best trainers it improves the average number of recognized letters from 6.8 to 7.13. In case of the worst trainers the improvement is from 2.93 to 3.33 letters per font. To investigate this effect further we have experimented with adding fuzzy data of various levels to the training set and then using the remaining fonts to test the results. Table 3 summarizes the effect of adding fuzzy fonts of various levels to Verdana and applying the network trained by a combined set to the remaining fonts. Here V represents Verdana, V + k represents Verdana and Verdana fuzzy level k combined. As previously, the results are reported in the number of correctly recognized letters. Similar results have been observed when other fonts have been experimented with.

It can be observed that adding fuzzy data to the plain font as a mechanism for adding diversity to the data set and thus trying to improve the recognition does not help. None of the pairs (plain font + fuzzy font) has results better than the plain font only. In addition, no particular pattern of interaction between the font pairs and the remaining fonts has been observed.

Next we checked that the fuzzy fonts are recognized correctly by the network trained using the original plain fonts. We run an experiment in which we have used all plain fonts together as the training set. Then we established what is the maximum level of "fuzziness" in the letter representations that this PNN will be able to correctly recognize by using the fuzzy data sets of increasing level as test data. We have found that the network correctly recognizes all fuzzy fonts up to level 6. This means that for each letter in each font, if each of the 196 bits was flipped with probability 6/100 (leading to approximately 12 bits flipped in each letter), the network was capable of recognizing it correctly. For some fonts (e.g. Brush, Colonna, Corsiva) even at fuzzy data of level 9 (about 18 bits flipped in each letter) the network worked correctly (recognizing all of the letters in the font).

Table 4: Training PNNs using all fonts but one and testing against this font.

	ALL	ALL + 1	ALL +2
Arial	19	18	18
Bookman	15	15	15
Brush	6	7	8
Colonna	11	9	9
Corsiva	14	14	15
Courier	6	6	6
Dresden	11	9	9
Haetten	13	7	8
Impact	9	5	5
Kino	7	6	7
Monocorsiva	4	5	4
New Century	12	12	12
School	16	16	16
Script	2	2	2
Tahoma	19	19	19
TNR	13	12	12
Verdana	19	24	24
TOTAL	196	186	189

In the next set of experiments all fonts but one were used to train the network. The network was then shown the one font that was excluded from the training data. This experiment was repeated for the training set consisting of plain fonts and the fuzzy fonts of level 1 and plain fonts combined with fuzzy fonts of levels 1 and 2. The results (number of correctly recognized letters in the test font) are summarized in Table 4. The total number of letters recognized in all fonts is also displayed.

When the data in Table 4 is compared with that in Table 1 a substantial improvement in the recognition of most fonts it can be observed. The only notable exception is Script, where no improvement occurred. However the overall results are rather disappointing as only Arial, Tahoma and Verdana are close to being correctly recognized by a network trained using the remaining fonts (and possibly their fuzzy representations). It can be also observed that in this case adding the fuzzy data sets does not help the network.

The final set of experiments involved two additional features of the NeuroShell environment. Iterative improvement and the application of genetic algorithms can be combined with the PNNs to improve their performance. Both options attempt at adjusting the weights connecting the nodes of the network to better match the data and add sensitivity to the network. As above, we used all fonts but one to train the network (and apply the iterative or genetic improvements to the training process). Then we used the excluded font as a test pattern. Both, the iterative and the genetic improvement techniques require that a subset of the data be removed from the training set. This subset is then used for the testing of the improvement during the training process. Table 5 summarizes the results for iterative (I) and genetic (G) improvements. The results are reported for 5, 10 and 20% of the data extracted and used as the "internal" testing set. In the case of genetic improvements we have used the smallest breeding pool of 20. This was due to the timing constraints. On a 400MHz Pentium II with 256 Mbytes of memory it took approximately 15-30 minutes to train each set. The same experiment with a breeding pool size of 300 had not completed training within two hours.

Table 5: Experimenting with iterative and genetic improvements.

	I 5%	I 10%	I 20%	G 5%	G 10%	G 20%
Arial	17	17	17	13	11	11
Bookman	15	15	13	14	17	14
Brush	5	5	4	2	2	2
Colonna	12	12	11	4	5	1
Corsiva	14	13	12	3	1	1
Courier	5	5	5	5	5	6
Dresden	10	9	9	3	2	3
Haetten	13	9	10	10	12	6
Impact	4	4	5	6	7	7
Kino	4	4	5	6	7	7
Monocorsiva	3	2	3	3	2	4
New Century	13	12	13	11	9	8
School	15	14	14	7	11	12
Script	1	1	0	3	0	1
Tahoma	18	18	16	19	17	17
TNR	11	11	10	7	8	6
Verdana	19	19	18	15	19	20
TOTAL	186	178	169	134	138	128

It is clear that, similarly to the results reported in [9], the usage of either iterative or genetic improvements does not improve the performance of PNNs. To the contrary, not only the results are worse, but also the PNN looses the initial capability of recognizing the fonts that the training set consisted of. While in the earlier cases (when interactive or genetic improvements were not used) the network was always able to correctly recognize its own training set, this is no longer the case. After the iterative/genetic process is applied, then the network can incorrectly classify even up to three elements from the training set.

4. Concluding Remarks

In this note we have discussed the performance of Probabilistic Neural Networks applied to the multifont recognition problem. We have found that their performance is rather uneven. While they can very well recognize fuzzy forms of the fonts that were used to train them, they have problems with properly generalizing the training data to correctly recognize a new font that was not presented earlier to the network. We have also found that the iterative and genetic improvement strategies do not seem to lead to the expected improvements in the performance of PNNs.

In the near future we plan to further investigate the problem. First, we will increase the size of the input data to about 25 fonts. We will select these fonts so that they will be as diverse as possible. We will also increase the size of the bit-map representation of the data to about 20x20. Using this data we will conclude our investigation of PNNs. We will then turn our attention to other NN architectures available in the NeuroShell environment. We will start our investigations with the Ward Nets and Kohonen's self-organizing networks.

Acknowledgements

Work of Sean Bowers was partially sponsored by the NASA Mississippi Space Grant Program. NeuroShell was purchased by the Department of Computer Science and Statistics with support from the Office of the Dean of the College of Science and Technology.

References

1. D.E. Rumelhardt, D.L. McClelland and the PDCP Research Group, Parallel Distributed Processing, MIT Press, Cambridge (1986).
2. C.G. Looney, Pattern Recognition Using Neural Networks, Oxford University Press, New York (1997).
3. D. Specht, in: Proceedings IEEE Conference on Neural Networks, Vol. 1, IEEE, Los Alamitos, (1989), p. 525.
4. D. Specht, Neural Networks, **3**, 109 (1990).
5. D. Specht, IEEE Transactions on Neural Networks, **1**, 111 (1990).
6. http://www.wardsystems.com
7. http://www.trajan-software.deman.co.uk
8. http://www.mathworks.com/products/neuralnet/
9. A.N. Fairchild and M. Paprzycki, in: Proceedings of the 14th CAM, UCO, Edmond, 161 (1998).

9th Int. Coll. on Differential Equations, pp. 319-324
D. Bainov (Ed)

A Nested then Coupled Mediterranean Sea Layer-Level Model

A. LOUISE PERKINS
Naval Research Laboratory, Stennis Space Center, MS 39529-5004, USA

Abstract: We couple a layered primitive equation coastal ocean model to an Ionian Sea Basin model which is itself nested within a full basin Mediterranean Sea model. An EOF-based coupling methodology, that includes some baroclinic forcing of the coastal model, is introduced to provide the Gulf of Sidra coastal model with "real-time" boundary conditions. The boundary conditions are designed to conserve either kinetic or potential energy, or both.
Key Words and Phrases: *Mediterranean Sea, Coupled Ocean Models.*

1. Introduction

Our basin-wide Mediterranean Sea model results suggest that one of three seasonally stable states are usually present in the Ionian Basin. These states are identified by the location of seasonally "permanent" sub-basin scale eddies. The location of these features appears to be a strong function of the wind-stress curl field over the Ionian Basin. While the inflow through the straits of Sicily appears to be the dominant driving mechanism for the Atlantic Ionian Stream, the location of these sub-permanent sub-basin scale eddies may help determine the eventual path of this incoming water. The proximity of these eddies to the shelf/slope region may significantly influence the shelf dynamics.

To study this interaction, we introduced a nested model within the Ionian Basin to capture these sub-basin scale eddies. We then use these higher resolution model results to force a coastal Gulf of Sidra model that contains realistic shelf/slope topography. In this paper we describe our boundary condition correction strategy for updating our coastal ocean level model from the deep water layer ocean model.

2. Related Research

[1] reports both a frontal partition of water masses at the shelf/slope and enhanced mixing along this same front due to several different dynamical factors. He also reports substantial reflection of deep water dynamics at greater than a few hours wave frequency.

Episodic eddies may impact shelf/slope flow as well. Vigorous cross-shelf/slope mixing may occur at the shelf/slope front due to frontal eddies capturing cross front water masses. For example, Gulf Stream Rings are responsible for a significant amount of the available kinetic and potential energy. And, as the exchange of shelf and slope water may be significant, the impinging of eddies onto a shelf/slope is expected to differ radically from the more benign quiescent flow generally seen there. For a horizontally isopycnal disturbance in a stratified fluid, such as occurs where isopycnal surfaces meet sloping topography, the finite difference approximation may induce an apparent geostrophic flow both in the upper and the lower layers. However this apparent flow is along isobaths, and hence should not impact the cross-shelf flow.

Eric Chaussinet (private communication), with a Rossby radius of order $50km$, and monthly time scales, used a model to show that the seat of kinetic energy is along the shelf break. He suggests that layer models are more natural for century long time scales since the dynamics and composition of bottom water can be maintained without numerical diffusion. But, as was discussed in Johnson Van Leer and Mooers [2], in their study of coastal upwelling off the Oregon coast, we need to resolve the vertical down to 5 meters and sample at least 3 times an hour in order to resolve the tidal, inertial and *cross-isobath* flows. This is not feasible for a layer model, and is why we introduce the coastal ocean model.

In [3] he stresses that eddy-shelf/slope interactions are coupled, but different, physical domains. The interaction between these different physical domains still being an open question, he suggests that one must address the interaction between both the eddy and the shelf/slope, and vis-a-versa. The dipole model presented there indicates that both initial distance from the topography and the change in height over the shelf/slope determine in large part the qualitative flow dynamics. However, the details are not yet well understood.

To model this region, we construct an idealized, but still realistic, experiment. We couple a deep and shallow water model, where the deep water model generates the eddy, and then helps maintain its location as it moves towards the topographic impediment. As it nears the shelf/slope, and strong coupling between the eddy and the topography occur, the eddy is allowed to interact with the shelf/slope in the shallow water model.

3. Shelf/Slope Physics

In [4] they examined wind-driven currents over an irregular coastline. Many researchers have examined mean flow over sloping shelves, typically following the lead of [5]. This work is built on a theory of small bottom topography, but appears to have broader applicability. For example, it has been extended to finite topography by [6].

The relevant physics may differ dependent upon which side of a body of water one is measuring, for the east and west coasts first order circulation are dominated by different dynamics. Further, the zeroth order topographic interactions along the shelf/slope region are different depending upon geological formulation of the shelf. Ocean ridges produce vastly different topographies than subductive processes.

Mean along-shore flow theory is still incomplete. Barotropic models have been utilized to demonstrate that alongshore pressure gradients do not penetrate up onto the shelf. However, the baroclinic effects, as well as the apparently important (but much slower) cross-shelf currents may play a significant role. A small across-shelf momentum flux due to wind forcing may precondition significant shelf dynamics.

A thermal front occurs at the shelf/slope break during the warmer months, and partitions the region into a deep water slope and a well mixed shelf area. This front alters horizontal mixing as it reflects water back into the deeper ocean.

4. Ocean Models

The Mediterranean Model of [7] is used in the deep ocean. For this implementation, the Naval Research Laboratory (NRL) Primitive Equation Ocean Model of Hurlburt and

Thompson [8], [9] and Wallcraft [10] is configured in a three-layer hydrodynamic finite-depth implementation. This model is then coupled to a finer resolution one-way nested Ionian Basin model. This finer resolution model is then coupled to the hybrid $\sigma - z$ coastal model developed by Paul Martin (personal communication). The hybrid coastal model was selected to allow us to vary the vertical coordinate system during the study to determine its impact on the model results. The layer model does mix between the two water masses, however, this may not be sufficient for predictive dynamics generating cross-frontal circulation [11]. Hence we couple the deep water layer model to a thermodynamic coastal level model.

P. Martin (personal communication) constructed a general ocean circulation model using vertical hybrid coordinates for the Naval Research Laboratory. We take the initial conditions for this coastal model from the nested Ionian Basin model. The method used to map from one set of prognostic variables to another is analogous to data assimilation methods for disparate data types.

The initial conditions are smoothed once in space with a 9 point filter before being assigned to the level model grid (this is true for the boundary conditions as well, where a 3 point filter is used).

5. Energy Sensitive Boundary Conditions

Because the layer and level models have distinct equations of state, strict conservation of mass between our coupled models is undesirable. Our boundary conditions are derived from a hydrostatically balanced model. However, the difference in mass between the two models introduces a hydrostatic imbalance, for which we must compensate. It is also desirable that the boundary condition treatment lead to a geostrophic current when both fluids are geostrophic, regardless of different model approximations. This requires that both models have consistent discrete pressure or velocity in any overlap areas.

From previous work, we expect the nominal maximum energy exchange across the shelf/slope region to be aligned with the free wave resonances. When the eddies bring a substantially higher temperature gradient with them, the first mode resonance is not significantly affected, however the next two modes are altered.

Momentum can also be exchanged across water mass separating fronts. [1] outlines which may be important, based on scaling arguments. But in this strongly non-linear regime, these scaling arguments may not always be applicable. However, neither can we assert that the numerical approximations for the non-linear dynamics accurately models this region. We hope only that a model of this region will prompt discussion of these issues, regardless of the model's shortcomings.

In atmospheric models, the need for conversions between kinetic and potential energy to cancel over the entire domain, forced specific discretization approximations as long ago as [12]. Spurious release of potential energy may generate eddy artifacts. Further, the geostrophic kinetic energy is dominated by mesoscale eddies. Hence the energy exchange between the coupled models is of critical importance.

We prescribe the variation in kinetic energy along the boundary, but compute it by combining the mass from the receiving coastal model with the tendency of the velocity values from the deep ocean basin model. Refer to the coastal model as model r and the deep water model as model l. Straightforward conservation of energy would net

$$PE_r + KE_r = PE_l + KE_l\,.$$

However, we prescribe only the change in energy across each model because of the mismatch between the Equations of State. The coastal ocean model velocity field is adjusted as a result. Conservation of energy, in an ambiguous sense, is then achieved by adjusting the velocity mean along the boundary. While not exact, such adjustments provide degrees of freedom that can be used to balance geostrophic flow or hydrostatic relations across the coupled boundary. An analogous approach can be used to prescribe potential energy along the coupled boundary, where sea surface height replaces velocity. Since the balance of both of these terms can not be known precisely, a combination of these boundary conditions can also be tuned for the coupled region. Since each individual model is in hydrostatic balance, and should model the geostrophic flow well, the latitude that exists within this formulation should permit the modeler to avoid excessive errors.

Under the assumption of a stratified fluid with buoyancy variations only along the isobaths, the across-isobath component of the internal pressure gradient vanishes. Hence no errors are introduced even though we have strong alongshore flow. However, when we have cross-isobath flow, it will have an error along the sigma coordinate surface, but at least this error will be unaffected by the alongisobath flow. So neither strong stratification, nor along isobathic flow should contribute to the discretization errors. However, across-isobath flow will contribute, and, as density surfaces tilt as they move into strong topography, the direction depending upon whether upwelling or downwelling is occurring, this will expose our model to errors.

6. Model Coupling in the Overlap Region

The NRL Layer model has density constant within a layer, netting an approximation of a vertically intergrated quantity, which is only a zeroth order approximation. For this reason, the level model is allowed to integrate the temperature and salinity quantities in the overlap area, with data assimilation the means by which the density information is transferred to the coastal level model. Because our overlap conditions do not update mass, this is consistent with the coupled boundary conditions. It would be nice to nudge water masses correctly, not just density. However, this information is not available from the deep water layer model directly. To compensate, we use an EOF fit to data in the region, based on dynamic height, which is positively correlated with water masses in this region. In the Mediterranean, the temperature and salinity profiles near the surface are critical. Most seasonal signals reside above the thermocline. The layer model integrated density does not include surface heating. The lack of surface heating in this layer model introduces a mismatch between data and model which forces the model to a wrong sea surface height that must be compensated for in the lower (deep) flows. To generate realistic temperature and salinity profiles we meld the integrated model density with real data. The dynamic height of the hydrodynamic layer model is first adjusted for such steric heating by computing the monthly mean anomalies from the model, which does not include steric heating, and adjusting them with the monthly mean from the data, which does include a zeroth order steric heating effect.

The resulting density plus adjusted dynamic height is then regressed to a temperature and salinity value (tracers) at predefined z levels.

The temperature and salinity profiles in the Gulf of Sidra are initialized from, and then forced by, the EOF projections of the density taken from the Ionian Basin model.

7. Conclusions

In the face of heterogeneous data fields, one must return to first principles to construct a boundary matching condition, considering the mathematical formulation and numerical discretization of these basic principles as guides, rather than absolutes. Still, one must be mindful of the effects of these approximations on the model outcome.

Up to one week experiments have been performed with varying degrees of boundary conservation and overlap nudging. Preliminary results appear to show an advantage to the near conservation scheme presented here. However, due to the strongly non-linear flow regime being modeled, we can only examine process studies focusing on the impact of differing topography on eddy propagation across a shelf slope region. We can state that while the layer model sustained eddies over shallow topography, the coastal model, with more realistic topography, dissipates them more rapidly. This is also seen in the available data.

Acknowledgments

This work was performed under the auspices of the Naval Research Laboratory and of the Office of Naval Research, Navy Ocean Modeling Program. The outer model runs are executed on a Cray C90 at the Naval Oceanographic Office (MSRCNAVO) located at the Stennis Space Center. The nested model runs are executed on the U.S. Army Corps of Engineers Waterways Experiment Station (CEWES) Cray YMP8. Computational time for these experiments has been provided by the High Performance Computing Initiative of the U.S. Department of Defense. ETOP05-based topography was supplied by the U.S. Naval Oceanographic Office. The authors would like to thank Mr. Joe Metzger and Mr. Rick Allard of the Naval Research Laboratory, and Mr. Bruce Hundermark of Sverdrup Technologies for their work in supplying the wind data, Dr. Alan Wallcraft of Planning Systems, Inc for the development of the NRL Layered Model.

References

1. J.M. Huthnance, Prog. Oceanog., **29**, 161–196 (1992).

2. W.R. Johnson, J.C. Van Leer and C.N.K. Mooers, Journal of Physical Oceanography, **5**, 556–574 (1976).

3. X. Wang, Interaction of an Eddy with a Continental Slope, Woods Hole Oceanographic Institution Massachusetts Institute of Technology Joint Program in Oceanography/Applied Ocean Science and Engineering Doctoral Dissertation, WHOI-92-40 (1992).

4. D.B. Haidvogel and K.H. Brink, Journal of Physical Oceanography, **16**, 2159–2171 (1986).

5. J.M. Huthnance, Estuarine Coastal Marine Science, **1**, 89–99 (1973).

6. J.W. Loder, Journal of Physical Oceanography, **10**, 1399–1416 (1980).

7. A.L. Perkins, D. W. Blake, L. F. Smedstad, G. W. Heburn and A. J. Wallcraft, Journal of Geophysical Research, (1997).

8. H.E. Hurlburt and J.D. Thompson, NORDA Technical Note 64, Naval Research Laboratory, Stennis Space Center, MS (1980).

9. H.E. Hurlburt and J.D. Thompson, in: Hydrodynamics of Semi-enclosed Seas, J.C.J. Nihoul (Ed.), Elsevier (1982) pp. 243–298.

10. A.J. Wallcraft, NOARL Report 35, Naval Research Laboratory, Stennis Space Center, MS (1991).

11. Dong-Ping Wang, Journal of Physical Oceanography, 1191–1199 (1984).

12. G.A. Corby, A. Gilchrist and R.L. Newson, Quart. J.R. Met. Soc., **98**, 809–832 (1972).

9th Int. Coll. on Differential Equations, pp. 325–330
D. Bainov (Ed)

One Phenomenon for the Wave Equation in $\mathbb{R}^3$

N.I. POPIVANOV
Department of Mathematics and Informatics, University of Sofia, 1164 Sofia, Bulgaria
nedyu@fmi.uni-sofia.bg
Ts.D. HRISTOV
Institute of Mathematics and Informatics, Bulgarian Academy of Sciences, 1113 Sofia, Bulgaria
tsvetan@math.bas.bg

Abstract: In this paper we investigate some boundary value problems for the wave equation, which are the three-dimensional analogues of the Darboux-problems (or Cauchy-Goursat problems) on the plane. It is well known that for the infinite number of smooth functions at the right hand side some of these problems do not have a classical solution. The reason of this fact is: for the same smooth functions in the right hand side there exist general solutions with a strong power-type singularity on the characteristic cone [1]. It is interesting, that this singularity is isolated only at the vertex of the characteristic cone and does not propagate along the cone. In the present paper we investigate in addition the right behavior of the singularity solution at the vertex of the characteristic cone. On the base of all these results we find that an analogous situation is true for the more general boundary value problems: there exist infinite number of smooth right hand sides functions, the corresponding general solutions for which are singular. Some weight a priori estimates are proved.
Key Words and Phrases: *Wave equation, boundary value problems, generalized solution, singular solutions, propagation of singularities.*
1991 Mathematics Subject Classification. 35L05, 35L20, 35D05, 35A20.

1. Introduction and Statement of the Problem

We investigate some boundary value problems for the wave equation in a simply connected region $\Omega \subset \mathbb{R}^3$, where, using $x_1 = \varrho\cos\varphi$, $x_2 = \varrho\sin\varphi$, t,

$$\Omega := \{(\varrho, \varphi, t) | t > 0,\ 0 \le \varphi < 2\pi,\ t < \varrho < 1 - t\}.$$

The domain Ω is bounded by the disc $S_0 := \{(\varrho, \varphi, t) | t = 0, \varrho < 1\}$ and the characteristics of the wave equation

$$S_1 := \{(\varrho, \varphi, t) | 0 \le \varphi < 2\pi, \varrho = 1 - t\},\quad S_2 := \{(\varrho, \varphi, t) | 0 \le \varphi < 2\pi, \varrho = t\}.$$

We seek sufficient conditions for the existence and uniqueness of generalized solution of the problem:

Problem P3. Find a solution of the wave equation

$$\Box u := \Delta_x u - u_{tt} = \frac{1}{\varrho}(\varrho u_\varrho)_\varrho + \frac{1}{\varrho^2}u_{\varphi\varphi} - u_{tt} = f \text{ in } \Omega, \tag{1.1}$$

which satisfies the boundary conditions

$$\underline{P3} \quad u|_{S_1} = 0, \quad [u_t + \alpha(\rho)u]|_{S_0} = 0 \quad , \tag{1.2}$$

with $\alpha \in C[0,1]$.

We observe that the adjoint problem is:

Problem P3*. Find a solution of the wave equation (1.1) in Ω, with the boundary conditions:

$$\underline{P3^*} \quad u|_{S_2} = 0, \quad [u_t + \alpha(\rho)u]|_{S_0} = 0 \quad . \tag{1.3}$$

M.H. Protter [2] formulated the next problems $P1$ and $P1^*$ in Ω.

Protter's Problems. Find a solution of the wave equation (1.1) with boundary conditions

$$\begin{array}{llll} \underline{P1} & u|_{S_0 \cup S_1} = 0 \quad , & \underline{P1^*} & u|_{S_0 \cup S_2} = 0 \, ; \\ \underline{P2} & u|_{S_1} = 0, u_t|_{S_0} = 0 \quad , & \underline{P2^*} & u|_{S_2} = 0, u_t|_{S_0} = 0 \, . \end{array} \tag{1.4}$$

We mention here that the boundary conditions in problem $P1^*$ (respectively $P2^*$) are the adjoint boundary conditions to problem $P1$ (respectively $P2$) for equation (1.1) in Ω. M.H. Protter [2] investigated problems $P1$ and $P1^*$ in Ω as multi-dimensional analogues of the Darboux problem on the plane. It is well known that the corresponding Darboux problems in $\mathbb{R}^2$ are well posed, whereas this is not true for the Protter's problems in $\mathbb{R}^3$. For known results about the problems (1.4) see papers of Popivanov, Schneider [3], [1] and the references cited there. For further publications in this area see [4–8]. S.A. Aldashev studied the Problems $P3$ and $P3^*$ in [9]. He claimed that the corresponding solution exists, but he didn't mention any possible singular solutions in [9]. In our opinion that approach need more careful treatment. It is interesting that A. Bazarbekov [10] gave another analogue of the classical Darboux problem in the same domain Ω. It is known now (see [11], [12]).

Theorem 1.1. *For all $n \in \mathbb{N}, n \geq 4; a_n, b_n$ arbitrary constants, the functions*

$$v_n(\varrho, \varphi, t) = t\varrho^{-n}[\varrho^2 - t^2]^{n-\frac{3}{2}}(a_n \cos n\varphi + b_n \sin n\varphi) \tag{1.5}$$

are classical solutions of the homogeneous problem $P1^*$ and the functions

$$w_n(\varrho, \varphi, t) = \varrho^{-n}[\varrho^2 - t^2]^{n-\frac{1}{2}}(a_n \cos n\varphi + b_n \sin n\varphi) \tag{1.6}$$

are classical solutions of the homogeneous problem $P2^*$.

We mention here the new paper of Khe Kan Cher [8] where some non-trivial solutions were found for the homogeneous Problems $P1^*$ and $P2^*$ but for Euler-Poisson-Darboux equation. It has a short connection with results of the Theorem 1.1. Theorem 1.1 shows that for the classical solvability of problem $P1$ the function f must at least be orthogonal to all functions (1.5) and in case of problem $P2$ orthogonal to all functions (1.6). Using Theorem 1.1 Popivanov, Schneider [1] proved the existence of some *generalized solutions* of Problems $P1$ and $P2$ which have at least power-type singularities at the vertex (0,0,0) of the cone S_2. For the homogeneous Problem $P3^*$ (instead of the case $\alpha \equiv 0$, i.e. Problem $P2^*$), we do not know some analogous solutions as (1.5) and (1.6). Anyway, we prove an existence of many singular solutions. We begin with the definition of solution with a possible singularity at (0,0,0).

Definition 1.1. *A function $u(\varrho, \varphi, t)$ is called a generalized solution of problem P3 in Ω, if*

i) $u \in C^1(\bar{\Omega} \setminus (0,0,0))$, satisfies the conditions (1.2) and

ii) the identity

$$\int_{\Omega} [u_t v_t - u_\varrho v_\varrho - \frac{1}{\varrho^2} u_\varphi v_\varphi - f v] \rho d\rho d\varphi dt = \int_{S_0} \alpha(\rho) u v \rho d\rho d\varphi$$

holds for all

$$v \in V_0 = \{v \in C^1(\bar{\Omega}) \ : [v_t + \alpha(\rho) v]|_{S_0} = 0, \ v \equiv 0 \text{ in a neighbourhood of } S_2\}.$$

2. Existence and Uniqueness Results

Theorem 2.1. *There exists at most one generalized solution of problem P3 in Ω.*

Remark 2.1. *We mention here that in 1960 P. Garabedian proved [13] the uniqueness of classical solution of the Problem P1. Existence and uniqueness results for the generalized solution of Problems P1 and P2 you could find in [3], [1].*

We begin here with the special, but main case of existence. The final result we will give in the Theorem 3.1.

Theorem 2.2. *Let the function $f \in C^\circ(\bar{\Omega})$ have a form*

$$f(\rho, \varphi, t) = \sum_{n=0}^{N} \{f_n^{(1)}(\rho, t) \cos n\varphi + f_n^{(2)}(\rho, t) \sin n\varphi\}. \tag{2.1}$$

Then there exists one and only one generalized solution of the Problem P3

$$u(\varrho, \varphi, t) \in C^1(\bar{\Omega} \setminus (0,.,0)). \tag{2.2}$$

3. A Priori Estimates

Let we denote: $M_\alpha := \sup |\alpha(\rho)|, \rho \in [0,1], \quad \Omega_\varepsilon := \Omega \cap \{\rho \geq \varepsilon\}$.

Theorem 3.1. *Suppose $f(x_1, x_2, t) \in C^1(\bar{\Omega})$ has a form*

$$f(\rho, \varphi, t) = \sum_{n=0}^{\infty} \{f_n^{(1)}(\rho, t) \cos n\varphi + f_n^{(2)}(\rho, t) \sin n\varphi\}$$

and for the Fourier coefficients $f_n^{(1)}(\rho, t)$ and $f_n^{(2)}(\rho, t)$ of the function f we have

$$\begin{aligned} \|f\|_{\exp(\varepsilon)} &:= \exp\left(\frac{1}{4\varepsilon^2}\right) \|f_0^{(1)}\|_{C^\circ(\bar{\Omega})} \\ &+ \sum_{n=1}^{\infty} \frac{1}{n^2} \exp\left(\frac{n^2}{\varepsilon^2}\right) \left(\|f_n^{(1)}\|_{C^\circ(\bar{\Omega})} + \|f_n^{(2)}\|_{C^\circ(\bar{\Omega})}\right) < \infty . \end{aligned} \tag{3.1}$$

Then there exist one and only one generalized solution $u \in C^1(\bar{\Omega} \setminus (0,0,0))$ of the problem P3 and the a priori estimate

$$\|u\|_{C^1(\bar{\Omega}_\varepsilon)} \leq 8 \exp(2M_\alpha) \|f\|_{\exp(\varepsilon)} \tag{3.2}$$

holds.

Remark 3.1. *The condition* (3.1) *is valid for each* $\varepsilon > 0$ *if there exists a function* ψ *with* $\psi(n) \to \infty$ *as* $n \to \infty$ *such that*

$$\sum_{n=1}^{\infty} \left(\|f_n^{(1)}\|_{C^0(\bar{\Omega})} + \|f_n^{(2)}\|_{C^0(\bar{\Omega})} \right) \frac{1}{n^2} \exp(n^2\psi(n)) < \infty. \tag{3.3}$$

Remark 3.2. *The a priori estimate for the smooth solutions of the Problem P1 in* Ω *(but for the wave equation with some necessary lower terms) one could find in G. Karatoprakliev [14]. First a priori estimates for the original Protter's Problems P1 and P2 about the wave equation in* $\mathbb{R}^3$ *was proved in [1] for the possible singular solutions.*

4. Singularities of the Generalized Solution

For Problems $P1$ and $P2$ Popivanov and Schneider announced in [12] that there exist some singular solutions for both wave equation and degenerate hyperbolic equation. The proofs of that results are in [1] and [3] respectively. S.A. Aldashev (after [12]) showed in [4] that there are some solutions with singularities to on the cone S_2 of order $\varepsilon^{-(n+m-2)}$ (or $\varepsilon^{-(n+m-1)}$) when the cones $S_{2\varepsilon} := \{\rho = t + \varepsilon\}$ approximate S_2, for the Problem $P1$ (or $P2$ respectively). His result is valid for the wave equation with m space variables (compare this result with the singularities (4.2), (4.3) in the next theorem).

Theorem 4.1. *For each function* $\alpha(\rho) \in C[0,1], \alpha(\rho) \geq 0$ *and* $n \in \mathbb{N}$, $n \geq 4$, *there exists a function*

$$f_n(x_1, x_2, t) := f_n^{(1)}(\varrho, t) \cos n\varphi \in C^{n-2}(\bar{\Omega}) \tag{4.1}$$

such that for the corresponding generalized solution *of the problem P3 from Theorem 2.2 we have*

$$\begin{gathered} u_n(x_1, x_2, t) := u_n(\varrho, \varphi, t) = u_n^{(1)}(\varrho, t) \cos n\varphi \in C^1(\bar{\Omega} \setminus (0,.,0)) \,, \\ u_n^{(1)}(\varrho, \varrho) \geq \varrho^{-n} \,. \end{gathered} \tag{4.2}$$

Theorem 4.2. *In the case* $\alpha(\rho) \equiv 0$, *for the generalized solution* $u_n(x_1, x_2, t)$ *from* (4.2) *we have in addition:*

$$u_n^{(1)}(\varrho, 0) \leq \varrho^{-n}, \tag{4.3}$$

i.e. in the case of Problem P2 the exact behavior of $u_n(x_1, x_2, t)$ *around (0,0,0) is* $(x_1^2 + x_2^2)^{-n/2}$.

It follows that these solutions have strong singularities at the vertex (0,0,0) of the cone S_2. These singularities of the generalized solutions do not propagate in the direction of the bicharacteristics of the characteristic cone. It is traditionally assumed that the wave equation with right-hand side sufficiently smooth in $\bar{\Omega}$ cannot have a solution with an isolated singular point (For results about the propagation of singularities for second order operators see L. Hörmander [15], Chapter 24.5.). But this rule is not without exceptions, as you can see in the results mentioned above. For the plane Darboux-Problem A.M. Nakhushev proved the analogous result [16].

Remark 4.1. *It is not so strange, that the norm* (3.1) *from the right-hand side of the estimate* (3.2) *blows up for* $\varepsilon = 0$. *The reason is that the singular generalized solutions* $u_n(x_1, x_2, t)$ *from Theorem 4.1 satisfies the same a priori estimate. We mention that we have proved the power-type singularity at* $(0,0,0)$ *for the special functions* $u_n(x_1, x_2, t)$. *It is possible that there exist some other solutions with some higher type of singularity. Anyway, if that singularity exists it is surely less than that exponential one from* (3.1), *according to Theorem 3.1.*

For another approach to Protter's problems $P1$ and $P2$ we refer to D. Edmunds, N. Popivanov [17]. To regularize the ill-posed Protter's Problems for the Tricomi equation in the hyperbolic part only, they have used a nonlocal regularization. For example, instead of the wave equation it can be considered a nonlocal equation

$$\Box\omega(\varrho,\varphi,t) - \frac{1}{\varrho^2}\omega_{\varphi\varphi}\left(\sqrt{\varrho t},\varphi,\sqrt{\varrho t}\right) = f(\varrho,\varphi,t) \tag{4.4}$$

with boundary conditions

$$\omega|_{S_0\cup S_1} = 0. \tag{4.5}$$

The new nonlocal problem (4.4), (4.5) is well posed and uniqueness and existence theorems are proved. It was shown that for some class of functions $f(\varrho,\varphi,t)$ the solutions of the problems $P1$ and (4.4), (4.5) coincide.

Acknowledgements. This research was partially supported by the Bulgarian NSF under Grant MM-512/95.

References

1. N.I. Popivanov and M. Schneider, J. Math. Anal. Appl., **194**, 50-77 (1995).
2. M.H. Protter, J. Rat. Mech. Anal., **3**, 435-446 (1954).
3. N.I. Popivanov and M. Schneider, J. Math. Anal. Appl., **175**, 537-579 (1993).
4. S.A. Aldashev, Ukrainian Math. J., **45**, 1456-1464 (1993).
5. S.A. Aldashev, Differ. Equations, **34**, 64–68 (1998).
6. Khe Kan Cher, Math. Notices of Jacutsk State Univ., **2**, 105-109 (1995).
7. Jong Duek Jeon, Khe Kan Cher, Ji Hyun Park, Yong Hee Jeon, Jong Bae Choi, J. Korean Math. Soc., **33**, 857-863 (1996).
8. Khe Kan Cher, Differ. Equations, **34**, 133–135 (1998).
9. S.A. Aldashev, Differ. Equations, **19**, No. 1, 1–6 (1983).
10. Ar. B. Bazarbekov and Ak.B. Bazarbekov, Differ. Equations, **30**, 741-748 (1994).
11. Tong Kwang-Chang, Science Record, New Series, **1**, 1-3 (1957).
12. N.I. Popivanov and M. Schneider, Compt. Rend. Acad. Bulg. Sci., **41**, No. 11, 7-9 (1988).
13. P.R. Garabedian, J. Math. Mech., **9**, 241-271 (1960).
14. G.D. Karatoprakliev, Differ. Equations, **18**, 49-53 (1982).

15. L. Hörmander, The Analysis of Linear Partial Differential Operators III., Springer-Verlag, Berlin -Heidelberg-New York-Tokyo (1985).

16. A.M. Nakhushev, Differ. Equations, **28**, 1445-1457 (1992).

17. D.E. Edmunds and N.I. Popivanov, SIAM J. Math. Anal., **29**, No. 1, 85-105 (1998).

9th Int. Coll. on Differential Equations, pp. 331-335
D. Bainov (Ed)

Blow Up of the Solutions for Quasilinear Hyperbolic Systems and Application to the Nonlinear Vibrating String Equation

PETER POPIVANOV
Institute of Mathematics, Bulgarian Academy of Sciences, Sofia 1113, Bulgaria
ANGELA SLAVOVA
Department of Mathematics, University of Mining and Geology, Sofia 1100, Bulgaria

Abstract: This paper deals with the blow up of the solutions of strictly hyperbolic quasilinear systems in the plane. An application to the Cauchy problem for the nonlinear string equation is given.
Key Words and Phrases: *quasilinear hyperbolic system, blow up of the solution, Cauchy problem*

1. Introduction and Formulation of the Main Results

This paper deals with blow up of the solutions of the Cauchy problem for several classes of quasilinear first order hyperbolic systems in the plane. Moreover, an application of the previous results to the nonlinear vibrating string equation is proposed.

There are a lot of papers and several books devoted to the subject. Because of the lack of space we shall mention only the papers of P.Lax [1], F. John [2] and S.Klainerman and A.Majda [3] as well as the monographs of L. Hörmander [4], S.Alinhac [5], Li Tatsien [6]. Recently the paper of Colombini-Del Santo [7] discussed the blow up problem for periodic initial data, while R.Manfrin announced at the conference in Seiffen (Germany) in April 1998 theorems on the blow up of the solutions of the Cauchy problem when the norms of the Cauchy data are not sufficiently small and they do not possess a limit for $x \to +\infty$.

We study here hyperbolic systems of the type:

$$\begin{aligned} \frac{\partial s}{\partial t} + \lambda(r,s)\frac{\partial s}{\partial x} &= 0, \\ \frac{\partial r}{\partial t} + \mu(r,s)\frac{\partial r}{\partial x} &= 0, \quad t > 0 \\ r|_{t=0} = r_0(x), \quad s|_{t=0} &= s_0(x), \end{aligned} \tag{1}$$

$\lambda(r,s) < \mu(r,s)$, $\lambda, \mu \in C^1$, and we assume that the Cauchy data $r_0, s_0 \in C^1(R^1)$ possess sufficiently small C^0 norms (uniform norms). We can expect blow up of the derivatives $s_x(t,x)$, $r_x(t,x)$ for a finite time $T > 0$. In this paper we shall consider two different cases.

Case 1.

$$(A) \qquad \frac{\partial \lambda}{\partial s} \geq 0, \ s_0'(x) \geq 0, \ \frac{\partial \mu}{\partial r} \geq 0.$$

It is shown in [6] that s_x can not blow up for a finite time $T > 0$. So a blow up is possible for r_x only. Evidently, s_0 is periodic if and only if $s_0 = const$.

We shall formulate our first result.

Theorem 1 *Assume that under condition (A) the following condition (B) is fulfilled:*

$$(B) \quad \int^{+\infty} \frac{\partial \mu}{\partial r}(r_0(\beta), s_0(\alpha))\, d\alpha = +\infty$$

and one can find a point $\beta \in R^1$ and such that $r_0'(\beta) < 0$.

Then r_x blows up in a finite time $T > 0$ supposing the uniform norms of the Cauchy data r_0, s_0 to be sufficiently small.

Remark 1. The integral in (B) is taken from an arbitrary finite α_0 till $+\infty$ and therefore the divergence of the integral (B) at $+\infty$ is important only. Assuming the Cauchy data to be $\varepsilon s_0(\alpha)$, $\varepsilon r_0(\beta)$, where $0 < \varepsilon \leq \varepsilon_0$ is a sufficiently small parameter we can rewrite the condition (B) in the following form:

$$\int^{+\infty} \frac{\partial \mu}{\partial r}(\varepsilon r_0(\beta), \varepsilon s_0(\alpha))\, d\alpha = +\infty.$$

This is a typical growth condition for $s_0(\alpha)$ at $+\infty$.

Case 2. Assume that $\frac{\partial \lambda}{\partial s} \geq 0$, $s_0'(x) \geq 0$ but $\frac{\partial \mu}{\partial r}$ is not obliged to conserve its sign near $(0,0)$. For the sake of simplicity we shall suppose that $\alpha s_0'(\alpha)_{\alpha \to +\infty} \to d$, $d \geq 0$.

In this situation the next result holds.

Theorem 2 *Suppose that in case 2 the following condition (C) is valid:*

$$(C) \quad \begin{cases} \int_0^\infty |\mu_r(r_0(\beta), s_0(\alpha))|\, d\alpha \leq const. \\ \beta^2 |r_0'(\beta)| \leq const., \forall \beta \leq 0. \end{cases}$$

Then for each Cauchy data r_0, s_0 having sufficiently small uniform norms there exists a unique global in $t \geq 0$ solution (s, r) of the hyperbolic system (1).

And now we propose our application of Theorem 1 to the study of the nonlinear vibrating string equation. We remind to the reader that this equation is written in the form

$$u_{tt} - (K(u_x))_x = 0, \tag{2}$$
$$u|_{t=0} = u_0(x), \quad u_t|_{t=0} = u_1(x).$$

The function $K \in C^2(R^1)$, $u_0(x) \in C^2(R^1)$, $u_1 \in C^1(R^1)$ and the uniform norms $||u_1||_{C^0(R^1)}$, $||u_0||_{C^1(R)}$ are supposed to be sufficiently small. Moreover, we suppose that

$$K(0) = 0, \;\; K' > 0 \text{ everywhere}, \;\; K'' \leq 0 \text{ everywhere}. \tag{3}$$

Introduce now the functions

$$s_0(x) = \frac{1}{2}\left(u_1(x) + \int_0^{u_0'} \sqrt{K'(\lambda)}\, d\lambda\right), \tag{4}$$
$$r_0(x) = \frac{1}{2}\left(u_1(x) - \int_0^{u_0'} \sqrt{K'(\lambda)}\, d\lambda\right).$$

Let $v = G(\nu) \in C^2$ be the inverse function of the strictly monotonically increasing function $\nu = \int_0^v \sqrt{K'(\lambda)}\, d\lambda \in C^2(R^1)$.

This is our last result.

Theorem 3 *Consider the Cauchy problem (2) with sufficiently small uniform norms* $||u_1||_{C^0(R^1)}$, $||u_0||_{C^1(R)}$ *and suppose that* $s_0' \geq 0$, $r_0'(\beta) < 0$ *at some point* $\beta \in R'$ *and*

$$(B') \quad \int^{+\infty} \frac{K''}{K'}(G(-r_0(\beta) + s_0(\alpha))d\alpha = -\infty.$$

Then the classical solution $u(t, x)$ *of (2) will develop a singularity in a finite time.*

Certainly, the condition (B') is rather implicit. Because of this reason we assume, moreover, that

(D) $K' > 0$ and $K'' \leq 0$ is not identically zero on any open interval of the real line.

Geometrically condition (D) means that the curve $\Gamma : t = K(x)$ is strictly monotonically increasing and (nonstrictly) concave. More precisely, at each point $(t_0, x_0) \in \Gamma$, in which $K''(x_0) = 0$, the corresponding tangential line l to Γ passing through (t_0, x_0) has a second order contact with Γ.

Corollary 1 *Consider Cauchy problem (2) in* $[0, +\infty) \times R_x$ *and assume that the condition (D) holds, the function* $s_0(x)$ *given by (4) is monotonically increasing and* $r_0(x)$ *is not monotonically increasing. Then the classical solution* $u(t, x)$ *of (2) with sufficiently small norms* $||u_1||_{C^0(R^1)}$, $||u_0||_{C^1(R)}$ *will develop singularity in a finite time* $T > 0$.

We shall prove in this Note only how to deduce Theorem 3 from Theorem 1.

2. Proof of Theorem 3

We reduce the second order nonlinear hyperbolic equation (2) via the change $v = u_x$, $w = u_t$ to the system

$$\begin{aligned} v_t &= w_x, \\ w_t &= K'(v)v_x. \end{aligned} \tag{5}$$

This is a strictly hyperbolic quasilinear homogeneous system with two distinct characteristic roots $\lambda = -\sqrt{K'(v)} < 0$, $\mu = \sqrt{K'(v)} > 0$.

Introduce now the Riemann invariants r, s:

$$\begin{aligned} 2r &= w - \int_0^v \sqrt{K'(\lambda)}d\lambda \\ 2s &= w + \int_0^v \sqrt{K'(\lambda)}d\lambda, \end{aligned} \tag{6}$$

hence,

$$\begin{aligned} r + s &= w, \\ s - r &= \int_0^v \sqrt{K'(\lambda)}d\lambda, \end{aligned} \tag{7}$$

and consider the strictly monotonically increasing function $\nu = \int_0^v \sqrt{K'(\lambda)}d\lambda$. As we know $v = G(\nu)$ is its inverse, i.e. (7) implies that $v = G(s - r)$. Then the system (5) reduces globally and in an equivalent way to the system

$$\begin{aligned} &\frac{\partial s}{\partial t} - k(s - r)\frac{\partial s}{\partial x} = 0, \\ &\frac{\partial r}{\partial t} + k(s - r)\frac{\partial r}{\partial x} = 0, \\ &s|_{t=0} = s_0, r|_{t=0} = r_0, \end{aligned} \tag{8}$$

where $k(\nu) = \sqrt{K'(G(\nu))}$, r_0, s_0 are given by (4).

Obviously, (8) is a system of the type (1) with $\lambda(r,s) = -k(s-r) < 0$, $\mu(r,s) = k(r-s) > 0 \Rightarrow \frac{\partial\lambda}{\partial s} = -k'(s-r)$, $\frac{\partial\mu}{\partial r} = -k'(s-r)$. Having in mind that $k'(\nu) = \frac{1}{2}\frac{K''(G(\nu))}{K'(G(\nu))}$ we see that we are in the case (A).

Condition (B) takes the form

$$-\int^{+\infty} k'(s_0(\alpha) - r_0(\beta))d\alpha = +\infty, \tag{9}$$

i.e. we obtain the condition (B'). Applying Theorem 1 to the system (8) we get the desired conclusion for the Cauchy problem (2).

Corollary is proved in the following way. As $s_0' \geq 0$ and s_0 is bounded then $s_0(\alpha)_{\alpha\to\infty} \to s_0(+\infty) \neq \infty$. If $k'(s_0(+\infty) - r_0(\beta)) \neq 0$ then the integral (9) is divergent. Assuming $k'(s_0(+\infty) - r_0(\beta)) = 0$ we apply the condition (D) and we conclude that near β one can find a point $\bar{\beta}$ and such that $r_0'(\bar{\beta}) < 0$, $k'(s_0(+\infty) - r_0(\bar{\beta})) \neq 0$. Thus everything is proved.

Remark 2. As it was shown, nonlinear vibrating string equation (2) was equivalently reduced to the system (8). Consider now the Cauchy problem (8) with the following initial data: $s|_{t=0} = \varepsilon s_0(x)$, $r|_{t=0} = \varepsilon r_0(x)$, where $\varepsilon > 0$ is a small parameter and the uniform norms of r_0, s_0 are bounded. It is interesting to estimate the life span time $T(\varepsilon)$ of the C^1 solution of the system (8). In this direction we shall formulate the following result from [6] for compactly supported and nontrivial initial functions s_0, r_0. So assume that $k \in C^1$, $k(0) > 0$ and there exists an integer $p \geq 1$ such that

$$k'(0) = \ldots = k^{(p-1)}(0) = 0, \ k^{(p)}(0) \neq 0.$$

Then one can find $\varepsilon_0 > 0$, so small that for any given $\varepsilon \in (0, \varepsilon_0]$ the C^1 solution (s,r) of the Cauchy problem (8) will blow up in a finite time. The life span time $T(\varepsilon)$ of the solution can be estimated in the next way:

$$C_1\varepsilon^{-p} \leq T(\varepsilon) \leq C_2\varepsilon^{-p},$$

where C_1, C_2 are constants independent of $\varepsilon > 0$.

Now it is classical (see [3]) that for sufficiently small, nontrivial periodic initial data the classical C^2 solution of the problem (2) develops singularity at finite time.

So the following open problem arise : to estimate the life span time of (8) under the condition (9).

Acknowledgements

The first author was partialy supported by the grant between Bulgarian Academy of Sciences and CNR- Italy "Nonlinear hyperbolic equations and applications to mechanics".

References

1. P.D.Lax, J. Math. Physics, **5** , 611- 613 (1964).

2. F. John, Comm. Pure Appl. Math., **27**, 235-268 (1974).

3. S.Klainerman and A.Majda, Comm. Pure Appl. Math., **33**, 241-263 (1980).

4. L.Hörmander, Nonlinear Hyperbolic Differential Equations, N 1988:2/ISSN 0327-845, Univ. of Lund, Sweden (1988).

5. S. Alinhac, Blow Up for Nonlinear Hyperbolic Equations, Birkhäuser, Boston (1995).

6. Li Ta-tsien, in: Research in Appl. Math., P.Ciarlet and J.L.Lions (Eds.), John Wiley and Sons (1994).

7. F.Colombini and D.Del Santo, Ann. Univ. Ferrara, **XLI**, 15-28 (1995).

8. A.Majda, Compressible Fluid Flow and Systems of Conservation Laws in Several Spaces Variables, Springer-Verlag, N.Y. (1984).

9. P.Popivanov and A.Slavova, (in preparation), (1998).

9th Int. Coll. on Differential Equations, pp. 337-342
D. Bainov (Ed)

Regularity of the First Derivatives of the Solutions of a Class of Linear Elliptic Equations of Second Order with Discontinuous Coefficients

MARIA ALESSANDRA RAGUSA
Dipartimento di Matematica, Università di Catania,
Viale A. Doria, 6, 95125 Catania, Italy
maragusa@dipmat.unict.it

Key Words and Phrases: *Linear elliptic operator in divergence form, Calderon-Zygmund kernel, a priori estimates, integral operators.*

Dedicated to the memory of my teacher Filippo Chiarenza, who I am greatly indebted to.

1. Introduction

In this lecture the author wish as to review some recent results on the $L^{p,\lambda}$ regularity of the derivatives of the solutions to a certain class of linear elliptic equations in divergence form with discontinuous coefficients. We assume that p is of any value in the range $]1;+\infty[$ and $0 < \lambda < n$. The crucial assumption on which we dwell more is that the coefficients belong to the Sarason class VMO of vanishing mean oscillation functions (see Section 2 for its definition). The classical method to obtain the estimates for the first derivatives of the solutions when the coefficients are continuous is, according to the author's knowledge, first to establish the same result for constant coefficient operators and then, via a perturbation argument, to extend it to the variable coefficient case. This technique of pointwise perturbation about the constant coefficients is known in literature as " Korn's trick " or " freezing method ". Some applications of the freezing method to second order and higher order elliptic equations with discontinuous coefficients are studied by C. B. Morrey and C. Simader in their books [1] and [2].

Let us assume the coefficients to be discontinuous, more precisely in the class VMO. The first step is to establish a representation formula for the first derivatives of the solutions directly in the case of variable coefficients. These derivatives are expressed by means of singular integral operators and commutators applied to the known term. These derivatives appear in the commutators whose norm can be made small if the coefficients have a small integral oscillation. The use of the class VMO is now clear. The method above mentioned is interesting for the author because it allows us to obtain estimates in the Morrey norm of the gradient of the solutions using only well known real analysis results and the properties of $L^{p,\lambda}$ spaces. This technique which uses explicit representation formulas for the derivatives of the solutions is associated with the names of A. Calderón and A. Zygmund because they examined this method closely, see e. g. the results contained in the pioneer papers [3] and [4]. It has been used by many authors in order to study the regularity problem in Hölder spaces, for example E. Levi and C. Miranda [5].

This method has been applied in the study of nondivergence form equations by several other authors, see e. g. the papers by A. Koshelev [6], D. Greco [7] and the book written by D. Gilbarg and N. Trudinger [8]. More recently it has been used in the noteworthy papers by Chiarenza, Frasca and Longo [9], [10] and the results I am going to deal with, follow the same procedure as the two above mentioned papers. The technique we are mentioning here has been applied also in the study of quasilinear elliptic equations by D. K. Palagachev in the significant paper [11].

2. Definitions and Assumptions

Let us suppose that in the following Ω be an open bounded set in $\mathbb{R}^n$, where $n \geq 3$, and f be a locally integrable function in Ω.

Definition 2.1. *Let $1 < p < \infty$, $0 < \lambda < n$. We call Morrey space $L^{p,\lambda}(\Omega)$ the set of those measurable f for which the following norm of f is finite*

$$\|f\|^p_{L^{p,\lambda}(\Omega)} \equiv \sup_{x\in\Omega,\ \rho>0} \frac{1}{\rho^\lambda} \int_{B_\rho(x)\cap\Omega} |f(y)|^p dy.$$

Definition 2.2. *(see [12]). A function f belongs to the John-Nirenberg space BMO of the functions of bounded mean oscillation if*

$$\sup_B \frac{1}{|B|}\int_B |f(x) - f_B|dx \equiv \|f\|_* < \infty$$

where we suppose that B ranges in the class of the balls contained in Ω and $f_B \equiv \frac{1}{|B|}\int_B f(x)dx$.

We denote the number $\|f\|_*$ as the norm of f in BMO (see the paper of U. Neri [13]).

Definition 2.3. *(see [14]). Let $f \in BMO$ and r is a positive number. We set*

$$\eta(r) \equiv \sup_{x\in\mathbb{R}^n,\ \rho\leq r} \frac{1}{|B_\rho|}\int_{B_\rho} |f(x) - f_{B_\rho}|dx$$

where B_ρ stands for a ball with radius ρ less than or equal to r. We say that $f \in BMO$ is in the space VMO if $\lim_{r\to 0^+} \eta(r) = 0$.

Moreover we set η_{ij} the VMO modulus of the function a_{ij} and $\eta(r) = (\sum_{i,j=1}^n \eta^2_{ij}(r))^{\frac{1}{2}}$. The space of vanishing mean oscillation functions just defined is the subspace of the functions in the John-Nirenberg space BMO whose BMO norm over a ball vanishes as the radius of the ball tends to zero. These VMO functions own many good properties not shared by general bounded mean oscillation functions, for example they can be approximated by smooth functions; moreover the bounded uniformly continuous functions belong to the Sarason class.

Definition 2.4. *Let us define $k : \mathbb{R}^n \setminus \{0\} \to \mathbb{R}$. We say that $k(x)$ is a Calderón-Zygmund kernel (C-Z kernel) if*
i) $k \in C^\infty(\mathbb{R}^n \setminus \{0\})$;
ii) $k(x)$ *is homogeneous of degree* $-n$;
iii) $\int_\Sigma k(x)\,dx = 0$, *where* $\Sigma = \{x \in \mathbb{R}^n : |x| = 1\}$.

Let us assume the following hypothesis

$$\exists p \in]1, +\infty[,\ \lambda \in]0, n[\text{ such that}$$

(2.1) $$f = (f_1, ..., f_n) \in [L^{p,\lambda}(\Omega)]^n$$

(2.2) $$a_{ij}(x) \in L^\infty(\Omega) \cap VMO \quad, \forall i,j = 1...n;$$

(2.3) $$a_{ij}(x) = a_{ji}(x), \quad \forall i,j = 1...n \quad \text{a.a. } x \in \Omega;$$

(2.4) $$\exists \tau > 0 : \ \tau^{-1}|\xi|^2 \leq a_{ij}(x)\xi_i\xi_j \leq \tau|\xi|^2 \quad \forall \xi \in \mathbb{R}^n, \text{ a.a. } x \in \Omega.$$

Let us consider the following elliptic second order equation in divergence form

(2.5) $$\mathcal{L}u \equiv -\sum_{i,j=1}^n (a_{ij}(x)u_{x_i})_{x_j} = div f(x) \quad \text{a. a. } x \in \Omega.$$

A function u is a solution of the above equation if $\forall i = 1, \dots, n$ and for some $1 < p < \infty$ we have $u, \partial_{x_i} u \in L^p(\Omega)$ and

$$\int_\Omega a_{ij}u_{x_i}\chi_{x_j}dx = -\int_\Omega f_i\chi_{x_i}dx \qquad \forall \chi \in C_0^\infty(\Omega).$$

Let us now define, for $a \in BMO$, $f \in L^p(\Omega)$ and $k(x, .)$ a Calderón–Zygmund kernel for a.a. $x \in \Omega$, the Calderon-Zygmund singular integral

$$Kf(x) = \int_\Omega k(x, x-y)f(y)dy,$$

and the following commutator between the operator K and the multiplication operator $a(x)$

$$C[a, f](x) = a(x)(Kf)(x) - K(af)(x) = \int_\Omega k(x, x-y)(a(x) - a(y))f(y)dy.$$

The last kind of operator was examined closely by many authors, for example by A. Calderon in [15]; R. Coifman, R. Rochberg and G. Weiss in [16].

3. Real Analysis Results

In this section we recall some results of singular integrals and commutators useful to the requested estimates.

Theorem 3.1. [17, Theorem 2.3]. *Let B be an open ball in $\mathbb{R}^n$, $f \in L^{p,\lambda}(B)$, $1 < p < \infty$, $0 < \lambda < n$, $a \in BMO(B)$. Let $k(x, z)$ be a real measurable function in $B \times (\mathbb{R}^n \setminus \{0\})$ such that*
i) $k(x, .)$ *is a Calderón–Zygmund kernel for a. a.* $x \in B$;
ii) $\max_{|j| \leq 2n} \left\| \frac{\partial^j}{\partial z^j} k(x, z) \right\|_{L^\infty(B \times \Sigma)} = M < +\infty$.

Let us define, for any $\varepsilon > 0$,

$$K_\varepsilon f(x) = \int\limits_{\substack{|x-y|>\varepsilon \\ x\in B}} k(x, x-y) f(y) dy, \quad C_\varepsilon[a, f](x) = \int\limits_{\substack{|x-y|>\varepsilon \\ x\in B}} k(x, x-y)(a(x) - a(y)) f(y) dy.$$

Then, there exist Kf, $C[a, f] \in L^{p,\lambda}(B)$ such that

$$\lim_{\varepsilon\to 0} \|K_\varepsilon f - Kf\|_{L^{p,\lambda}(B)} = 0 \quad \text{and} \quad \lim_{\varepsilon\to 0} \|C_\varepsilon[a, f] - C[a, f]\|_{L^{p,\lambda}(B)} = 0.$$

Moreover there exists a constant $c = c(n, p, \lambda, M)$ such that

$$\|Kf\|_{L^{p,\lambda}(B)} \le c\|f\|_{L^{p,\lambda}(B)} \quad \text{and} \quad \|C[a, f]\|_{L^{p,\lambda}(B)} \le c\|a\|_* \|f\|_{L^{p,\lambda}(B)}.$$

The proof of this theorem follows from classical results about the expansion of the kernel into spherical harmonics (see e.g. [18]). Therefore as a consequence of the above theorem the next result follows.

Theorem 3.2. [17, Theorem 2.4]. *Let us suppose $a \in VMO \cap L^\infty(\mathbb{R}^n)$ and $k(x, z)$ satisfying the hypothesis of Theorem 3.1. Then for any $\epsilon > 0$, there exists $\rho_0 > 0$ such that for any ball B_r of radius $r \in]0, \rho_0[$ and $f \in L^{p,\lambda}(B_r)$, with $1 < p < \infty$ and $0 < \lambda < n$, we have*

$$\|C[a, f]\|_{L^{p,\lambda}(B_r)} \le c\,\epsilon\, \|f\|_{L^{p,\lambda}(B_r)}. \tag{3.1}$$

Theorem 3.3. [19, Lemma 2.8] *Let Ω an open subset of $\mathbb{R}^n$, $0 < \alpha < n$, $1 < \nu < \infty$, $0 < \beta < n$ such that $\beta + \alpha p < n$ and $\zeta \in C^\infty(\mathbb{R}^n \setminus \{0\})$ an homogeneous function of degree $-\alpha$. Then, for every $\delta \in L^{\nu,\beta}(\Omega)$ the operator*

$$\Lambda\,\delta(x) = \int\limits_\Omega \zeta(x-y)\,\delta(y) dy$$

is a.e. defined, belongs to $L^{p,\lambda}(\Omega)$ where $\frac{1}{p} = \frac{1}{\nu} - \frac{\alpha}{n}$, $\lambda = \beta\frac{p}{\nu}$ and there exists $c = c(\nu, \mu, \alpha) > 0$ such that

$$\|\Lambda\delta\|_{L^{p,\lambda}(\Omega)} \le c\|\delta\|_{L^{\nu,\beta}(\Omega)}.$$

4. Regularity

To express the first derivatives of a solution u of the above divergence form equations we need to define the following functions (see e. g. the books of Miranda [5] and Hellwig [20])

$$\Gamma(x, t) = \frac{1}{n(2-n)\omega_n\sqrt{\det\{a_{ij}(x)\}}} \left(\sum_{i,j=1}^n A_{ij}(x) t_i t_j\right)^{(2-n)/2}$$

$$\Gamma_i(x, t) = \frac{\partial}{\partial t_i}\Gamma(x, t), \qquad \forall i = 1, \dots, n$$

and

$$\Gamma_{ij}(x, t) = \frac{\partial}{\partial t_i \partial t_j}\Gamma(x, t), \qquad \forall i, j = 1, \dots, n.$$

Moreover we set

$$M = \max_{i,j=1,\dots,n} \max_{|\alpha|\le 2n} \left\| \frac{\partial^\alpha \Gamma_{ij}(x,t)}{\partial t^\alpha} \right\|_{L^\infty(\Omega\times\Sigma)}$$

for almost all $x \in B$ and $\forall t \in \mathbb{R}^n \setminus \{0\}$, where $A_{ij}(x)$ stand for the entries of the inverse matrix of the matrix $\{a_{ij}(x)\}_{i,j=1,\dots,n}$, and ω_n is the measure of the unit ball in $\mathbb{R}^n$.

The functions $\Gamma_{ij}(x,t)$ are Calderón–Zygmund kernels in the t variable for all x, as they are the first derivatives of an homogeneous function of degree $1-n$.

In the following theorem we use the integral representation formula for the first derivatives of a solution u of (2.5).

Theorem 4.1. *Let $a_{ij}(x)$, $\forall i,j = 1,\dots,n$, satisfying* (2.2), (2.3) *and* (2.4).

Let also u be a solution of $\mathcal{L}u = div\, f$ with $u \in L^p(\Omega)$, $\partial_{x_i} u \in L^p(\Omega)$ $\forall i = 1,\dots,n$, and $f \in \left[L^{p,\lambda}(\Omega)\right]^n$ for $0 < \lambda < n$ and $1 < p < \infty$.

Then for any compact set $K \subset \Omega$ there exists $c = c(n,p,\lambda,\tau,\eta,K,\Omega)$ such that

$$\|\partial_{x_i} u\|_{L^{p,\lambda}(K)} \le c\left(\|u\|_{L^2(\Omega)} + \|f\|_{L^{p,\lambda}(\Omega)}\right)$$

for every $i = 1,\dots,n$ $0 < \lambda < n$, $2 < p < +\infty$.

Proof. Let $r, R \in \mathbb{R}^+$, $r < R$ and $\phi \in C^\infty(\mathbb{R})$ a standard cut-off function such that for every $B_R \subset \Omega$

$$\phi(x) = 1 \quad in\, B_r \qquad \phi(x) = 0 \ , \forall x \notin B_R.$$

Then if u is a solution of (2.5) and $v = \phi u$ we have

$$L(v) = div\, G + g \tag{4.1}$$

with G and g supported in B_R, and the first derivatives of the function v can be expressed as the following representation formula

(4.2)

$$\partial_{x_i}(\phi u)(x) = \sum_{h,j=1}^{n} \text{P.V.} \int_{\mathbb{R}^n} \Gamma_{ij}(x, x-y)\left\{(a_{jh}(x) - a_{jh}(y))\partial_{x_h}(\phi u)(y) - \right.$$
$$\left. -G_j(y)\right\} dy - \int_{\mathbb{R}^n} \Gamma_i(x,x-y)\, g(y) dy + \sum_{h=1}^{n} c_{ih}(x) G_h(x)\,, \text{ for a. e. } x \in \mathbb{R}^n$$

where $c_{ih} = \int_{|t|=1} \Gamma_i(x,t)\, t_h d\sigma_t$.

From the above formula we obtain

$$\partial_{x_i} u \in L^{p,\lambda}(K)\,, \forall i = 1,\dots,n;$$

and also

$$\|\partial_{x_i} u\|_{L^{p,\lambda}(K)} \le c\left(\|u\|_{L^{p,\lambda}(\Omega)} + \|f\|_{L^{p,\lambda}(\Omega)} + \|\partial_{x_i} u\|_{L^{q,\mu}(\Omega)}\right)$$

where $q = p_*$ is such that $\frac{1}{p_*} = \frac{1}{p} + \frac{1}{n}$ and $\mu = \frac{\lambda p_*}{p}$.

Using the continuously imbedding of $L^{p,\lambda}(\Omega)$ into $L^{p',\lambda'}(\Omega)$ for $p' \le p$ and $\lambda' \le \lambda\frac{p'}{p} + n(1 - \frac{p'}{p})$ the results contained in [21] and the classical $L^{2,\lambda}$ theory we obtain the requested inequality. □

We are now ready to obtain the Hölder regularity for the first derivatives of the equation (2.5) as an immediate consequence of the above result.

Theorem 4.2. *Let* $0 < \lambda < n$, $2 < p < \infty$, *such that* $p > n - \lambda$, a_{ij}, $\forall i, j = 1, \dots, n$ *such that* (2.2), (2.3) *and* (2.4) *holds true.*

Then if $u \in L^p(\Omega)$ *is a solution of* $\mathcal{L}u = \operatorname{div} f$ *with* $\partial_{x_i} u \in L^p(\Omega)$ *for every* $i=1, \dots, n$ *and* $f \in \left[L^{p,\lambda}(\Omega)\right]^n$ *we have that for any compact set* $K \subset \Omega$ *there exists* $c = c(n, p, \lambda, \tau, \eta, K, \Omega)$ *such that* $u \in C^{(0,\alpha)}(K)$ *and the following formula is true*

$$\|u\|_{C^{(0,\alpha)}(K)} \leq c\left(\|f\|_{L^{p,\lambda}(\Omega)} + \|u\|_{L^2\Omega}\right)$$

with $\alpha = 1 - \frac{n}{p} + \frac{\lambda}{p}$.

References

1. C. B. Morrey, Multiple Integrals in the Calculus of Variations, Springer Verlag (1966).
2. C. Simader, On Dirichlet's Boundary Value Problem, Springer Verlag (1972).
3. A. Calderón, Bull. Amer. Math. Soc., **72**, 427–465 (1966).
4. A. Calderón and A. Zygmund, Acta Mathematica, **88**, 85–139 (1952).
5. C. Miranda, Partial Differential Equations of Elliptic Type, Springer Verlag (1970).
6. A. Koshelev, Mat. Sb., **38**, 359–372 (1956).
7. D. Greco, Ricerche Mat., **5**, 126–149 (1956).
8. D. Gilbarg and N. Trudinger, Elliptic Partial Differential Equations of Second Order, Springer Verlag (1983).
9. F. Chiarenza, M. Frasca and P. Longo, Ricerche Mat., **40**, 149–168 (1991).
10. F. Chiarenza, M. Frasca and P. Longo, Trans. Amer. Math. Soc., **336**, 841–853 (1993).
11. D. K. Palagachev, Trans. Amer. Math. Soc., **347**, 2481–2493 (1995).
12. F. John and L. Nirenberg, Commun. Pure Appl. Math., **14**, 415–426 (1961).
13. U. Neri, Studia Math., **41**, 63–75 (1977).
14. D. Sarason, Trans. Amer. Math. Soc., **207**, 391–405 (1975).
15. A. Calderón, Proc. Nat. Acad. Sci. USA, **53**, 1092–1099 (1965).
16. R. Coifman, R. Rochberg and G. Weiss, Ann. Math., **103**, 611–635 (1976).
17. G. Di Fazio and M.A. Ragusa, J. Funct. Anal., **112**, 241–256 (1993).
18. A. Calderón, Integrales Singulares y sus Aplicaciones a Ecuaciones Diferenciales Hyperbolicas, Cursos y Seminarios de Matematicas 3, Univ de Buenos Aires (1960).
19. S. Polidoro and M.A.Ragusa, Manuscripta Math., **96**, 371–392 (1998).
20. G. Hellwig, Partial Differential Equations, Blaisdell Publishing Co. (1964).
21. F. Chiarenza and M. Frasca, Rend. Mat. Appl., **7**, 273–279 (1987).

9th Int. Coll. on Differential Equations, pp. 343-351
D. Bainov (Ed)

Fractal Geometry for Natural-Looking Tree Generation

FARID RAIDAN AND ISSAM A.R. MOGHRABI
Natural Science Division, Lebanese American University, P.O.Box 13-5053, Beirut, Lebanon
e-mail: imoghrbi@lau.edu.lb

Abstract: Fractal geometry is a way to represent natural looking objects. Of the most famous natural objects generated by fractal are trees. We address in this work several modified popular techniques for generating random naturally looking trees with non-uniformity in the pattern of generation so as to allow for irregularities is shape. We stress statistical rather than absolute self-similarity.

1. Introduction

Sharp boundaries or smooth shapes for real entities reflect a map model or geometric bias rather than an appropriate model for nature. Fractal geometry is used to represent those models that are not smooth in shape as in the case of natural phenomena, such as coastlines. Benoit Mandelbrot recognized that the relationship between large scale structure and small scale detail is an important aspect of natural phenomena. He gave the name fractals to objects that exhibit increasing details as one zooms in closer. Mandelbrot raised an example question "How long is the coastline of Great Britain?". At first sight this question may seem trivial. Given a map one can sit down with a ruler and soon come up with a value for the length. The problem is that repeating the operation with a larger scale map yields a greater estimate of the length (Fig. 1). If one actually went to the

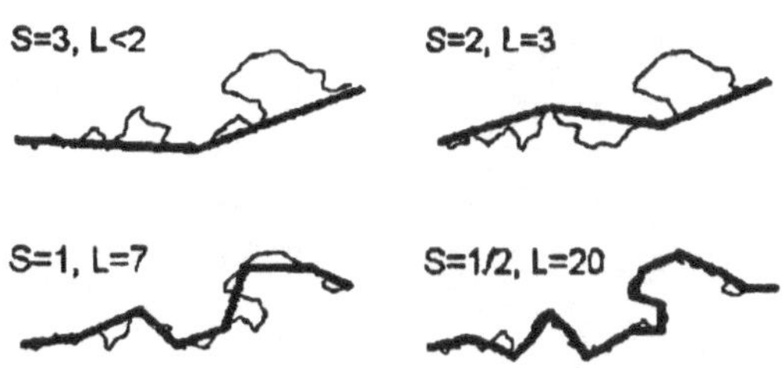

Figure 1: Using sticks of different size S to estimate the length L of the coast-line.

coast and measured them directly, then still greater estimates would result. It turns out that as the scale of measurement decreases the estimated length increases without limit.

In discussing measurement, scale can be characterized in terms of a measuring stick of a particular length: the finer the scale, the shorter the stick. Thus at any particular scale, we can think of a curve as being represented by a sequence of sticks (Fig. 1), all of the appropriate length, joined end-to-end. Clearly, any feature shorter than the stick will vanish from a map constructed this way. Of course, no one actually makes map by laying sticks on the ground, but the stick analogy reflects the sorts of distortions that are inevitably produced by the limited resolution of the photographs, or by the thickness of the pen being used in drafting.

An important difference between fractal curves and idealized curves that are normally applied to natural processes is that fractal curves are nowhere differentiable. Fractals can be characterized by the way in which representation of their structure changes with changing scale. So it is important to realize that true fractals are an idealization. No curve or surface in the real world is a true fractal; real objects are produced by processes that act only over a finite range of scales only.

Some words used within fractals:

Iterate: to repeat an operation, generally using the last result of that operation as the input.

Self-similarity: a similar appearance at all scales. More about this is shown later.

Fractal dimension: the dimension of a fractal in general is not a whole number, not an integer. Also, more on this after some more examples of fractals.

Replacement rule: in going from one stage of construction of a fractal to the next, one graphical object is replaced with another.

Fractals can be generated using recursive algorithms as well as other algorithms such as the grammar based model which is called L-system.

2. Self-Similarity

Self-similarity is symmetry across different scales; these are patterns within patterns. Fractals are geometric shapes that are equally complex in their details as in their overall form. That is, if the small scale detail resembles the large scale detail, the object is said to be self-similar. An appreciation for these concepts is best approached by seeing how particular types of objects can be created.

(a) Cantor dust:

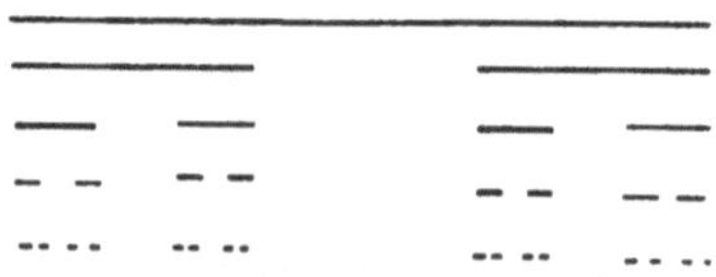

Figure 2: Cantor dust

Consider a line segment split into 3 equal parts with the middle section discarded. An indefinitely long continued process of splitting the line segments will in the end produce a set of very small aligned segments or points (whose length is tending to zero) which are called **Cantor dust** (Fig. 2).

Such a process of **pattern generation** has two components:

i. An initiator (for Cantor dust this is the line segment).

ii. A replacement rule (two other segments) or pattern generator.

At step n, the set consists of 2^n segments. The segment length is "13^{-n}", where 1 is the unit length of the initiator. Thus, for the third level of application of the recursive process, if the original line was 4 units, then each segment piece will be 4/9. For the replacement rule, its self-similarity ratio is 1/3, identifying the subdivision of the original line into 3 pieces.

(b) Sierpinski triangle:

To make a fractal from a triangle, draw lines connecting the midpoints of the sides and cut out the center. Take the result and do again and again (Fig. 3).

Figure 3:

The above fractals look the same over all ranges of scale. This is called "self-similarity".

3. Fractal Dimension

Associated with the notion of self-similarity is the notion of fractal dimension. The notion of fractal dimension provides a way to measure how rough fractal curves are. The dimension of a fractal is not an integer. So, a fractal curve (an one-dimensional object in a plane that has two dimensions) has a fractal dimension that lies between 1 and 2. Likewise, a fractal surface has a dimension between 2 and 3. The value depends on how the fractal is constructed. The closer the dimension of a fractal is to its possible upper limit which is the dimension of the space in which it is embedded, the rougher, i.e., the more filling of that space it is (Fig. 4).

To compute fractal dimensions, examine first some properties of objects whose dimension is known. For example: a line segment is 1D; if it is divided into N equal parts, each part looks like the original scaled down by a factor of $N = N^{1/1}$. A square is 2D: if it's divided into N parts, each part looks like the original scaled down by a factor of

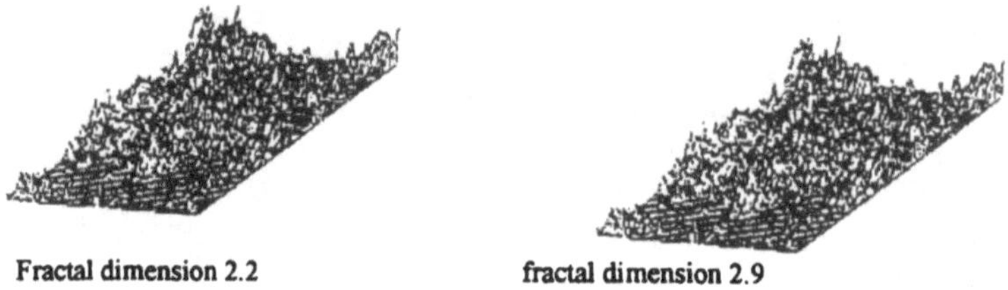

Figure 4:

$\sqrt{N} = N^{1/2}$. For example, a square divides into 9 subsquares, each looks like the original scaled down by a factor of 1/3.

Consider now the fractal, the von Koch curve. When it's divided into 4 pieces (the pieces associated with original 4 segments in Fig. 3), each resulting piece looks like the original scaled down by a factor of 3. Let d be the dimension of the von Koch curve where $4^{1/d} = 3 \Longrightarrow d = \log(4)/\log(3) = 1.26\ldots$ which is not an integer.

4. Statistical Self-Similarity

To create more natural looking shapes, that is involving variety, use randomization. This is known as stochastic fractals.

Example 1: starting with a line segment lying on the x-axis (Fig. 5a). Subdivide

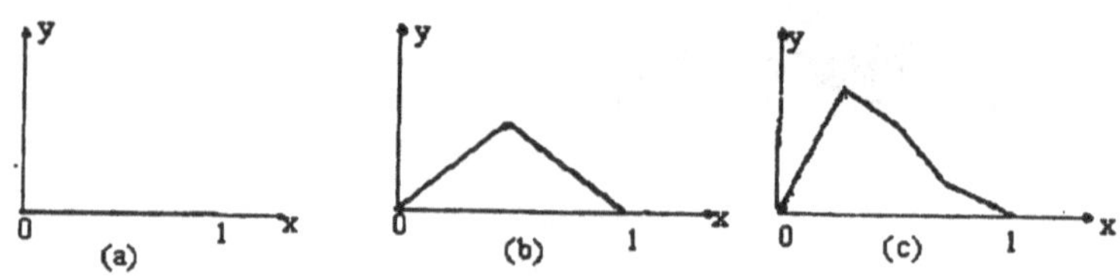

Figure 5:

the line into 2 halves and then move the mid-point some distance in the y direction (Fig. 5b). To continue subdividing each segment, compute a new value for the mid-point of the segment from (X_j, Y_j) to (X_{j+1}, Y_{j+1}) as follows: $X_{\text{new}} = 1/2(X_j + X_{j+1})$, $Y_{\text{new}} = 1/2(Y_j + Y_{j+1}) + P(X_{j+1} - X_j)^{*}R(X_{\text{new}})$; where P is a function determining the extent of the perturbation in terms of the size of the line being perturbed ($P(s)$ could be s or s^a or 2^{-s}), and $R(\,)$ is a random number between 0 and 1 selected on the basis of X_{new} (Fig. 5b).

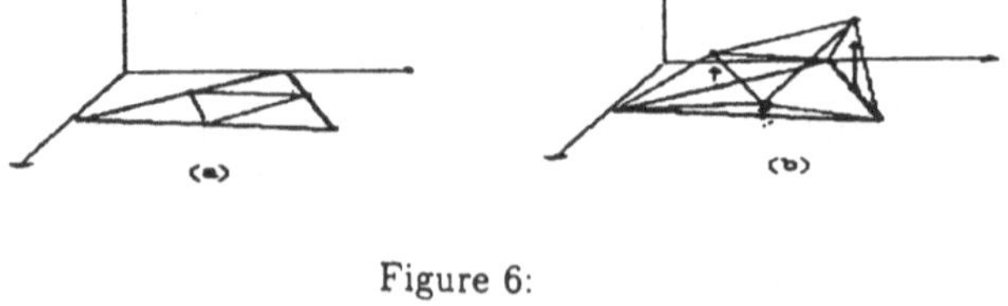

Figure 6:

This process can be used to modify 2D shapes in the following fashion. Starting with a triangle, mark the midpoints of each edge in Fig. 6a. The y coordinate is then modified in the manner described above, so that the result looks like in Fig. 6b. This process when iterated produces realistic-looking mountains.

Example 2: adding a little randomness to the Sierpinski triangle can produce something very natural looking. Change the rule so that instead of using the exact midpoints of the sides of the triangle, take a point at random around the midpoint. The self-similarity will be statistical rather than absolute. Constrain the randomness by making the point some where within the circle centered on the midpoint whose diameter is half the length of the side. Connect these points and the corners. The heavy lines are the starting triangle. In the second figure below, the heavy lines are result of the last iteration. The third figure is the result of clearing the construction and making the corners black.

These results are extremely suggestive for modeling natural forms, since many natural objects seem to exhibit striking self-similarity. Mountains have peaks and smaller peaks and rock which all look similar; trees have limbs and branches, which again look similar. Hence, modeling self-similarity of statistical self-similarity at some scale seems to be a way to generate appealing-looking models of natural phenomena. When an object has been generated recursively through enough steps that all further changes happen at well below pixel resolution, there is no need to continue.

5. L–Systems

L-systems are sets of rules and symbols (also known as "formal grammars") that model growth processes. The name "L-system" is short for "Lindenmayer System". A simple L-system contains four elements:

1. **VARIABLES** are symbols denoting elements that are replaced.

2. **CONSTANTS** are symbols denoting elements that remain fixed. e.g. The expression <subject> <verb> <predicate>
   consists of grammatical variables. Each variable may be replaced by constants (English words or phrases) to produce sentences in English, such as "The cat sat on the mat" or "The dog ate the bone".

3. **RULES** ("syntax") define how the variables are to be replaced by constants or other variables, e.g., in the above example

   $$< \text{subject} > \longrightarrow \text{the cat}$$

   would be one such rule.

4. **START** words are expressions defining how the system begins. e.g. the above examples from English might start from a single variable
<sentence>

Example: Fibonacci numbers:
Consider the simple grammar, defined as follows

Variables	:	A B
Constants	:	none
Start	:	A
Rules	:	A ⟶ B
		B ⟶ A B

This L-system produces the following sequence of strings ...

Stage 0	:	A
Stage 1	:	B
Stage 2	:	AB
Stage 3	:	BAB
Stage 4	:	ABBAB
Stage 5	:	BABABBAB
Stage 6	:	ABBABBABABBAB
Stage 7	:	BABABBABABBABBABABBAB

If the length of each string is counted, the famous Fibonacci sequence of numbers will be obtained:

1 1 2 3 5 8 13 21 34....

The power of L-systems comes when we assign meaning to the symbols and rules. For instance, the symbols might represent branches of a growing tree and the rules denote the way these symbols are to be used.

The languages are described by the grammar consisting of a collection of productions, all of which are applied at once. A typical example is the grammar with variables A and B, constants "[" and "]", and two production rules:

1. A ⟶ AA

2. B ⟶ A[B]AA[B]

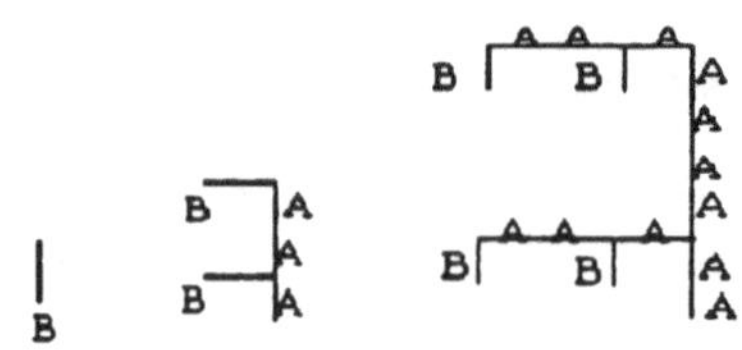

Figure 7:

Starting from axiom A, the first few generations are A, AA, AAAA, ... Starting from axiom B, the first few generations are:

1. B
2. A[B]AA[B]
3. AA[A[B]AA[B]]AAAA[A[B]AA[B]]
4. ...

Suppose that a word in the language represents a sequence of segments in a graph structure and that the bracketed portions represents portions that branch from the symbol preceding them; then the associated pictures are shown in Fig. 7.

Adding two more constants "(" and ")", and changing the second production to:

2. B $\longrightarrow$ A[B]AA(B)

Then starting from axiom B, the first few generations are:

1. B
2. A[B]AA(B)
3. AA[A[B]AA(B)]AAAA(A[B]AA(B))
4. ...

Now, suppose that square brackets denote a left branch and parenthesis denote a right branch, then the associates pictures are shown in Fig. 8.

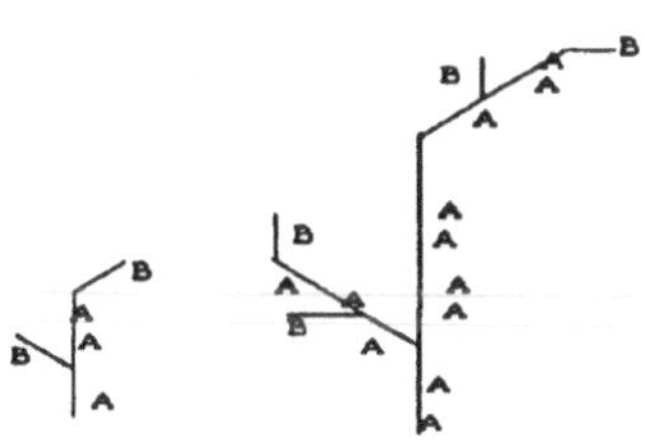

Figure 8:

By progressing to later generations in such a language, graph structures representing complex patterns will be produced. These graph structures have a sort of self-similarity, in that the pattern described by the n-th-generation is contained (repeatedly, in this case) in the $(n+1)$-th-generation word.

Choosing varying branching angles for different depth branches, and varying thickness for lines (or even cylinders) representing the segments gives different results; drawing a "leaf" or "flower" at each terminal node of the tree further enhances the picture. The grammar itself has no inherent geometric content, so using a grammar-based model requires both a grammar and a geometric interpretation of the language. At some point, additional features should be added to the grammar or to the interpretation of a word in it. The productions of the grammar are applied probabilistically rather than deterministically. In this model, start as before with a single stem. At the tip of this stem is a *bud*, which can undergo several transitions: it may die, it may flower and die, it may sleep some period of time, or it may become an *internode,* a segment of the plant between buds.

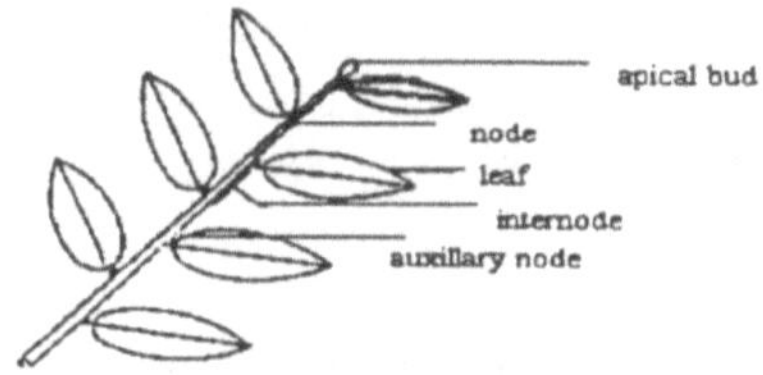

Figure 9:

The process of becoming an internode is added, and the end of the internode becomes an *apical bud* (a bud at the very end of the sequence of internodes) see Fig. 9.

Each of the buds in the resulting object can then undergo similar transitions. Suppose the initial segment of the tree is of *order-1,* define the order of all other internodes inductively: Internodes generated from the apical bud of an order-i internode are also of order-i; those generated from auxiliary buds of an order-i internode are of order (i+1).

Thus the entire trunk of a tree is of order-1, the limbs are order-2, the branches on those limbs are of order-3, and so on (Fig. 10). The placement of auxiliary buds on a sequence of order-i internodes may occur in different ways, and the angles at which the order (i+1) internodes (if any) branch out from order-i auxiliary buds also determine the shape of the plant. There are some anomalies in the tree growth, in which the behavior of a collection of order (i+1) internodes is not standard, but instead resembles that of some lower order (called reiteration), and this is too must be modeled. Finally converting this description into an actual image requires a model for the shapes of its various components: an order-1 internode may be large tapered cylinder, and an order-7 internode may be a small green line, for example. The sole requirement is that there must be a leaf at each auxiliary node.

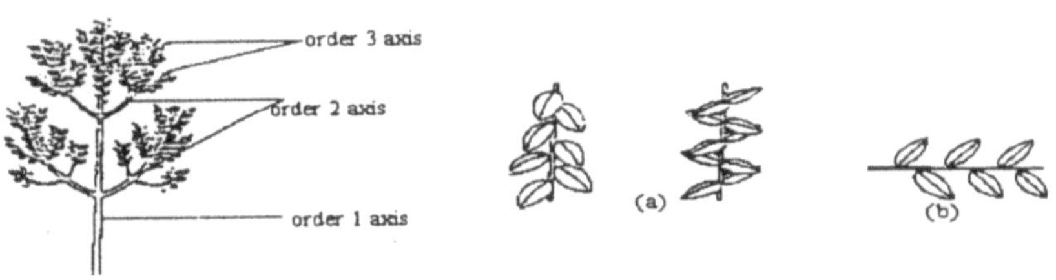

Figure 10:

Varying the values for the probabilities and the angles can produce a wide variety of extremely convincing tree models. The correct choices for these parameters depend on knowledge of plant biology or on the modeler's artistic eye; by using the wrong values, they can also generate plants bearing no resemblance at all to anything real.

6. Conclusion

Fractal geometry proves to be more adequate than Euclidean geometry for generating naturally looking objects modeling natural phenomena. Stochastic fractals (statistical self-similarity) create more "natural looking" shapes than exact self-similarity since the former involves randomization which is closer to nature because nature is not exactly self-similar. Fractals are recursive in nature; however, in producing trees for example, a similar approach could be used which is called graftals (L-grammar). This method can be used in any domain in which the object being modeled exhibit sufficient regularity.

Appendix

Sample Results

This appendix shows some sample trees generated using the package.

References

1. J.A. Ford and P.J. Cooper, Chaos and Fractals in Numerical Computation, Science and Engineering, **5**, (1991).

2. Foley, Van Dam, Feiner and Hughes, Computer Graphics: Principles and Practice, 2-nd edition, (1993), pp. 1020–1030.

3. M. McGuire, In Eye for Fractals, (1991).

4. D.G. Green, L-Systems, Environmental and Information Sciences, Charles Sturt University (1993).

5. D.G. Green, Fractals and Scale, Environmental and Information Sciences, Charles Sturt University (1993).

9th Int. Coll. on Differential Equations, pp. 353-359
D. Bainov (Ed)

Exact Solutions of the Navier-Stokes Equations

K.B. RANGER AND R.A. ROSS
Department of Mathematics, University of Toronto, 100 St.George St., Toronto, Ont. M5S 3G3, Canada

Abstract: This paper discusses two new exact solutions of the Navier-Stokes equations. In the first part of the paper the flow is time-independent and axisymmetric, and the fluid is viscous and incompressible. The flow is generated by a semi-infinite line source perpendicular to an infinite plane boundary, on which the no-slip boundary conditions are satisfied. The second part of the paper investigates the motion of a solid sphere which is both translating and rotating with time-dependent decay in a viscous incompressible fluid, relative to a uniform flow and a steady rigid body rotation of the fluid at infinity.
Key Words and Phrases: *Fluid Mechanics, Navier-Stokes equations.*

1. Exact Solution of the Time-Independent Navier-Stokes Equations

The Navier-Stokes Equations written in terms of spherical polar coordinates are

$$u\frac{\partial u}{\partial r} + \frac{v}{r}\frac{\partial u}{\partial \theta} - \frac{v^2}{r} = -\frac{1}{\rho}\frac{\partial p}{\partial r} + \sigma\left(\nabla^2 u - \frac{2u}{r^2} - \frac{2}{r^2}\frac{\partial v}{\partial \theta} - \frac{2v\cot\theta}{r^2}\right) \tag{1}$$

$$u\frac{\partial v}{\partial r} + \frac{v}{r}\frac{\partial v}{\partial \theta} + \frac{uv}{r} = -\frac{1}{\rho r}\frac{\partial p}{\partial \theta} + \sigma\left(\nabla^2 v + \frac{2}{r^2}\frac{\partial u}{\partial \theta} - \frac{v}{r^2\sin^2\theta}\right) \tag{2}$$

$$\frac{1}{r^2}\frac{\partial}{\partial r}(r^2 u) + \frac{1}{r\sin\theta}\frac{\partial}{\partial \theta}(v\sin\theta) = 0 \tag{3}$$

where

$$\nabla^2 = \frac{1}{r^2}\frac{\partial}{\partial r}\left(r^2\frac{\partial}{\partial r}\right) + \frac{1}{r^2\sin\theta}\frac{\partial}{\partial \theta}\left(\sin\theta\frac{\partial}{\partial \theta}\right) \qquad [1]$$

p = pressure
ρ = density
σ = kinematic viscosity
u = r–component of velocity
v = θ–component of velocity

and we assume that the flow is axisymmetric, so that the azimuthal component of velocity is zero. Because the flow is axisymmetric, there exists a Stokes stream function ψ, and the velocity components u and v are

$$u = \frac{1}{r^2\sin\theta}\frac{\partial\psi}{\partial\theta}, \qquad v = -\frac{1}{r\sin\theta}\frac{\partial\psi}{\partial r}.$$

Then equation (3) is identically satisfied. We solve equations (1) and (2) by writing $\psi = \sigma r f(\theta)$. Then

$$u = \frac{\sigma}{r\sin\theta} f'(\theta)\ , \qquad v = -\frac{\sigma}{r\sin\theta} f(\theta)\ .$$

If we set $x = \cos\theta$, it can be shown [1], that equations (1) and (2) are satisfied if f satisfies the Riccati equation

$$\frac{df}{dx} + \frac{2xf}{1-x^2} - \frac{f^2}{2(1-x^2)} = \frac{2(ax^2+bx+c)}{1-x^2} \tag{4}$$

where a, b, c are arbitrary constants. To reduce (4) to a second order linear differential equation, set

$$f(x) = -2(1-x^2)\frac{\phi'}{\phi}\ .$$

Then

$$(1-x^2)\phi'' + (ax^2+bx+c)\phi = 0. \tag{5}$$

Equation (5) is of Fuchsian type, with $x = \pm 1, \infty$ as regular singular points. If one takes $a = b = c = 0$, we find

$$f = \frac{2\sin^2\theta}{1+d-\cos\theta}$$

where d is an arbitrary constant. [1]

This solution, known as the Landau source, has been extensively studied. A more general solution can be found by taking $b = 0$, and a and c non-zero. Then two linearly independent solutions of (5) are

$$\phi_1(x) = \sqrt{1-x^2}\, P_\nu^\mu(x)\ , \qquad \phi_2(x) = \sqrt{1-x^2}\, Q_\nu^\mu(x)$$

where $P_\nu^\mu(x)$, $Q_\nu^\mu(x)$ are Legendre functions, and $\mu = \sqrt{1-a-c}$, $\nu = -\frac{1}{2} + \sqrt{1-4a}$. We wish to obtain the solution for flow in a cone of semi-vertical angle α. This requires that we impose the no-slip boundary conditions

$$f(\alpha) = f'(\alpha) = 0$$

on the surface of the cone. If we take $\phi = P_\nu^\mu(\cos\theta)\sin\theta$ (rejecting $Q_\nu^\mu(\cos\theta)$ because of a singularity at $\theta = 0$) we find

$$f(\alpha) = 2(1-\mu)\cos\alpha - 2g(\mu,\nu)\frac{P_\nu^{\mu-1}(\cos\alpha)}{P_\nu^\mu(\cos\alpha)}$$

$$f'(\alpha) = -2(1-\mu)\sin\alpha - 4g(\mu,\nu)\cos\alpha\frac{P_\nu^{\mu-1}(\cos\alpha)}{P_\nu^\mu(\cos\alpha)} +$$
$$2\sin\alpha g(\mu,\nu)g(\mu-1,\nu)\frac{P_\nu^{\mu-2}(\cos\alpha)}{P_\nu^\mu(\cos\alpha)} - 2[g(\mu,\nu)]^2\sin\alpha\left[\frac{P_\nu^{\mu-1}(\cos\alpha)}{P_\nu^\mu(\cos\alpha)}\right]^2$$

where $g(\mu,\nu) = (\nu+\mu)(\nu-\mu+1)$. These boundary conditions require that we solve $f(\alpha) = f'(\alpha) = 0$ for μ and ν. The simplest case to consider is $\alpha = \pi/2$: the cone then degenerates into the right half space. The boundary conditions then become

$$2g(\mu,\nu)P_\nu^{\mu-1}(0) = 0$$

$$-2(1-\mu) + g(\mu,\nu)g(\mu-1,\nu)\frac{P_\nu^{\mu-2}(0)}{P_\nu^\mu(0)} = 2[g(\mu,\nu)]^2\left[\frac{P_\nu^{\mu-1}(0)}{P_\nu^\mu(0)}\right]^2 .$$

However

$$P_\nu^\mu(0) = \frac{2^\mu\sqrt{\pi}}{\Gamma\left(1+\frac{\nu}{2}-\frac{\mu}{2}\right)\Gamma\left(\frac{1}{2}-\frac{\nu}{2}-\frac{\mu}{2}\right)}. \qquad [2]$$

The two boundary conditions then become, using properties of the Γ function,

$$(\mu+\nu)(\nu-\mu+1) = \mu - 1$$

$$2-\mu-\nu = -2n \qquad \textit{where } n = 0,1,2,3,\ldots.$$

Solving gives

$$\mu = \frac{4n^2+10n+7}{4n+5}, \qquad \nu = \frac{4n^2+8n+3}{4n+5}$$

and the problem has a solution for each value of n, giving rise to an infinity of solutions. The functions $\phi_n(\theta) = \sin\theta P_\nu^\mu(\cos\theta)$ can be shown to satisfy the differential equation

$$\frac{d}{d\theta}\left(\frac{1}{\sin\theta}\frac{d\phi_n}{d\theta}\right) - \lambda_n\frac{\cos^2\theta}{\sin^3\theta}\phi_n = 0$$

where

$$\lambda_n = \frac{\left(n+\frac{1}{2}\right)(n+1)\left(n+\frac{3}{2}\right)(n+2)}{\left(n+\frac{5}{4}\right)^2} \qquad n = 1,2,3\ldots.$$

For small values of θ, the r and θ components of velocity are approximately

$$u = \frac{-2\sigma(1-\mu)}{r} \qquad v = \frac{-2\sigma(1-\mu)}{r}\cot\theta\ .$$

As a result there is a line source on the x-axis ($\theta = 0$). It can be shown that the rate of outflow per unit length of the line source is

$$4\pi\sigma(\mu-1) = \frac{4\pi\sigma}{4n+5}(4n^2+6n+2)\ .$$

The fluid from this line source flows out into the semi-infinite region, in the presence of the plane barrier at $\theta = \frac{\pi}{2}$.

2. Exact Solution of the Time-Dependent Navier-Stokes Equations

The time-dependent motion of a viscous incompressible fluid is governed by the Navier-Stokes equations

$$\frac{\partial\vec{q}}{\partial t} + (\vec{q}\cdot\nabla)\vec{q} = -\frac{1}{\rho_0}\nabla p + \nu\nabla^2\vec{q}$$

$$\nabla\cdot\vec{q} = 0$$

where

$\vec{q} = \vec{q}(x,y,z,t)$ is the fluid velocity
$p = p(x,y,z,t)$ is the pressure
ρ_0 is the (constant) density and
ν is the kinematic viscosity.

Let ρ, ϕ, z be cylindrical polar coordinates, where $x = \rho\cos\phi$, $y = \rho\sin\phi$, $z = z$ and $\vec{\rho}$, $\vec{\phi}$, $\vec{k}$ are unit vectors in the ρ, ϕ, z directions respectively.

Consider axisymmetric fluid motion, with swirl, as given by

$$\vec{q} = \left(\vec{Q} + V\vec{k}\right)e^{-\alpha^2\nu t} + U\vec{k} + \omega\rho\vec{\phi}$$

where U, V are constant speeds, ω is a constant angular velocity, and α is a constant with dimensions 1/length.

Now an axisymmetric solution of the Beltrami equation

$$\nabla \times \vec{Q} = \alpha\vec{Q}$$

has been discussed in [3], and can be written as

$$\vec{Q} = \nabla \times \left(\frac{\chi}{\rho}\vec{\phi}\right) + \frac{\alpha\chi}{\rho}\vec{\phi}$$

where χ satisfies $(L + \alpha^2)\chi = 0$

$$L = \frac{\partial^2}{\partial z^2} + \frac{\partial^2}{\partial \rho^2} - \frac{1}{\rho}\frac{\partial}{\partial \rho}\,.$$

Clearly $\nabla \cdot \vec{Q} = 0$ automatically. If we write $\vec{Q} = F\vec{k} + G\vec{\rho} + H\vec{\phi}$, where $F = F(\rho, z, t)$, $G = G(\rho, z, t)$, $H = H(\rho, z, t)$, then F, G, H are given in terms of χ by

$$F = \frac{1}{\rho}\frac{\partial\chi}{\partial\rho}, \qquad G = -\frac{1}{\rho}\frac{\partial\chi}{\partial z}, \qquad H = \frac{\alpha\chi}{\rho}\,.$$

Now $\nabla\times\nabla\times\vec{Q} = \alpha\nabla\times\vec{Q} = \alpha^2\vec{Q}$. Since $\nabla\cdot\vec{Q} = 0$, $\nabla\times\nabla\times\vec{Q} = \nabla(\nabla\cdot\vec{Q}) - \nabla^2\vec{Q} = -\nabla^2\vec{Q}$. So $(\nabla^2 + \alpha^2)\vec{Q} = 0$. Also by routine calculation

$$\begin{aligned}\omega(\vec{Q}\cdot\nabla)(\rho\vec{\phi}) + \omega(\rho\vec{\phi}\cdot\nabla)\vec{Q} &= 2\omega(G\vec{\phi} - H\vec{\rho}) \\ &= -\frac{2\omega}{\alpha}\left[\frac{\partial\vec{Q}}{\partial z} - \nabla\left(\frac{1}{\rho}\frac{\partial\chi}{\partial\rho}\right)\right]\,.\end{aligned}$$

The convection terms in the Navier-Stokes equations are

$$\begin{aligned}\frac{\partial\vec{q}}{\partial t} + (\vec{q}\cdot\nabla)\vec{q} &= \frac{\partial\vec{Q}}{\partial t}e^{-\alpha^2\nu t} - \alpha^2\nu(\vec{Q} + V\vec{k})e^{-\alpha^2\nu t} \\ &\quad + \left[\left(\vec{Q} + V\vec{k}\right)e^{-\alpha^2\nu t} + U\vec{k} + \omega\rho\vec{\phi}\right]\cdot\nabla\left[\left(\vec{Q} + V\vec{k}\right)e^{-\alpha^2\nu t} + U\vec{k} + \omega\rho\vec{\phi}\right] \\ &= \left[V\frac{\partial\vec{Q}}{\partial z} + \frac{\partial\vec{Q}}{\partial t}e^{\alpha^2\nu t}\right]e^{-2\alpha^2\nu t} + (\vec{Q}\cdot\nabla)\vec{Q}e^{-2\alpha^2\nu t} \\ &\quad -\alpha^2\nu(\vec{Q} + V\vec{k})e^{-\alpha^2\nu t} + U\frac{\partial\vec{Q}}{\partial z}e^{-\alpha^2\nu t} \\ &\quad +\omega(\vec{Q}\cdot\nabla)(\rho\vec{\phi})e^{-\alpha^2\nu t} + \omega\rho(\vec{Q}\cdot\nabla)\vec{Q}e^{-\alpha^2\nu t} + \omega^2\rho(\vec{\phi}\cdot\nabla)(\rho\vec{\phi})\,.\end{aligned}$$

The Navier-Stokes equations become

$$\begin{aligned}&\left(V\frac{\partial\vec{Q}}{\partial z} + \frac{\partial\vec{Q}}{\partial t}e^{\alpha^2\nu t}\right)e^{-2\alpha^2\nu t} + \left(U - \frac{2\omega}{\alpha}\right)\frac{\partial\vec{Q}}{\partial z}e^{-\alpha^2\nu t} - \alpha^2\nu Q e^{-\alpha^2\nu t} \\ &+\nabla\left[\frac{1}{2}Q^2e^{-2\alpha^2\nu t} + \frac{2\omega}{\alpha\rho}\frac{\partial\chi}{\partial\rho}e^{-\alpha^2\nu t} - \alpha^2\nu z V e^{-\alpha^2\nu t} - \frac{\omega^2\rho^2}{2} + \frac{p}{\rho_0}\right] \\ &-\nu\nabla^2[\vec{Q} + V\vec{k}]e^{-\alpha^2\nu t} - \nu\nabla^2(\rho\omega\vec{\phi}) = 0\,.\end{aligned}$$

Since $\nabla \times (\nabla \times \rho\vec{\phi}) = 0$ and $-\nu\nabla^2(\rho\omega\vec{\phi}) = 0$ we have

$$\left[V\frac{\partial\vec{Q}}{\partial z} + \frac{\partial\vec{Q}}{\partial t}e^{\alpha^2\nu t}\right]e^{-2\alpha^2\nu t} + \left(U - \frac{2\omega}{\alpha}\right)\frac{\partial\vec{Q}}{\partial z}e^{-\alpha^2\nu t}$$
$$+\nabla\left[\frac{1}{2}Q^2e^{-2\alpha^2\nu t} + \frac{2\omega}{\alpha\rho}\frac{\partial\chi}{\partial\rho}e^{-\alpha^2\nu t} - \alpha^2\nu zVe^{-\alpha^2\nu t} - \frac{\omega^2\rho^2}{2} + \frac{p}{\rho_0}\right] = 0 .$$

To satisfy this equation set

$$\frac{1}{2}Q^2e^{-2\alpha^2\nu t} + \frac{2\omega}{\alpha\rho}\frac{\partial\chi}{\partial\rho}e^{-\alpha^2\nu t} - \alpha^2\nu zVe^{-\alpha^2\nu t} - \frac{\omega^2\rho^2}{2} + \frac{p}{\rho_0} = K$$

(This equation defines the pressure p)

$$U = \frac{2\omega}{\alpha}, \qquad V\frac{\partial\vec{Q}}{\partial z} + \frac{\partial\vec{Q}}{\partial t}e^{\alpha^2\nu t} = 0 .$$

In the latter equation we find

$$\vec{Q} = F\vec{k} + G\vec{\rho} + H\vec{\phi}$$

where

$$F = F(z', \rho), \quad G = G(z', \rho), \quad H = H(z', \rho)$$

$$z' = z + V\frac{e^{-\alpha^2\nu t}}{\alpha^2\nu}, \quad \text{and} \quad \nabla \times \vec{Q} = \alpha\vec{Q} .$$

Now consider the motion of a solid sphere of radius a translating with decaying velocity $Ve^{-\alpha^2\nu t}\vec{k}$, and rotating with angular velocity $\omega' e^{-\alpha^2\nu t}\vec{k}$, where ω' is a constant. This motion is relative to a uniform stream of fluid, with velocity $(U + Ve^{-\alpha^2\nu t})\vec{k}$, and a rigid body rotation of the fluid with angular velocity $\omega = \frac{\alpha U}{2}$. At time t the equation of the surface of the sphere is

$$x^2 + y^2 + z'^2 = x^2 + y^2 + \left[z + \frac{Ve^{-\alpha^2\nu t}}{\alpha^2\nu}\right]^2 = a^2 .$$

Now define spherical coordinates (r, θ, ϕ) in the moving frame by

$$z' = z + \frac{Ve^{-\alpha^2\nu t}}{\alpha^2\nu} = r\cos\theta , \quad \rho = \sqrt{x^2 + y^2} = r\sin\theta .$$

Then

$$\vec{Q} = \nabla \times \left(\frac{\chi\vec{\iota}_\phi}{r\sin\theta}\right) + \alpha\frac{\chi\vec{\iota}_\phi}{r\sin\theta}$$
$$= \frac{1}{r^2\sin\theta}\frac{\partial\chi}{\partial\theta}\vec{\iota}_r - \frac{1}{r\sin\theta}\frac{\partial\chi}{\partial r}\vec{\iota}_\theta + \frac{\alpha\chi}{r\sin\theta}\vec{\iota}_\phi$$

where χ satisfies

$$\frac{\partial^2\chi}{\partial r^2} + \frac{\sin\theta}{r^2}\frac{\partial}{\partial\theta}\left(\frac{1}{\sin\theta}\frac{\partial\chi}{\partial\theta}\right) + \alpha^2\chi = 0 .$$

A general solution for χ is

$$\chi = r^{\frac{1}{2}}\left\{a_n J_{n+\frac{1}{2}}(\alpha r) + b_n J_{-n-\frac{1}{2}}(\alpha r)\right\}(P_{n-1}(\cos\theta) - P_{n+1}(\cos\theta))$$

where $J_{\pm(n+\frac{1}{2})}(\alpha r)$ are Bessel functions and

$P_n(\cos\theta)$ is a Legendre polynomial.

The no-slip boundary conditions on the sphere are

$$\vec{Q} = V\vec{k} + \omega' a \sin\theta \vec{\iota}_\phi\,, \quad \text{at } r = a\,, \qquad |\vec{Q}| \to 0 \quad \text{as } r \to \infty\,.$$

In terms of χ, the boundary conditions are

$$\frac{1}{a^2\sin\theta}\frac{\partial\chi}{\partial\theta} = V\cos\theta\,, \quad -\frac{1}{a\sin\theta}\frac{\partial\chi}{\partial r} = -V\sin\theta\,, \quad \frac{\alpha\chi}{a\sin\theta} = \omega' a\sin\theta$$

and $\frac{\chi}{r\sin\theta} \to 0$ as $r \to \infty$.

A suitable solution for χ is

$$\chi = \left\{a_1\left(\alpha\cos\alpha r - \frac{\sin\alpha r}{r}\right) + b_1\left(\alpha\sin\alpha r + \frac{\cos\alpha r}{r}\right)\right\}\sin^2\theta\,.$$

The condition at infinity is automatically satisfied, and the three boundary conditions become

$$\begin{aligned} a_1\left(\alpha\cos\alpha - \frac{\sin\alpha a}{a}\right) + b_1\left(\frac{\cos\alpha a}{a} + a\sin\alpha a\right) &= \frac{Va^2}{2} \\ a_1\left(\sin\alpha a\left(\frac{1}{a^2} - \alpha^2\right) - \frac{\alpha}{a}\cos\alpha a\right) + b_1\left[\left(\alpha^2 - \frac{1}{a^2}\right)\cos\alpha a - \frac{\alpha}{a}\sin\alpha a\right] &= Va \end{aligned}$$

and $\frac{2\omega'}{V} = a$.

Solving gives

$$\begin{aligned} a_1 &= \frac{Va}{2\alpha^3}\left\{a\left[\left(\alpha^2 - \frac{1}{a^2}\right)\cos\alpha a - \frac{\alpha}{a}\sin\alpha a\right] - \frac{2\cos\alpha a}{a} - 2\alpha\sin\alpha a\right\} \\ b_1 &= -\frac{Va}{2\alpha^3}\left\{a\left[\left(\frac{1}{a^2} - \alpha^2\right)\sin\alpha a - \frac{\alpha}{a}\cos\alpha a\right] - 2\alpha\cos\alpha a + \frac{2}{a}\sin\alpha a\right\} \\ \chi &= \frac{Va}{2\alpha^3}\sin^2\theta G(r), \quad \text{where} \\ G(r) &= \left\{\left[a\left(\alpha^2 - \frac{3}{a^2}\right)\cos\alpha a - 3\alpha\sin\alpha a\right]\left[2\cos\alpha r - \frac{\sin\alpha r}{r}\right]\right. \\ &\quad \left. - \left[a\left(\frac{3}{a^2} - \alpha^2\right)\sin\alpha a - 3\alpha\cos\alpha a\right]\left[\alpha\sin\alpha r + \frac{\cos\alpha r}{r}\right]\right\}\,. \end{aligned}$$

In terms of the unit vectors for the spherical coordinates moving with the sphere, the velocity $\vec{q}$ can be written as $\vec{q} = \vec{q}_r\vec{\iota}_r + q_\theta\vec{\iota}_\theta + q_\phi\vec{\iota}_\phi$ where

$$\begin{aligned} q_r &= U\cos\theta + Ve^{-\alpha^2\nu t}\cos\theta\left\{1 + \frac{a}{r^2\alpha^3}G(r)\right\} \\ q_\theta &= -U\sin\theta - Ve^{-\alpha^2\nu t}\sin\theta \times \\ &\quad \left\{1 + \frac{a}{2r\alpha^3}\left[a\left(\alpha^2 - \frac{3}{a^2}\right)\cos\alpha a - 3\alpha\sin\alpha a\right]\left[\frac{\sin\alpha r}{r^2} - \frac{\alpha\cos\alpha r}{r} - \alpha^2\sin\alpha r\right]\right. \\ &\quad \left. - \frac{a}{2r\alpha^3}\left[a\left(\frac{3}{a^2} - \alpha^2\right)\sin\alpha a - 3\alpha\cos\alpha a\right]\left[\alpha^2\cos\alpha r - \frac{\cos\alpha r}{r^2} - \frac{\alpha\sin\alpha r}{r}\right]\right\} \\ q_\phi &= \omega r\sin\theta + \frac{Vae^{-\alpha^2\nu t}\sin\theta}{2\alpha^2}G(r). \end{aligned}$$

The force on the sphere can be found from the stress vector. We find

$$\vec{F} = -\frac{8\pi}{3} V \mu a^3 \alpha^2 e^{-\alpha^2 \nu t} \vec{k} .$$

References

1. G.B. Whitham, in: Laminar Boundary Layers, L. Rosenhead, (Ed.), Oxford, London (1963) pp. 134–151.
2. A. Erdélyi, in: Higher Transcendental Functions, Vol. 1, McGraw-Hill, New York (1953) p.145.
3. S. Weinbaum and V. O'Brien, Phys. Fluids, **10**, 1438–1477 (1967).

9th Int. Coll. on Differential Equations, pp. 361–368
D. Bainov (Ed)

Remarks on the N-Body Problem in Celestial Mechanics

A. RAUH
Fachbereich Physik, Carl von Ossietzky Universität, D-26111 Oldenburg, Germany

Abstract: The Nekhoroshev theory is discussed in view of the restrictions implied by exponential rather than power law estimates. It is pleaded that, in applied N-body problems of celestial mechanics with given smallness parameter, power law estimates should be preferred. The advantages are discussed for the three-body problem Sun-Earth-Jupiter. Quantitative results are given for the first order resonances.
Key Words and Phrases: *Nekhoroshev theorem, planetary three-body problem.*

1. Introduction

The following contribution relates to the Nekhoroshev theorem [1] of nearly integrable Hamiltonian systems which we write as

$$H = H_0(\mathbf{I}) + \epsilon H_1(\mathbf{I}, \varphi) \tag{1}$$

with action variables $\mathbf{I} \in \mathbb{R}^n$ and angle variables $\varphi \in \mathbf{T}^n$. For $\epsilon = 0$ the components I_j, $j = 1, 2, ...n$ are constants of motion and every trajectory describes a quasi-periodic motion on the n-torus specified by the initial conditions.

In the case $\epsilon \neq 0$ one has the two general theorems of Kolmogorov-Arnold-Moser (KAM) [2] and of Nekhoroshev [1]. Whereas KAM predicts the survival of invariant tori in a Cantor like subset of phase space, the latter theorem establishes nearly integrable properties in compact domains, but only over a finite time interval.

The Nekhoroshev theorem appears particularly attractive for applications to the N-body problem of celestial mechanics, because it both allows for compact subsets of initial conditions and holds for time intervals T which increase exponentially with $\epsilon \to 0$. Evidently, the objects of celestial mechanics have a finite origin and live time, both of the order of a couple of billion years.

In the Newtonian model of planetary motion, H_0 and H_1, in general, refer to the unperturbed Kepler ellipses and to the gravitational interaction potential, respectively. In the case of the 3-body problem Sun-Earth-Jupiter e.g. the smallness parameter ϵ may be identified with the relative magnitude $|H_1/E_0| \approx 10^{-4}$ where E_0 denotes the energy of the Earth orbit. The question arises whether the Nekhoroshev theorem is valid up to such, relatively large, values of ϵ. As a matter of fact, the bounds of ϵ as inferred at first from the original paper [1], turned out to be ridiculously small [3], see also remarks in [4]. Meanwhile, the estimates were much improved, in particular by work of Lochak [5], see also [6] and references therein. Based on the latter, Niedermann [6], in a special application to the 3-body problem of Sun-Jupiter-Saturn, succeeded in corroborating the validity range of the theorem up to the bound $|\epsilon| \approx 10^{-13}$. And yet this bound is still too small as compared with the realistic ϵ parameter.

As we shall discuss in the next Section, the exponential form of the Nekhoroshev time $T \propto \exp[1/\epsilon^\alpha]$ implies severe restrictions on ϵ. Clearly, as compared to power law estimates, namely $T \propto 1/\epsilon^\beta$, the exponential form attains its significance only in the asymptotic limit $\epsilon \to 0$. This, however, is irrelevant in specific problems with a given finite ϵ. In order to relax the constraints on ϵ, we plead for renouncing exponential in favour of power law estimates.

When the machinery of the Nekhoroshev theorem is applied to the 3-body problem of Sun-Earth-Jupiter, after only two consecutive canonical transformations one arrives at a remainder of order ϵ^3 which, maybe, is already small enough to be negligible for the relevant time interval of about 2 billion years.

In Section 3 the program of power law estimates will be outlined. However, the application to the Sun-Earth-Jupiter problem will remain fragmentary here, and specific results are given only on the first order resonances. As compared to a previous investigation [7] which was based on standard estimates of the analyticity domain, the numerical calculation of the resonant Fourier amplitudes of H_1 will bring a gain of about 6 decimals in accuracy making Nekhoroshev's geometrical reasoning redundant in the given case.

As should be remarked, in reference [7] we had adopted a nonconvential scheme of canonical transformations where the degenerate variables were treated as time dependent parameters. Unfortunately, the statement below eq.(77) of [7] that the time dependent degenerate variables behave adiabatically also to second order in ϵ is wrong, as was first pointed out by Niedermann [8]. We cannot therefore recommend that canonical scheme.

By choosing the 3-body problem of Sun-Earth-Jupiter we have in mind a geocentric definition of stability: Can we understand why the energy flux from the sun, characterized by the solar constant S_0, was stable long enough to allow for the evolution of life over a time span of 1-2 billion years? The yearly average of S_0 is proportional to $(a^2\sqrt{1-e^2})^{-1}$ where a and e denote the semi-mayor axis and excentricity of the Earth orbit, respectively. The problem here amounts to estimate the gravitational effect upon the Earth elements a and e due to the other planets. In terms of the magnitude ϵ as defined further above, among the planets, Jupiter perturbs the Earth-Moon system most. We remark that in the mathematical literature the notion stability of the solar system in general refers to the possibility of escapes and collisions, see e.g. [9], and [10]. In this case Sun, Jupiter and Saturn are considered as the three dominant bodies.

2. Nekhoroshev Theory

In this Section we give a brief, unavoidably oversimplified, sketch of the basic ideas. The main purpose is to demonstrate the constraints on ϵ by aiming at exponential estimates rather than at power law forms. In the version of Benettin, Galgani, and Giorgilli [4], henceforth abbreviated by (BGG), the Nekhoroshev theorem reads

$$\| \mathbf{I}(t) - \mathbf{I}(0) \|_G \leq c_0\epsilon^\beta \quad \text{for} \quad t \in (0,T); \quad T := t_0\epsilon^{-1/2} \exp[1/\epsilon^\alpha] \tag{2}$$

where $\mathbf{I}(0) \in G$ is a compact domain, $\alpha, \beta, c_0, t_0 > 0$, and $\| \dots \|_G$ is the supremum norm over G. Apart from the assumptions of analyticity of $H(\mathbf{I}, \varphi)$ and of small enough $|\epsilon|$, (BGG) proved the theorem for nondegenerate H_0 and under the condition that the real symmetric matrix A defined as

$$A_{ij} := \frac{\partial^2 H_0}{\partial I_i \partial I_j}; \quad i,j = 1,2,...n \tag{3}$$

is positive definite for $\mathbf{I} \in G$.

The **first step** in the proof of the theorem is to cutoff the Fourier series representation of H_1

$$H_1 = \sum_{\mathbf{k}\in\mathbf{Z}^n} H_{\mathbf{k}}(\mathbf{I}) \exp[i\mathbf{k}\cdot\varphi] \tag{4}$$

at some integer K with $|\mathbf{k}| := |k_1| + |k_2| + ...|k_n| \leq K$ i.e.

$$H_1 = H_1' + H_1''; \quad H_1' = \sum_{|\mathbf{k}|\leq K} H_{\mathbf{k}} \exp[i\mathbf{k}\cdot\varphi], \quad H_1'' = \sum_{|\mathbf{k}|>K} H_{\mathbf{k}} \exp[i\mathbf{k}\cdot\varphi]. \tag{5}$$

From the canonical equations $\dot{\mathbf{I}} = -\partial H/\partial\varphi$ one finds the a priori estimate

$$|I_j(t) - I_j(0)| < [\sup_{(\mathbf{I},\varphi)\in G\otimes\mathbf{T}^n} |\frac{\partial H_1'}{\partial\varphi_j}| + r_K]\, t; \quad j = 1,2,...n, \tag{6}$$

where due to the analyticity of H_1, the remainder r_K decays exponentially as

$$r_K := r_0 K^n \exp[-\sigma K], \quad r_0, \sigma > 0. \tag{7}$$

This remainder can be neglected if the time interval considered is small compared to $T^* := 1/r_k$. In Section 3 we will choose the Fourier cutoff K such that $T^* \gg 2 * 10^9$ years. The exponential form of the theorem requires in addition the following connection between K and ϵ

$$r_K = r_0 K^n \exp[-\sigma K] \propto \exp[-1/\epsilon^\alpha]. \tag{8}$$

Clearly, in the limit $\epsilon \to 0$ the cutoff K is unbounded. This, on the other hand, makes the number of possible resonant zones unbounded too, which, as we will see later, will impose a self-consistency condition upon the exponent α.

In the **second step** the truncated Fourier series H_1' is split into resonant H_{res} and nonresonant H_{nonres} terms. The former are connected with small denominators and are defined in such a way that H_{nonres} can be removed, to first order in ϵ, by means of a near identity canonical transformation. Accordingly the set of resonant Fourier amplitudes is defined as

$$R := \{\mathbf{k}; \quad |\omega\cdot\mathbf{k}| < d_0\epsilon^\gamma, \quad \mathbf{I}\in G, \quad |\mathbf{k}| \leq K\}, \tag{9}$$

where $\omega_j = \partial H_0/\partial I_j, \ j = 1,2,...n$, and we have

$$H_1' = H_{res} + H_{nonres}; \ H_{res} = \sum_{\mathbf{k}\in R} H_{\mathbf{k}}(\mathbf{I}) \exp[i\mathbf{k}\cdot\varphi], \ H_{nonres} = \sum_{\substack{\mathbf{k}\notin R \\ |\mathbf{k}|\leq K}} H_{\mathbf{k}}(\mathbf{I}) \exp[i\mathbf{k}\cdot\varphi]. \tag{10}$$

As a consequence, in the canonical transformation for H_{nonres}, only denominators appear with $|\omega\cdot\mathbf{k}| \geq d_0\epsilon^\gamma$, and for the near identity property one needs the necessary condition $0 \leq \gamma < 1$.

In the **third step**, H_{nonres} is successively transformed to higher orders. In each order one gets additional resonances and a new set of nonresonant Fourier amplitudes. If H_0 is nondegenerate, this procedure can be done in a superconvergent way. For instance, at the

n-th iteration we can generate the canonical transformation $(\mathbf{I}, \varphi) \to (\mathbf{J}, \psi)$ by a mixed variable function $S(\mathbf{J}, \varphi; \epsilon)$ as

$$J_j = I_j - \epsilon^{2^n} \frac{\partial S(\mathbf{J}, \varphi; \epsilon)}{\partial \varphi_j}; \qquad \psi_j = \varphi_j + \epsilon^{2^n} \frac{\partial S(\mathbf{J}, \varphi; \epsilon)}{\partial J_j}, \quad j = 1, 2, ...n. \tag{11}$$

After N iterations the transformed Hamiltonian attains the following form

$$H_0 + \epsilon H_1' \quad \to \quad H_0^{(N)}(\mathbf{J}; \epsilon) + \epsilon H_{res}^{(N)}(\mathbf{J}, \psi; \epsilon) + \epsilon^{2^N} \rho^{(N)}(\mathbf{J}, \psi; \epsilon) \tag{12}$$

where the last term is considered as a remainder, which should be of the order of the remainder (8), i.e.

$$\epsilon^{2^N} \propto \exp[-1/\epsilon^{\alpha}], \quad \text{or} \quad \mathrm{N} \propto |\mathrm{Log}[\epsilon]| \quad \text{for} \quad \epsilon \to 0. \tag{13}$$

In the **fourth step** we come to the hard core of the theory: estimating the effect of the resonant Fourier amplitudes. Let me point out some basic ideas in a very simplified way. Suppose the initial vector $\mathbf{I}^* := \mathbf{I}(0)$ lies in the intersection of r resonance zones

$$|\mathbf{k}^{(\rho)} \cdot \omega| < d_0 \epsilon^{\gamma}; \qquad \rho = 1, 2, ...r, \quad \mathbf{I} \in G. \tag{14}$$

Suppose further that, in the time interval considered, the drift $\Delta\mathbf{I}(t)$ due to the resonances is sufficiently small. Then, in the canonical equations we can approximately take the Fourier amplitudes at the constant vector $\mathbf{I} = \mathbf{I}^*$ as follows

$$\dot{\mathbf{I}} = -i \sum_{\rho=1}^{r} \mathbf{k}^{(\rho)} H_{\mathbf{k}^{(\rho)}}(\mathbf{I}^*) \exp[i\mathbf{k}^{(\rho)} \cdot \varphi]. \tag{15}$$

As a consequence, the drift $\Delta\mathbf{I}(t)$ lies in the space spanned by the vectors $\mathbf{k}^{(\rho)}$ with

$$\Delta\mathbf{I}(t) = a^{(1)}(t)\mathbf{k}^{(1)} + a^{(2)}(t)\mathbf{k}^{(2)} + ...a^{(r)}(t)\mathbf{k}^{(r)}. \tag{16}$$

The idea proposed by Nekhoroshev was to estimate $\| \Delta\mathbf{I}(t) \|$ geometrically from the maximal diameter of the domains (14), $r = 1, 2, ...n$, without having information on the coefficients $a^{(\rho)}(t)$, $\rho = 1, 2, ...r$.

In the simplest case, $r = 1$, $\mathbf{I}^*$ lies in one resonance zone, say $\mathbf{k}^{(1)}$. We choose the point $\mathbf{I}^*$ on the resonance manifold $\mathbf{k}^{(1)} \cdot \omega(\mathbf{I}^*) = 0$ which is contained in the resonance zone. In linear approximation and using the matrix A defined in (3) one obtains

$$|\mathbf{k}^{(1)} \cdot (A\Delta\mathbf{I})| < d_0 \epsilon^{\gamma} \tag{17}$$

where the matrix A maps $\Delta\mathbf{I}$ to the vector $(A\Delta\mathbf{I})$. Since by (16) $\Delta\mathbf{I}$ is parallel to $\mathbf{k}^{(1)}$ and because A is positive, we get immediately

$$\lambda_{min} k^{(1)} \| \Delta\mathbf{I} \|_e \leq |\mathbf{k}^{(1)} \cdot (A\Delta\mathbf{I})| < d_0 \epsilon^{\gamma} \tag{18}$$

where λ_{min} is the smallest eigenvalue of A and $\| ... \|_e$ denotes the euclidean norm. For the euclidean length, $k^{(1)}$, one has $k^{(1)} \geq 1$ because $\mathbf{k}^{(1)}$ has integer components. Therefore, $\| \Delta\mathbf{I} \|_e$ has an upper bound vanishing with $\epsilon \to 0$.

In the more general case of an r-fold resonance with

$$|\mathbf{k}^{(\rho)} \cdot (A\Delta\mathbf{I})| < d_0 \epsilon^{\gamma} \quad \text{for} \quad \rho = 1, 2, ...r, \tag{19}$$

only the projection $\mathbf{w}_r$ of $(A\Delta\mathbf{I})$ onto the r vectors $\mathbf{k}^{(\rho)}$ enters. We now introduce an orthogonal transformation S such that the transformed vectors $k_i^{\rho'}$ have zero components for $i = r+1, r+2, ...n$. The analogous property holds for $\Delta\mathbf{I}' = (S\Delta\mathbf{I})$ which by (16) lies in the same r-dimensional subspace. We then obtain for the transformed projection $\mathbf{w}'_r := A'\Delta\mathbf{I}' = SA\Delta\mathbf{I}$

$$\lambda_{min}(r) \parallel \Delta\mathbf{I}' \parallel_e \leq \left[\sum_{\rho=1}^{r}(A'\Delta\mathbf{I}')_\rho(A'\Delta\mathbf{I}')_\rho\right]^{1/2} = \parallel \Delta\mathbf{w}'_r \parallel_e \tag{20}$$

where $\lambda_{min}(r)$ is the smallest eigenvalue of the $r \times r$ submatrix $A'_{\rho,\rho'}$.

To proceed, we use a lemma proved in [4]: The conditions (16) imply

$$\parallel \mathbf{w}'_r \parallel_e \leq d_0 r K^{r-1}\epsilon^\gamma \tag{21}$$

where $K \geq \parallel \mathbf{k}^{(\rho)'} \parallel_e = \parallel \mathbf{k}^{(\rho)} \parallel_e$ is the Fourier cutoff. Thus one arrives at the upper bound

$$\parallel \Delta\mathbf{I}' \parallel_e \equiv \parallel \Delta\mathbf{I} \parallel_e \leq \frac{1}{\lambda_{min}(r)} d_0 r K^{r-1}\epsilon^\gamma. \tag{22}$$

By Sturm's separation theorem, $\lambda_{min}(r) \geq \lambda_{min}(r+1)$, see e.g. [11], in the inequality above, one can replace $\lambda_{min}(r)$ by $\lambda_{min}(n)$.

But according to (8) the cutoff $K \propto 1/(\sigma\epsilon^\alpha)$ is unbounded for $\epsilon \to 0$. Therefore, in order that the drift $\parallel \Delta\mathbf{I} \parallel_e$ vanishes in the limit $\epsilon \to 0$, we have to impose the self-consistency condition

$$K^{r-1}\epsilon^\gamma \leq e_0\epsilon^\delta; \qquad e_0, \delta > 0, \quad r = 1, 2, ...n, \tag{23}$$

which asymptotically for $\epsilon \to 0$ can be fulfilled if

$$0 < \alpha \leq \frac{\gamma - \delta}{n-1}; \qquad \gamma > \delta. \tag{24}$$

In the case $r = n$, the inequality (23) can be related to the volume occupied by the resonance zones in the action space $\mathbf{I} \in G$. To this end one observes that the factor ϵ^γ is proportional to the volume of a single resonance zone whereas K^{n-1} is proportional to the maximal possible number of resonance zones, i.e. to the number of different directions of the integer vectors $|\mathbf{k}| \leq K$. Therefore, (23) tells that the overall volume of the resonance domains vanishes with $\epsilon \to 0$. This argument, however, does not solve the problem of overlapping resonance zones. For the latter see [1], [4], [5] and [6].

3. Towards Power Law Estimates in the Sun-Earth-Jupiter Problem

As should become clear in this Section, the replacement of exponential by power law estimates significantly relaxes the conditions on the Fourier cutoff K, the number of iterated canonical transformations N, and the smallness parameter ϵ.

We eliminate the motion of the Sun by using Jacobian coordinates with the Earth vector $\mathbf{r}$ taken from the center of the Sun and the origin of the Jupiter vector $\tilde{\mathbf{r}}$ located in the Sun-Earth center of mass. The latter is at $\kappa\mathbf{r}$ with $\kappa = m/(m_0 + m) \approx 3 \times 10^{-6}$,

m_0 and m are the mass of Sun and Earth, respectively. The corresponding Hamiltonian can be written as, see e.g. [7]

$$H = H_0 + H_1; \; H_0 = -\frac{\mu\gamma^2}{2I_3^2} - \frac{\tilde{\mu}\tilde{\gamma}^2}{2\tilde{I}_3^2}, \; H_1 = \gamma_2 \left[\frac{1}{\tilde{r}} - \frac{1}{|\tilde{\mathbf{r}} + \kappa\mathbf{r}|}\right] + \gamma_{12}\left[\frac{1}{\tilde{r}} - \frac{1}{|\tilde{\mathbf{r}} - \overline{\kappa}\mathbf{r}|}\right] \quad (25)$$

where tilde refers to Jupiter magnitudes; furthermore, $\overline{\kappa} = 1 - \kappa$, $\mu = m_0 m/(m_0 + m)$, $\tilde{\mu} = (m_0 + m)\tilde{m}/(m_0 + m + \tilde{m})$, $\gamma = Gm_0 m$, $\gamma_2 = Gm_0\tilde{m}$, $\gamma_{12} = Gm\tilde{m}$ with G the gravitational constant. The unperturbed Hamiltonian H_0 depends on the two action variables I_3 and $\tilde{I}_3$ only and is therefore degenerate with respect to $I_1, I_2, \tilde{I}_1, \tilde{I}_2$. The vectors $\mathbf{r}$ and $\tilde{\mathbf{r}}$ have to be expressed in terms of the action angle variables $(\mathbf{I}, \tilde{\mathbf{I}}) \in \mathbb{R}^6$ and $(\varphi, \tilde{\varphi}) \in \mathbf{T}^6$, see e.g. [7].

It is convenient to scale $H, (\mathbf{I}, \tilde{\mathbf{I}}), t$ by the initial data of the Earth, namely $E_0 = \gamma/(2a(0))$, $I_3(0)$, and $T_0 = \pi I_3(0)/E_0 =$ one year, respectively. Using the same notation, one has e.g. the dimensionless canonical equations $\dot{\mathbf{I}} = -\pi\partial H_1/\partial\varphi$.

When H_1 is Taylor expanded with respect to $r/\tilde{r} \approx 1/5$, the zero and first order terms cancel out exactly. Furthermore, due to the smallness of κ, in the first term of H_1, we can neglect the third and higher orders in $r/\tilde{r}$, whereas the second Taylor coefficient has magnitude $\approx 10^{-10}$ and turns out to be irrelevant further below for the calculation of the resonant Fourier amplitudes. Due to the canceling in H_1, the effective smallness parameter $\epsilon = 2\pi(\gamma_{12}/\gamma)\,(r/\tilde{r})^3 \approx 4.3 \times 10^{-5}$.

We now try to eliminate the dependence on the nondegenerate (fast) angles φ_3 and $\tilde{\varphi}_3$ by means of two consecutive canonical transformations $(\mathbf{I}, \tilde{\mathbf{I}}, \varphi, \tilde{\varphi}) \to (\mathbf{I}', \tilde{\mathbf{I}}', \varphi', \tilde{\varphi}') \to (\mathbf{I}'', \tilde{\mathbf{I}}'', \varphi'', \tilde{\varphi}'')$. To this end we represent H_1 by the Fourier series

$$H_1 = \sum_{(k_3, \tilde{k}_3) \in \mathbf{Z}^2} H_{k_3, \tilde{k}_3} \exp[ik_3\varphi_3 + i\tilde{k}_3\tilde{\varphi}_3] \quad (26)$$

where the amplitudes $H_{k_3,\tilde{k}_3}$, apart from the action variables, depend on the four degenerate (slow) angles.

We define the resonance zones generously, i.e. independently of ϵ, by

$$R := \{(k_3, \tilde{k}_3); |k_3\omega(I_3) + \tilde{k}_3\tilde{\omega}(\tilde{I}_3)| < \tilde{\omega}; \quad (I_3, \tilde{I}_3) \in G, \quad |k_3|, |\tilde{k}_3| \le 13\} \quad (27)$$

where the domain G has to be determined a posteriori; $\omega = \partial H_0/(\partial I_3)$ and $\tilde{\omega} = \partial H_0/(\partial \tilde{I}_3)$ are the orbit frequencies of Earth and Jupiter, respectively. If the semi-mayor axis of the Earth $a = a(0)(1 + p)$ has allowance $|p| \le 0.05$, then by the conservation of energy and angular momentum, we have $11.0 < \omega/\tilde{\omega} < 12.9$, see [7]. As a consequence, there are only three resonances $R = \{(1, -11),\ (1, -12),\ (1, -13)\}$. The cutoff $|k_3|, |\tilde{k}_3| \le 13$ is inferred from numerical calculations of the Fourier coefficients.

The resonant Fourier amplitudes are calculated from

$$H_{k_3,\tilde{k}_3} = \frac{1}{(2\pi)^2}\int_0^{2\pi} d\varphi_3 \int_0^{2\pi} d\tilde{\varphi}_3 \hat{H}_1(\mathbf{r}, \tilde{\mathbf{r}}) \exp[-ik_3\varphi_3 - i\tilde{k}_3\tilde{\varphi}_3] \quad (28)$$

where $\hat{H}_1$ is dimensionless and given as, see discussion further above,

$$\hat{H}_1 = 2\pi\frac{\gamma_{12}}{\gamma}a(0)\left[\frac{1}{\tilde{r}} - \frac{1}{|\tilde{\mathbf{r}} - \overline{\kappa}\mathbf{r}|} + \overline{\kappa}\left(\frac{d}{d\overline{\kappa}}\frac{1}{|\tilde{\mathbf{r}} - \overline{\kappa}\mathbf{r}|}\right)_{\overline{\kappa}=0}\right]. \quad (29)$$

For the numerical calculation of the double integral it is convenient to substitute the mean anomalies $(\varphi_3, \tilde{\varphi}_3)$ by the excentric anomalies $(E, \tilde{E})$. Thus, in the exponents of

Table: Absolute magnitude of the first order resonance amplitudes of the Sun-Earth-Jupiter problem for coplanar orbits

$(k_3, \tilde{k}_3)$	$(1, -11)$	$(1, -12)$	(1,-13)
$\lvert H_{k_3,\tilde{k}_3}\rvert$	8.141×10^{-16}	6.416×10^{-17}	5.072×10^{-18}

the integrand we use the Kepler equation and set $\varphi_3 = E - e\sin[E]$, $\tilde{\varphi}_3 = \tilde{E} - \tilde{e}\sin[\tilde{E}]$. Furthermore, we use the standard relations $r = a(1 - e\cos[E]), \tilde{r} = \tilde{a}(1 - \tilde{e}\cos[\tilde{E}])$, see e.g. [12]. In terms of the true anomalies $\theta, \tilde{\theta}$ we can write

$$\mathbf{r} \cdot \tilde{\mathbf{r}} = r\tilde{r}\left(c_{11}\cos[\theta]\cos[\tilde{\theta}] + c_{12}\cos[\theta]\sin[\tilde{\theta}] + c_{21}\sin[\theta]\cos[\tilde{\theta}] + c_{22}\sin[\theta]\sin[\tilde{\theta}]\right) \quad (30)$$

where the coefficients c_{ik} do not depend on any anomaly variable. Finally, we replace $\theta, \tilde{\theta}$ by $E, \tilde{E}$ through the standard relations

$$\cos[\theta] = (\cos[E] - e)/(1 - e\cos[E]); \quad \sin[\theta] = \sqrt{1 - e^2}\sin[E]/(1 - e\cos[E]), \quad (31)$$

and through the analogous relations for $\tilde{\theta}$.

In view of the small mutual inclination, at present epoch, we assume coplanar orbits. As it turns out, the integrals become maximal in absolute magnitude when the apsidal lines are parallel with the Sun located between the perihelion points. Indeed, for given semi-mayor and semi-minor axes, this mutual orientation of the ellipses leads to the smallest distance between the two planets.

The numerical integration was done within MATHEMATICA using the trapezoidal rule which turned out to be more efficient than the Simpson method. The number of discretization points was $30 \times k_3$ and $30 \times \tilde{k}_3$ for the E and $\tilde{E}$ integrations, respectively; the numerical evaluation was carried out with 35 digits precision, and with the present epoch parameters of $e, \tilde{e}$, and $\overline{\kappa a}/\tilde{a}$ specified by rational numbers. For results see the table above.

As to the work left for the future, some few comments are in order. After two canonical transformations, the remainder is proportinal to $\epsilon^3 \approx 10^{-13}$. However, this term contains derivatives with respect to the degenerate action variables which give rise to singularities at zero inclinations and excentricities. This is the price for using action-angle rather than cartesian coordinates. If the excentricities become not much smaller than 10^{-2} within the time interval of 2 billion years, then the remainder is increased by a factor of order 100 which may still be sufficient to neglect it. The magnitudes of the inclinations, on the other hand, depend on the orientation of the coordinate axes which may be chosen suitably. If both the remainder and the resonant Fourier amplitudes can be neglected, one is left with a problem of four degrees of freedom. The natural perturbation parameter is then the quotient $a/\tilde{a} \approx 1/5$ instead of ϵ which is used up for the elimination of the fast angles φ_3 and $\tilde{\varphi}_3$.

Conclusions

In the case of a three-body problem of celestial mechanics, it was shown exemplarily that power law estimates allow for a rather low Fourier cutoff as compared with Nekhoroshev type exponential estimates. The low cutoff gave rise to only three resonance zones. The direct numerical integration showed that the first order resonance amplitudes are negligible provided the values of the excentricities are not too far from present epoch data. Moreover, the numerical values turned out to be smaller by a factor 10^{-6} as compared with previous results [7] based on standard estimates of the analyticity domain in the complex plane.

Acknowledgement

The author is very thankful to the organizers of the conference for their hospitality, especially to Drumi D. Bainov and Emil Minchev. He is also indepted to Andreas Spille for a critical reading of the manuscript.

References

1. N.N. Nekhoroshev, Russ. Mathem. Surv., **32**, 1-65 (1977).
2. A.N. Kolmogorov, Dokl. Akadem. Nauk SSSR, **98**, 527-530 (1954).

   V.I. Arnold, Usp. Mat. Nauk, **18**, 9-36 (1963); English transl.: Russ. Mathem. Surv. **18**, 9-36 (1963).

   J. Moser, Nachr. Akad. Wiss. Gött. II. Math.-Phys. Kl., 1-20 (1962)
3. A. Mikschl and T. Zachrau, Zur Anwendbarkeit des Nekhoroshev-Theorems auf ein planetares Dreikörperproblem, Diplomarbeit, Universität Oldenburg (1990), (unpublished).
4. G. Benettin, L. Galgani, and A. Giorgilli, Celest. Mech., **37**, 1-25 (1985).
5. L. Lochak, Nonlinearity, **6**, 885-904 (1993).
6. L. Niederman, Nonlinearity, **9**, 1703-1751 (1996).
7. A. Rauh, Celest. Mech., **55**, 161-206 (1993); Ibidem: Erratum, 415-415.
8. L. Niederman, (private communication).
9. V.I. Arnold, V.V. Kozlov and A.I. Neishtadt, Mathematical Aspects of Classical and Celestial Mechanics, Springer-Verlag, Berlin (1993).
10. F. Diacu and Ph. Holmes, Celestial Encounters, Princeton Univers. Press, Princeton (1996).
11. R. Bellman, Introduction to Matrix Analysis, Second Ed., TATA McGraw-Hill, New Delhi (1974).
12. A.E. Roy, Orbital Motion, Adam Hilger, Bristol (1978).

9th Int. Coll. on Differential Equations, pp. 369–372
D. Bainov (Ed)

Toeplitz-Theorem for Set-Valued Sequences in Banach-Spaces

RICHARD RÖDLER
University of Trier, Fachbereich IV - Mathematik, D-54286 Trier, Germany
e-mail: roedler@uni-trier.de

Abstract: The aim of this paper is the theorem of Toeplitz [1] for set-valued sequences in Banach-spaces. We consider the so called Mosco-convergence (see [2]). The characterization of Mosco-regularity of a limitation matrix is the same as for regularity.

1. Introduction

We consider infinite lower triangular matrices $A = [a_{n,k}]_{n,k \in \mathbb{N}_0} \in [0,\infty[^{\mathbb{N}_0 \times \mathbb{N}_0}$ and set-valued sequences $x = \{x_n : n \in \mathbb{N}_0\}$ in $\mathcal{P}_c(\mathcal{X}) = \{M \subset \mathcal{X} : M$ is bounded, closed and convex$\}$ where $(\mathcal{X}, \|\cdot\|)$ is a reflexive separable real or complex Banach-space. Further we consider the Mosco-convergence (see Mosco [2]). We will show that for every Mosco-convergent sequence $x = \{x_n : n \in \mathbb{N}_0\}$ in $\mathcal{P}_c(\mathcal{X})$ to the limit $\lim x_n$ the set-valued sequence of A-transforms $\sigma^A(x) := \{\sigma_n^A(x) : n \in \mathbb{N}_0\} := \{\sum_{k=0}^{n} a_{n,k} x_k : n \in \mathbb{N}_0\}$ is Mosco-convergent to the limit $\lim x_n$, if and only if the conditions of the Toeplitz theorem are fullfilled (see Peyerimhoff [1], Zeller and Beekmann [3]).

2. Basic Definitions and a Result

Definitions 2.1

(1.) Let $(\mathcal{X}, \|\cdot\|)$ be a reflexive and separable real or complex Banach-space and $x = \{x_n : n \in \mathbb{N}_0\}$ a set-valued sequence in $\mathcal{P}_c(\mathcal{X}) := \{M \subset \mathcal{X} : M$ is bounded, closed and convex$\}$. We call x *Mosco − convergent* to the limit $\lim x_n$, if

$$\begin{aligned} s - \liminf x_n &= \{\xi \in \mathcal{X} : \forall n \in \mathbb{N}\ \exists \xi_n \in x_n : \lim_{n\to\infty} \| \xi_n - \xi \| = 0\} \qquad (2.1) \\ &= w - \limsup x_n \\ &= \{\xi \in \mathcal{X} : \exists \xi_k \in x_{n_k} : \xi_k \xrightarrow{w} \xi \text{ (converges weakly)}\} \end{aligned}$$

holds.

(2.) For every $M \subset \mathcal{X}$ we write $| M | := \sup_{\mu \in M} | \mu |$.

Theorem 2.2 (Toeplitz) *An infinite matrix $A = [a_{n,k}]_{n,k\in \mathbb{N}_0} \in \mathbb{C}^{\mathbb{N}_0\times \mathbb{N}_0}$ is regular if and only if*

$$\begin{aligned} \lim_{n\to\infty} a_{n,k} &= 0 \quad \text{for all } k \in \mathbb{N}_0; \\ \lim_{n\to\infty} \sum_{k=0}^{\infty} a_{n,k} &= 1 \quad \text{and} \\ \sup_{n\in \mathbb{N}_0} \sum_{k=0}^{\infty} | a_{n,k} | &< \infty \end{aligned} \tag{2.2}$$

hold.

3. The Main Result

At first we prove that for all Mosco-convergent sequences $x = \{x_n : n \in \mathbb{N}_0\}$ in $\mathcal{P}_c(\mathcal{X})$ the limit $\lim x_n$ is also in $\mathcal{P}_c(\mathcal{X})$.

Lemma 3.1 *Let $(\mathcal{X}, \|\cdot\|)$ be a real or complex separable and reflexive Banach-space and $x = \{x_n : n \in \mathbb{N}_0\}$ a Mosco-convergent set-valued sequence in $\mathcal{P}_c(\mathcal{X})$ to the limit $\lim x_n$. Then*

$$\lim x_n \in \mathcal{P}_c(\mathcal{X}) \tag{3.1}$$

holds.

Proof. Because of our assumptions we have $\lim x_n = s - \liminf x_n$ and $x_n \in \mathcal{P}_c(\mathcal{X})$ for all $n \in \mathbb{N}_0$. Therefore we get

$$| \lim x_n | = \sup_{\xi\in \lim x_n} | \xi | = \sup_{\xi \in s-\liminf x_n} | \xi | < \infty. \tag{3.2}$$

Let $\{\xi^{(k)} : k \in \mathbb{N}_0\}$ be a convergent sequence in $(\mathcal{X}, \|\cdot\|)$ to the limit ξ. Then we can conclude the relation:

$$\forall \xi^{(k)} \in \lim x_n \; \exists \xi_n^{(k)} \in x_n \; (n \in \mathbb{N}_0) \text{ and } \lim_{n\to\infty} \| \xi_n^{(k)} - \xi^{(k)} \| = 0. \tag{3.3}$$

For the sequence $\{\xi_k^{(k)} : k \in \mathbb{N}_0\}$ in $(\mathcal{X}, \|\cdot\|)$ we get

$$\xi_k^{(k)} \in x_k \; (k \in \mathbb{N}_0) \quad \text{and} \quad \lim_{k\to\infty} \| \xi_k^{(k)} - \xi \| = 0 \tag{3.4}$$

and therefore $\xi \in s - \liminf x_n = \lim x_n$.

Finally let $\xi, \eta \in \lim x_n$ and $\lambda \in [0, 1]$ be arbitrary choosen. We have to show that $\varphi := \lambda\xi + (1 - \lambda)\eta \in \lim x_n$ holds. From our assumptions we get that for all $n \in \mathbb{N}_0$ there exist $\xi_n, \eta_n \in \mathcal{X}$ such that

$$\lim_{n\to\infty} \| \xi_n - \xi \| = 0 \text{ and } \lim_{n\to\infty} \| \eta_n - \eta \| = 0 \tag{3.5}$$

hold. Because of $x_n \in \mathcal{P}_c(\mathcal{X})$ for all $n \in \mathbb{N}_0$, we can conclude

$$\varphi_n = \lambda\xi_n + (1 - \lambda)\eta_n \in x_n \quad \text{for all } n \in \mathbb{N}_0. \tag{3.6}$$

Further we get

$$\lim_{n\to\infty} \| \varphi_n - \varphi \| = \lim_{n\to\infty} \| \lambda\xi_n + (1 - \lambda)\eta_n - \lambda\xi - (1 - \lambda)\eta \| = 0 \tag{3.7}$$

and therefore we have $\varphi \in s - \liminf x_n = \lim x_n$. q.e.d.

Definition 3.2 Let $A = [a_{n,k}]_{n,k\in \mathbb{N}_0} \in \mathbb{C}^{\mathbb{N}_0\times \mathbb{N}_0}$ be an infinite matrix and $(\mathcal{X}, \|\cdot\|)$ be a real or complex separable and reflexive Banach-space. We call A *Mosco-regular*, if for all Mosco-convergent set-valued sequences $x = \{x_n : n \in \mathbb{N}_0\}$ in $\mathcal{P}_c(\mathcal{X})$ the sequence of A-transforms $\{\sum_{k=0}^{\infty} a_{n,k}x_n : n \in \mathbb{N}_0\}$ exists and is Mosco-convergent to the limit $\lim x_n$.

Now we are able to formulate and prove our main result.

Theorem 3.3 (Toeplitz) *Let $A = [a_{n,k}]_{n,k\in N_0} \in [0,\infty[^{N_0\times N_0}$ be an infinite lower triangular matrix with $\sum_{k=0}^{n} a_{n,k} = 1$ for all $n \in \mathbb{N}_0$. A is Mosco-regular if and only if*

$$\lim_{n\to\infty} a_{n,k} = 0 \quad \textit{for all} \quad k \in \mathbb{N}_0 \tag{3.8}$$

holds.

Proof. Because of the single-valued case e.g. $x_n = \{\xi_n\}$ for all $n \in \mathbb{N}_0$ and the classical theorem of Toeplitz the condition (3.8) is necessary. On the other hand let $x = \{x_n : n \in \mathbb{N}_0\}$ be a Mosco-convergent sequence in $\mathcal{P}_c(\mathcal{X})$. The matrix A is a lower triangular matrix and therefore $\sigma_n^A(x) = \sum_{k=0}^{n} a_{n,k}x_k$ is defined for all $n \in \mathbb{N}_0$ (see [4], p. 155, no. (1.2) and (1.3)). The assumptions $\sum_{k=0}^{n} a_{n,k} = 1$ and $a_{n,k} \geq 0$ imply $\sigma_n^A(x) \in \mathcal{P}_c(\mathcal{X})$ for all $n \in \mathbb{N}_0$.

Let $\xi \in \lim x_n$ be arbitrary chosen then we get $\xi_n \in x_n$ such that $\lim_{n\to\infty} \| \xi_n - \xi \| = 0$. Under our conditions the matrix A is regular in the classical sense and therefore we can conclude

$$\eta_n := \sum_{k=0}^{n} a_{n,k}\xi_k \in \sigma_n^A(x) \text{ for all } n \in \mathbb{N}_0 \text{ and} \tag{3.9}$$

$$\lim_{n\to\infty} \| \eta_n - \xi \| = 0. \tag{3.10}$$

From this we get $\xi \in s - \liminf \sigma_n^A(x)$, e.g.

$$\lim x_n \subset s - \liminf \sigma_n^A(x). \tag{3.11}$$

Because of $s - \liminf \sigma_n^A(x) \subset w - \limsup \sigma_n^A(x)$ it is sufficient to show that $w - \limsup \sigma_n^A(x) \subset \lim x_n$.

Let now $\xi \in w - \limsup \sigma_n^A(x)$ be arbitrary chosen. By definition we get a sequence $\xi_k \in \sigma_{n_k}^A(x)$ for all $k \in \mathbb{N}_0$ such that the relation

$$\xi_k \xrightarrow{w} \xi \quad (k \to \infty) \tag{3.12}$$

holds. Therefore there exists a sequence $\eta = \{\eta_j : j \in \mathbb{N}_0\}$ such that

$$\xi_k = \sum_{j=0}^{n_k} a_{n_k,j}\eta_j \quad (\eta_j \in x_j \text{ for all } j \in \mathbb{N}_0). \tag{3.13}$$

Because of our assumptions on A we get the relation

$$\xi \in \overline{co}(D_w(\eta)), \tag{3.14}$$

where $\overline{co}(D_w(\eta))$ denotes closed convex hull of the weak density points of η. The set-valued sequence $x = \{x_n : n \in \mathbb{N}_0\}$ is Mosco-convergent to $\lim x_n$. From this we get

$$D_w(\eta) \subset \lim x_n. \tag{3.15}$$

Together with lemma (3.1) we have $\lim x_n \in \mathcal{P}_c(\mathcal{X})$ and therefore we can conclude

$$\overline{co}(D_w(\eta)) \subset \overline{co}(\lim x_n) = \lim x_n. \tag{3.16}$$

If we put everything together we have shown that $\xi \in \lim x_n$, e.g.

$$w - \limsup \sigma_n^A(x) \subset \lim x_n \tag{3.17}$$

holds. q.e.d.

References

1. A. Peyerimhoff, Lectures on Summability, Springer-Verlag (1969).
2. U. Mosco, Advances in Math., **3**, 518–535 (1969).
3. K. Zeller and W. Beekmann, Theorie der Limitierungsverfahren, Springer-Verlag (1970).
4. F. Hiai and H. Umegaki, J. Multivariate Anal., **7**, 149–182 (1977).

9th Int. Coll. on Differential Equations, pp. 373–378
D. Bainov (Ed)

Mixed Problem for a Damped Quasilinear Wave Equation

L. P. SAN GIL JUTUCA
Dept. de Matemática e Estatística, Uni-Rio, Rio de Janeiro, RJ, Brasil
rsangil@iq.ufrj.br
M. MILLA MIRANDA
Instituto de Matemática, UFRJ, Rio de Janeiro, RJ, Brasil
milla@dmm.im.ufrj.br

Abstract: This paper is concerned with the existence, uniqueness and exponential decay of solutions for a mixed homogeneous boundary conditions for the damped Kirchhoff equation

$$\frac{\partial^2 u}{\partial t^2} - M\Big(\int_\Omega |\nabla u|^2 d\Omega\Big)\,\Delta u + \alpha \frac{\partial u}{\partial t} = 0\,,$$

where M is a function satisfying $M(\lambda) \geq m_0 > 0$, Ω is an open bounded set of $\mathbb{R}^n$ $(n \geq 1)$ and α is a positive constant.
Key Words and Phrases: *Mixed Problem, Galerkin method, Kirchhoff equation.*

1. Introduction

The goal of this paper is to establish the existence, uniqueness and the exponential decay of solutions for the following initial-boundary value problem:

$$u'' - M\Big(\int_\Omega |\nabla u|^2 d\Omega\Big)\,\Delta u + \alpha u' = 0 \ \text{ in } \ \Omega \times [0, \infty), \tag{1.1}$$

$$u = 0 \ \text{ on } \ \Gamma_0 \times [0, \infty), \tag{1.2}$$

$$\frac{\partial u}{\partial \nu} + \delta(x) u' = 0 \ \text{ on } \ \Gamma_1 \times [0, \infty), \tag{1.3}$$

$$u(x, 0) = u^0(x), \ \ u'(x, 0) = u^1(x) \ \text{ in } \ \Omega\,, \tag{1.4}$$

where, as usually, u' denotes the derivative of u with respect to the time. Here Ω is an open bounded set of $\mathbb{R}^n$ with boundary Γ of class C^3, M is a function of class C^1 on $[0, \infty[$ which satisfies $M(\lambda) \geq m_0 > 0$, δ is a non-negative function which belongs to $W^{1,\infty}(\Gamma)$ and α is a positive constant. Also $\{\Gamma_0, \Gamma_1\}$ is a partition of Γ such that the measures of Γ_0 and Γ_1 are positives, Γ_1 is open and the intersection of $\overline{\Gamma}_0$ with $\overline{\Gamma}_1$ is non empty. The vector ν denotes the outward unit normal to the boundary Γ.

We note that if $\alpha = 0$, that is, if it is not considered a friction force, equation (1.1) is a generalization of the Kirchhoff's model [1] introduced by him to study the small transversal vibrations of an elastic stretched string.

On the existence of global solutions for equation (1.1), with null Dirichlet boundary conditions and some smallness conditions or analyticity on initial data, we cite among others, the papers of Pohozaev [2], Bernstein [3], Arosio-Spagnolo [4], J. L. Lions [5], Nishihara [6], Brito [7, 8] and Hosoya-Yamada [9]. When $\alpha = 0$ the problem (1.1) - (1.4) in the case of one space dimension was studied by Tucsnack [10] and Ono [11]. Still for $\alpha = 0$ we cite the papers of Milla Miranda and Medeiros [12] which worked with $M(\lambda(t)) = \mu(t)$, Milla Miranda and San Gil Jutuca [13] and I. Lasiecka-Ong [14].

Our main contribution is to prove the existence of global solutions of problem (1.1) - (1.4) when $u^0 \in H^2(\Omega)$, $u^1 \in H^1(\Omega)$ and $\delta(x) \geq 0$. We also find a "hidden" trace regularity and the asymptotic behavior of the solutions of (1.1) - (1.4).

To prove the global existence of solutions we apply the Galerkin procedure and consider a suitable smallness of the initial data u^0 and u^1. The exponential decay of solutions is obtained by applying a method introduced by Komornik and Zuazua [15].

This article is organized as follows: In Section 2 we give the notation and the basic results. In Section 3 we make an outline of the proof of existence of global solutions. In Section 4 we prove the exponential decay.

2. Notation and Basic Results

We consider the Hilbert space

$$V = \{v \in H^1(\Omega); v = 0 \text{ on } \Gamma_0\}$$

equipped with the scalar product and the norm given by:

$$((u, v)) = \sum_{i=1}^{n} \int_\Omega \frac{\partial u}{\partial x_i}(x) \frac{\partial v}{\partial x_i}(x)\, dx \text{ and } \|v\| = \left(\sum_{i=1}^{n} \int_\Omega \left(\frac{\partial v}{\partial x_i}(x) \right)^2 dx \right)^{1/2}.$$

The scalar product and norm of $L^2(\Omega)$ are represented by $(.,.)$ and $|.|$, respectively.

Let W be the space of functions $u : \Omega \to \mathbb{R}$ such that $u \in V$, $\Delta u \in L^2(\Omega)$ and there exists $g_u \in H^{1/2}(\Gamma)$ which satisfies $g_u \equiv 0$ on Γ_0 and

$$(-\Delta u, v) = ((u, v)) - (g_u, v)_{L^2(\Gamma)}, \quad \text{for all } v \in V.$$

Notice that g_u is unique. So, we introduce on W the norm

$$\|u\|_W = \left(|\Delta u|^2 + \|g_u\|^2_{H^{1/2}(\Gamma)}\right)^{1/2}.$$

We also note that W is a separable Hilbert space and W is compactly embedding into V.

Proposition 2.1 *The space W is dense in V.*

The proof of this Proposition, based on the density of $\mathcal{D}(-\Delta)$ in V, is reasonably straightforward. It is obtained by applying similar arguments to the used in [12].

3. Existence of Global Solutions

We recall that

$$M(\lambda) \geq m_0 > 0 \qquad m_0 \text{ constant}.$$

In order to state the main result, we introduce the constants

$$N_1 = \frac{1}{m_0}\left[M(\|u^0\|^2)|\Delta u^0| + \delta|u^1|\right]^2 + \alpha^2|u^1|^2\widehat{M}(\|u^0\|^2),$$

$$N_2 = \frac{1}{m_0}\left[|u_1|^2 + \widehat{M}(\|u^0\|^2)\right],$$

$$N_3 = \max\left\{\left|\frac{dM}{d\lambda}(\lambda)\right|;\ 0 \leq \lambda \leq \frac{1}{m_0}\left[|u^1|^2 + \widehat{M}(\|u^0\|^2)\right]\right\},$$

where

$$\widehat{M}(\lambda) = \int_0^\lambda M(\sigma)\,d\sigma\,.$$

We have.

Theorem 3.1 *Let $u^0 \in V$ with $\Delta u^0 \in L^2(\Omega)$ and $u^1 \in V$ such that*

$$(-\Delta u^0, v) = ((u^0, v)) + (\delta u^1, v)_{L^2(\Gamma)}, \quad \text{for all } v \in V\,,$$

$$N_1 N_2 N_3 < \frac{\alpha}{2} \min\left\{m_0, m_0^2\right\}.$$

Then there exists a unique function $u : \Omega\times]0,\infty[\to \mathbb{R}$ in the class

$$u \in L^\infty_{loc}(0,\infty; W), \quad u' \in L^\infty_{loc}(0,\infty; V), \quad u'' \in L^\infty_{loc}(0,\infty; L^2(\Omega))\,,$$

$$\delta^{1/2} u'' \in L^2_{loc}(0,\infty; L^2(\Gamma_1))\,,$$

satisfying the equation

$$u'' - M(|u(.)|^2)\Delta u + |u'|^\rho u' = 0 \ \text{ in } \ L^\infty_{loc}(0,\infty; L^2(\Omega))$$

and the initial conditions

$$u(0) = u^0, \quad u'(0) = u^1\,.$$

Furthermore u verifies

$$\frac{\partial u}{\partial \nu} + \delta u' = 0 \quad \text{in} \quad L^\infty_{loc}(0,\infty; H^{1/2}(\Gamma_1))\,,$$

$$\frac{\partial u'}{\partial \nu} + \delta u'' = 0 \quad \text{in} \quad H^{-1}_{loc}(0,\infty; L^2(\Gamma_1))\,.$$

The next result is important in the proof of Theorem 3.1.

Lemma 3.1 *Let us suppose that $u^0 \in V$, $\Delta u^0 \in L^2(\Omega)$ and $u^1 \in V$ with*

$$(-\Delta u^0, v) = ((u^0, v)) + (\delta u^1, v)_{L^2(\Gamma)}, \quad \text{for all } v \in V.$$

Let $\varepsilon > 0$. Then there exist w and z in W such that

$$(-\Delta w, v) = ((w, v)) + (\delta z, v)_{L^2(\Gamma)}, \quad \text{for all } v \in V,$$

$$\|w - u^0\|_W < \varepsilon, \quad \|z - u^1\| < \varepsilon.$$

Proof. Fix $\varepsilon > 0$. By Proposition 2.1, there exists $z \in W$ such that $\|z - u^1\| \le \varepsilon$. Let w be the solution of the variational problem

$$\left|\begin{array}{l} w \in V \\ ((w, v)) = (-\Delta u^0, v) + (\delta z, v)_{L^2(\Gamma)} \quad \text{for all } v \in V. \end{array}\right.$$

Then $\Delta w = \Delta u^0$; moreover

$$\begin{aligned} \|w - u^0\|_W^2 &= |\Delta w - \Delta u^0|^2 + \|\delta(z - u^1)\|^2_{H^{1/2}(\Gamma)} \\ &\le C\|z - u^1\|^2_{H^{1/2}(\Gamma)} \\ &\le C_1\|z - u^1\|^2 \le C_1\varepsilon^2 \end{aligned}$$

where C_1 is a positive constant that depends only of δ and Ω. Thus $w,\ z \in W$ and

$$(-\Delta w, v) = ((w, v)) + (\delta z, v)_{L^2(\Gamma)}, \quad \text{for all } v \in V.$$

◇

The proof of Theorem 3.1, obtained by the Galerkin method, is organized as follows:

Step 1. We construct a special basis for the space W by using of Lemma 3.1. In fact, from $u^0 \in W$ and $u^1 \in V$ satisfying the hypotheses of theorem, we found sequences (u_ℓ^0) and u_ℓ^1 in W such that:

$$u_\ell^0 \to u^0 \quad \text{strongly in } W$$

$$u_\ell^1 \to u^1 \quad \text{strongly in } V$$

$$(-\Delta u_\ell^0, v) = ((u_\ell^0, v)) + (\delta u_\ell^1, v)_{L^2(\Gamma)}, \quad \text{for all } v \in V.$$

From above sequences, for each $\ell \in \mathbb{N}$ fixed, we consider the vectors (u_ℓ^0) and u_ℓ^1. So, to obtain the special basis of W we proceed in the following way: first, we determine a orthonormal basis w_k^ℓ of the subspace of W spanned by u_ℓ^0 and u_ℓ^1 (ℓ fixed). Thus $k = 1$ or $k = 1, 2$. Then by the orthonormalization process, we complete (w_k^ℓ) just to obtain a basis of W.

Step 2. We construct the approximate problems $(P_{\ell m})$ of the problem (1.1) - (1.4) and determine the respective approximate solutions $(u_{\ell m})$.

Step 3. We obtain a priori estimates in the space $L^\infty(0, \infty; W)$ and $L^\infty(0, \infty; V)$ for $(u_{\ell m})$ and $(u'_{\ell m})$, respectively. These estimates permit us to pass to the limit in the nonlinear part of the equations of $(P_{\ell m})$.

Step 4. We show the identities in the boundary for the solution u of problem (1.1) - (1.4), making use of Remark 2.1 and of results contained in Milla Miranda [16]. ◇

4. Asymptotic Behavior

In this section we shall study the energy decay of solution of (1.1) - (1.4) as $t \to \infty$. First of all let us note that the energy

$$E(t) = \frac{1}{2}\left\{|u'(t)|^2 + \widehat{M}(\|u(t)\|^2)\right\},$$

does not increase because

$$E'(t) = -M(\|u(t)\|^2)\int_{\Gamma_1} \delta(x)(u'(t))^2 d\Gamma - \alpha|u'(t)|^2. \tag{4.1}$$

In order to establish the result which assures the exponential decay of $E(t)$, we assume that

$$\widehat{M}(\lambda) \geq M(\lambda)\,\lambda. \tag{4.2}$$

So, we have:

Theorem 4.1 *Let u^0, u^1 satisfy the assumptions of Theorem 3.1. Then there exists a positive constant η such that*

$$E(t) \leq 3\,E(0)\,e^{-\eta t}, \quad \textit{for all } t \geq 0. \tag{4.3}$$

Proof. Let

$$\rho(t) = (u'(t), u(t)) + \frac{\alpha}{2}|u(t)|^2 + \frac{m_0}{2}\|\delta^{1/2}u(t)\|^2_{L^2(\Gamma_1)}, \tag{4.4}$$

and constants C_1, C_2 and C_3 such that

$$\|v\|^2_{L^2(\Gamma_1)} \le C_1 \|v\|^2_{H^{1/2}(\Gamma_1)}, \tag{4.5}$$

$$\|v\|^2_{H^{1/2}(\Gamma_1)} \le C_2 \|v\|^2, \tag{4.6}$$

$$|v|^2 \le C_3 \|v\|^2. \tag{4.7}$$

Thus, by using (4.5) – (4.7), from (4.4) it follows that

$$|\rho(t)| \le \frac{|u'(t)|^2}{2} + C_3(1+\alpha)\frac{\|u(t)\|^2}{2} + \frac{m_0}{2}\|\delta\|_{L^\infty(\Gamma_1)} C_1 C_2 \|u(t)\|^2. \tag{4.8}$$

Furthermore, if we consider $\varepsilon > 0$ such that

$$\varepsilon \le \frac{1}{2}\min\left\{\frac{1}{\max\left\{1, \frac{C_3(1+\alpha)}{m_0} + \|\delta\|_{L^\infty(\Gamma_1)} C_1 C_2\right\}}, \frac{4\alpha}{3}\right\}, \tag{4.9}$$

and E_ε defined by

$$E_\varepsilon(t) = E(t) + \varepsilon\,\rho(t), \tag{4.10}$$

from (4.7) – (4.9) we get

$$E(t) + \varepsilon\left[\frac{|u'(t)|^2}{2} + \left(\frac{C_3(1+\alpha)}{m_0} + \|\delta\|_{L^\infty(\Gamma_1)} C_1 C_2\right)\frac{m_0}{2}\|u(t)\|^2\right].$$

Due to (4.8), from above inequality we can conclude that

$$E_\varepsilon(t) \le \frac{3}{2}E(t). \tag{4.11}$$

We also note that the relation (4.9) implies

$$E_\varepsilon(t) \ge E(t) - \varepsilon\,|\rho(t)|,$$

whence by use of (4.8) and (4.10), we obtain

$$\frac{E(t)}{2} \le E_\varepsilon(t) \le \frac{3\,E(t)}{2}. \tag{4.12}$$

In order to obtain $E'_\varepsilon(t)$ we now calcule the derivative of ρ with respect to time. So,

$$\begin{aligned}\rho'(t) &= \left(M(\|u(t)\|^2)\Delta u(t), u(t)\right) + |u'(t)|^2 + m_0\left(\delta^{1/2}u'(t), u(t)\right)_{L^2(\Gamma_1)} \\ &\le -M(\|u(t)\|^2)\|u(t)\|^2 + |u'(t)|^2.\end{aligned}$$

Thus, from above inequality, (4.2) and (4.3) we can write

$$E'_\varepsilon(t) \le -\varepsilon\left(\frac{1}{2}|u'(t)|^2 + \frac{1}{2}\widehat{M}(\|u(t\|^2)\right) = -\varepsilon\,E(t), \quad \text{for all } t \ge 0.$$

By using (4.12) and the above inequality, we have

$$E'_\varepsilon(t) \le -\frac{2}{3}\varepsilon\,E_\varepsilon(t) \quad \text{for all } t \ge 0,$$

which implies that

$$E_\varepsilon(t) \le E_\varepsilon(0)e^{-\frac{2}{3}t}, \quad \text{for all } t \ge 0.$$

By the use of (4.12) once again, we obtain (4.3). ◇

References

1. G. Kirchhoff, Vorlesungen über Mechanik, Tauber, Leipzig (1883).
2. S.I. Pohozaev, Math. Sbornik, **25**, 145-158 (1975).
3. S. Bernstein, Izv. Acad. Nauk SSSR, Ser. Math., **4**, 17-26 (1940).
4. A. Arosio and S. Spagnolo, in: Nonlinear Partial Differential Equations and Their Applications, H. Brezis and J. L. Lions (Eds.), College de France Seminar, Vol.6, Pitman, London (1984).
5. J.L. Lions, in: Contemporary Developments in Continuum Mechanics and Partial Differential Equations, G. M. De la Penha and L. A. Medeiros (Eds.), North-Holland, London (1978).
6. K. Nishihara, Nonlinear Anal., **8**, 623-636 (1984).
7. E.H. Brito, Appl. Anal., **13**, 219-233 (1982).
8. E.H. Brito, Nonlinear Anal., **8**, 1489-1496 (1984).
9. M. Hosoya and Y. Yamada, J. Fac. Univ. Tokyo, **38**, 239-250 (1991).
10. M. Tucsnak, Diff. Integral Equations, **6**, 925-935 (1993).
11. K. Ono, Kyushu Journal of Mathematics, **48**, 265-281 (1994).
12. M. Milla Miranda and L.A. Medeiros, Rev. de Mat. Apl., Univ. de Chile, **17**, 47-73 (1996).
13. M. Milla Miranda and L.P. San Gil Jutuca, (preprint).
14. I. Lasiecka and J. Ong, (preprint).
15. V. Komornik and E. Zuazua, J. Math. Pure et Appl., **69**, 33-54 (1990).
16. M. Milla Miranda, Bol. Soc. Paran. Matemática, **11**, 131-157 (1990).
17. Y. Yamada, Nagoya Math. J., **87**, 17-39 (1982).

9th Int. Coll. on Differential Equations, pp. 379–384
D. Bainov (Ed)

Stability of Integral Manifolds of Neutral Functional Differential Equations

ZHOUDE SHAO
Department of Mathematics, Millersville University, Millersville, PA 17551, USA

Abstract: Assume that an autonomous neutral functional differential equation of the form $\frac{d}{dt}D(x_t) = f(x_t)$ has a $(k+1)$- parameter family of periodic solutions. It is shown that the integral manifold $\mathfrak{M}$ determined by these periodic solutions is orbitally asymptotically stable with asymptotic phase under appropriate assumptions. This extends the corresponding results for ordinary and retarded functional differential equations.
Key Words and Phrases: *Integral manifolds, periodic orbits, orbital stability.*

1. Introduction

Suppose that $r \geq 0$ is a given real number, and $\mathbb{R}^n$ is the n-dimensional vector space with norm $|\cdot|$. Let $C = C([-r,0],\mathbb{R}^n)$ be the Banach space of all the continuous functions mapping the interval $[-r,0]$ into $\mathbb{R}^n$ with norm $|\varphi|_C = \max_{-r\leq\theta\leq 0}|\varphi(\theta)|$, $\varphi \in C$. If $t_0 \in \mathbb{R}$, $\sigma \geq 0$, and $x : [t_0 - r, t_0 + \sigma] \to \mathbb{R}^n$ is a continuous function, for any $t \in [t_0, t_0+\sigma]$, $x_t \in C$ is defined by $x_t(\theta) = x(t+\theta)$, $\theta \in [-r, 0]$.

We consider an autonomous neutral functional differential equation of the form

$$\frac{d}{dt}D(x_t) = f(x_t). \tag{1.1}$$

Throughout this paper, it is assumed that $f : C \to \mathbb{R}^n$ is twice continuously Fréchet differentiable on C and the operator $D : C \to \mathbb{R}^n$ is linear, stable, and atomic at 0.

Suppose that (1.1) has a $(k+1)$- parameter family of periodic solutions, $\{p_t(\lambda),\ \lambda \in \Lambda\}$, and the minimal period of $p_t(\lambda)$ is $\omega(\lambda) > 0$, $\lambda \in \Lambda$, where Λ is an open set in $\mathbb{R}^k$. It is further assumed that the functions $\omega(\lambda)$ and $p_t(\lambda)$ are C^2 in $\lambda \in \Lambda$.

Let $\mathfrak{M} = \{p_t(\lambda) : 0 \leq t \leq \omega(\lambda), \lambda \in \Lambda\}$. Then $\mathfrak{M}$ is an integral manifold of (1.1) and we are mainly interested in the stability property of $\mathfrak{M}$. For ordinary differential equations, it is proved that $\mathfrak{M}$ is orbitally asymptotically stable with asymptotic phase under appropriate assumptions, see Hale and Stokes [1]. The result was extended to retarded functional differential equations by Hale [2] and Stokes [3,4]; and to neutral functional differential equations in case that Λ consists of a single point by Shao [5]. In addition, the behavior of solutions of (1.1) near constant solutions was studied by Hale, see, for example, Hale [6,7]. See Hale and Lunel [8] for more references.

The qualitative study of solutions of neutral functional differential equations is extremely difficult. This is mainly due to the fact that the solution operator does not smooth with time t. To overcome this (lack of smoothness) difficulty, we use a result obtained in Shao [9] which states that, for initial functions in a dense set of the phase space, the corresponding solutions of (1.1) are actually continuously differentiable in t on their interval of existence. The denseness of the set of such initial functions implies that, if a certain estimate or property is obtained for the solutions which are continuously differentiable, it maybe possible to show that the same estimate or property holds for all solutions.

This paper is organized as follows. In Section 2, we present some preliminary results on linear variational equation of (1.1) associated with $p_t(\lambda)$, and the differentiability result of solutions of (1.1). These results will be presented in the form which is needed in later sections. Section 3 is devoted to the existence of local stable manifolds. Finally, in Section 4, the main result is stated and proved.

2. Preliminary Results

For any fixed $\lambda \in \Lambda$, performing the change of variable $y_t = x_t - p_t(\lambda)$ on (1.1), we obtain that

$$\frac{d}{dt}D(y_t) = L(t, y_t, \lambda) + F(t, y_t, \lambda), \tag{2.1}$$

where L is the linear operator given by $L(t,\varphi) = f'_\varphi(p_t(\lambda))\varphi$, and F is defined by $F(t,\varphi,\lambda) = f(p_t(\lambda)+\varphi) - f(p_t(\lambda)) - f'_\varphi(p_t(\lambda))\varphi$. One can easily see that F is continuously Fréchet differentiable and satisfies

$$F(t,0,\lambda) = 0, \qquad F'_\varphi(t,0,\lambda) = 0. \tag{2.2}$$

Since $p_t(\lambda)$ is periodic with period $\omega(\lambda)$, $\{p_t(\lambda) : 0 \le t \le \omega(\lambda)\}$, for each fixed λ, is a compact set in C. This implies that

$$|F(t,\varphi_1,\lambda) - F(t,\varphi_2,\lambda)| \le l(\max\{|\varphi_1|_C, |\varphi_2|_C\}, \lambda)|\varphi_1 - \varphi_2|_C \tag{2.3}$$

for all $t \in \mathbb{R}$ and φ_1, φ_2 in a neighborhood of 0 in C, where $l : [0,\infty) \times \Lambda$ is continuous and satisfies that $l(0,\lambda) = 0$. Because of (2.2) and (2.3), we can regard (2.1) as a perturbation of the linear periodic equation

$$\frac{d}{dt}D(y_t) = L(t, y_t, \lambda) \tag{2.4}$$

near the solution $y \equiv 0$. Since $\frac{\partial}{\partial t}p_t(\lambda)$, $\frac{\partial}{\partial \lambda_1}p_t(\lambda), \cdots, \frac{\partial}{\partial \lambda_k}p_t(\lambda)$ are periodic solutions of (2.4) with period $\omega(\lambda)$, 1 is a characteristic multiplier of (2.4). For the definition and properties of characteristic multipliers, see Hale and Lunel [8] and Stokes [10].

For and $s \in \mathbb{R}$, $t \ge s$, let $T(t,s,\lambda)$ be the solution operator of (2.4), that is, $T(t,s,\lambda)\varphi = y_t(s,\varphi,\lambda)$, where $y_t(s,\varphi,\lambda)$ is the solution of (2.4) through (s,φ). Denote $T^0(s,\lambda) = T(s+\omega(\lambda), s, \lambda)$. The following two lemmas provide some preliminary results related to (2.4) and (2.1), most of which are corollaries of proven results tailored in the form needed in later sections. See Cruz and Hale [11, 12] and Hale and Lunel [8] for more general results.

Lemma 2.1. *Assume that (1). 1 is a characteristic multiplier of multiplicity $(k+1)$ of (2.4), and $\frac{\partial}{\partial t}p_t(\lambda), \frac{\partial}{\partial \lambda_1}p_t(\lambda), \cdots, \frac{\partial}{\partial \lambda_k}p_t(\lambda)$ are linearly independent. (2). All the remaining characteristic multipliers of (2.4) have absolute values less than 1. Then, for any $s \in R$, there is a $k+1$ dimensional subspace $P(s,\lambda)$ of C and a closed subspace $Q(s,\lambda)$ of C such that*

(1). $C = P(s,\lambda) \oplus Q(s,\lambda)$.

(2). $T^0(s,\lambda)P(s,\lambda) = P(s,\lambda)$, $T^0(s,\lambda)Q(s,\lambda) \subset Q(s,\lambda)$.

(3). $\sigma(T^0(s,\lambda)\,|_{P(s,\lambda)}) = \{1\}$, $\sigma(T^0(s,\lambda)\,|_{Q(s,\lambda)}) = \sigma(T^0(s,\lambda)) - \{1\}$.

(4). There are constants $M = M(\lambda) \ge 1$ and $\alpha = \alpha(\lambda) > 0$ such that, for any $\varphi \in P(s,\lambda), \psi \in Q(s,\lambda)$, the solution $y_t(s,\varphi,\lambda)$ and $y_t(s,\psi,\lambda)$ of (2.4) through (s,φ) and (s,ψ), respectively, satisfy that

$$|y_t(s,\varphi,\lambda)|_C \le M\,|\varphi| \qquad \text{for all } t, s \in \mathbb{R}$$

$$|y_t(s,\psi,\lambda)|_C \le M\,|\psi|_C e^{-\alpha(t-s)} \qquad \text{for all } t \ge s \in \mathbb{R}.$$

For any $s \in \mathbb{R}$, let $X(t,s,\lambda)$ denote the fundamental solution of (2.4) and let $K(t,\cdot,\lambda) : [s,t] \to C$ is defined by

$$K(t,\tau,\lambda) = \int_s^\tau X(t+\theta,\alpha,\lambda)d\alpha.$$

Then the solution $y_t = y_t(s,\varphi,\lambda)$ of (2.1) through (s,φ) can be written as

$$y_t = T(t,s,\lambda)\varphi + \int_0^t d[K(t,\tau,\lambda)]F(\tau,y_\tau,\lambda)d\tau.$$

We will let $\pi_{P(s,\lambda)} : C \to P(s,\lambda)$, and $\pi_{Q(s,\lambda)} : C \to Q(s,\lambda)$ represent the canonical projections of C onto $P(s,\lambda)$ and $Q(s,\lambda)$, respectively.

Lemma 2.2. *Under the assumptions of Lemma 2.1, the solution y_t of (2.1) through (s,φ) can be written as $y_t = y_t^P + y_t^Q$, where $y_t^P = \pi_{P(t,\lambda)}y_t \in P(t,\lambda)$ and $y_t^Q = \pi_{Q(t,\lambda)}y_t \in Q(t,\lambda)$ are given by the following formulas*

$$\begin{aligned} y_t^P &= T(t,s,\lambda)\pi_{P(s,\lambda)}\varphi + \int_s^t d[\pi_{P(\tau,\lambda)}K(t,\tau,\lambda)]F(\tau,y_\tau,\lambda)d\tau \\ y_t^Q &= T(t,s,\lambda)\pi_{Q(s,\lambda)}\varphi + \int_s^t d[\pi_{Q(\tau,\lambda)}K(t,\tau,\lambda)]F(\tau,y_\tau,\lambda)d\tau. \end{aligned}$$

Moreover, there are constants $B = B(\lambda)$ and $\alpha = \alpha(\lambda) > 0$ such that

(1). $Var_{[s,t)}[\pi_{P(\cdot,\lambda)}K(t,\cdot,\lambda)] \le B$ for all s, $t \in (-\infty,\infty)$.

(2). $Var_{[s,t)}[\pi_{Q(\cdot,\lambda)}K(t,\cdot,\lambda)] \le Be^{-\alpha(t-s)}$ for all $t \ge s$.

The following differentiabilility result of solutions of (1.1) can be found in Shao [9].

Lemma 2.3. *Assume that f is local Lipschitz continuous on C. Then the solution of (1.1) through $(0,\varphi)$ for any $\varphi \in S = \{\varphi \in C : \dot\varphi \in C, D(\dot\varphi) = f(\phi)\}$ is continuously differentiable on its interval of existence. Moreover, the set S is dense in C.*

3. Local Stable Manifolds

In this section, we consider the nonlinear equation (2.1). Based on the results given in Section 2, we can show the existence of local stable manifolds of (2.1) near the solution $y(t) \equiv 0$.

Lemma 3.1. *Assume that the assumptions of Lemma 2.2 hold. Then, for any $\varphi \in C$, if $z^*(\varphi,\lambda) \in C([-r,\infty),\mathbb{R}^n)$ is a solution of the integral equation*

$$\begin{aligned} z(t) = T(t,0,\lambda)\varphi &+ \int_0^t d[\pi_{Q(s,\lambda)}K(t,s,\lambda)]F(s,z_s,\lambda)ds \qquad (3.1) \\ &- \int_t^\infty d[\pi_{P(s,\lambda)}K(t,s,\lambda)]F(s,z_s,\lambda)ds, \qquad t \ge 0 \end{aligned}$$

$$z(t) = \varphi(t) - \int_0^\infty d[\pi_{P(s,\lambda)}K(t,s,\lambda)]F(s,z_s,\lambda)ds, \quad -r \le t \le 0\,, \qquad (3.2)$$

then $z_t^*(\varphi,\lambda)$ is also a solution of (2.1).

The proof of this Lemma uses standard arguments, see, for example, Hale and Lunel [8] (Lemma 1.1 in Section 10.1) or Stokes [3] (Lemma 3.3).

Lemma 3.2. *Under the assumptions of Lemma 2.2, for any given constant β, $0<\beta<\alpha$, there is a constant $\eta_0=\eta_0(\lambda)>0$ such that, for any $\varphi \in \mathcal{U}_Q(\eta_0)=\{\varphi \in Q(0,\lambda): |\varphi|_C \le \eta_0\}$, there is a solution $z^*(\varphi,\lambda) \in C([-r,\infty),\mathbb{R}^n)$ of (3.1) and (3.2) satisfying*

$$|z^*(\varphi,\lambda)(t)| \le \rho(\eta_0,\lambda)e^{-\beta t}, \qquad t \ge 0,$$

where $\rho:[0,\infty)\times\Lambda \to [0,\infty)$ is a continuous function with $\rho(0,\lambda)=0$. Furthermore, $H(\cdot,\lambda):\mathcal{U}_Q(\eta_0)\to C$ defined by

$$H(\varphi,\lambda)(\cdot)=\int_0^\infty d[\pi_{P(s,\lambda)}K(\cdot,s,\lambda)]F(s,z_s^*(\varphi,\lambda),\lambda)ds$$

is continuously Fréchet differentiable and satisfies that $H'_\varphi(0,\lambda)=0:Q(0,\lambda)\to C$.

We would like to remark that, for a fixed compact subset of Λ, $\eta_0=\eta_0(\lambda)$ can be chosen to be uniformly bounded away from 0.

Outline of the Proof. For any $\delta>0$, we define

$$\begin{aligned} Z(\delta) &= \{z \in C([-r,\infty),\mathbb{R}^n): \ |z(t)| \le \delta e^{-\beta t}, \ t\ge 0, \ |z_0|\le\delta\} \\ \bar{Z}(\delta) &= \{z \in C([-r,\infty),\mathbb{R}^n): \ |\bar{z}(t)| \le \delta, \ t \ge -r\}. \end{aligned}$$

Then $Z(\delta)$ and $\bar{Z}(\delta)$ are closed bounded convex subsets in the Banach space $C([-r,\infty),\mathbb{R}^n)$ with the uniform topology. Moreover, there is a one-to-one correspondence h: $Z(\delta)\to\bar{Z}(\delta)$, defined by, for any $z\in Z(\delta)$,

$$h(z)=\bar{z}\in\bar{Z}(\delta): \quad \bar{z}(t)=z(t)e^{\beta t}, \ t\ge 0; \ \bar{z}_0=z_0.$$

For any $\varphi\in C$ and $\lambda\in\Lambda$, let $A(\varphi,\cdot,\lambda):Z(\delta)\to C([-r,\infty),\mathbb{R}^n)$ be the operator defined by the right-hand sides of (3.1) and (3.2), and let $B(\varphi,\cdot,\lambda):\bar{Z}(\delta)\to C([-r,\infty),\mathbb{R}^n)$ be the operator defined as $B=h\circ A(\varphi,\cdot,\lambda)\circ h^{-1}$. Similarly as in the proof presented in Stokes [3, 4] (Lemma 3.4 in Stokes [3], for example), using (2.2), (2.3), and Lemma 2.2, we can prove that there are constants $\eta_0=\eta_0(\lambda)>0$ and δ_0 such that $B(\varphi,\cdot,\lambda):\bar{Z}(\delta_0)\to\bar{Z}(\delta_0)$ is a uniform contraction operator with respect to $\varphi\in\mathcal{U}_Q(\eta_0)$. This implies that $B(\varphi,\cdot,\lambda):\bar{Z}(\delta_0)\to\bar{Z}(\delta_0)$ has a unique fixed point $\bar{z}^*(\varphi,\lambda)\in\bar{Z}(\delta_0)$. Therefore, $z^*(\varphi,\lambda)=h^{-1}(\bar{z}^*(\varphi,\lambda))\in Z(\delta_0)$ is a unique fixed point of $A(\varphi,\cdot,\lambda):Z(\delta_0)\to Z(\delta_0)$. Hence $z^*(\varphi,\lambda)$ is a solution of (3.1) and (3.2). Using (2.2) and (2.3), we can derive that $z^*(0,\lambda)=0$ and the existence of the function $\rho(\eta_0,\lambda)$.

To show that H is continuously Fréchet differentiable, we first show that $z^*(\varphi,\lambda)$ is continuously Fréchet differentiable on $\mathcal{U}_Q(\eta_0)$. In fact, it can be shown that $B(\varphi,\bar{z},\lambda)$ is continuously Fréchet differentiable with respect to $(\varphi,\bar{z},\lambda)\in\mathcal{U}_Q(\eta_0)\times\bar{Z}(\rho(\delta_0))\times\Lambda$. This implies that, by Theorem 3.2 in Chapter 0 of Hale [13], $\bar{z}_\varphi^{*\prime}(\varphi,\lambda)$ exists and is continuous. From this it is easy to show that $z_\varphi^{*\prime}(\varphi,\lambda)$ exists and is continuous. Finally, by the definition of H, we can easily show that H is Fréchet differentiable on $\mathcal{U}_Q(\eta_0)$, $H'_\varphi(\varphi,\lambda)$ is continuous, and $H'_\varphi(0,\lambda)=0:Q(0,\lambda)\to C$.

Corollary 3.1. *Under the assumptions in Lemma 3.2, if we let $R(\cdot,\lambda)=I-H(\cdot,\lambda):\mathcal{U}_Q(\eta_0)\to C$, then we have*

(1). $R(\cdot,\lambda):\mathcal{U}_Q(\eta_0)\to R(\mathcal{U}_Q(\eta_0))$ is a homeomorphism.

(2). $R(\cdot,\lambda)$ is continuously Fréchet differentiable and $R'_\varphi(0,\lambda)=I:Q(0,\lambda)\to C$;

(3). If y_t is a solution of (2.1) through $(0,\varphi)$ with $\varphi\in R(\mathcal{U}_Q(\eta_0,\lambda))$, then

$$|y_t| \le \rho(\eta_0,\lambda)e^{-\beta t}, \qquad t\ge 0.$$

Corollary 3.2. *Let $\pi = \pi_{Q(0,\lambda)} : C \to Q(0,\lambda)$ be the canonical projection, and $H_1(\cdot,\lambda)$, $R_1(\cdot,\lambda) : \mathcal{U}_Q(\eta_0) \to Q(0,\lambda)$ be defined by $H_1(\varphi,\lambda) = \pi H(\varphi,\lambda)$, and $R_1(\varphi,\lambda) = \varphi - H_1(\varphi,\lambda) = \pi R(\varphi,\lambda)$. Then there are open neighborhoods of 0 in $Q(0,\lambda)$, W_1 and W_2, such that $R_1(\cdot,\lambda) : W_1 \to W_2$ is a diffeomorphism. Moreover, $R_1(0,\lambda) = 0$.*

The conclusions of these corollaries follow directly from the implicit function theorem and the inverse function theorem.

4. Main Result

This section is devoted to the proof of our main result which is stated below. Recall that $\mathfrak{M} = \{p_t(\lambda), t \in (-\infty,\infty), \lambda \in \lambda\}$. For the definitions related to orbital stability of $\mathfrak{M}$, see Stokes [4].

Theorem 1. *Assume that (1). 1 is a characteristic multiplier of multiplicity $(k+1)$ of (2.4), and $\frac{\partial}{\partial t}p_t(\lambda), \frac{\partial}{\partial \lambda_1}p_t(\lambda), \cdots, \frac{\partial}{\partial \lambda_k}p_t(\lambda)$ are linearly independent. (2). All the remaining characteristic multipliers of (2.4) have absolute values less than 1. Then the integral manifold $\mathfrak{M}$ is orbitally asymptotically stable with asymptotic phase.*

Proof. For any precompact subset Λ^* of Λ with $\bar{\Lambda}^* \subset \Lambda$, let $\mathfrak{M}^* = \{p_t(\lambda) : t \in (-\infty,\infty), \lambda \in \Lambda^*\}$. By Corollary 3.1, all we need to show is that there is a $\delta > 0$ such that, if $\mathrm{dist}(\varphi, \mathfrak{M}^*) < \delta$, then, for some $c \in \mathbb{R}$ and $\lambda \in \Lambda$, we have

$$x_c(\varphi) - p_0(\lambda) \in R(\mathcal{U}_Q(\eta_0),\lambda)), \tag{4.1}$$

where R and $\mathcal{U}_Q(\eta_0)$ are as in Corollary 3.3. By the compactness of $\bar{\Lambda}^*$, we need only show that (4.1) is true in the case that $\bar{\Lambda}^*$ consists of a single point, say $\bar{\Lambda}^* = \{\lambda_0\}$. This is equivalent to show that there is a $\delta > 0$ such that, if $\mathrm{dist}(\varphi, p_0(\lambda_0)) < \delta$, then, for some $c \in \mathbb{R}$ and $\lambda \in \Lambda$, (4.1) is true. For any $\delta > 0$ and $\zeta > 0$, set

$$V_{\delta,\zeta} \stackrel{\text{def}}{=} (\omega(\lambda_0) - \zeta, \omega(\lambda_0) + \zeta) \times \{|\varphi - p_0|_C < \delta\} \times \{|\lambda - \lambda_0| < \zeta\}.$$

By the continuity of solutions, we can choose $\delta > 0$ and $\zeta > 0$ so small that for any $(t,\varphi,\lambda) \in V_{\delta,\zeta}$, $|x_t(\varphi) - p_0(\lambda)|_C$ is small enough such that

$$\pi(x_t(\varphi) - p_0(\lambda)) \in \pi R(U_Q(\eta_0),\lambda) = R_1(\mathcal{U}_Q(\eta_0),\lambda),$$

where π, R, R_1 are as in Corollary 3.1 and Corollary 3.2, and $x_t(\varphi)$ is the solution of (1.1) through $(0,\varphi)$. The relation above implies that there exists a $\psi(t,\varphi,\lambda)$ such that $R_1(\psi(t,\varphi,\lambda),\lambda) = \pi(x_t(\varphi) - p_0(\lambda))$. Therefore, the proof would be completed if we can show that there are $c \in (\omega(\lambda_0) - \zeta, \omega(\lambda_0) + \zeta)$ and $\lambda \in \{|\lambda - \lambda_0| < \zeta\}$ such that

$$\pi^\perp(x_c(\varphi) - p_0(\lambda)) = \pi^\perp R(\psi(t,\lambda,\varphi),\lambda),$$

where $\pi^\perp = I - \pi = \pi_{P(0,\lambda)}$.

Let G be the function defined by

$$G(t,\varphi,\lambda) = \pi^\perp(x_t(\varphi) - p_0(\lambda)) - \pi^\perp R(\psi(t,\varphi,\lambda),\lambda)$$

on $V_{\delta,\zeta}$. It is easy to see that $G(t,\varphi,\lambda)$ is continuous on $V_{\delta,\zeta}$. Also, one can verify, by the properties of R and R_1, that $\psi(\omega(\lambda_0), p_0(\lambda_0), \lambda_0) = 0$. This implies that $G(\omega(\lambda_0), p_0(\lambda_0), \lambda_0) = 0$. If we further restrict that $\varphi \in S$, where S is defined in Lemma 2.3, we also have that $G(t,\varphi,\lambda)$ is continuously differentiable with respect to $t \in (\omega(\lambda_0) - \zeta, \omega(\lambda_0) + \zeta)$. In fact, if $\varphi \in S$, Lemma 2.3 says that $x_t(\varphi)$ is continuously differentiable with respect to t. Since π and $\pi^\perp$ are both bounded linear operators, by the properties of R and R_1, we can show that $G'_t(t,\varphi,\lambda)$ is continuous for $(t,\varphi,\lambda)\in V^S_{\delta,\lambda}$, where

$$V^S_{\delta,\zeta} \stackrel{\text{def}}{=} (\omega(\lambda_0)-\zeta,\omega(\lambda_0)+\zeta)\times\{|\varphi-p_0(\lambda)|_C<\delta\}\cap S\times\{\lambda:|\lambda-\lambda_0|<\zeta\}.$$

Furthermore, it is clear that

$$\frac{\partial}{\partial t}(x_t(\varphi)-p_0(\lambda))|_{I_0}=\dot{x}_{\omega(\lambda_0)}(p_0(\lambda_0))=\frac{\partial}{\partial t}p_0(\lambda_0) \quad \textit{and} \quad \frac{\partial}{\partial\lambda_i}(x_t(\varphi)-p_0(\lambda))|_{I_0}=\frac{\partial}{\partial\lambda_i}p_0(\lambda_0),$$

where, for simplicity, we have set $I_0=(\omega(\lambda_0),p_0(\lambda_0),\lambda_0)$. This implies that

$$\frac{\partial}{\partial t}\pi^{\perp}(x_t(\varphi)-p_0(\lambda))|_{I_0}=\frac{\partial}{\partial t}p_0(\lambda_0) \quad\text{and}\quad \frac{\partial}{\partial t}\pi(x_t(\varphi)-p_0)|_{I_0}=0.$$

$$\frac{\partial}{\partial\lambda_i}\pi^{\perp}(x_t(\varphi)-p_0(\lambda))|_{I_0}=\frac{\partial}{\partial\lambda_i}p_0(\lambda_0) \quad\text{and}\quad \frac{\partial}{\partial\lambda_i}\pi(x_t(\varphi)-p_0(\lambda_0))|_{I_0}=0.$$

Therefore, $\frac{\partial}{\partial t}\psi(t,\varphi,\lambda)|_{I_0}=0$. By the definition of G, we can show that $G'_t(t,\varphi,\lambda)|_{I_0}=\frac{\partial}{\partial t}p_0(\lambda_0)$, and $G'_{\lambda_i}(t,\varphi,\lambda)|_{I_0}=\frac{\partial}{\partial\lambda_i}p_0(\lambda_0)$ for $k=1,\ 2,\ \cdots,\ k$. Since $\frac{\partial}{\partial t}p_0(\lambda_0)$, $\frac{\partial}{\partial\lambda_1}p_0(\lambda_0)$, $\cdots$, $\frac{\partial}{\partial\lambda_k}p_0(\lambda_0)$ form a basis for the space $P(0,\lambda)$, by the implicit function theorem, we may assume that the $\delta>0$ and $\zeta>0$ has been chosen so small such that, for any for any $\varphi\in\{|\varphi-p_0|<\delta\}\cap S$, there is an $c=c(\varphi)\in(\omega(\lambda_0)-\zeta,\omega(\lambda_0)+\zeta)$ and $\lambda=\lambda(\varphi)\in\{|\lambda-\lambda_0|<\zeta\}$ such that

$$G(c(\varphi),\varphi,\lambda(\varphi))=0.$$

This proves that (4.1) holds provided that $\varphi\in S$.

Since S is dense in C, for any $\varphi\in C$ with $|\varphi-p_0(\lambda_0)|_C<\delta$, we can find a sequence $\{\varphi_n\}\subset S$ satisfying that $|\varphi_n-p_0|_C<\delta$ and $\varphi_n\to\varphi$ as $n\to\infty$. Let $c_n=c(\varphi_n)$ and $\lambda_n=\lambda(\varphi_n)$ be such that $G(c_n,\varphi_n,\lambda_n)=0$. Since $\{c_n\}$ and $\{\lambda_n\}$ are bounded sequences, we may assume that $c_n\to c_0$ and $\lambda_n\to\lambda_0$ as $n\to\infty$. By the continuity of G, we obtain that

$$0=\lim_{n\to\infty}G(c_n,\varphi_n,\lambda_n)=G(c_0,\varphi,\lambda_0),$$

i.e., (4.1) hold for $\varphi\in C$, $|\varphi-p_0(\lambda_0)|_C<\delta$. This completes the proof.

References

1. J.K. Hale and A.P. Stokes, Arch. Rat. Mech. Anal., **2**, 133-170 (1960).
2. J.K. Hale, J. Differential Equations, **7**, 126-183 (1970).
3. A.P. Stokes, in: Contri. Differential Equations, **3**, 121-139 (1963).
4. A.P. Stokes, J. Differential Equations, **9**, 405-419 (1971).
5. Z. Shao, preprint, (1998).
6. J.K. Hale, J. Differential Equations, **10**, 59-82 (1971).
7. J.K. Hale, J. Differential Equations, **15**, 278-294 (1974).
8. J.K. Hale and S.V. Lunel, Introduction to Functional Differential Equations, Applied Math. Sci., Vol. 99, Springer-Verlag, New York (1993).
9. Z. Shao, Annals of Mathematics (Shanghai), **8A**, 394-402 (1987).
10. A.P. Stokes, Proc. Nat. Acad. of Sci. U.S.A., **48**, 1330-1334 (1962).
11. M.A. Cruz and J.K. Hale, J. Differential Equations, **7**, 334-355 (1970).
12. M.A. Cruz and J.K. Hale, Math. Anal. Appl., **34**, 267-285 (1971).
13. J.K. Hale, Ordinary Differential Equations, Wiley-Interscience, New York (1969).

9th Int. Coll. on Differential Equations, pp. 385-392
D. Bainov (Ed)

Numerical Accuracy of Dynamical System Integration

KOICHIRO SHIDA
Musashi Institute of Technology, 1-28-1 Tamazutumi, Setagaya, 158-8557, Japan
KAZUO HARAOKA
TOSHIO KAWAI
Chitose Institute of Science and Technology, Japan

Abstract: Assume hormonic oscillator $\ddot{x} = -x$ and the initial conditions $x(0) = 1$, and $\dot{x} = 0$. Finite difference symplectic scheme contains angular velocity, which is the second order of h. The amplitude error is not a problem because it is strictly constant for harmonic oscillator, and more generally, the energy is held within a certain range. But the accuracy can be very easily degraded to first order by careless initial condition.

1. Introduction

Molecular Dynamics does time integration. The accuracy of the result becomes a problem by the selection of time integration method. Also attention is necessary to the handling of the initial condition. Often, a strict initial condition depend on the time integration method.

In this paper, we will discuss about two points in the linear systems. First of all, we will discuss Stoermer method [1, 2] as the time integration method. Stoermer method is sometimes so called "Verlet method." It is the simplest case of the symplectic method [3, 4, 5]. The accuracy of this method is the 2nd order of the time step h theoretically. However, there are many cases it is not calculated with the accuracy because of its bad initial conditions. We will see one body harmonic oscillator as an example. We recognize that the Stoermer method is sensitive to an initial condition. Two bodies oscillator problem is also calculated. It is also sensitive as above. Then we will give the proper initial condition in the n body problem.

Next we study "velocity Verlet method." It is more accurate than the original Verlet method. We applied it to the system that has friction proportional to the velocity. In this case, we discovered that the better result of accuracy may be obtained than theoretical prediction. It is very strange result. We considered the reason.

Then we will conclude this paper.

2. Stoermer Method

2.1. General Descpition

The Stoermer method is obtained the next two steps: expanding of the integrand by time, and eliminating the terms of the 3rd or higher order. The general description is as follows,

$$\frac{X_{n+1} - 2X_n + X_{n-1}}{h^2} = F_n, \tag{1}$$

where X_n means the state vector at n-th time steps, h is the time step and F_n is the force. Then you can easily replace as follows,

$$X_{n+1} = 2X_n - X_{n-1} + h^2 F_n. \tag{2}$$

Using Eq.(2), the states are determined gradually under the initial condition X_{-1} and X_0.

2.2. One Body Oscillator

In this section, we will apply the Stoermer method to one body harmonic oscillator. The equation of motion is,

$$\ddot{x} = -\omega^2 x, \tag{3}$$

where x is the position of the mass point. Assume the following initial condition,

$$x(t=0) = x_0, \tag{4}$$
$$v(t=0) = v_0, \tag{5}$$

where v is the velocity and given by $v = \dot{x}$. Analytical solution of Eq.(3) is,

$$x(t) = x_0 \cos \omega t + v_0/\omega \sin \omega t \tag{6}$$
$$= Re(x_0 + v_0/i\omega) \exp i\omega t \tag{7}$$
$$= Re(a \exp i\omega t), \tag{8}$$

where ω is the angular velocity and i is the imaginary unit. The coefficient "a" is amplitude and determined by the initial conditions.

Deference equation for Eq.(3) is,

$$x_{n+1} - 2x_n + x_{n-1} = -\omega^2 h^2 x_n, \tag{9}$$

and Stoermer expression is,

$$x_{n+1} = 2x_n - x_{n-1} - \omega^2 h^2 x_n. \tag{10}$$

The characteristic equation of Eq.(10) is,

$$\lambda^2 - (2 - h^2)\lambda + 1 = 0, \tag{11}$$

and the eigenvalues of Eq.(11) is easily obtained as,

$$\lambda_{\pm} = ((2 - \omega h) \pm \sqrt{4\omega^2 h^2 - \omega^4 h^4})/2. \tag{12}$$

The time step h must be smaller than 2 for numerical stability. The absolute value of λ is unity, therefore another expression of the eigenvalues with a real positive value Ω,

$$\lambda_{\pm} = \exp i\Omega h. \tag{13}$$

Putting this result into Eq.(11), we have another expression of the equation,

$$\Omega h/2 = \sin^{-1} \omega h/2. \tag{14}$$

Of course, $\Omega = \omega$ at the limit of $h \to 0$.

The solution of Eq.(10) is,

$$x = A/2 \exp i\Omega hn + B/2 \exp(-i\Omega hn), \tag{15}$$

where A and B are determined by the initial conditions. Typically initial velocity v_0 and initial position x_0 are given. But Stoermer method need two initial positions. Then formal notation of the required initial position by Stoermer method instead of velocity is x_{-1}, where $x_{-1} = x(-\Omega h)$. Then exact form of A is,

$$A = x_0 \exp(i\Omega h/2) \frac{\sin(\Omega h/2)}{(\sin \Omega h)/2} + \frac{x_0 - x_{-1}}{i \sin \Omega h}. \tag{16}$$

At last, we reached at the exact solution of the Stoermer method,

$$x_n = Re(A \exp i\Omega hn). \tag{17}$$

Let us compare the two positions which are expressed by Eq.(8) and Eq.(17). The expression $x(nh)$ and x_n in each equation means the same time. The constants Ω and A correspond to ω and a respectively. It means that Stoermer method calculates the slightly different problem.

We will discuss the detail of the error. Let us set $A = a + \delta a$ and $\Omega = \omega + \delta\omega$, where $\delta a = O(h)$ and $\delta\omega = O(h^2)$. Then,

$$\begin{aligned} error(t = nh) &= x_n - x(nh) & (18)\\ &= Re[A \exp(i\Omega nh)] - Re[a \exp(i\omega nh)] \\ &= Re[a \exp(i\omega nh)(\exp(i\delta\omega nh) - 1) + \delta a \exp(i\omega nh) \exp(i\delta\omega nh)] \\ &= Re[x(nh)(i\delta\omega nh + \delta a/a)]. & (19) \end{aligned}$$

In Eq.(19), there are two types of errors, one is time dependent and the other is not. Time dependent error is estimated as $\omega = O(h^2)$. This is the reason why Stoermer is called 2nd order accuracy of time integration. We use Taylor expansion in the above analysis, so this estimate is effective under the next criterion,

$$\delta\omega t << 1, or \quad \omega t << 24/\omega^2 h^2.$$

There is another error term in Eq.(19), $\delta a/a$. You can notice that the calculation accuracy becomes also worse if δa is worse than h^2, . In other words, the accuracy of the Stoermer integration depends on the accuracy of assumed initial condition x_{-1}. The analytical value of x_{-1} is given as,

$$\begin{aligned} x_{-1} &= x_0 \cos(-\omega h) + v_0/\omega \sin(\omega h) & (20)\\ &= x_0 - v_0 h - \frac{(\omega h)^2 x_0}{2} + O(h^3). & (21) \end{aligned}$$

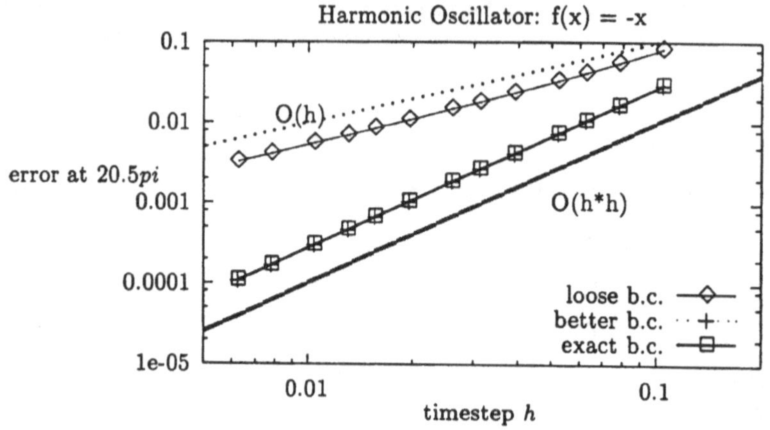

Figure 1: The *error* vs. h at $t = 20.5\pi$. The dashed line is the error with the accurate initial condition, $x_{-1} = \cos(-h)$, and the solid line is the looser, $x_{-1} = x_0$.

When $v_0 = 0$, we often employ the loose initial guess, $x_{-1} = x_0$. From Eq.(21), we see that the loose assumed initial condition causes the worse calculation result. Using strict x_{-1}, δa becomes $-1/2i\omega h x_0$.

Now we will show the calculation results. We assume one body harmonic oscillator given in Eq.(3) under the initial conditions $\omega = 1$, $x_0 = 1$, and $v_0 = 0$. Analytical solution of this system is

$$x(t) = \cos(t). \tag{22}$$

Figure 1 shows the *error* vs h graph. Parameter is time step h and observed at $t = 20.5\pi$ (see Appendix). Strict position is origin. Calculation result is analyzed and the numerical error δx at 20.5π is proportional to the h for loose condition and proportional h^2 (for exact condition) . The theory and calculation agree well.

2.3. N-Body General Theory

The above theory is for the most simplest system. But it is expandable to many bodies system. In general, the equation of motion for linear system is given as,

$$\ddot{X} = -AX, \tag{23}$$

where X is a state vector and A is a matrix. If A is positive definite, Stoermer method is easily applied to the system,

$$X_{n+1} - (2 - Ah^2)X_n + X_{n-1} = 0. \tag{24}$$

Because of the characteristics of A, Eq.(24) can be rewritten to diagonal form,

$$Y_{n+1} - (2 - Dh^2)Y_n + Y_{n-1} = 0, \tag{25}$$

and the m-th component of Eq.(25) is,

$$y^m_{n+1} - (2 - (\omega^m)^2 h^2)y^m_n + y^m_{n-1} = 0, \tag{26}$$

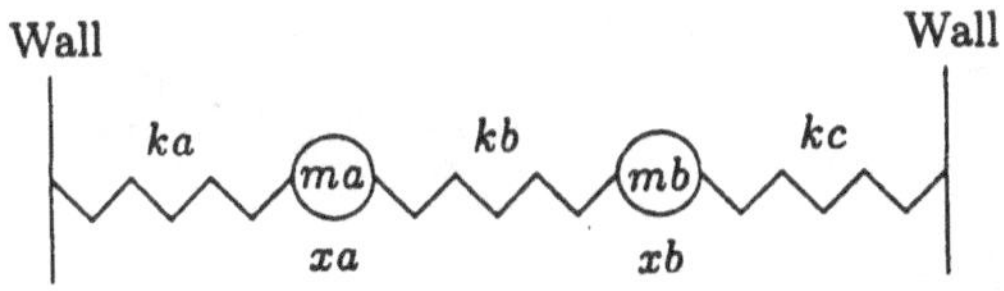

Figure 2: Three springs and two mass points system.

where superscript m means the m-th mode. This is similar to Eq.(9). All components are able to be described in the same way. So Stoermer method for linear many body system solves the very close problem to the original. The angular velocity $\omega^1, \omega^2, \ldots$ changed to $\Omega^1, \Omega^2, \ldots$. Another important point is that the alternative initial condition is given the next expression,

$$X_{-1} = X_0 - V_0 h + 1/2X"_0 h^2 + O(h^3), \tag{27}$$

where V_0 is initial condition for velocity and $\ddot{X}_0 = -AX_0$. Of course if you select the traditional initial condition, such as $X_{-1} = X_0 - V_0 h$, the result is going worse.

2.4. Two Body Harmonic Chain

In this section, we confirm the last conclusion, treating three springs and two mass points system. Figure 2 shows the detail.

In the Fig. 2, k_i are spring constants and ma and mb are mass, xa and xb are positions from the balanced points. For easy handling, spring constants, masses, and natural length of springs are assumed unity. The system is described as follows,

$$\ddot{xa} = -xa + (xb - xa), \tag{28}$$
$$\ddot{xb} = -(xb - xa) - xb, \tag{29}$$

where initial conditions are,

$$xa(0) = xa_0, \quad xb(0) = xb_0, \tag{30}$$
$$va(0) = 0, \quad vb(0) = 0. \tag{31}$$

Theoretical solution of this problem is;

$$xa = A\cos(t) + B\cos(\sqrt{3}t), \tag{32}$$
$$xb = A\cos(t) - B\cos(\sqrt{3}t), \tag{33}$$

where $A = xa_0 + xb_0$ and $B = xa_0 - xb_0$.

Figure 3 shows the Stoermer result. The solid line means the better initial conditions and the dashed means the loose initial conditions. In this case, the system has two modes so the error behavior seems strange, but the numerical result shows good agreement with the analysis.

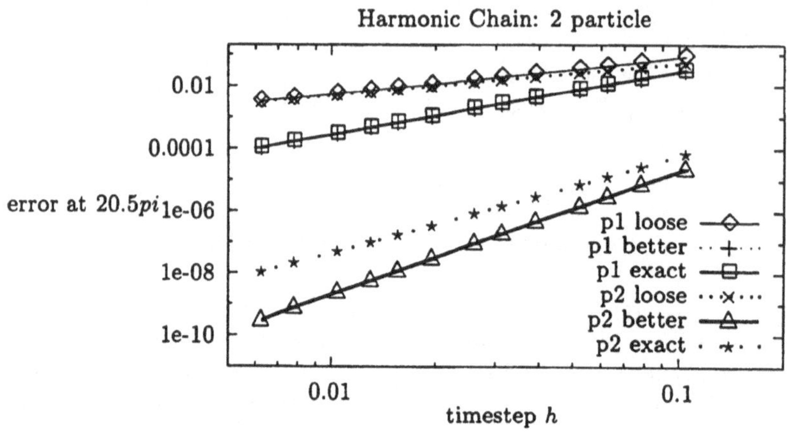

Figure 3: The *error* vs. h at $t = 20.5\pi$ for the 2body system. The graphs of better and exat initial condition overlap each other.

3. Velocity Verlet Method

Velocity Verlet is a very useful time integration method except for one weak point. When the force depends on the velocity we cannot calculate explicitly. We employ iteration method instead of implicit calculation for computing time.

Standard velocity Verlet is,

$$x_{n+1} = x_n + hv_n + h^2/2F(x_n), \tag{34}$$
$$v_{n+1} = v_n + h/2(F(x_n) + F(x_{n+1})). \tag{35}$$

3.1. Iterate Velocity Verlet Method

Assuming $F = F(v(t))$, Eq.(35) is changed into,

$$v_{n+1} = v_n + h/2(F(v_n) + F(v_{n+1})). \tag{36}$$

From Eq.(36), the iteration is defined as the next recipe,

$$V^i = v_n + h/2(F(v_n) + F(V^{i-1})), \tag{37}$$

where V^i is the value at i-th iteration. Of course $V^0 = v_n$ and $V^\infty = v_{n+1}$. We should confirm that it reaches a limit.

3.2. Application Velocity Verlet Method to Frictional System

Now we applied the above method to the simplest case,

$$f = -v, \tag{38}$$

where $v(0) = v_0$. Analytical solution of this problem is,

$$v(t) = v_0(\exp -t). \tag{39}$$

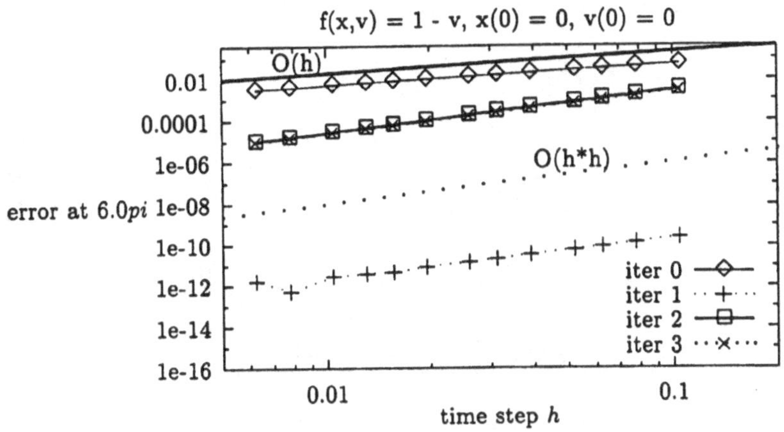

Figure 4: The error of position vs. the number of iteration at 6π. One iteration gives the best result [6].

This means that there is friction which is linear to the velocity. The velocity is determined as,

$$v_{n+1} = v_n + h/2(-v_n - v_{n+1}). \tag{40}$$

Exact solution of the above equation is,

$$v_{n+1} = (2-h)/(2+h)v_n, \tag{41}$$

and after i time iterations, we get

$$V^i = v_n(1 - h + h^2/2 - h^3/4.....). \tag{42}$$

In this case, a limiting value exists and is,

$$V^\infty = -v_n + 2hv_n \lim_{n->\infty}(1-(-h/2)^n)/(1-(-h/2)) \tag{43}$$

$$= v_n(2-h)/(2+h). \tag{44}$$

Then the limit of V^∞ equals to the Eq.(41).

Now we will show the calculation results in Fig. 4. The vertical axis means the error between theoretical estimate and horizontal axis is time step h. Observation is made at $t = 6\pi$, and circle is using V^1, triangle is V^2,,,,,. The suprising result came out!! One iteration gives extremely accurate result.

3.3. Discussion

Let's estimate the theoretical propagation operator during time step h. From Eq.(39),

$$v_{n+1} = \exp(-h)v_n \tag{45}$$

$$= (1 - h + h^2/2 - h^3/3! + \ldots)v_n. \tag{46}$$

From Eq.(42), velocity Verlet propagation follows

$$V^i = v_n(1 - h + h^2/2 - h^3/4\ldots). \tag{47}$$

And from Eq.(34), exact form of x_{nh} is easily given as,

$$x_n = x_{n-1} + hV_{n-1}(1 - \frac{h}{2}) \quad (48)$$

$$= x_{n-2} + (hV_{n-2} + hV_{n-1})(1 - \frac{h}{2}) \quad (49)$$

$$\dots$$

$$= x_0 + h(1 - \frac{h}{2}) \sum_{k=0}^{n} V_k. \quad (50)$$

Putting Eq. (47) into Eq. (50), we get the expression of the position at n-th step,

$$x_n = x_0 + h(1 - \frac{h}{2}) \frac{1 - (1 - h + (h^2/2))^n}{1 - (1 - h + (h^2/2))} \quad (51)$$

$$= x_0 + 1 - (1 - h + \frac{h^2}{2})^n. \quad (52)$$

The limiting value of x_{nh} at $n \to \infty$ is $x_0 + 1$ which is the strict position.

Appendix

Sometimes, under the particular initial condition or observing time, the dominant error term is disappeared by chance. See the Eq.(19), if the observing time is $\pi, 2\pi, 3\pi....\delta a$ term is 0. Of course, we can decide x_{-1} to δa is 0, but it is over speculation.

References

1. D.J. Hardy and D.I. Okunbor, J. Chem. Phys, **102**, 8978-8982 (1995).
2. M. Tuckerman, B.J. Berne and G.J. Martyna, J. Chem. Phys., **97**, 1990-2001 (1992).
3. H. Kinoshita, H. Yoshida and H. Nakai, Celest. Mech., **50**, 59-71 (1991).
4. H. Yoshida, Phys. Lett. A, **150**, 262-268 (1990).
5. H. Yoshida, Celest. Mech., **56**, 27-43 (1993).
6. K. Shida, K. Haraoka and T. Kawai, Meeting Abst. of the Phys. Soc. Jpn., **53**, 918 (1998).

9th Int. Coll. on Differential Equations, pp. 393–400
D. Bainov (Ed)

Approximations of a General Class of Quantum-Mechanical One-Dimensional Point Interactions by Local Self-Adjoint Interactions

T. SHIGEHARA, H. MIZOGUCHI, T. MISHIMA
Department of Information and Computer Sciences, Saitama University, Shimo-Okubo, Urawa, Saitama 338-8570, Japan
e-mail: sigehara@ics.saitama-u.ac.jp
TAKSU CHEON
Laboratory of Physics, Kochi University of Technology, Tosa Yamada, Kochi 782-8502, Japan
and
Theory Division, High Energy Accelerator Research Organization (KEK), Tsukuba, Ibaraki 305-0801, Japan
e-mail: cheon@mech.kochi-tech.ac.jp

Abstract: We first construct in the small-size limit of a local potential the one-dimensional zero-range force (ε potential) which induces such a boundary condition that the wave function has continuous first derivative on the right and left, but a jump proportional to the first derivative. By using the ε potential, we realize all the boundary conditions allowable in one-dimensional point interactions under the time reversal symmetry. Numerical evidence for the existence of the ε potential is shown by using a realistic finite-range potential. The scattering property of the ε potential is compared to that of the conventional δ potential.

1. Introduction

One-dimensional point interaction is a simple but useful tool for examining the effect of small impurities on low-energy particle motion in quantum mechanics. A well-known example is Kronig-Penney model [1] (potential consisting of a periodic array of δ functions), which has been widely considered as a standard reference model in solid state physics for more than sixty years. In two and three dimensions, the point interaction is nothing but an extended object of the usual δ potential. However the phenomena are much richer in one dimension. This is due to the fact that whereas the symmetric operator $A = -\Delta \lceil C_0^\infty(\mathbf{R}^d \backslash \{0\})$ has deficiency indices $(1,1)$ in spatial dimension $d = 2, 3$, it has deficiency indices $(2,2)$ in dimension $d = 1$. As a result, the operator A has in one dimension a four-parameter family of self-adjoint extensions [2], characterized by the connection condition [3]

$$\begin{pmatrix} \varphi'(+0) \\ \varphi(+0) \end{pmatrix} = \mathcal{V} \begin{pmatrix} \varphi'(-0) \\ \varphi(-0) \end{pmatrix}, \tag{1}$$

where $\mathcal{V} = e^{i\theta}\mathcal{U}$ with $\theta \in [0, 2\pi)$, $\mathcal{U} \in SL(2, \mathbf{R})$. The condition (1) covers the usual δ potential (of strength v), the transfer matrix of which is given by

$$\mathcal{V} = \mathcal{V}_\delta(v) \equiv \begin{pmatrix} 1 & v \\ 0 & 1 \end{pmatrix}. \tag{2}$$

Another important example is a so-called δ' potential, which leaves the derivative of the wave function $\varphi'(x)$ continuous, but makes $\varphi(x)$ itself discontinuous [4];

$$\mathcal{V} = \mathcal{V}_\varepsilon(u) \equiv \begin{pmatrix} 1 & 0 \\ u & 1 \end{pmatrix}. \tag{3}$$

We denote the contact interaction with the transfer matrix (3) by $\varepsilon(x)$ in this paper.

Except for δ, the connection condition (1) appears at first sight to be unnatural for quantum mechanics, and some efforts have been made in order to seek the physical Hamiltonians which embody the condition (1). Šeba has observed [5] that a certain two-parameter family is obtained by non-local pseudopotentials with ε potential. Chernoff and Hughes have shown [6] that a three-parameter family disjoint from Šeba's class corresponds to local pseudopotentials involving δ^2. Carreau has succeeded in realizing a general class in the small-size limit of non-hermitian operators [7]. In spite of considerable works, existing approximations do not seem to have much relevance to the experimentally realizable systems. Our main purpose is to rectify this situation. We study the cases with time reversal symmetry in this paper and show that the connection condition (1) can be realized by the perturbation of a local potential with a simple form. More precisely, it is shown in case of $\mathcal{V} \in SL(2, \mathbf{R})$ ($\theta = 0$) that a potential consisting of neighboring δ functions with disappearing distances can give rise to all possible connection conditions, if one renormalizes the coupling strengths of the δ's according to the distances in an appropriate manner. We start in Sect.2 by constructing ε potential in the small separation limit of three nearby δ's with suitably renormalized strengths according to the distance. By using ε and usual δ, the general connection condition (1) is realized in Sect.3. We give in Sect.4 numerical evidence for the ε potential by using a finite-range interaction composed of three local short-range potentials with the coupling strengths adjusted in an appropriate manner. Some elementary results for the scattering problem are presented in Sect.5, where a dual role of the δ and ε potentials is emphasized. The current work is summarized and some remarks are given in Sect.6.

2. Realization of Epsilon Potential

We begin with the Schrödinger equation in one-dimensional free space; $-\varphi''(x) = k^2\varphi(x)$. Here k is the wave number of the particle. In a vector notation $\Psi(x) = {}^t(\varphi'(x)\ \varphi(x))$, the equation is rewritten as

$$\Psi'(x) = \mathcal{H}(k)\Psi(x), \qquad \mathcal{H}(k) = \begin{pmatrix} 0 & -k^2 \\ 1 & 0 \end{pmatrix}. \tag{4}$$

The solution of Eq.(4) is given by

$$\Psi(x) = \mathcal{G}(k; x - x_0)\Psi(x_0), \qquad \mathcal{G}(k; x) \equiv e^{\mathcal{H}(k)x} = \begin{pmatrix} \cos kx & -k \sin kx \\ \frac{1}{k}\sin kx & \cos kx \end{pmatrix}. \tag{5}$$

Since $Tr\mathcal{H}(k) = 0$, $\mathcal{G}(k; x) \in SL(2, \mathbf{R})$, namely $\det \mathcal{G}(k; x) = 1$.

In order to realize the ε potential of strength u, we consider three nearby δ potentials which are located with equal distance a [8];

$$V(a;x) = v_1\delta(x+a) + v_0\delta(x) + v_1\delta(x-a). \tag{6}$$

With these three δ's, the connection between $x = -a-0$ and $x = a+0$ is given by

$$\Psi(a+0) = \mathcal{V}_{\varepsilon;a}(k)\Psi(-a-0), \tag{7}$$

where

$$\mathcal{V}_{\varepsilon;a}(k) = \mathcal{V}_\delta(v_1)\mathcal{G}(k;a)\mathcal{V}_\delta(v_0)\mathcal{G}(k;a)\mathcal{V}_\delta(v_1). \tag{8}$$

Note that $\mathcal{V}_{\varepsilon;a}(k) \in SL(2,\mathbf{R})$, since each matrix on RHS is in $SL(2,\mathbf{R})$. Since

$$\mathcal{G}(k;a) \simeq \begin{pmatrix} 1 & -k^2a \\ a & 1 \end{pmatrix} \tag{9}$$

at low energy $ka \ll 1$, we obtain

$$[\mathcal{V}_{\varepsilon;a}(k)]_{11} = [\mathcal{V}_{\varepsilon;a}(k)]_{22} = 1 + (v_0 + 2v_1 + v_0v_1a - k^2a)a, \tag{10}$$

$$[\mathcal{V}_{\varepsilon;a}(k)]_{12} = v_0 + 2v_1 + 2v_1a(v_0 + v_1 - k^2a) + v_0v_1^2a^2 - 2k^2a, \tag{11}$$

$$[\mathcal{V}_{\varepsilon;a}(k)]_{21} = 2a + v_0a^2. \tag{12}$$

In the limit $a \longrightarrow +0$ with v_0, v_1 constant, we simply obtain the connection condition for δ potential of strength $v_0 + 2v_1$. On the other hand, Eq.(12) shows that if the coupling strength v_0 changes as

$$v_0(a) \simeq \frac{u}{a^2} \tag{13}$$

for small a, one obtains $\lim_{a\longrightarrow+0}[\mathcal{V}_{\varepsilon;a}(k)]_{21} = u$. Furthermore, if

$$v_1(a) \simeq \frac{2}{u} - \frac{1}{a} \tag{14}$$

for small a, the diagonal element (10) converges into one in the small a limit. We thus obtain $\lim_{a\longrightarrow+0}\mathcal{V}_{\varepsilon;a}(k) = \mathcal{V}_\varepsilon(u)$ with Eqs.(13) and (14). (Note that $\lim_{a\longrightarrow+0}[\mathcal{V}_{\varepsilon;a}(k)]_{12} = 0$, since $\mathcal{V}_{\varepsilon;a}(k) \in SL(2,\mathbf{R})$.) This indicates that one realizes the ε potential of strength u in the small distance limit of the three δ potentials (6) by changing the strengths according to Eqs.(13) and (14) (See Fig.1). Though both the strengths v_0, v_1 diverge in the small a limit, $\lim_{a\longrightarrow+0}V(a;x)$ is nonsingular by construction.

It is easy to see in a similar manner that two nearby δ's are insufficient to construct the transfer matrix (3), even if the coupling strengths are renormalized according to the distance between them. This indicates that the ε potential has little to do with the derivative of Dirac's δ function, which is approximated by two nearby δ's of the strengths with the same magnitude but opposite sign each other (See also [9]).

3. Realization of the General Cases

By using the ε potential, one can realize the connection condition characterized by transfer matrix

$$\mathcal{V} = \begin{pmatrix} \alpha & \beta \\ \gamma & \delta \end{pmatrix}, \qquad \alpha\delta - \beta\gamma = 1, \quad \alpha,\beta,\gamma,\delta \in \mathbf{R}. \tag{15}$$

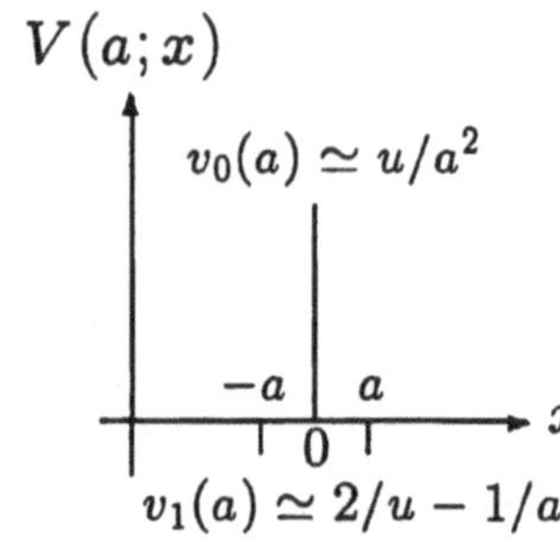

Figure 1: Approximation of ε potential of strength u by three nearby δ potentials

The following matrix decompositions serve for our purpose;

$$\begin{pmatrix} \alpha & \beta \\ \gamma & \delta \end{pmatrix} = \begin{cases} \mathcal{V}_\delta(\frac{\alpha-1}{\gamma})\mathcal{V}_\varepsilon(\gamma)\mathcal{V}_\delta(\frac{\delta-1}{\gamma}) & \text{for } \gamma \neq 0, \\ \mathcal{V}_\varepsilon(\frac{\delta-1}{\beta})\mathcal{V}_\delta(\beta)\mathcal{V}_\varepsilon(\frac{\alpha-1}{\beta}) & \text{for } \beta \neq 0. \end{cases} \tag{16}$$

Each matrix on RHS takes a form of either δ or ε potential. This implies that one can realize general connection condition by using the potential

$$V(x) = \begin{cases} \lim_{a\to+0} \left(\frac{\delta-1}{\gamma}\delta(x+a) + \gamma\varepsilon(x) + \frac{\alpha-1}{\gamma}\delta(x-a) \right) & \text{for } \gamma \neq 0, \\ \lim_{a\to+0} \left(\frac{\alpha-1}{\beta}\varepsilon(x+a) + \beta\delta(x) + \frac{\delta-1}{\beta}\varepsilon(x-a) \right) & \text{for } \beta \neq 0. \end{cases} \tag{17}$$

In case of $\beta = \gamma = 0$, one can use the decomposition

$$\begin{pmatrix} \pm|\alpha| & 0 \\ 0 & \pm|\delta| \end{pmatrix} = \mathcal{V}_\delta(\rho)\mathcal{V}_\varepsilon(-\frac{1}{\rho})\mathcal{V}_\delta(\rho)\mathcal{V}_\delta(\mp\frac{1}{\rho})\mathcal{V}_\varepsilon(\pm\rho)\mathcal{V}_\delta(\mp\frac{1}{\rho}) \tag{18}$$

with $\rho = \sqrt{|\alpha|}$, which is also written in terms of the transfer matrices for the δ and ε potentials.

4. Numerical Evidence for Epsilon Potential

The manner to construct the ε potential in Sect.2 is somewhat beyond our intuition, since it has been obtained in the zero-separation limit of three zero-range forces. In this section, we give a numerical example which shows that it works even with finite-range forces with a finite-size separation. For this purpose, we approximate the δ function by a step-like function of small but finite width h;

$$\delta_h(x) = \begin{cases} \frac{1}{h}, & |x| < \frac{h}{2}, \\ 0, & \text{otherwise}. \end{cases} \tag{19}$$

We consider on the interval $[X_1, X_2]$ ($X_1 < 0 < X_2$) the Schrödinger equation $-\varphi''(x) + V(x)\varphi(x) = k^2\varphi(x)$ with the potential

$$V(x) = v_1(a)\delta_{h_1}(x+a) + v_0(a)\delta_{h_0}(x) + v_1(a)\delta_{h_1}(x-a). \tag{20}$$

The wave propagation on $[X_1, X_2]$ is described by

$$\begin{aligned}\tilde{\mathcal{G}}(k) = &\mathcal{G}(k; X_2 - a - \frac{h_1}{2})\mathcal{G}(K_1; h_1)\mathcal{G}(k; a - \frac{h_0 + h_1}{2})\mathcal{G}(K_0; h_0) \\ &\times \mathcal{G}(k; a - \frac{h_0 + h_1}{2})\mathcal{G}(K_1; h_1)\mathcal{G}(k; -a - \frac{h_1}{2} - X_1)\end{aligned} \tag{21}$$

with $K_0 = \sqrt{k^2 - v_0(a)/h_0}$ and $K_1 = \sqrt{k^2 - v_1(a)/h_1}$. We assume the Dirichlet boundary condition so that $\varphi(X_1) = \varphi(X_2) = 0$. In this case, the eigenvalue is determined by

$$\left[\tilde{\mathcal{G}}(k_n)\right]_{21} = 0. \tag{22}$$

The associated (not necessarily normalized) eigenfunction is given, for example in the

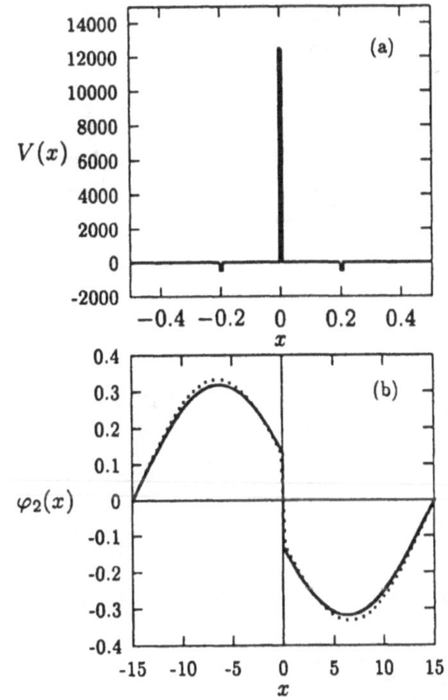

Figure 2: (a) Finite-range approximation of ε potential by three neighboring short-range well potentials. (b) The broken line shows the wave function for the second eigenstate ($k_2 = 0.185527$) for the potential in (a). The parameter is taken as $X_2 = -X_1 = 15.0$, $u = 5$, $a = 0.2$ and $h_0 = h_1 = 0.01$. For comparison, the approximated wave function by using the ε potential ($k_2 = 0.181097$) is exhibited by the solid line.

region $x \in [-a + h_1/2, -h_0/2]$ by the second component of

$$\begin{pmatrix} \varphi_n'(x) \\ \varphi_n(x) \end{pmatrix} = \mathcal{G}(k; x + a - \frac{h_1}{2})\mathcal{G}(K_1; h_1)\mathcal{G}(k; -a - \frac{h_1}{2} - X_1)\begin{pmatrix} 1 \\ 0 \end{pmatrix} \tag{23}$$

together with $k = k_n$. The eigenfunction in the other regions is obtained in a similar manner.

In the numerical calculations, we take $X_2 = -X_1 = 15.0$, $u = 5$, $a = 0.2$ and $h_0 = h_1 = 0.01$. The potential strengths v_0, v_1 are determined by Eqs.(13) and (14), respectively. Fig.2(a) illustrates the assumed potential (20). The broken curve in Fig.2(b) shows the wave function for the second state ($k_2 = 0.185527$). One can observe abrupt change of the wave function around the potential, which is successfully reproduced in the corresponding eigenfunction approximated by using the ε potential (solid line).

5. Scattering Properties of Epsilon Potential

In this section, we briefly discuss the scattering properties of the ε potential and point out a dual role between the δ and ε potentials.

We start by putting a general point interaction at the origin on x-axis. The wave function can be written as

$$\varphi(x) = \begin{cases} A(k)e^{ikx} + B(k)e^{-ikx}, & x < 0, \\ e^{ikx}, & x > 0. \end{cases} \tag{24}$$

Here we assume the incident wave comes from minus infinity and as a result, the wave function has only an outgoing component for $x > 0$, whose amplitude is normalized to one. The connection condition in Eq.(1) is written as

$$\begin{pmatrix} ik \\ 1 \end{pmatrix} = \mathcal{V} \begin{pmatrix} ik & -ik \\ 1 & 1 \end{pmatrix} \begin{pmatrix} A(k) \\ B(k) \end{pmatrix}, \tag{25}$$

leading to

$$\begin{pmatrix} A(k) \\ B(k) \end{pmatrix} = \frac{1}{2ik} \begin{pmatrix} 1 & ik \\ -1 & ik \end{pmatrix} \mathcal{V}^{-1} \begin{pmatrix} ik \\ 1 \end{pmatrix}. \tag{26}$$

The transmission and reflection coefficients are calculated by

$$T(k) = \left| \frac{1}{A(k)} \right|^2, \qquad R(k) = \left| \frac{B(k)}{A(k)} \right|^2, \tag{27}$$

respectively. In case of $\mathcal{V} = \mathcal{V}_\delta(v)$, Eq.(27) reproduces a well-known result

$$T_\delta(k) = \frac{k^2}{k^2 + (v/2)^2}, \quad R_\delta(k) = \frac{(v/2)^2}{k^2 + (v/2)^2}. \tag{28}$$

For $\mathcal{V} = \mathcal{V}_\varepsilon(u)$, we obtain

$$T_\varepsilon(k) = \frac{(2/u)^2}{k^2 + (2/u)^2}, \quad R_\varepsilon(k) = \frac{k^2}{k^2 + (2/u)^2}, \tag{29}$$

which are shown in Fig.3. One can observe that the transmission and reflection coefficients take an opposite role in the strong (resp. weak) δ and weak (resp. strong) ε potentials each other. More precisely, if $vu = 4$, $T_\delta(k) = R_\varepsilon(k)$ and $R_\delta(k) = T_\varepsilon(k)$ hold. This is related to a nontrivial duality between one-dimensional fermionic and bosonic many particle systems interacting through a contact zero-range force [10]. We also recognize that if $u = v$, $T_\delta(k) = T_\varepsilon(1/k)$ and $R_\delta(k) = R_\varepsilon(1/k)$ are satisfied. This implies that the low (resp. high) energy dynamics with ε potential is described by the high (resp. low) energy dynamics with δ potential.

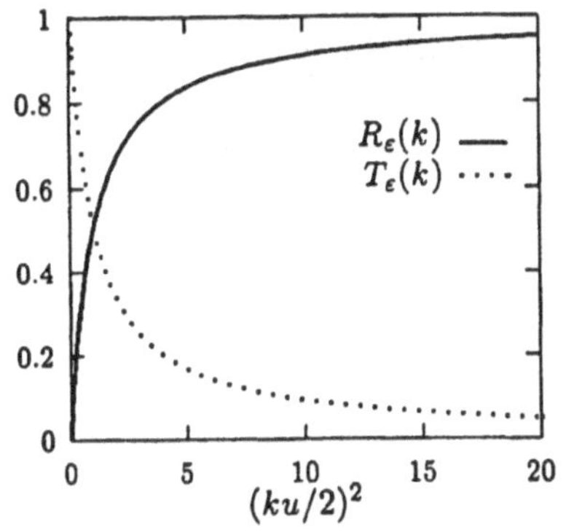

Figure 3: The transmission and reflection coefficients for the ε potential.

6. Summary and Remark

We have constructed a general class of one-dimensional zero-range point interactions in the small separation limit of several nearby δ functions. In particular, the ε potential where the derivative of the wave function is continuous but the wave function itself is discontinuous, is obtained by using three δ potentials with suitably renormalized strengths according to the distance. Numerical results show that such behavior of wave function is observed even at finite-size and finite-distance level. It is realized from the scattering properties that the δ and ε potentials have a reversed nature each other.

As another good approximation of the ε potential which is possibly realized in experiment, geometric scatterer (multichannel scatterer) is proposed in [11].

The spectral properties for the ε potential have been extensively studied in [2, 4, 11, 12].

We are grateful to Prof. P. Exner for useful discussions and comments.

References

1. R. de L. Kronig and W. G. Penney, Proc. Roy. Soc. (London), **130A**, 499 (1931).
2. S. Albeverio, F. Gesztesy, R. Høegh-Krohn and H. Holden, Solvable Models in Quantum Mechanics, Springer, New York (1988).
3. F. Gesztesy and W. Kirsch, J. Reine Angew. Math., **362**, 28 (1985).
4. A. Grossmann, R. Høegh-Krohn and M. Mebkhout, J. Math. Phys., **21**, 2376 (1980).
5. P. Šeba, Czech. J. Phys., **B36**, 667 (1986).
6. P. R. Chernoff and R. J. Hughes, J. Funct. Anal., **111**, 97 (1993).
7. M. Carreau, J. Phys., **A26**, 427 (1993).
8. T. Cheon and T. Shigehara, Phys. Lett., **A243**, 111 (1998).

9. P. Šeba, Rep. Math. Phys., **24**, 111 (1986).

10. T. Cheon and T. Shigehara, LANL preprint quant-ph/9806041.

11. J. E. Avron, P. Exner and Y. Last, Phys. Rev. Lett., **72**, 896 (1994).

12. Yu. N. Demkov and V. N. Ostrovskii, Zero Range Potentials and Their Applications in Atomic Physics, Plenum, New York (1988).

9th Int. Coll. on Differential Equations, pp. 401-406
D. Bainov (Ed)

An Exponentially-Fitted Finite Difference Method for the Numerical Solution of Differential Equations with Engineering Applications

T.E. SIMOS AND G. PAPAKALIATAKIS
School of Engineering, Department of Civil Engineering, University of Thrace, GR-67 100 Xanthi, Greece

Abstract: An exponentially-fitted finite difference method the numerical solution of fourth-order differential equations with engineering applications is developed in this paper. Numerical and theoretical results show that this new approach is more efficient compared with previously derived methods.
1991 Mathematics Subject Classification. 65L05.

1. Introduction

Several problems in Engineering and especially in plate deflection theory can be expanded in terms of the following boundary-value problem

$$Dy^{(4)}(x) + ky(x) = q \tag{1}$$
$$y\left(\pm\frac{L}{2}\right) = y^{(2)}\left(\pm\frac{L}{2}\right) = 0$$

where L, D, k and q are used to present the physical and mechanical characteristics of the problem (for details see [1]-[10]). For example one of the most known problems of this type from the theory of plates is the problem of bending of a plate on an elastic foundation (see [6] pp. 30-32 and 466-532).

This problem is a part of the following general class of fourth order boundary value problems

$$y^{(4)}(x) + p(x)y(x) = q(x), \; A \leq x \leq B \tag{2}$$
$$y(A) = S_1, \; y(B) = S_2, \; y''(A) = S_3, \; y''(B) = S_4.$$

The analytical solution of the above problem cannot be found, for all $p(x)$ and $q(x)$.

For the numerical solution of the general problem (2) some numerical methods have been proposed [1]-[5], [8]-[9], [10].

The purpose of this paper is to introduce a new approach, the exponential fitting, to constuct methods for the above problem (1) especially in the case which we have periodic and/or oscillating solutions. This is the case when $p(x)$ is a slowly varying function over the region of interest. In section 2 we will develop the basic theory for the exponentially-fitted methods. A simple exponentially-fitted finite difference method has been constructed in section 3. Finally, in section 4 the application of the developed method to the problem (1) has been presented and extended numerical results has been produced to show the efficiency of the new approach.

2. Exponential Multistep Methods

For the numerical solution of the initial value problem

$$y^{(r)} = f(x,y),\ y^{(j)}(A) = 0,\ j = 0,1,\ldots,r-1 \tag{3}$$

multistep methods of the form

$$\sum_{i=0}^{k} a_i y_{n+i} = h^r \sum_{i=0}^{k} b_i f(x_{n+i}, y_{n+i}) \tag{4}$$

can be used over the equally spaced intervals $\{x_i\}_{i=0}^{k} \in [A,B]$ and $h = | x_{i+1} - x_i |$, $i = 0,\ldots,k-1$.

The method (4) is associated with the operator

$$L(x) = \sum_{i=0}^{k} [a_i z(x+ih) - h^r b_i z^{(r)}(x+ih)] \tag{5}$$

where z is a continuously differentiable function.

Definition 1 *The multistep method (4) is called algebraic (or exponential) of order p if the associated linear operator L vanishes for any linear combination of the linearly independent functions $1, x, x^2, \ldots, x^{p+r-1}$ (or $exp(v_0 x), exp(v_1 x), \ldots, exp(v_{p+r-1}x)$ where $v_i, i = 0,1,\ldots,p+r-1$ are real or complex numbers).*

Remark 1 *(see [11] and [12]) If $v_i = v$ for $i = 0,1,\ldots,n,\ n \leq p+r-1$ then the operator L vanishes for any linear combination of*

$$\{exp(vx), x exp(vx), x^2 exp(vx), \ldots, x^n exp(vx), exp(v_{n+1}x), \ldots, exp(v_{p+r-1}x)\}. \tag{6}$$

Remark 2 *(see [11] and [12]) Every exponential multistep method of order p corresponds in a unique way, to an algebraic multistep method of order p (by setting $v_i = 0$ for all i).*

Lemma 1 *(For proof see [11] and [13]) Consider an operator L of the form (5). With $v \in C,\ h \in \mathcal{R},\ n \geq r$ if $v = 0$, and $n \geq 1$ otherwise, then we have*

$$L[x^m exp(vx)] = 0,\ m = 0,1,\ldots,n-1,\ L[x^n exp(vx)] \neq 0 \tag{7}$$

if and only if the function φ has a zero of exact multiplicity s at $exp(vh)$, where $s = n$ if $v \neq 0$, and $s = n - r$ if $v = 0$, $\varphi(w) = \rho(w)/log^r w - \sigma(w)$, $\rho(w) = \sum_{i=0}^{k} a_i w^i$ and $\sigma(w) = \sum_{i=0}^{k} b_i w^i$.

Proposition 1 *(For proof see [14] and [12]) Consider an operator L with*

$$L[exp(\pm v_i x)] = 0,\ j = 0,1,\ldots,k \leq \frac{p+r-1}{2} \tag{8}$$

then for given a_i and p with $a_i = (-1)^r a_{k-i}$ there is a unique set of b_i such that $b_i = b_{k-i}$.

3. The Derivation of Exponentially - Fitted Methods for General Problems

Consider the construction of an exponentially-fitted multistep method (4) which exactly integrates the set of functions $\{exp(\pm v_j x)\}_{j=0}^{k}$. We will use this for the numerical solution of the general problem (3).

From Lemma 1 we obtain the equations

$$\rho[exp(\pm v_j h)] - (\pm v_j h)^r \sigma[exp(\pm v_j h)] = 0 \tag{9}$$

or equivalently

$$\sum_{i=0}^{k}[a_i exp(\pm v_j h) - (\pm v_j h)^r b_i exp(\pm v_j h)] = 0, \; j = 0, 1, \ldots, n \tag{10}$$

where $n \leq k$ and $a_i,\ b_i,\ i = 0, \ldots, k$ are the coefficients of the multistep method (4).

We investigate here the case where k is a positive number. Then, from Proposition 1 we have a set of k equations:

$$a_i = (-1)^r a_{k-i},\ b_i = b_{k-i},\ i = 0, 1, \ldots, k. \tag{11}$$

We let $a_k = 1$, which is the convention adopted for all families of known multistep methods. Then (10) and (11) give the following system of equations

$$2\sum_{i=1}^{k/2-1} a_i sinh\left[\left(\frac{k}{2} - i\right) w_j\right]$$

$$+a_{k/2} - w_j^r \left[2\sum_{i=0}^{k/2-1} b_i cosh\left[\left(\frac{k}{2} - i\right) w_j\right] + b_{\frac{k}{2}}\right] = -2sinh\left(\frac{kw_j}{2}\right), \text{ for } r = 1, 3, 5, \ldots \tag{12}$$

$$2\sum_{i=1}^{k/2-1} a_i cosh\left[\left(\frac{k}{2} - i\right) w_j\right]$$

$$+a_{k/2} - w_j^r \left[2\sum_{i=0}^{k/2-1} b_i cosh\left[\left(\frac{k}{2} - i\right) w_j\right] + b_{\frac{k}{2}}\right] = -2cosh\left(\frac{kw_j}{2}\right), \text{ for } r = 2, 4, 6, \ldots \tag{13}$$

where $w_j = v_j h$ and $j = 0, 1, \ldots, k$.

We now prove that the system of equations either (i) has a unique solution when $w_i \neq \pm w_j$, or (ii) lead to so called 'indeterminate' expressions of the form $\left(\frac{0}{0}\right)$ when $w_i = \pm w_j$ for some i and j.

Let $X(w)$ and $Y(w)$ $(w = vh)$ be the matrices of the unknown coefficients in the systems of equations (12) and (13) respectively. Consider case (i). In order to make the matrices $X(w)$ (or $Y(w)$) singular then their columns would have to be linearly dependent. The elements in a row consist of terms like $\cosh Mw_j$, $\sinh Nw_j$ and powers of w_j. The multiple angle hyperbolic functions can be expressed in terms of powers of $\cosh w_j$, $\sinh x_j$ and their products. These with powers of w_j form a linearly independent set of functions. Therefore the columns cannot be linearly dependent. Hence in this case $detX(w) \neq 0$ (or $detY(w) \neq 0$). Thus the system of equations (12) and (13) has a unique solution.

Consider case (ii). Here we simply have two rows of the matrix of coefficients the same and hence $detX(w) = 0$ (or $detY(w) = 0$). Similarly we have the right hand side of two of the equations in (10) or (11) the same so that the numerator determinant which is formed when a column of $X(w)$ (or $Y(w)$) is replaced by the right hand column will also give two identical rows. Hence the numerator determinant is 0. In these cases **L'Hospital's rule** must be used to evaluate these expressions.

In the present paper we investigate the case $r = 4$.

4. Derivation of the new exponentially-fitted method

Consider the four-step method

$$\overline{y}_n = y_n - ah^4(y_{n+2}^{(4)} - 2y_n^{(4)} + y_{n-2}^{(4)}) \tag{14}$$

$$\begin{aligned} y_{n+2} - 4y_{n+1} + 6y_n - 4y_{n-1} + y_{n-2} = h^4(b_0 y_{n+2}^{(4)} + b_1 y_{n+1}^{(4)} + \\ + b_2\overline{y}_n^{(4)} + b_1 y_{n-1}^{(4)} + b_0 y_{n-2}^{(4)}) \end{aligned} \tag{15}$$

where $y_k^{(4)} = -p(x_k, y_k) + q(x_k)$, with $k = n-2(1)n+2$.

We require that the method (15) should integrate exactly any linear combination of the functions:

$$\{1, x, x^2, x^3, x^4, x^5, x^6, x^7, \exp(\pm vx)\}. \tag{16}$$

To construct a method of the form (15) which integrates exactly the functions (16), we require that the method (15) integrates exactly (see [14], [12]):

$$\{1, x, \exp(\pm v_0 x), \exp(\pm v_1 x), \exp(\pm v_2 x), \exp(\pm v_3 x)\} \tag{17}$$

and then put:

$$v_0 = v_1 = v_2 = 0 \text{ and } v_3 = v. \tag{18}$$

The method (15) integrates exactly the functions 1, x. Demanding that (15) integrates (17) exactly, we obtain the following system of equations for b_0, b_1, a and b

$$\begin{aligned} 2b_0 w^4 cosh(2w) + b_2 w^4 + 2b_2 a w^8 - 2b_2 a w^8 cosh(2w) \\ +2b_1 w^4 cosh(w) = 6 - 8cosh(w) + 2cosh(2w) \end{aligned} \tag{19}$$

where $w_j = v_j h, j = 0, 1, 2, 3$.

Solving for b_i, $i = 0(1)2$ and a we obtain the coefficients of the new method. The formulae obtained above are subject to heavy cancellations for small values of $w = vh$. For this reason series expansions of the coeffcients obtained we the above procedure must be produced.

If $w = i\phi$, then the family of methods (15) is exact for any linear combination of the functions:

$$\{1, x, x^2, x^3, x^4, x^5, x^6, x^7, sin(\phi x), cos(\phi x)\}. \tag{20}$$

Applying the Taylor series expansions of $y_{n+2}, y_{n+1}, y_n, y_{n-1}, y_{n-2}, y_{n+2}^{(4)}, y_{n+1}^{(4)}, y_n^{(4)}, y_{n-1}^{(4)}$ and $y_{n-2}^{(4)}$ in (15) we have the following result for the local truncation error $T(h)$:

$$T(h) = h^{10}\left(\frac{1}{3024}y_n^{(10)} + \frac{1501}{29030400}y_n^{(6)} + \frac{61}{241920}v^4 y_n^{(6)}\right) \tag{21}$$

i.e. we have an eighth order method.

At the boundaries we use the formulae introduced in [15].

5. Numerical Illustration

To test the efficiency of the proposed new approach we have applied the new method to a problem which occurs in the plate deflection theory.

5.1. A problem which occurs in plate deflection theory

We solve the problem (1) for

$$D = 1,\ k = 64 - 0.01x^4,\ q = 1,\ L = 10. \tag{22}$$

There is no analytical solution to this problem. So we obtain our reference solution (which we call "exact solution") using the high order method of Usmani [2] with $h = \frac{1}{32}$.

In Table 1 we present the maximum absolute errors for various choices of step size. We note that the new method is mentioned in the Table as Method [b] and the method of Usmani [2] as Method [a].

	Maximum Absolute errors	
h	Method [a]	Method [b]
$\frac{1}{2}$	2.810^{-5}	7.210^{-11}
$\frac{1}{4}$	1.510^{-7}	6.110^{-13}
$\frac{1}{8}$	1.510^{-9}	0
$\frac{1}{16}$	2.410^{-11}	0

Table 1: Comparison of maximum absolute errors $\mid E^{calculated} - E^{exact} \mid$ for the test problem (22) produced by the high order method of Usmani [2] (presented as Method [a]) and the new method (presented as Method [b]).

6. Conclusions

In this paper a new procedure to obtain numerical methods for the numerical solution of some fourth order boundary value problems with engineering applications has been developed. This approach is based on exponential fitting. Based on this, we have produced a method which integrates exactly any linear combination of $\{1, x, x^2, x^3, x^4, x^5, x^6, x^7, \exp(\pm vx)\}$. The numerical results indicate that the new method is much more accurate than the high order method of Usmani [2].

References

1. R.A. Usmani and M.J. Marsden, J. Engrg. Math., **9**, 1-10 (1975).

2. R.A. Usmani, J. Inst. Maths. Applics., **20**, 331-333 (1977).

3. R.A. Usmani, Math. Comput., **32**, 1087-1096 (1978).

4. R.A. Usmani and S.A. Warsi, Comput. Maths. with Appls. **6**, 205-211 (1980).

5. R.A. Usmani, J. Comput. Appl. Math., **6**, 93-98 (1980).

6. S.P. Timoshenko and S. Woinowsky-Kreger, Theory of Plate and Shells, McGraw-Hill, New York (1959).

7. C.P.Katti, Math. Comput., **35**, 1117-1119 (1980).

8. M.K. Jain, S.R. Iyengar and I.S.V. Saldanha, J. Engrg. Math., **11**, 373-380 (1977).

9. M.M. Chawla and C.P. Katti, BIT, **19**, 27-33 (1979).

10. M. Van Daele, G. Vanden Berghe and H. De Meyer, J. Comput. Appl. Math., **51**, 383-394 (1994).

11. T. Lyche, Numerische Mathematik, **10**, 65-75 (1972).

12. T.E. Simos, Numerical Solution of Ordinary Differential Equations with Periodical Solution, Doctoral Dissertation, National Technical University of Athens (1990).

13. P. Henrici, Discrete Variable Methods in Ordinary Differential Equations, John Wiley and Sons, New York (1962).

14. A.D. Raptis, Bulletin of the Greek Mathematical Society, **25**, 113-126 (1984).

15. G. Papakaliatakis and T.E. Simos, in: Proceedings of the V-th International Colloquium on Numerical Analysis, August 13-17, 1996, Plovdiv, Bulgaria, E. Minchev (Ed.), Academic Publications (1996) pp. 123-132.

9th Int. Coll. on Differential Equations, pp. 407-410
D. Bainov (Ed)

L^∞-Estimates for Strong Solutions to Quasilinear Elliptic Equations

LUBOMIRA G. SOFTOVA
Department of Mathematics, Higher Military Transport School, 158 "Geo Milev" Street, 1574 Sofia, Bulgaria
e-mail: luba@mail.vvtu.bg

Abstract: By the aid of Aleksandrov–Pucci maximum principle for linear elliptic operators, we derive L^∞-a priori estimate and uniqueness for strong solutions to the Dirichlet problem for quasilinear strictly elliptic equations with Carathéodory's coefficients. The results obtained compose a preliminary step towards the study of strong solvability (in Sobolev or Morrey spaces) of boundary value problems for quasilinear elliptic operators with discontinuous coefficients.

Key Words and Phrases: *Aleksandrov–Pucci maximum principle, strong solutions, quasilinear elliptic equations.*

1991 Mathematics Subject Classification. 35J65, 35B50, 35R05.

Let $\Omega \subset \mathbb{R}^n$ be a bounded domain with Lipschitz boundary $\partial\Omega$, and consider in Ω the quasilinear operator

$$Qu \equiv a^{ij}(x,u,Du)D_{ij}u + b(x,u,Du)$$

where the usual summation convention on repeated indices is adopted. The coefficients $a^{ij}(x,z,p)$ and $b(x,z,p)$ are supposed to be defined on $\Omega \times \mathbb{R} \times \mathbb{R}^n$ with values in $\mathbb{R}$, and to satisfy the *Carathéodory* condition, i.e., they are measurable functions on x for all $(z,p) \in \mathbb{R} \times \mathbb{R}^n$ and continuous on (z,p) for almost all $x \in \Omega$. We will suppose that Q is a strictly elliptic operator, that is,

$$a^{ij}(x,z,p)\xi^i\xi^j \geq \lambda(x,z,p)|\xi|^2 \quad \text{for a.a. } x \in \Omega, \quad \forall \xi \in \mathbb{R}^n, \quad \lambda(x,z,p) > 0. \tag{1}$$

Concerning the function $b(x,z,p)$, we impose the next conditions:

$$\begin{cases} \text{the nonlinear operator } u \longrightarrow b(x,u,Du) \text{ is a continuous} \\ \text{operator from } W^{2,n}(\Omega) \text{ into } L^n(\Omega), \end{cases} \tag{2}$$

and

$$\begin{cases} \dfrac{\operatorname{sign} z.b(x,z,p)}{\lambda(x,z,p)} \leq \mu_1(x)|p| + \mu_2(x) \quad \text{for a. a. } x \in \Omega, \quad \forall (z,p) \in \mathbb{R} \times \mathbb{R}^n, \\ \mu_1,\ \mu_2 \geq 0, \qquad \mu_1,\ \mu_2 \in L^n(\Omega). \end{cases} \tag{3}$$

Remark 1. The validity of (2) can be ensured by imposing growth conditions on $b(x,z,p)$ with respect to (z,p).

For example, using continuity of the imbedding $W^{2,n}(\Omega) \hookrightarrow W^{1,q}(\Omega) \cap C^0(\overline{\Omega})$ for each $q \geq 1$, the Nemitskii operator $u(x) \to b(x,u(x),Du(x))$ is continuous, i.e., (2) holds, if:

$$\begin{cases} |b(x,z,p)| \leq b_1(|z|)\Big(b_2(x) + |p|^q\Big) & \text{for arbitrary } q \geq 1, \\ \text{where } b_1(\cdot) \text{ is a continuous function and } & b_2(x) \in L^n(\Omega) \end{cases}$$

(see Theorem 16. 11 in [1]).

Our first result establishes a upper estimate for strong subsolutions to the equation $Qu = 0$.

Theorem 1. *Suppose the validity of* (1), (2) *and* (3) *and let* $u \in W^{2,n}(\Omega)$ *satisfy*

$$Qu \equiv a^{ij}(x,u(x),Du(x))D_{ij}u + b(x,u(x),Du(x)) \geq 0 \qquad \textit{almost everywhere in } \Omega.$$

Then

$$\sup_{\Omega} u \leq \sup_{\partial\Omega} u^+ + C\|\mu_2\|_{L^n(\Omega)} \tag{4}$$

where $u^+(x) = \max\{u(x),0\}$ *and* $C = C\Big(n, \|\mu_1\|_{L^n(\Omega)}, \operatorname{diam}\Omega\Big)$.

Proof. In view of (3), we have for almost all $x \in \Omega^+ = \{x \in \Omega : \ u(x) > 0\}$ that

$$\begin{aligned} \frac{b(x,u(x),Du(x))}{\lambda(x,u(x),Du(x))} &\leq \mu_1(x)|Du(x)| + \mu_2(x) \\ &\leq \mu_1(x).\text{sign}\,(D_iu(x))D_iu(x) + \mu_2(x). \end{aligned}$$

Thus,

$$\bar{a}^{ij}(x)D_{ij}u(x) + \bar{b}^i(x)D_iu(x) \geq -\mu_2(x) \qquad \text{for a. a. } x \in \Omega^+,$$

where $\bar{a}^{ij}(x) = \dfrac{a^{ij}(x,u(x),Du(x))}{\lambda(x,u(x),Du(x))}$, $\bar{a}^{ij}\xi^i\xi^j \geq |\xi|^2$ and $\bar{b}^i(x) = \mu_1(x).\text{sign}\,(D_iu(x)) \in L^n(\Omega)$. Since $u \in W^{2,n}(\Omega)$, it follows from the Sobolev imbedding theorem that $u \in C^0(\overline{\Omega})$ and Aleksandrov-Pucci maximum principle [2, Theorem 9.1] asserts

$$\sup_{\Omega^+} u \leq \sup_{\partial\Omega^+} u^+ + C\|\mu_2\|_{L^n(\Omega)}$$

with C depending on the same quantities as in (4). The continuity of $u(x)$ and the fact that $\sup_{\Omega} u \leq \sup_{\Omega^+} u$ complete the proof. □

Corollary 1. *Under the assumptions of Theorem 1, let* $u \in W^{2,n}(\Omega)$ *be a strong solution of the equation*

$$a^{ij}(x,u(x),Du(x))D_{ij}u + b(x,u(x),Du(x)) = 0 \qquad x \in \Omega.$$

Then

$$\sup_{\Omega} |u| \leq \sup_{\partial\Omega} |u| + C\|\mu_2\|_{L^n(\Omega)}. \tag{5}$$

Proof. It follows from Theorem 1 that

$$u(x) \le \sup_\Omega u \le \sup_{\partial\Omega} u^+ + C\|\mu_2\|_{L^n(\Omega)}, \quad x \in \Omega. \tag{6}$$

In order to derive an estimate for u from below, we consider the function $v = -u$ which satisfies

$$a^{ij}(x, -v, -Dv)D_{ij}v + \bar{b}(x, v, Dv) \ge 0 \quad \text{a.e. in } \Omega.$$

Here $\bar{b}(x,z,p) = -b(x,-z,-p)$. It is clear that

$$\operatorname{sign} z.\bar{b}(x,z,p) = \operatorname{sign}\{-z\}.b(x,-z,-p) \le \mu_1(x)|p| + \mu_2(x).$$

Applying Theorem 1 to the function v, we obtain

$$\sup_\Omega v \le \sup_{\partial\Omega} v^+ + C\|\mu_2\|_{L^n(\Omega)}$$

and therefore

$$\begin{aligned} -\inf_\Omega u &\le -\inf_{\partial\Omega} u^- + C|\mu_2\|_{L^n(\Omega)} \\ &\le \sup_{\partial\Omega}|u| + C\|\mu_2\|_{L^n(\Omega)}, \end{aligned}$$

whence

$$u(x) \ge \inf_\Omega u \ge -\sup_{\partial\Omega}|u| - C\|\mu_2\|_{L^n(\Omega)}.$$

The last inequality combined with (6) proves the desired estimate (5). □

Remark 2. The above results remain true if one supposes $u \in C^0(\overline{\Omega}) \cap W^{2,n}_{\rm loc}(\Omega)$.

Additional requirements on the coefficients of the operator Q lead to uniqueness result for the strong solutions of the Dirichlet problem

$$\begin{cases} a^{ij}(x, Du)D_{ij}u(x) + b(x,u,Du) = 0, \\ u \in W_0^{1,n}(\Omega) \cap W^{2,n}(\Omega). \end{cases} \tag{7}$$

Theorem 2. *Suppose that the coefficients a^{ij} of the operator Q satisfy* (1), *they are independent of u and let*

$$\left|a^{ij}(x,p') - a^{ij}(x,p'')\right| \le \tilde{a}(x)|p'-p''| \quad \textit{for a.a. } x \in \Omega, \quad i,j = 1,\dots,n, \tag{8}$$

where $\tilde{a} \in L^\infty(\Omega)$. Assume further that the function $b(x,z,p)$ is non-increasing in z and

$$\left|b(x,z,p') - b(x,z,p'')\right| \le \tilde{b}(x,z)|p'-p''| \quad \textit{for a.a. } x \in \Omega, \tag{9}$$

where $\sup_{|z|\le M}\tilde{b}(\cdot,z) \in L^n(\Omega)$.

Then, if u, $v \in W_0^{1,n}(\Omega) \cap W^{2,n}(\Omega)$ solve the Dirichlet problem (7), *one has $u \equiv v$.*

Proof. Denote $w = u - v$ and consider the set $\Omega^+ = \{x \in \Omega : \ w(x) > 0\}$. We have

$$b(x,u(x),Du(x)) \le b(x,v(x),Du(x)) \qquad \text{for a.a. } x \in \Omega^+,$$

as consequence of the monotonicity of $b(x,z,p)$ with respect to z. Thus,

$$A^{ij}(x)D_{ij}w + B^k(x)D_kw \geq 0 \qquad \text{a.e. in } \Omega^+, \tag{10}$$

where

$$\begin{aligned} A^{ij}(x) &= a^{ij}(x, Dv(x)),\\ B^k(x) &= \sum_{i,j=1}^{n} D_{ij}u(x)\left(\int_0^1 \frac{\partial a^{ij}}{\partial p_k}(x, s(Du-Dv)+Dv)ds\right)\\ &\quad + \int_0^1 \frac{\partial b}{\partial p_k}(x, v, s(Du-Dv)+Dv)ds. \end{aligned}$$

We should note that the derivatives $\partial a^{ij}/\partial p_k$ and $\partial b/\partial p_k$ exist by virtue of the Lipschitz continuity with respect to p (cf. (8) and (9)) and the Rademacher theorem. Further, (8) and (9) imply $B^k \in L^n(\Omega^+)$ in (10). Therefore, the Aleksandrov-Pucci maximum principle yields

$$\sup_{x\in\Omega^+} w(x) \leq \sup_{x\in\partial\Omega^+} w(x) = 0,$$

that is, $\Omega^+ \equiv \emptyset$ whence $w(x) \leq 0$ for $x \in \Omega$.

Proceeding in the same manner, one obtains $w(x) \geq 0$ for $x \in \Omega$, and this completes the proof of Theorem 2. □

References

1. S. Fučik and A. Kufner, Nonlinear Differential Equations, Elsevier Sci. Publ. Co., New York (1980).

2. D. Gilbarg and N. S. Trudinger, Elliptic Partial Differential Equations of Second Order, 2nd edition, Springer–Verlag, Berlin (1983).

9th Int. Coll. on Differential Equations, pp. 411–415
D. Bainov (Ed)

On Convolutional Decompositions of the Generalized Inverse Gaussian Distributions and the Student t-Distributions

KATSUO TAKANO
Department of Mathematics, Faculty of Science, Ibaraki University, Mito-city, Ibaraki 310, Japan

1991 Mathematics Subject Classification. 44A10, 45E10, 60E7.

1. Introduction

The probability distribution function $F(x)$ is called a generalized gamma convolution if $F(x)$ satisfies the following (a), (b), (c).

- (a) $F(x)$ is concentrated on $[0, \infty)$.
- (b) The Laplace transform of $F(x)$ can be written as

$$\zeta(s) = \exp[-as - \int_0^\infty \log(1 + \frac{s}{w})dU(w)], a \geq 0, \tag{1}$$

where

- (c) the function $U(w)$ is nondecreasing on $(0, \infty)$ and
- 1) $U(0) = 0$, 2) $\int_0^1 |\log w| \, dU(w) < \infty$, 3) $\int_1^\infty 1/w \, dU(w) < \infty$.

The generalized gamma convolution is infinitely divisible and a self-decomposable distribution (cf. [1], [2]). F. W. Steutel [3] found firstly that an infinitely divisible distribution, which is concentrated on $[0, \infty)$, satisfies an integral equation and this fact provides us fruitful analytical ways developed in differential and integral equations for study of the infinite divisibility of probability distribution. This author became aware from a convolutional interal equation satisfied by an infinitely divisible probability density that convolutional decompositions can be obtained for the generalized inverse Gaussian distributions and the Student t-distributions. These facts will be shown in this paper. The author thinks that Theorem 2 may be useful in analysis like the theory of singular integrals in weighted L^p spaces.

2. Convolutional Decompositions of the Generalized Inverse Gaussian Distributions

The Laplace transform of the generalized gamma convolution can be written as

$$\zeta(s) = \exp[-as + \int_{(0,\infty)} (e^{-sx} - 1)\frac{1}{x}k(x)dx], \tag{2}$$

where

$$k(x) = \int_{(0,\infty)} e^{-xw} dU(w) \geq 0, \ x > 0.$$

The condition (c) 1),2),3) is equivalent to the following condition,

$$k(x) = 0, \ (x \leq 0), \ \int_{(0,1]} k(x)dx < \infty, \ \int_1^\infty \frac{k(x)}{x} dx < \infty. \tag{3}$$

For simplicity we suppose $a = 0$. Then

$$\phi(t) = \exp[\int_{(0,\infty)} (e^{itx} - 1)\frac{1}{x}(\int_0^\infty e^{-xw} dU(w))dx]. \tag{4}$$

Suppose $k(+0) = U(+\infty) > 0$ in what follows. By using the result of [4, Theorem 2.1] we can obtain a following

Remark 1 *The distribution function $F(x)$ of the generalized gamma convolution with the characteristic function (4) has a continuous density $f(x)$ on $(0,\infty)$ and $f(x)$ satisfies an integral equation,*

$$xf(x) = \int_{(0,x)} f(x-u)k(u)du, \ x > 0, \tag{5}$$

where $k(x) = \int_{(0,\infty)} e^{-xw} dU(w) \geq 0, \ x > 0$.

Let $\alpha > 0$ and consider a generalized inverse Gaussian distribution with density

$$\begin{aligned} g(x) &= \frac{1}{\Gamma(\alpha)} x^{-\alpha-1} e^{-1/(4x)}, \ x > 0, \\ &= 0, \ x \leq 0. \end{aligned} \tag{6}$$

The Laplace transform of $g(x)$ is as follows:

$$\zeta(s) = \frac{1}{4^\alpha \Gamma(\alpha)} \int_0^\infty x^{-\alpha-1} e^{-(1/(4x)+sx)} dx = \frac{2}{2^\alpha \Gamma(\alpha)} (\sqrt{s})^\alpha K_\alpha(\sqrt{s}), \tag{7}$$

where $K_\alpha(z)$ denotes the Bessel function of third kind. It is known in [5], [6] that the generalized inverse Gaussian distribution is a generalized Gamma convolution. As seen from (5), then by the inverse Laplace transform of

$$-\frac{\zeta'(s)}{\zeta(s)} = \frac{K_{\alpha-1}(\sqrt{s})}{2\sqrt{s}K_\alpha(\sqrt{s})} = \int_0^\infty e^{-sx} k(x)dx \tag{8}$$

we can get the function of $k(t)$, that is,

$$k(x) = \int_0^\infty e^{-xt} \frac{1}{\pi^2 t (J_\alpha^2(\sqrt{t}) + Y_\alpha^2(\sqrt{t}))} dt. \tag{9}$$

Here $J_\alpha(t)$ and $Y_\alpha(t)$ are the Bessel functions of the first kind and of second kind, respectively, with order α. The function $k(x)$ is rewritten as

$$k(x) = \int_0^\infty e^{-xv^2} \frac{2dv}{\pi^2 v (J_\alpha^2(v) + Y_\alpha^2(v))}, \ x > 0. \tag{10}$$

From the above result we can obtain the following

Theorem 1 *If $\alpha > \beta > 0$ there exists a probability density function $f(\alpha, \beta; x)$ of a generalized Gamma convolution with characteristic function*

$$\phi(t) = \exp[\int_{(0,\infty)} (e^{itx} - 1) \cdot \frac{1}{x}(\int_0^\infty e^{-xv^2} \frac{2}{\pi^2 v}(\frac{1}{J_\beta^2(v) + Y_\beta^2(v)} - \frac{1}{J_\alpha^2(v) + Y_\alpha^2(v)})dv)dx], \quad (11)$$

and the convolutional equation

$$\frac{1}{\Gamma(\beta)} x^{-\beta-1} e^{-1/(4x)} = \int_0^x \frac{1}{\Gamma(\alpha)}(x-y)^{-\alpha-1} e^{-1/4(x-y)} f(\alpha, \beta; y) dy \quad (12)$$

holds for $x > 0$.

Proof. Consider the characteristic functions of the both sides of (12). From the formula in G. N. Watson's Book [7, Chap. XIII p.444] we have

$$\frac{1}{4}\pi^2(J_\nu^2(x) + Y_\nu^2(x)) = \int_0^\infty K_0(2x \sinh T)(e^{-2\nu T} + e^{2\nu T})dT,$$

and this function is increasing in ν. Therefore we see that

$$\frac{1}{J_\beta^2(v) + Y_\beta^2(v)} - \frac{1}{J_\alpha^2(v) + Y_\alpha^2(v)}$$

is positive on $(0, \infty)$ and by (4) and (10) the following measure

$$\frac{1}{|x|}(\int_0^\infty \frac{2}{\pi^2 v}(\frac{1}{J_\beta^2(v) + Y_\beta^2(v)} - \frac{1}{J_\alpha^2(v) + Y_\alpha^2(v)})e^{-xv^2} dv)dx$$

is a Levy measure of the probability distribution with the density function $f(\alpha, \beta; x)$ and $\phi(t)$ is a characteristic function of a certain generalized Gamma convolution.

3. Convolutional Decompositions of the Student t-Distribution

Consider a probability distribution with density

$$f(x) = \frac{\Gamma(\alpha + 1/2)}{\sqrt{\pi}\,\Gamma(\alpha)} \frac{1}{(1 + x^2)^{\alpha+1/2}}, \quad (\alpha > 0, \ -\infty < x < \infty).$$

If $2\alpha = n$ is a positive integer $f(x)$ is the density of the Student t-distribution with degree n of freedom. This density is written such as

$$f(x) = \int_0^\infty \frac{1}{\sqrt{2\pi t}} e^{-x^2/(2t)} \frac{1}{2^\alpha \Gamma(\alpha)} t^{-\alpha-1} e^{-1/(2t)} dt. \quad (13)$$

From the section 2, the density

$$g(x) = \frac{1}{2^\alpha \Gamma(\alpha)} x^{-\alpha-1} e^{-1/(2x)}$$

is a generalized gamma convolution with the density

$$U'(w) = \frac{1}{\pi^2 w (J_\alpha^2(\sqrt{2w}) + Y_\alpha^2(\sqrt{2w}))}, \quad w > 0. \quad (14)$$

Hence we obtain a canonical representation of the probability distribution with the density function $f(x)$,

$$\begin{aligned}\phi(v) &= \int_0^\infty e^{-tv^2/2}\frac{1}{2^\alpha\Gamma(\alpha)}t^{-\alpha-1}e^{-1/(2t)}dt = \exp[-\int_{(0,\infty)}\log(1+\frac{v^2}{2w})dU(w) \quad (15)\\ &= \exp[\int_{(0,\infty)}(\int_{\mathbf{R}-\{0\}}(e^{itu}-1)\frac{1}{|u|}e^{-\sqrt{2w}|u|}du)dU(w)]\\ &= \exp[\int_{\mathbf{R}-\{0\}}(e^{itv}-1)\frac{1}{|v|}(\int_{(0,\infty)}e^{-\sqrt{2w}|v|}dU(w))dv], \quad (16)\end{aligned}$$

namely

$$\phi(t) = \exp[\int_{R-\{0\}}(e^{itx}-1-\frac{itx}{1+x^2})\frac{k(x)}{x}dx],$$

where

$$\begin{aligned}k(x) &= (sgn\ x)\int_0^\infty e^{-|x|\sqrt{2t}}\frac{1}{\pi^2 t(J_\alpha^2(\sqrt{2t})+Y_\alpha^2(\sqrt{2t}))}dt\\ &= (sgn\ x)\int_0^\infty e^{-|x|v}\frac{2}{\pi^2 v(J_\alpha^2(v)+Y_\alpha^2(v))}dv,\ x\neq 0. \quad (17)\end{aligned}$$

From the fact that $1/\{\pi^2 v(J_\alpha^2(v)+Y_\alpha^2(v))\}\to 1/\pi$ as $v\to\infty$, it follows that $k(+0) = -k(-0) = U(+\infty) = \infty$.

Theorem 2 *If $\alpha > \beta > 0$ there exists a probability density function $h(\alpha,\beta;x)$ with characteristic function*

$$\begin{aligned}&\phi(t) = \exp[\int_{R-\{0\}}(e^{itx}-1-\frac{itx}{1+|x|^2})\\ &\cdot(\frac{1}{|x|}\int_0^\infty\frac{2}{\pi^2 v}(\frac{1}{J_\beta^2(v)+Y_\beta^2(v)}-\frac{1}{\pi v(J_\alpha^2(v)+Y_\alpha^2(v))})e^{-v|x|}dv)dx], \quad (18)\end{aligned}$$

and the convolutional equation

$$\frac{\Gamma(\beta+1/2)}{\pi^{1/2}\Gamma(\beta)}\frac{1}{(1+x^2)^{\beta+1/2}} = \int_{-\infty}^\infty\frac{\Gamma(\alpha+1/2)}{\pi^{1/2}\Gamma(\alpha)}\frac{1}{(1+(x-y)^2)^{\alpha+1/2}}h(\alpha,\beta;y)dy \quad (19)$$

holds.

Proof. As the same in the proof of Theorem 2, consider the characteristic functions of the both sides of (19). Then we see that

$$\frac{1}{|x|}(\int_0^\infty\frac{2}{\pi^2 v}(\frac{1}{J_\beta^2(v)+Y_\beta^2(v)}-\frac{1}{J_\alpha^2(v)+Y_\alpha^2(v)})e^{-v|x|}dv)dx$$

is a positive measure on the real line minus zero and so, $\phi(t)$ is a characteristic function of a certain density function $h(\alpha,\beta;x)$ and we obtain the convolutional equation (19). q.e.d.

In a subsequent paper an extension of the above Theorem 2 to the d-dimensional probability density,

$$\frac{\Gamma(\alpha+d/2)}{\pi^{d/2}\Gamma(\alpha)}\frac{1}{(1+|x|^2)^{\alpha+d/2}},\ x\in\mathbf{R}^d,\ \alpha>0,$$

will be shown.

References

1. L. Bondesson, Generalized Gamma Convolutions and Related Classes of Distributions and Densities, Lecture Note in Statistics, 76, Springer-Verlag (1992).
2. O. Thorin, Scand. Acturial. J., 31-40 (1977).
3. F. W. Steutel, Math. Centre Tracts, Math. Centre, Amsterdam, **33**, (1970).
4. K. Sato and M. Yamazato, Z. Wahrscheinlichkeitstheorie verw. Gebiete, **43**, 273–308 (1978).
5. E. Grosswald, Z. Wahrscheinlichkeitstheorie verw. Gebiete, **36**, 103–109 (1976).
6. C. Halgreen, Z. Wahrscheinlich-keitstheorie verw. Gebiete, **47**, 13–17 (1979).
7. G. N. Watson, A Treatise on the Theory of Bessel Functions, Second Edition, Cambridge University Press (1966).
8. M. Abramowitz and I. A. Stegun, Handbook of Mathematical Functions, New York, Dover (1970).
9. L. Bondesson, Probab. Th. Rel. Fields, **85**, 181–194 (1990).
10. A. Diédhiou, J. Math. Anal. Appl., **220**, 42–64 (1998).
11. K. Takano, Bull. Fac. Sci. Ibaraki Univ., Math., **21**, 29–41 (1989), (Correction and Addendum, ibid. **22**).

9th Int. Coll. on Differential Equations, pp. 417–422
D. Bainov (Ed)

The Complete Ducks Not Satisfying S^1

KIYOYUKI TCHIZAWA
Dept. of Mathematics, Musashi Institute of Technology, 1-28-1 Tamazutsumi Setagaya-Ku Tokyo, 158, Japan

Abstract: In this paper, we will introduce a complete (and an incomplete) duck solution; a singular limit of the duck solution for a parameter. Then, we will give the explicit form of the solution which does not have the property of S^1 under certain conditions.
Key Words and Phrases: *constrained system, singular perturbation, duck solution, winding number.*
1991 Mathematics Subject Classification. 34A34,34A47,34C35.

1. Introduction

In 1990, Benoit [1] showed how to construct the explicit duck solutions (or simply ducks). Furthermore, he showed, in his paper, that if the difference of each the winding numbers associated with the ducks is more than 3/2, there exists a duck which is not S^1 (S^1 is a smoothness class in nonstandard analysis). However, this solution has not been constructed explicitly yet. In this paper, we would construct the above solution in this framework. In Section 2 and Section 3, introducing a parameter in the differential equations, an incomplete duck and a complete duck without S^1 are constructed explicitly as a singular limit of the duck for the parameter. The explicit form of the complete duck is the exact solution in the first approximation of the "local model" ([1], [2]).

2. Preliminaries

Let consider a constrained system (2.1):

$$
\begin{aligned}
dx/dt &= f(x, y, z, u), \\
dy/dt &= g(x, y, z, u), \\
h(x, y, z, u) &= 0,
\end{aligned}
\tag{2.1}
$$

where u is a parameter (any fixed) and f,g,h are defined in $R^3 \times R^1$. Furthermore, let consider the singular perturbation problem of the system (2.1):

$$
\begin{aligned}
dx/dt &= f(x, y, z, u), \\
dy/dt &= g(x, y, z, u), \\
\epsilon dz/dt &= h(x, y, z, u),
\end{aligned}
\tag{2.2}
$$

where ϵ is infinitesimally small.

We assume that the system (2.1) satisfies the following conditions $(A1) - (A5)$:

$(A1)$ f and g are of class C^1 and h is of class C^2.

$(A2)$ The set $S = \{(x, y, z) \in R^3 : h(x, y, z, u) = 0\}$ is a 2-dimensional differentiable manifold and the set S intersects the set

$T = \{(x,y,z) \in R^3 : \partial h(x,y,z,u)/\partial z = 0\}$ transversely so that the set $PL = \{(x,y,z) \in S \cap T\}$ is a 1-dimensional differentiable manifold.

($A3$) Either the value of f or that of g is nonzero at any point $p \in PL$.

Let $(x(t,u), y(t,u), z(t,u))$ be a solution of (2.1). By differentiating $h(x,y,z,u)$ with respect to the time t, the following equation holds:

$$h_x(x,y,z,u)f(x,y,z,u) + h_y(x,y,z,u)g(x,y,z,u) + h_z(x,y,z,u)dz/dt = 0, \tag{2.3}$$

where $h_i(x,y,z,u) = \partial h(x,y,z,u)/\partial i$, $i = x,y,z$. The above system (2.1) becomes the following system:

$$\begin{aligned} dx/dt &= f(x,y,z,u), \\ dy/dt &= g(x,y,z,u), \\ dz/dt &= -\{h_x(x,y,z,u)f(x,y,z,u) + \\ &\quad h_y(x,y,z,u)g(x,y,z,u)\}/h_z(x,y,z,u), \end{aligned} \tag{2.4}$$

where $(x,y,z) \in S \setminus PL$. The system (2.1) coincides with the system (2.4) at any point $p \in S \setminus PL$. In order to study the system (2.4), let consider the following system:

$$\begin{aligned} dx/dt &= -h_z(x,y,z,u)f(x,y,z,u), \\ dy/dt &= -h_z(x,y,z,u)g(x,y,z,u), \\ dz/dt &= h_x(x,y,z,u)f(x,y,z,u) + h_y(x,y,z,u)g(x,y,z,u). \end{aligned} \tag{2.5}$$

As the system (2.5) is well defined at any point of R^3, it is well defined indeed at any point of PL. The solutions of (2.4) coincide with those of (2.1) on $S \setminus PL$ except the velocity when they start from the same initial points.

($A4$) For any $(x,y,z) \in S$, the implicit function theorem holds;

$$h_y(x,y,z,u) \neq 0, h_x(x,y,z,u) \neq 0, \tag{2.6}$$

that is, the surface S can be expressed by using $y = \varphi(x,z,u)$ or $x = \psi(y,z,u)$ in the neighborhood of PL. Let $y = \varphi(x,z,u)$ exist, then the projected system, which restricts the system (2.5) is obtained:

$$\begin{aligned} dx/dt &= -h_z(x,\varphi(x,z,u),z,u)f(x,\varphi(x,z,u),z,u), \\ dz/dt &= h_x(x,\varphi(x,z,u),z,u)f(x,\varphi(x,z,u),z,u) + \\ &\quad h_y(x,\varphi(x,z,u),z,u)g(x,\varphi(x,\varphi(x,z,u),z,u). \end{aligned} \tag{2.7}$$

($A5$) All the singular points of (2.7) are nondegenerate, the matrix induced from the linearized system of (2.6) at a singular point has two nonzero eigenvalues. Note that all the points contained in $PS = \{(x,y,z) \in PL : dz/dt = 0\}$, which are called *pseudo singular points* are the singular points of (2.7).

Definition 2.1. Let $p \in PS$ and $\mu_1(u)$, $\mu_2(u)$ be two eigenvalues of the linearized system of (2.7), then the point p is called *pseudo singular saddle* if $\mu_1(u) < 0 < \mu_2(u)$ and called *pseudo singular node* if $\mu_1(u) < \mu_2(u) < 0$ or $\mu_1(u) > \mu_2(u) > 0$.

Definition 2.2. A solution $(x(t,u), y(t,u), z(t,u))$ of the system (2.2) is called *a duck*, if there exist standard $t_1 < t_0 < t_2$ such that

(1) ${}^*(x(t_0,u), y(t_0,u), z(t_0,u)) \in S$, where ${}^*(X)$ denotes the standard part of X,

(2) for $t \in (t_1,t_0)$ the segment of the trajectory $(x(t,u), y(t,u), z(t,u))$ is infinitesimally close to the attracting part of the slow curves,

(3) for $t \in (t_0,t_2)$, it is infinitesimally close to the repelling part of the slow curves, and

(4) the attracting and repelling parts of the trajectory are not infinitesimally small.

Definition 2.3. Let E be a set in R^3. We call a point p is a δ- micro-galaxy of E when the distance from p to E is less than $exp(-n/\delta)$, where n is some positive integer and $\delta = \epsilon/\alpha^2$ (α is infinitesimally small).

Definition 2.4. Let θ is an angle of the polar coordinate after changing the coordinates in the "local model" such as the orbit passing through the pseudo singular point becomes the z-axis itself as the below. See [1], [3]. Then, the winding number $N(\psi)$ of a duck ψ is defined as follows:

$$N(\psi) = (1/2\psi) \int_{\psi} d\theta, \tag{2.8}$$

where ψ is contained partially in the δ-micro- galaxy of γ_μ.

Theorem 2.1. (Benoit) *In the system* (2.1), *if the following two conditions at a pseudo singular saddle or node point;*

(1) $f(O,u) \simeq h(O,u) \simeq h_y(O,u) \simeq h_z(O,u) \simeq 0$,

(2) $g(O,u) \not\simeq 0, h_x(O,u) \not\simeq 0, h_{zz}(O,u) \not\simeq 0$, *where* $O = (0,0,0) \in PS$,

are satisfied, the explicit duck solutions $\gamma_{\mu_i(u)}$ *in the first approximation of the local model can be constructed:*

$$\gamma_{\mu_i(u)}(t) = (-\mu_i(u)^2 t^2 - \delta\mu_i(u), t, \mu_i(u)t) \quad (i = 1, 2), \tag{2.9}$$

wehere δ *is an infinitesimally small constant.*

The above Definition 2.3 is based on the following fact. If ϵ is fixed arbitrarily and $\gamma(t)$ is a duck near $\gamma_{\mu(u)}(t)$ is within $exp(-n/\delta)$ in some neighborhood of the pseudo singular point. See [4].

In the system (2.2), under the conditions (1) and (2) in the Theorem 2.1, making the following coordinate transformations (2.10) and (2.11) successively;

$$\begin{pmatrix} x \\ y \\ z \end{pmatrix} = \begin{pmatrix} \alpha^2\tilde{x} \\ \alpha\tilde{y} \\ \alpha\tilde{z} \end{pmatrix}, (\alpha \simeq 0, \epsilon/\alpha^2 \simeq 0) \tag{2.10}$$

$$\begin{pmatrix} X \\ Y \\ Z \end{pmatrix} = \begin{pmatrix} h_x(0,u)h_{zz}(0,u)\tilde{x}/2 + (h_{yy}(0,u)h_{zz}(0,u) - h_{yz}(0,u)^2)\tilde{y}^2/4 \\ \tilde{y}/g(0,u) \\ -h_{yz}(0,u)\tilde{y}/2 - h_{zz}(0,u)\tilde{z}/2 \end{pmatrix}, \tag{2.11}$$

the following local model (2.12) is obtained:

$$\begin{aligned} dX/dt &= pY + qZ + \xi(X,Y,Z,u), \\ dY/dt &= 1 + \eta(X,Y,Z,u), \\ \delta dZ/dt &= -(Z^2 + X) + \zeta(X,Y,Z,u), \end{aligned} \tag{2.12}$$

where

$$\begin{aligned} p &= g(0,u)h_x(0,u)(f_y(0,u)h_{zz}(0,u) - f_z(0,u)h_{yz}(0,u))/2 \\ &\quad + g(0,u)^2(h_{yy}(0,u)h_{zz}(0,u) - h_{yz}(0,u)^2)/2, \\ q &= -h_x(0,u)f_z(0,u), \\ \delta &= \epsilon/\alpha^2. \end{aligned} \tag{2.13}$$

Here $\xi(X,Y,Z,u)$,$\eta(X,Y,Z,u)$ and $\zeta(X,Y,Z,u)$ are infinitesimal when X,Y and Z are limited. Note that the solutions (2.9) are in the first approximation system on (2.12).

By applying the following transformations of the coordinates as mentioned above, in Definition 2.4, successively;

(2.14)
$$u = X + Z^2 + \delta\mu,$$
$$v = Y - Z/\mu,$$
$$z = Z,$$

(2.15)
$$u = r cos\theta,$$
$$v = r sin\theta,$$

the Hermite equation (2.16) is obtained. This equation associated with $\gamma_{\mu_i(u)}$ $(i = 1,2)$ is the following:

(2.16)
$$\delta\ddot{z} - \tau\dot{z} + K_i z = 0, \dot{z} = dz/d\tau, t = \tau/\alpha, (i = 1,2),$$

where K_i is a positive integer and $K_1 = 1 + \mu_2(u)/\mu_1(u)$, $K_2 = 1 + \mu_1(u)/\mu_2(u)$. See [1].

It is said that a duck $\psi(t)$ has *a jump* if the shadow of it contains a vertical segment and that $\psi(t)$ is *long* if it is in an infinitesimally small neighborhood at the pseudo singular point. It can be proved that if ψ is not long, the standard part of the winding number $N(\psi_i)$ associated with μ_i is an integer. If the pseudo singular point is node, it is positive. If the point is saddle, it needs some conditions such as K_i is positive. The relation between $N(\psi_i)$ and K_i $(i = 1,2)$ is as follows.

Theorem 2.2. (Benoit) *If the duck ψ_1, which is not long has 2 jumps, $N(\psi_1) \approx -[K_1/2]$, and if the duck ψ_2 has 2 jumps, $N(\psi_2) \approx 0$.*

3. Complete and Incomplete Ducks

In this section, the main theorems of this paper; the sufficient condition for the existence of the complete duck solution are described explicitly.

Definition 3.1. In the system (2.12), if for any parameter u,
it satisfies the conditions $(A1)$-$(A5)$ and has a duck,
then the solution is called *a complete duck*.

Definition 3.2. In the system (2.12), if the following (1) and (2):

(1) for any parameter u except some limited u_0,
it satisfies the conditions $(A1)$-$(A5)$ and has a duck,

(2) for the parameter $(u = u_0)$,
it satisfies the conditions $(A1)$-$(A5)$ but the solution is not a duck,

are established, then it is called *an incomplete duck*.

Definition 3.3. A solution $\psi(x,u)$ is called S^1 *at a*,
if there exists a real number b such that

(3.1)
$$\frac{\psi(x,u) - \psi(y,u)}{x - y} \approx b,$$

for any x, y $(x \approx a, y \approx a)$.
A duck is called an S^1*duck* if it is S^1 in some neighborhood of the pseudo singular point.

Theorem 3.1. (Benoit) *In the first approximation of the local model* (2.12), *if $\mu_1(u)/\mu_2(u)$ is positive (> 3) but no an integer, then all the S^1 ducks are exponentially close to one of the two explicit ducks and there exists non S^1 ducks.*

In the local model, we assume that

$$f_y(0,u) = g_u(0,u) = h_{yz}(0,u) = h_{yyu}(0,u) = h_{zzu}(0,u) = 0, \tag{3.2}$$

and that the following (1) or (2):
(1) $h_x(0,u) = O(u)$ and $f_z(0,u) = O(1)$,
(2) $f_z(0,u) = O(u)$ and $h_x(0,u) = O(1)$,
where all the coefficients of higher order (more than 2) for u is negligible. In other words, we assume that only the coefficient q can take a unlimited number ($q = \tilde{q}u + o(1)$, a constant $\tilde{q} \not\simeq 0$). Then, blowing up only the variable Z again;

$$Z = (1/u)\zeta, \tag{3.3}$$

the first approximation of the local model becomes the following:

$$\begin{aligned} dX/dt &= pY + \tilde{q}\zeta, \\ dY/dt &= 1, \\ (\delta/u)d\zeta/dt &= -(\zeta^2/u^2 + X), \end{aligned} \tag{3.4}$$

where $\tilde{q}$ is limited (does not contain u) and $(\delta/u) \simeq 0$. The explicit solutions in the system (3.4) are

$$\gamma_{\mu_i(u)}(t) = (-\mu_i(u)^2t^2 - u\delta\mu_i(u), u^2t, u^2\mu_i(u)t) \quad (i = 1, 2), \tag{3.5}$$

where $\mu_1(u)$, $\mu_2(u)$ ($\mu_1(u) > \mu_2(u)$) are the solutions of the characteristic equation of the system (3.4) in case $\delta/u \simeq 0$.
The above system satifies the conditions ($A1$)-($A5$) and the solutions (3.5) satisfy the condition (1) and one of the solutions satisfies the condition (2) in Definition 3.2 when $u = u_0 = \omega = 1/\epsilon$. Since $\mu_2(u) \simeq -1/2\epsilon$ as $u = 1/\epsilon$ and for the first component of (3.5), the following

$$\frac{-(1/2\epsilon)^2(2\epsilon)^2 + (1/2\epsilon)^2(\epsilon)^2}{2\epsilon - \epsilon} = -3/4\epsilon, \tag{3.6}$$

is established. In this state, the winding number $N(\psi_2)$ associated with μ_2 is unlimited and the other $N(\psi_1)$ associated with μ_1 is infinitesimal. Then, this duck may be almost tangent to the X-axis when u tends to $1/\epsilon$.

Now, let $v = 1/u$. Then $\partial_u = -v^2\partial_v$ holds and then the following conditions are assumed; $f(x,y,z,u) = \tilde{f}(x,y,z,v) \in C^3$, $g(x,y,z,u) = \tilde{g}(x,y,z,v) \in C^1$ and $h(x,y,z,u) = \tilde{h}(x,y,z,v) \in C^3$ at almost everywhere but $v = v_0 = 0$.

Theorem 3.2. *In the first approximation of the local model, under the condition* (3.2) *and $\tilde{h}_{x,v}(0,v)\tilde{f}_{zv}(0,v) = 0$, if either the assumption* (1) *or* (2) ;

(1) $\tilde{f}_z(0,v) = 0$, *and* $\tilde{h}_x(0,v)\tilde{f}_{zvv}(0,v) = 0$,

(2) $\tilde{h}_x(0,v) = 0$, *and* $\tilde{h}_{xvv}(0,v)\tilde{f}_z(0,v) = 0$,

where all the coefficients of higher order (more than 2) for v is negligible, is satisfied, then this system has an incomplete duck.

Proof. From the above assumptions, the relation $q = -h_x(0,u)f_z(0,u) = \tilde{q}u$ holds. Differentiating the both side of the equation by the parameter v, we can directly lead to the conclusion. □

By blowing up the variable X again, like the same way above, the complete duck which is not S^1 is obtained.

Theorem 3.3. *In the equation* (3.4), *we assume that*

$$\tilde{f}_y(0,v) = \tilde{h}_{yz}(0,v) = 0, \quad \tilde{g}(0,v) = O(1) \tag{3.7}$$

and moreover, if either the condition $\tilde{h}_{yy}(0,v) = 0$, *or* $\tilde{h}_{zz}(0,v) = 0$, *is satisfied, that is;* $p = c_1 v^2 + o(1)$ *($c_1 \not\simeq 0$ is limited) and* $\tilde{q} = c_2 v^2 + o(1)$ *($c_2 \not\simeq 0$ is limited), then there exists a complete duck which is not* S^1 *in this system* (3.4).

Proof. Blowing up the variable X such as

$$X = v^2\xi, \tag{3.8}$$

the equation (3.4) becomes

$$\begin{aligned} d\xi/dt &= (1/v^2)pY + (1/v^2)\tilde{q}\zeta, \\ dY/dt &= 1, \\ (\delta/v)d\zeta/dt &= -(\zeta^2+\xi), \end{aligned} \tag{3.9}$$

where $\delta/v \simeq 0$.
Therefore, the explicit ducks $\tilde{\gamma}_{\mu_i(u)}$ of (3.9) are described as follows:

$$\tilde{\gamma}_{\mu_i(v)}(t) = (-\mu_i(v)^2t^2 - \delta\mu_i(v)/v, t, \mu_i(v)t) \quad (i = 1,2), \tag{3.10}$$

where $\mu_i(v)$ are the eigenvalues of the linearized system in the constrained system of (3.9). In this situation, the constrained surface $S = \{(\xi, Y, \zeta) : \xi + \zeta^2 = 0\}$ is invariant for the parameter v. Furthermore, one of the ducks tangents to the ξ-axis at the pseudo singular point when v tends to $\epsilon^{1/2}$. So, this duck is not S^1. □

Acknowledgement

I am grateful to Prof. E.Benoit for stimulating discussions and useful suggestions at CIRM in Marseille and in Nice. I would like to thank Prof. Marc and Francine Diener for discussing these related problems in nonstandard analysis.

References

1. E. Benoit, Publ. Math. IHES, **72**, 63–91 (1990).
2. N. Kakiuchi and K. Tchizawa, preprint.
3. N. Kakiuchi and K. Tchizawa, J. Differential Equations, **141**, 327–339 (1997).
4. A.K.Zvonkin and M.A.Shubin, Russian Math. Surveys, **39**, 69–131 (1984).
5. V.I. Arnold, Dynamical Systems V, Springer-Verlag (1994).
6. E. Benoit, Asterisque, **109–110**, 55-71 (1983).
7. J. Callot, Bifurcation du Portrait de Phase pur des Equations Differentielles Lineaires du Second Order Ayant pouy Type L'equation d'Hermite, These d'Etat, Strasbourg (1981).
8. A.I. Neishtadt, Differential Equations, **23**, 2060–2067 (1987); II, ibid **24**, 226–233 (1988).
9. E. Nelson, Bull. Amer. Math. Soc., **83**, 1165–1198 (1977).
10. K. Tchizawa, preprint.

9th Int. Coll. on Differential Equations, pp. 423–426
D. Bainov (Ed)

Oscillation Criteria for Linear Differential Equations

BETTY TRAVIS
Department of Mathematics and Statistics, The University of Texas at San Antonio, San Antonio, Texas, 78249, USA
e-mail: btravis@utsa.edu

Abstract: Oscillation and nonoscillation of certain classes of linear differential equations are discussed. Sufficient conditions based on the coefficients are examined.

1. Introduction

Consider the differential equation

$$(r(t)y')' + p(t)y = 0 \tag{1.1}$$

where $p(t)$ and $r(t)$ are continuous functions, $r(t) > 0$ and $p(t) \geq 0$ on some interval $[a, \infty)$. A classical result, first discovered by Wintner [1] for the case $r(t) = 1$ and later generalized by Leighton [2], states that if

$$\int_a^\infty \frac{1}{r(t)}dt = \infty = \int_a^\infty p(t)dt,$$

then all solutions of (1.1) are oscillatory, i.e. the set of zeros of every solution is unbounded above. In [3] the author extended this result to nth order equations of this form. Here, for the purpose of clarification, we further discuss the cases $n = 2$ and 3. The result in [3] is a natural extension of the Wintner-Leighton Theorem, which has had many extensions (see [4], [5], and [6]). However, the author has been unable to find an extension of this result for arbitrary n. Thus, we felt that a simple proof of such an elegant theorem should be made more accessible even if it were known.

Consider the nth order differential equation

$$(r_n(t)(\dots(r_2(t)(r_1(t)y)')'\dots)')' + p(t)y = 0. \tag{1.2}$$

By letting $y_1 = r_1(t)y$, then equation (1.2) can be written as the system

$$\begin{aligned} y_1' &= \frac{1}{r_2}y_2 \\ y_2' &= \frac{1}{r_3}y_3 \\ &\vdots \\ y_n' &= \frac{-p}{r_1}y \end{aligned}$$

The following theorem was established in [7] and [3].

Theorem 1 *Assume that $r_k(t)$ and $p(t)$ are continuous functions such that $r_k(t) > 0$, $p(t) \geq 0$ and*

$$(A) \qquad \int_0^\infty \frac{1}{r_k(t)} dt = \infty = \int_0^\infty \frac{p(t)}{r_1(t)} dt, \quad 2 \leq k \leq n.$$

If n is even, then all solutions of (1.2) oscillate. If n is odd, then (1.2) has a nonoscillatory solution $y(t)$ such that $(-1)^{k-1} y_k(t) > 0$ for all $t \geq 0$ and $1 \leq k \leq n$. Furthermore, when n is odd, every solution that vanishes once is oscillatory.

Remark. When n is odd, it was shown in [3] that the conclusion of the above theorem follows under the weaker requirement that $r_k(t) > 0$, $1 \leq k \leq n$, and $p(t) \geq 0$, $p(t) \neq 0$ on any interval of the form $[a, \infty)$.

2. Special Cases Involving $n = 2, 3$

For $n = 2$, our extension reduces to equations of the form

$$(2.1) \qquad (r_2(t)(r_1(t)y)')' + p(t)y = 0,$$

where $r_1(t), r_2(t), p(t) \geq 0$ and they are all continuous on $[a, \infty)$. Theorem 1 states that if

$$\int_a^\infty \frac{1}{r_2(t)} dt = \infty = \int_a^\infty \frac{p(t)}{r_1(t)} dt,$$

then all solutions of (2.1) are oscillatory. This is precisely the Wintner-Leighton Theorem if we let $r_1(t) \equiv 1$. Thus, even for $n = 2$ the Wintner-Leighton Theorem is a special case of our result.

Example 2.1. Consider the differential equation

$$((t+1)(e^{-t}y)')' + |\sin t| y = 0$$

on $[0, \infty)$. Clearly,

$$\int_0^\infty \frac{1}{t+1} dt = \infty = \int_0^\infty \frac{|sint|}{e^{-t}} dt.$$

Hence, by our theorem, all solutions oscillate. However, the Wintner-Leighton Theorem is not applicable. We note that by letting $y_1 = r_1 y$ and $r_2 y_1' = y_2$, we can write the above differential equation as a system

$$\begin{pmatrix} y_1' \\ y_2' \end{pmatrix} = \begin{pmatrix} 0 & \frac{1}{t+1} \\ \frac{|\sin t|}{e^{-t}} & 0 \end{pmatrix} \begin{pmatrix} y_1 \\ y_2 \end{pmatrix}.$$

Thus, the existence and uniqueness and other basic properties hold.

For $n = 3$, we consider equations of the form

$$(2.2) \qquad (r_3(t)(r_2(t)(r_1(t)y)')')' + p(t)y = 0,$$

where $r_k(t) > 0$, $1 \leq k \leq 3$, $p(t) \geq 0$, and all are continuous. Applying Theorem 1, it follows that if

$$\int_0^\infty \frac{1}{r_k(t)} dt = \infty = \int_0^\infty \frac{p(t)}{r_1(t)} dt, \quad k = 1, 2, 3,$$

then (2.2) has a nonoscillatory solution $y(t)$ such that $y > 0$, $(r_1y)' < 0$ and $(r_2(r_1y)')' > 0$ for all t, $t \geq 0$. It is further shown that every nontrivial solution that vanishes at some point is oscillatory. Thus, the two linearly independent solutions $y_1(t)$ and $y_2(t)$ satisfying the initial values

$$\begin{aligned} y_1(0) = 0, \quad y_1'(1) = 0, \quad y_1''(0) = 1 \\ y_2(0) = 0, \quad y_2'(1) = 1, \quad y_2''(0) = 0 \end{aligned}$$

are both oscillatory.

First we consider the following simple example:

$$y''' + y = 0. \tag{2.3}$$

Our theorem implies that (2.3) has a solution $y(t)$ such that $y(t) > 0$, $y'(t) < 0$ and $y''(t) > 0$. Solving equation (2.3), we see that $y(t) = e^{-t}$ is such a solution. Further, $y_1(t) = e^{\frac{t}{2}}\sin(\frac{\sqrt{3}}{2}t)$ and $y_2(t) = e^{\frac{t}{2}}\cos(\frac{\sqrt{3}}{2}t)$ are two linearly independent oscillatory solutions.

Example 2.2 Consider the differential equation

$$y''' + |\sin t| y = 0 \tag{2.4}$$

on $[0, \infty)$. Here $p(t) = |\sin t|$ and $r_k(t) = 1$, $1 \leq k \leq 3$. Hence,

$$\int_0^\infty p(t)dt = \infty = \int_0^\infty \frac{1}{r_k(t)}dt, \quad k = 1, 2, 3.$$

This implies that equation (2.4) has a nonoscillatory solution $y(t)$ such that $y(t) > 0$, $y'(t) < 0$ and $y''(t) > 0$. Further it has two linearly independent oscillatory solutions. By letting $y_1 = r_1y$, $y_2 = r_2y_1'$ and $y_3 = r_3y_2'$, we can write equation (2.4) as the system

$$\begin{pmatrix} y_1' \\ y_2' \\ y_3' \end{pmatrix} = \begin{pmatrix} 0 & \frac{1}{r_2} & 0 \\ 0 & 0 & \frac{1}{r_3} \\ \frac{-p}{r_1} & 0 & 0 \end{pmatrix} \begin{pmatrix} y_1 \\ y_2 \\ y_3 \end{pmatrix}.$$

References

1. A. Wintner, Quart. Appl. Math., **7**, (1949).
2. W. Leighton, J. London Math. Soc., **27**, (1952).
3. B. Travis, Applied Mathematics and Computation, (in press).
4. M. Gregus, Third Order Linear Differential Equations, D. Reidel Publishing Co. (1987).
5. C. Swanson, Comparison and Oscillation Theory of Linear Differential Equations, Academic Press, New York (1968).
6. D. Willet, Advances in Math., **3**, (1969).
7. B. Travis and R. Navarro, Dynamics of Continuous, Impulsive, and Discrete Systems, **4**, (1998).

8. J. Barret, Advances in Math., **3**, (1969).

9. W. Coppel, Disconjugacy, Lecture Notes in Mathematics, **220**, Springer-Verlag, Berlin (1971).

10. G. Polya, Trans. Amer. Math. Soc., **24**, (1924).

11. W. T. Reid, Ordinary Differential Equations, John Wiley & Sons (1971).

9th Int. Coll. on Differential Equations, pp. 427–430
D. Bainov (Ed)

Connection Between Differential Equations and Differential Geometry

GR. TSAGAS AND G. DIMOU
Division of Mathematics, Faculty of Technology, Department of Mathematics and Physics, Aristotle University of Thessaloniki, 54006 Thessaloniki, Greece

Abstract: There is a strong connection between Differential Equations and Differential Geometry. The aim of the present paper is to give this connection through of some problems.
Key Words and Phrases: *Differential equations, surface, system of differential equations, developable surface, minimal surface and rectilinear congruence.*
1991 Mathematics Subject Classification. 35F30, 53A05.

1. Introduction

It is known that the solution of many problems in Differential Geometry is reduced to find out solution of differential equations (ordinary or partial) or a systems of differential equations (ordinary or partial).

The aim of present talk is to give some problems from the Classical Differential Geometry whose solutions are reduced to solutions of differential equations or a system of differential equations.

The whole paper has four paragraphs. Each of them is analyzed as follows.

The first paragraph is the introduction.

Some basic problems of Classical Differential Geometry are given in the second paragraph. The third paragraph deals with developable surfaces.

Finally, the minimal surfaces are studied in the last paragraph.

2. Some Basic Problems of Classical Differential Geometry

Let S be a surface in IR^3 which is covered by a patch with local coordinate system (u,v). We consider an orthogonal coordinate system Oxyz in IR^3. The surface S can be represented by a vector equation

$$S \to \overrightarrow{r} = \overrightarrow{r}(u,v) = x(u,v)\overrightarrow{i} + y(u,v)\overrightarrow{j} + z(u,v)\overrightarrow{k} \tag{1}$$

To the surface S we can associate six functions defined by

$$E = \left(\frac{\partial \overrightarrow{r}}{\partial u}\right)^2 = x_u^2 + y_u^2 + z_u^2, F = \frac{\partial \overrightarrow{r}}{\partial u}\frac{\partial \overrightarrow{r}}{\partial u} = x_u x_v + y_u y_v + z_u z_v,$$

$$G = \left(\frac{\partial \overrightarrow{r}}{\partial v}\right)^2 = x_v^2 + y_v^2 + z_v^2 \tag{2}$$

$$L = \vec{r}_{u^2} \cdot \vec{\ell_o} = \frac{[\vec{r}_{uv} \vec{r}_u \vec{r}_v]}{\sqrt{EG - F^2}} = \frac{1}{\sqrt{EG - F^2}} \begin{vmatrix} x_{u^2} & y_{u^2} & z_{u^2} \\ x_u & y_u & z_u \\ x_v & y_v & z_v \end{vmatrix} \tag{3}$$

$$M = \vec{r}_{uv} \cdot \vec{\ell}_o = \frac{[\vec{r}_{uv} \vec{r}_u \vec{r}_v]}{\sqrt{EG - F^2}} = \frac{1}{\sqrt{EG - F^2}} \begin{vmatrix} x_{uv} & y_{uv} & z_{uv} \\ x_u & y_u & z_u \\ x_v & y_v & z_v \end{vmatrix} \tag{4}$$

$$N = \vec{r}_{v^2} \cdot \vec{\ell}_o = \frac{[\vec{r}_{v^2} \vec{r}_u \vec{r}_v]}{\sqrt{EG - F^2}} = \frac{1}{\sqrt{EG - F^2}} \begin{vmatrix} x_{v^2} & y_{v^2} & z_{v^2} \\ x_u & y_u & z_u \\ x_v & y_v & z_v \end{vmatrix} \tag{5}$$

The functions (E, F, G) and (L, M, N) are called fundamental magnitudes of the first and second order respectively of the surface S. These functions of u and v satisfy the system of partial differential equations

$$L_v - M_u = L\Gamma^1_{12} + M(\Gamma^2_{12} - \Gamma^1_{11}) - N\Gamma^2_{11} \tag{6}$$

$$M_v - N_u = L\Gamma^1_{12} + M(\Gamma^2_{12} - \Gamma^1_{11}) - N\Gamma^2_{11} \tag{7}$$

$$LN - M^2 = F\left\{\left(\Gamma^2_{22}\right)_n - \left(\Gamma^2_{12}\right)_v + \Gamma^1_{22}\Gamma^2_{11} - \Gamma^1_{22}\Gamma^2_{12}\right\}$$

$$+E\left\{\left(r^2_{12}\right)_v - \Gamma^1_{22}\Gamma^1_{11} + r^2_{22}r^1_{12} - r^1_{12}r^1_{12} - r^2_{12}r^2_{12}\right\} \tag{8}$$

where Γ^i_{jk} $i, j, k = 1, 2$ are the Christoffel's symbols and defined by

$$\Gamma^1_{11} = \frac{GE_u - 2FF_u + FE_v}{2(EG - F^2)} \qquad \Gamma^2_{11} = \frac{2EF_u - 2EE_v + FE_u}{2(EG - F^2)} \tag{9}$$

$$\Gamma^1_{12} = \frac{GE_v - FG_u}{2(EG - F^2)} \qquad \Gamma^2_{12} = \frac{EG_u - FE_v}{2(EG - F^2)} \tag{10}$$

$$\Gamma^1_{22} = \frac{2GF_v - GG_u - FG_v}{2(EG - F^2)} \qquad \Gamma^2_{22} = \frac{EG_v - 2FF_v - FG_u}{2(EG - F^2)} \tag{11}$$

The (6) and (7) are called Mannardi Codazzi equations and (8) Gauss equation. The basic theorem for the surfaces is the following

Theorem 1 *(Bonnet). Let E, F, G, L, M, N be differentiable functions in an open set $V \subset IR^2(u, v)$ with $E > 0$ and $G > 0$. Assume that the given functions satisfy formally Gauss and Mannardi Codazzi equation and that $EG - F^2 > 0$. Then for every $q \in V$ there exists a neighborhood $U \subset N$ of q and a diffeomorphism $t : U \to t(U) \subset IR^3$ such that the regular surface $t(V)$ has E, F, G and L, M, N as the fundamental magnitudes of the first and second order respectively. If U is connected and if*

$$\ell : U \to \ell(U) \subset IR^3$$

is another diffeomorphism satisfying the same conditions, then there exists a translation T and a proper linear orthogonal transformation P in IR^3 such that

$$\ell = ToPot$$

Remark 2 *The above theorem states in order that the six functions E(u,v), F(u, v), G(u, v) and L(u, v), M(u, v), N(u, v) determine a surface in IR^3 they must satisfy a system of partial differential equations which are (6), (7) and (8). If we require more properties for the surface, then these functions satisfy more partial differential equations. We give two problems in this direction.*

3. Developable Surfaces

Remark 3 *A surface S in IR^3 can be represented by the function*

$$S \to z = f(x,y) \tag{12}$$

In this case the fundamental magnitudes of the first and of second order S take the form

$$E = 1+p^2, \quad F = pq, \quad G = 1+q^2, \quad L = \frac{r}{w}, \quad M = \frac{s}{w}, \quad N = \frac{t}{w} \tag{13}$$

where

$$p = \frac{\partial f}{\partial x}, \quad q = \frac{\partial f}{\partial y}, \quad w = \sqrt{1+p^2+q^2}, \quad r = \frac{\partial^2 f}{\partial x^2}, \quad s = \frac{\partial^2 f}{\partial y^2}, \quad t = \frac{\partial^2 f}{\partial x \partial y} \tag{14}$$

Problem 4 *Determine all the developable surfaces.*

Solution 5 *A surface is a developable if it is isometric to a plane. If E, F, G and L, M, N are the fundamental magnitudes of the first and second order of this surface, then these satisfy except the equations (6), (7), (8) and the equation*

$$LN - M^2 = 0 \tag{15}$$

Therefore the determination of all developable surfaces in IR^3 is reduced to the solution of equations (6), (7), (8) and (15).

If the surface is defined by (12), then the determination of developable surfaces is reduced to solve the partial differential equation

$$rs - t^2 = 0 \Longleftrightarrow \frac{\partial^2 f}{\partial x^2}\frac{\partial^2 f}{\partial y^2} - \left(\frac{\partial^2 f}{\partial x \partial y}\right)^2 = 0 \tag{16}$$

Example 6 *A cylinder and cone are developable surfaces. We give only two such surfaces of revolution.*

Proposition 7 *Every developable surface S can be generated, except cylinder and cone, by the tangents of a non-plane curve in IR^3.*

Proof. A curve C in IR^3 can be represented by the vector equation

$$\overrightarrow{r} = \overrightarrow{r}(s) = x(s)\overrightarrow{i} + y(s)\overrightarrow{j} + z(s)\overrightarrow{k} \tag{17}$$

where s is the arc length of the curve C.

The developable surface is defined by the equation

$$\begin{aligned}\overrightarrow{R} = \overrightarrow{r}(s,v) = \overrightarrow{r}(s) + w\,\dot{\overrightarrow{r}}\,(s) = [(x(s) + w\,\dot{x}\,(s)]\,\overrightarrow{i} \\ + [(y(s) + w\,\dot{y}\,(s)]\,\overrightarrow{j} + [z(s) + w\,\dot{z}\,(s)]\,\overrightarrow{k}\end{aligned} \tag{18}$$

□

Technological Applications of the Developable Surfaces

It is important in the industry and the technology when we use a piece of plane from leather or plastic or iron plate or another material and we would like to change the shape of this in the surface without to destroy the strength of this material.

In this case the surface must be a developable surface, because this is mapped isometrically into a plane. Therefore this surface has the same inner strength as the piece of plane which took the shape of developable surface.

Problem 8 *WILLMORE'S PROBLEM. Construct a shoe using one peace of leather.*

Solution 9 *Thirty years ago Professor Willmore from the University of Durham had worked on this problem. This project has financed by a very big shoe industry in U.K. The problem was solved by Professor Willmore using developable surfaces. The company had a profit in one year 10.000.000 sterling's.*

4. Minimal Surfaces

Problem 10 *Determine of minimal surfaces solution. We consider a non-plane closed curve c. The surface which comes through c and has the minimal surface.*

Solution 11 *This minimal surface is characterized by the equation*

$$EN - 2FM + LN = 0 \tag{19}$$

Solution 12 *Hence the determination of the minimal surfaces is reduced to solve the system of partial differential equation (6), (7), (8) and (19).*

If the surface s is defined by the equation (12), then this is a minimal surface if f satisfies the partial differential equation

$$r(1 + q2) - 2spq + t(1 + p2) = 0 \tag{20}$$

Therefore when the curve c is given we try to find a solution of (20) with boundary conditions which are the given curve C.

Technological Applications of the Minimal Surfaces

The theory of minimal surfaces has many applications in industries and technology. We refer an example.

Example 13 *We consider the sassi of a car and we like to construct a surface coming from the sassi and having minimal area.*

Solution 14 *The sassi is a given curve in three dimensional space and we like to determine the surface through this curve and having the minimal area, that is a minimal surface.*

We require this condition in order the resistance of the car, when moving in the air, is minimal. This implies that the car spend less oil for the movement.

References

1. M. Gûrses, Lett. Math. Phys., **44** , 1-8 (1998).
2. J.-F. Hwang, Pacific J. Math., **183**, 91–102 (1998).
3. R. Koch, in: Proc. of the Congress of Geometry (Thessaloniki, 1987), Aristotle Univ. Thessaloniki (1998) pp. 82–98.
4. E. Kreyszig, SIAM J. Math. Anal., **25**, 174–178 (1994).
5. B. Opozda, Results Math., **27**, 97–104 (1995).
6. G.P. Pirola, Pacific J. Math., **183**, 333–357 (1998).
7. V. Rangachari, J. Bihar Math. Soc., **7**, 45-48 (1983).
8. V. Rengachari, J. Bihar Math. Soc., **8**, 46–48 (1984).
9. Gr. Tsagas, Ch. Christophoridou, As. Synefaki, in: New Frontiers in Algebras, Groups and Geometries (Monteroduni, 1995), Ser/ New Front. Adv. Math. Ist. Ric. Base, Hadronic Press, Palm Harbor, FL (1996) pp. 317–326.

9th Int. Coll. on Differential Equations, pp. 431–438
D. Bainov (Ed)

Errors in Iteration Points in Oscillatory State for Systems of Algebraic Equations

KUMIKO TSUJI
Kyushu Teikyo Junior College, 4-3-124 Shin-Katsutachi-Machi Ohmuta 836, Japan

Abstract: For system of the algebraic equations which is solved by Newton's iteration process, the round-off errors of iteration points are analyzed. The equation which determines the unknown round-off error is obtained for the iteration point under the condition that the iteration point is oscillate after some period. The equation is a system of algebraic equations with the unknown of round-off errors. The main term which determines the condition for ill or well depends on Lipschitz matrix. The condition becomes better as spectral radiuses of the Lipschitz matrices are smaller than 1. The sub term is sum of the product series of Lipschitz matrix whose coefficients are the round-off errors of numerical iteration process on computer at numerical oscillating points.
Key Words and Phrases: *round-off error, Fortran, Chebyshev polynomial, floating-point addition, floating-point multiplication, Newton's method.*

1. Introduction

The problem of round-off error arises when the error is computed practically. In [1] and [2], a zero point is computed by the Newton's iteration process $w_{n+1} = f(w_n)$ for the Chebyshev polynomial $T(w)$. But the Newton's iteration process turns to $w_{n+1} = g(w_n)$ $(n = 1, 2, 3, \dots)$ in practical computation. Here $g(w) = w \ominus H \otimes \tilde{T}_s(w)$. $H = 1 \oslash \tilde{T}_s'(w_1) \otimes 0.5$. $\tilde{T}_s(w)$ is computed residual of $T(w)$ by single precision. $\tilde{T}_s'(w_1)$ is computed derivative of $T(w)$ at $w = w_1$. $\ominus$, $\otimes$ and $\oslash$ denote floating-point subtraction, multiplication and division. In [1], assuming that the iteration points are in an oscillatory state after some iterations, the equation is obtained, which determines the round-off error of each iteration point in the oscillatory state with period m i.e. $w_{m+n_0} = w_{n_0}$ *for some* $m \geq 1$. This equation is called an OSI equation. By this OSI equation the round-off error of zero point is calculated for the algebraic equation. In [2], it is analyzed that OSI-equation is effective to calculate round-off error for ill condition case based on the singularity theory in [3]. In the present paper, it is shown that the OSI equation for the algebraic equation is applicable also systems of algebraic equations and the round-off error of the solution is determined by OSI equation.

1.1. Existence theorem for system of algebraic equations

Definition 1.1. *For any vector x, with components $x_1, x_2, \dots, x_n$, a vector with components $| x_1 |, | x_2 |, \dots, | x_n |$, is called a magnitude-vector of x and is denoted by*

$\sigma(x)$.

In [4], Urabe gives an existence theorem for nonlinear equations, which is related in the following proposition. Using this proposition, absolute value of round-off error of numerical solution $\bar{x}$ is calculated for the system of algebraic equations. Further, an existence of unique solution of the algebraic equations is obtained.

Proposition 1.1. *Let $F(x) = 0$ be a given real system of equations where x and $F(x)$ are vectors of the same dimension and $F(x)$ is a continuously differentiable function of x defined in some region D of the x-space. Assume that*

1. *$F(x) = 0$ has an approximate solution $\bar{x}$ for which the determinant of the Jacobian matrix $J(x)$ of $F(x)$ with respect to x does not vanish.*

2. *There are a positive constant vector δ and a non-negative matrix κ such that*

   (a) The spectral radius of κ is less than 1.

   (b) $D_\delta = \{\sigma(x - \bar{x}) \le \delta\} \subset D$.

   (c) $\sigma[I - HJ(x)] \le \kappa$ for any $x \in D_\delta$.

   (d) $(I - \kappa)^{-1}\sigma(HF(\bar{x})) \le \delta$, where H is the inverse matrix $J^{-1}(\bar{x}) : H = J^{-1}(\bar{x})$.

Then

1. *The Newton iteration process*

$$x_{n+1} = x_n - HF(x_n)\ (n = 0, 1, 2, \dots)$$

   with $x_0 = \bar{x}$ yields a fundamental sequence $\{x_n\}$ $(n = 0, 1, 2, \dots)$ in D_δ and we have $\sigma(x_n - \bar{x}) \le (I - \kappa)^{-1}\sigma(HF(\bar{x}))$ $(n = 0, 1, 2, \dots)$.

2. *The above fundamental sequence $\{x_n\}$ $(n = 0, 1, 2, \dots)$ is convergent in D namely, there is an $\hat{x} \in D$ such that $\sigma(\hat{x} - \bar{x}) \to 0$ as $n \to \infty$, then $\hat{x}$ is a unique solution of $F(x) = 0$ in D_δ.*

3. *Let ϵ is the error of $\bar{x}$ i.e. $\epsilon = \hat{x} - \bar{x}$. Then $\sigma(\epsilon) \le (I - \kappa)^{-1}\sigma(HF(\bar{x}))$.*

The magnitude-vector of error ϵ is estimated in the above proposition, however our error by OSI equation is calculated with signs and promotes the accuracy of numerical solution.

2. OSI Equation for System of Algebraic Equations

Definition 2.1. ***Lipschitz matrix*** *for (α, β) is defined as the divided difference of $f(x)$ with deviation $\alpha - \beta$ and denoted by $\chi(\alpha, \beta)$:*

$$\chi(\alpha, \beta) = I - H \int_0^1 J[\beta + \theta(\alpha - \beta)]d\theta.$$

Lipschitz matrix for $(x_1 + \epsilon_1, x_n)$ is denoted as $\kappa(\epsilon_1, x_n)$:

$$\kappa(\epsilon_1, x_n) = \chi(x_1 + \epsilon_1, x_n) = \chi(\hat{x}, x_n).$$

Definition 2.2. *In the present paper, theoretical Newton iteration process* $w_{n+1} = f(w_n)$ $(n = 1, 2, \dots)$ *and numerical Newton iteration process* $w_{n+1} = g(w_n)$ $(n = 1, 2, \dots)$ *are considered, where* $f(x) = x - HF(x)$, $g(x) = x \ominus H \otimes \tilde{F}(x)$. *and* $\tilde{F}(x)$ *is the computed floating-point vector of* $F(x)$.

Iteration round-off error *is defined as the round-off error of* $f(x)$:
$R(x) = f(x) - g(x)$.

Then $R(x) = -H(F(x) - \tilde{F}(x)) + r_M(-H, \tilde{F}(x)) + r_A(x, -H \otimes \tilde{F}(x))$. since

$$f(x) - g(x) = x - HF(x) - (x - H \otimes \tilde{F}(x) + x - H \otimes \tilde{F}(x) - x \ominus H \otimes \tilde{F}(x)$$

$$= -H(F(x) - \tilde{F}(x)) + r_M(-H, \tilde{F}(x)) + r_A(x, -H \otimes \tilde{F}(x)).$$

Here $r_A(x, y)$ is the round-off error of floating-point addition for vector x and y, and $r_M(H, y)$ is the round-off error of floating-point multiplication for matrix H and vector y. In [5], program to calculate the round-off errors of floating-point additions are obtained depending on the floating-point arithmetic algorithm.

Theorem 2.1. *Let* E *and* B *be* N*-dimensional Euclidean spaces. Let a function* $F(x)$ *map an open set* D *of* E *into* B.

1. *Let* $F(x) = 0$ *be a given real system of equations where* x *and* $F(x)$ *are vectors of the same dimension and* $F(x)$ *is a continuously differentiable function.*
2. *Assume that* $F(x) = 0$ *has an approximate solution* $\bar{x}$ *for which Jacobian matrix* $J(x)$ *with respect to* x *does not vanish. Put* $H = J(\bar{x})$.
3. *Suppose the iterative process* $w_{n+1} = g(w_n)$ $(n = 1, 2, \dots)$ *with* w_1 *yields a sequence* $\{w_n\}$ $(n = 1, \dots, m+1)$ *in oscillatory state such that* $w_1 \doteq w_{m+1}$.
4. *There is a unique solution* $\hat{x}$ *for* $F(x) = 0$.
5. *Let Lipschitz matrix* $\kappa(\epsilon_1, w_n)$ *be the divided difference of* $f(x)$ *with deviation* $(w_1 + \epsilon_1) - w_n$ *such that* $\kappa(\epsilon_1, w_n) = I - H \int_0^1 J[w_n + \theta\{(\epsilon_1 + w_1) - w_n\}]d\theta$.

Then the error ϵ_1 *satisfies OSI equation:*

$$(I - \Pi_{j=1}^{m} \kappa(\epsilon_1, w_j))\epsilon_1 = \sum_{i=1}^{m} \{\Pi_{j=i+1}^{m} \kappa(\epsilon_1, w_j)\} R(w_i).$$

Proof. Put $\epsilon_n = \hat{x} - w_n$. $\epsilon_2 = \hat{x} - w_1 = f(\hat{x}) - g(w_1) = f(\hat{x}) - f(w_1) + R(w_1) = \kappa(\epsilon_1, w_1)\epsilon_1 + R(w_1)$

$$\epsilon_3 = \hat{x} - w_3 = \kappa(\epsilon_1, w_2)\epsilon_2 + R(w_2) = \kappa(\epsilon_1, w_1)\kappa(\epsilon_1, w_2)\epsilon_1 + \kappa(\epsilon_1, w_2)R(w_1) + R(w_2).$$

Assumption for mathematical induction:

$$\epsilon_m = \Pi_{j=1}^{m-1} \kappa(\epsilon_1, w_j)\epsilon_1 + \sum_{i=1}^{m-1} \{\Pi_{j=i+1}^{m-1} \kappa(\epsilon_1, w_j)\} R(w_i).$$

Then ϵ_{m+1} satisfies

$$\epsilon_{m+1} = \kappa(\epsilon_1, w_m)\epsilon_m + R(w_m)$$

$$= \Pi_{j=1}^{m}\kappa(\epsilon_1, w_m)\epsilon_1 + \sum_{i=1}^{m-1}\{\Pi_{j=i+1}^{m}\kappa(\epsilon_1, w_j)\}R(w_i) + R(w_m)$$

$$= \Pi_{j=1}^{m}\kappa(\epsilon_1, w_j)\epsilon_1 + \sum_{i=1}^{m}\{\Pi_{j=i+1}^{m}\kappa(\epsilon_1, w_j)\}R(w_i).$$

Since $w_{m+1} = w_1$ by oscillation, $\epsilon_{m+1} = \epsilon_1$ and so

$$\epsilon_1 = \Pi_{j=1}^{m}\kappa(\epsilon_1, w_j)\epsilon_1 + \sum_{i=1}^{m}\{\Pi_{j=i+1}^{m}\kappa(\epsilon_1, w_j)\}R(w_i).$$

□

3. Sufficient Condition for Oscillation

Utilizing the theorem in [4], sufficient condition that iteration points are shut up in the sphere with radius 2δ is obtained for δ in the above proposition. If the iteration points are shut up in small sphere, iteration points become oscillatory state at some stage. But there is the problem from what stage, iteration points are oscillatory.

Theorem 3.1. *Assume that*

1. *Let $F(x) = 0$ be a given real system of equations where x and $F(x)$ are vectors of the same dimension and $F(x)$ is a continuously differentiable function of x defined in some region D of the x-space.*
2. *$F(x) = 0$ has an approximate solution $\bar{x}$ for which the determinant of the Jacobian matrix $J(x)$ of $F(x)$ with respect to x does not vanish.*
3. *There are a positive constant δ and a non-negative constant $\kappa < 1$ such that*
   (a) *$D_{2\delta} = \{\sigma\{x - \bar{x}\} \leq 2\delta\} \subset D$.*
   (b) *Spectral radius $I - HJ(x)$ is bounded by κ on D.*
   (c) *$(I - \kappa)^{-1}\sigma(HF(\bar{x})) \leq \delta$, where H is a matrix such that $H = J^{-1}(\bar{x})$.*
   (d) *Round-off error of $\sigma(R(x))$ is also bounded by vector η such that $(I-\kappa)^{-1}\eta \leq \delta$.*

Then

1. *Exact solution exists and the error of $\bar{x}$ satisfies*

$$\{H\int_0^1 J(\bar{x} + \theta\epsilon)d\theta\}\epsilon = -HF(\bar{x}).$$

2. *w_n are included in $D_{2\delta}$ and become OSI from some stage $1 + n_0$ with period m : $w_{1+n_0} = w_{1+n_0+m}$.*
3. *Spectral radius of $\kappa(\epsilon_1, x_j)$ i.e. the maximum in absolute value of eigenvalues of $\kappa(\epsilon, x_j)$ is bounded by κ on D.*

4. Round-off error $\epsilon_1 = \hat{x} - x_{1+n_0}$ satisfies our OSI equation.

$$(I - \Pi_{j=1}^{m}\kappa(\epsilon_1, x_j))\epsilon_1 = \sum_{i=1}^{m}\{\Pi_{j=i+1}^{m}\kappa(\epsilon_1, x_{j+n_0})\}R(x_{i+n_0}).$$

Proof. 1.Put $\zeta_{n+1} = x_{n+1} - x_n$. $\zeta_1 = f(\bar{x}) - \bar{x} = \bar{x} - HF(\bar{x}) - \bar{x} = -HF(\bar{x})$. $\zeta_{n+1} = x_{n+1} - x_n = f(x_n) - f(x_{n-1}) = \chi(x_n, x_{n-1})\zeta_n$.

$$\zeta_n = \Pi_{i=1}^{n}\chi(x_i, x_{i-1})\zeta_1 = -\Pi_{i=1}^{n}\chi(x_i, x_{i-1})HF(\bar{x}).$$

Put $\xi_n = x_n - \bar{x}$.

$$\xi_n = \sum_{j=1}^{n}\zeta_j = -\sum_{j=1}^{n}\Pi_{i=1}^{j}\chi(x_i, x_{i-1})HF(\bar{x}).$$

$$\sigma(\xi_n) \leq (I - \kappa)^{-1}\sigma(HF(\bar{x})) \leq \delta.$$

x_n is included in D_δ. Since $\{x_n\}$ is fundamental sequence in D, there exists the exact solution $\hat{x}$. $\epsilon_1 = \hat{x} - x_1$ satisfies

$$\epsilon = -\sum_{j=1}^{\infty}\Pi_{i=1}^{j}\chi(x_i, x_{i-1})HF(\bar{x}).$$

where the right hand is bounded by $(I - \kappa)^{-1}\sigma(HF(\bar{x})) \leq \delta$. ϵ satisfies

$$\epsilon - \zeta_1 = \epsilon + (\bar{x} - x_1) = \hat{x} - x_1 = \phi(\hat{x}) - \phi(\bar{x}) = \chi(\bar{x} + \epsilon, \bar{x})\epsilon,$$

where $\zeta_1 = -LF(\bar{x})$ and $I - \chi(\bar{x} + \epsilon, \bar{x}) = L\int_0^1 J(\bar{x} + \theta\epsilon]d\theta$.

2. By the proof of theorem for OSI equation,

$$\epsilon_{m+1} = \Pi_{j=1}^{m}\kappa(\epsilon_1, w_j)\epsilon_1 + \sum_{i=1}^{m}\{\Pi_{j=i+1}^{m}\kappa(\epsilon_1, w_j)\}R(w_i).$$

$$\sigma(\epsilon_{m+1}) \leq \sigma(I - \kappa)^{-1}\eta \leq \delta \text{ for all } m.$$

w_n are included in $\{\sigma(\hat{x} - w) \leq \sigma\}$. $\sigma(w_n - \bar{x}) \leq \sigma(\epsilon_n + \hat{x} - \bar{x}) \leq 2\delta$. w_n are included in $D_{2\delta}$.

Therefore w_n are bounded in $D_{2\delta}$ and oscillate. Since $w_{m+1+n_0} = w_{1+n_0}$ by oscillation, $\epsilon_{m+1} = \epsilon_1$ and so

$$\epsilon_1 = \Pi_{j=1}^{m}\kappa(\epsilon_1, w_{j+n_0})\epsilon_1 + \sum_{i=1}^{m}\{\Pi_{j=i+1}^{m}\kappa(\epsilon_1, w_{j+n_0})\}R(w_{i+n_0}).$$

□

4. Necessary and Sufficient Condition for Oscillatory State

In the following theorem, relation of iteration round-off error $R(x)$ and $HF(x)$ is obtained for equation $F(x) = 0$ when iteration points are OSI state. Let $r_A(w, v)$ denote the vector round-off error of vector floating-point sum $w \oplus v$: $r_A(w, v) = w + v - w \oplus v$. Put $v_{i+n_0} = -H \otimes \tilde{F}(w_{i+n_0})$.

Theorem 4.1. *$\{w_{i+n_0}\}$ be iteration points in oscillatory state if and only if one of following condition is satisfied:*

1. *The sum of the round-off errors for the floating-point additions $w_{i+n_0} \oplus v_{i+n_0}$ and sum of numerical vector values v_{i+n_0} coincide: $\sum_{i=1}^{m} r_A(w_{i+n_0}, v_{i+n_0}) = \sum_{i=1}^{m} v_{i+n_0}$.*

2. *The sum of iteration round-off errors and product of H and sum of residuals $F(w_{i+n_0})$ coincide: $\sum_{i=1}^{m} R(w_{i+n_0}) = -H \sum_{i=1}^{m} F(w_{i+n_0})$.*

Proof. $w_{m+1+n_0} = w_1$ if and only if 1, since

$$\sum_{i=1}^{m} w_{i+1+n_0} = \sum_{i=1}^{m} w_{i+n_0} + v_i - r_A(w_{i+n_0}, v_i); \, w_{m+1+n_0} = w_{1+n_0} + \sum_{i=1}^{m} (v_i - r_A(w_{i+n_0}, v_i))$$

$w_{m+1+n_0} = w_1$ if and only if 2, since

$$\sum_{i=1}^{m} R(w_{i+n_0}) = \sum_{i=1}^{m} (w_{i+n_0} - HF(w_{i+n_0}) - w_{i+1+n_0});$$

$$\sum_{i=1}^{m} R(w_{i+n_0}) = w_{1+n_0} - w_{m+1+n_0} - \sum_{i=1}^{m} HF(w_{i+n_0}).$$

□

5. Equivalence of L-equation and OSI Equation for Stationary Iteration Point

Equation $(H \int_0^1 J(\bar{x} + \theta\epsilon))d\theta\epsilon = -HF(\bar{x})$ is called L-equation, since it is made by linear approximation. OSI-equation is made by practical phenomenon. But following theorem shows that those coincide if iteration point becomes stationary i.e. $m = 1$. Proof is performed putting $m = 1$ in the above theorem for OSI case.

Theorem 5.1. *Let $\bar{x} = w_1$. Let iteration points become stationary at 2 stage i.e. $w_2 = w_1$. Then $r_A(w_{1+n_0}, -H \otimes \tilde{F}(w_{1+n_0})) = -H \otimes \tilde{F}(w_{1+n_0})$, $R(w_1) = -HF(w_1)$, $I - \kappa(\epsilon_1, w_1) = H \int_0^1 J(\bar{x} + \theta\epsilon)d\theta$, and L-equation and OSI equation coincide.*

For $m > 1$, OSI equation and L-equation are different from each other. By OSI equation, the singularity of $F(x)$ is reduced, since main term $(I - \Pi_{i=1}^{m} \kappa(\epsilon_1, w_i))$ approaches to unit matrix.

6. Numerical Example to Compute a Collocation Point for Chebyshev Polynomial

Chebyshev polynomial $T(w)$ is the polynomial as follows: $T(w) = 524288w^{20} - 2621440w^{18} + 5570560w^{16} - 6553600w^{14} + 4659200w^{12} - 2050048w^{10} + 54120w^8 - 84480w^6 + 6600w^4 - 200w^2 + 1$. In [2], it is shown the catastrophic cancellation arises in the sum of coefficients in Chebyshev polynomial $T(w)$ for single precision and so computed value $\tilde{T}(w)$ has large round-off errors as w near to 1. Computed value $\tilde{T}(w) = 0$ nevertheless $\tilde{T}(w)$ is large value $\tilde{T}(w) = -.FE9E8C(0)$ for $w = .FD0D88$. By our OSI equation in quadruple precision

(28 Hexadecimal digit floating-point number), the error is obtained 22 Hexadecimal digits correctly. By linear approximation, correct digits of error are not obtained at all even computed quadruple precision. In the present paper, it is shown the case that the iteration points do not converge to fundamental sequence by the round-off errors. Theoretically it is guaranteed that iteration points become fundamental sequence since κ is less than 1. Only 2 digits $-.A6$ are stationary in iteration points. The iteration points become oscillatory with period $m = 122$ as the following table shows.

Oscillating iteration points from w_{81} up to w_{203}, where $w_{81} = w_{203}$

81	$-0.A646B0$	82	-0.A6580B	83	-0.A63AA9	84	-0.A64F67
85	-0.A63F88	86	-0.A63253	87	-0.A6387D	88	-0.A630AF
89	-0.A6370A	90	-0.A625AF	91	-0.A63ACE	92	-0.A651F2
93	-0.A642D5	94	-0.A63CD1	95	-0.A64863	96	-0.A63EDF
97	-0.A6448A	98	-0.A6409E	99	-0.A63533	100	-0.A63E41
101	-0.A65423	102	-0.A63DC1	103	-0.A642BD	104	-0.A64597
105	-0.A63D98	106	-0.A65017	107	-0.A6428A	108	-0.A64B98
109	-0.A64483	110	-0.A646FC	111	-0.A64213	112	-0.A650CD
113	-0.A630F1	114	-0.A657FD	115	-0.A63B2D	116	-0.A63036
117	-0.A649E6	118	-0.A62CEF	119	-0.A637C7	120	-0.A6386C
121	-0.A64FB7	122	-0.A63EEE	123	-0.A63DE6	124	-0.A64551
125	-0.A63D8D	126	-0.A64484	127	-0.A63C37	128	-0.A62EB4
129	-0.A63D47	130	-0.A64096	131	-0.A642CB	132	-0.A62BBA
133	-0.A62E02	134	-0.A6441E	135	-0.A65978	136	-0.A64F4E
137	-0.A62C67	138	-0.A64B27	139	-0.A643C4	140	-0.A640F3
141	-0.A635D5	142	-0.A64472	143	-0.A640C0	144	-0.A6447A
145	-0.A64852	146	-0.A647A2	147	-0.A64E68	148	-0.A6477A
149	-0.A6446E	150	-0.A64939	151	-0.A62A5B	152	-0.A64CA5
153	-0.A642A2	154	-0.A63643	155	-0.A636C1	156	-0.A634D9
157	-0.A650EF	158	-0.A63685	159	-0.A62CE3	160	-0.A62D44
161	-0.A655A5	162	-0.A63759	163	-0.A6344D	164	-0.A6488C
165	-0.A62E2B	166	-0.A638F0	167	-0.A63EB9	168	-0.A65793
169	-0.A635BD	170	-0.A6347B	171	-0.A637FB	172	-0.A6443C
173	-0.A642B6	174	-0.A63CF6	175	-0.A64139	176	-0.A64267
177	-0.A641CB	178	-0.A6390A	179	-0.A64826	180	-0.A632A4
181	-0.A64ABB	182	-0.A64D95	183	-0.A634CE	184	-0.A63B9E
185	-0.A653AB	186	-0.A638D5	187	-0.A63124	188	-0.A64691
189	-0.A63CA2	190	-0.A64798	191	-0.A64705	192	-0.A64C32
193	-0.A63F5E	194	-0.A64530	195	-0.A63390	196	-0.A644E1
197	-0.A63639	198	-0.A63115	199	-0.A64B7E	200	-0.A644D4
201	-0.A6502B	202	$-0.A63CAE$	203	$-0.A646B0$		

The zero $cos\frac{29\pi}{40}$ is computed by the intrinsic function in quadruple precision. Its digits are correct 28 Hexadecimal digits as follows and denoted by $\tilde{x}$:

$$\tilde{x} = cos\frac{29\pi}{40} = -0.A6423A3007776FD278C1DCC49236.$$

1. Let $T(z - \delta z)$ be the residual at the approximate zero $z - \delta z$. Let $T'(z)$ be the derivative at exact zero $z = cos\ (2r + 1)\pi/40$. $T(z - \delta z)$ is approximated linearly by $-T'(z)\delta z$ at zero point z. The error is estimated by using residual and derivative as $\delta\tilde{z} = -\frac{T(z-\delta z)}{T'(z)}$. This method is called a LA (linear approximation). By augmenting $\delta\tilde{z}_n$ to w_n, corrected zero point $w_n + \delta\tilde{z}_n$ is considered and the accuracy's of $w_n + \delta\tilde{z}_n$ are compared. The number of correct digits in $w_n + \delta\tilde{z}_n$ increases more as the distance of iteration point w_n and exact solution is shorter. The correct digits in corrected zero

$w_n + \delta\tilde{z}_n$ varies from 9 digits in $w_{111} + \delta\tilde{z}_{111}$ for $w_{111} = -0.A64213$ to 5 digits in $w_{90} + \delta\tilde{z}_{90}$ for $w_{90} = -0.A625AF$:

$$w_{111} + \delta\tilde{z}_{111} = -0.A6423A300\underline{4189A26};$$

$$w_{90} + \delta\tilde{z}_{90} = -0.A64238\underline{3DEA7884}$$

(Underlined digits are incorrect digits). 2. For our OSI equation, the number of correct digits in $w_n + \delta\epsilon_n$ do not depend on distance of iteration point w_n and exact zero. The correct digits are same 25 digits both for w_{111} and for w_{90} :

$$w_{111} + \epsilon_{111} = -0.A6423A3007776FD278C1DCC48\underline{FFE},$$

$$w_{90} + \epsilon_{90} = -0.A6423A3007776FD278C1DCC48\underline{462}$$

3. Accuracy of our error by OSI equation is extraordinary superior than the error by LA. Correct digits in $w_n + \epsilon_n$ are more than those in $w_n + \delta\tilde{z}_n$ by $15 \sim 21$. The maximum number of correct digits is 26 for $w_{181} = -0.A64ABB$:

$$w_{181} + \epsilon_{181} = -0.A6423A3007776FD278C1DCC492\underline{78}.$$

$w_{181} + \epsilon_{181}$ attains the highest accuracy:

$$\min_{n=81\sim 202} |\ \tilde{x} - w_n - \epsilon_n\ | = .420000(-26) = |\ \tilde{x} - w_{181} - \epsilon_{181}\ | .$$

The minimum number of correct digits is 24 for $w_{185} = -.A653AB$:

$$w_{185} + \epsilon_{185} = -0.A6423A3007776FD278C1DCC4\underline{6AFB}.$$

$w_{185} + \epsilon_{185}$ attains the lowest accuracy:

$$\max_{n=81\sim 202} |\ \tilde{x} - w_n - \epsilon_n\ | = .273B00(-24) = |\ \tilde{x} - w_{185} - \epsilon_{185}\ | .$$

4. The corrected zero $w_n + \epsilon_n$ by augmenting digits of ϵ_n is correct in $24 \sim 26$ digits for OSI points w_n ($n = 81 \sim 202$). The number of correct digits varies only 2 in $w_n + \epsilon_n$ whereas 4 in $w_n + \delta\tilde{z}_n$. $w_n + \epsilon_n$ are the uniform accuracy for OSI points not depending on the distance between w_n and the exact zero point.

References

1. K. Tsuji, in: Proc. 7th Int.Coll.on Differential Equations, D. Bainov (Ed.), VSP, Utrecht, (1997) 421–428.

2. K. Tsuji, in: Proceedings of the First ISAAC, Kluwer Publisher (Netherlands) (1998) (in press).

3. J.H. Wilkinson, Notes on Applied Science No. 32, National Physical Laboratory, 46–47 (1963).

4. M. Urabe, Lec. Notes in Computer Sci. **29**, Interval Mathematics, Springer-Verlag, 99–117 (1975).

5. K. Tsuji, Mem. of Fac. of Sci., Kyushu Univ., Ser.A, **39**, No.2, 209–225 (1985).

9th Int. Coll. on Differential Equations, pp. 439-442
D. Bainov (Ed)

Equicontinuity Domains and Runge Domains

MIKI TSUJI
Graduate School of Mathematics, Kyushu University, Ropponmatsu, Chuo-ku Fukuoka 810-0044, Japan
Fax : +81-92-726-4846
E-mail: tsuji@rcmath.rc.kyushu-u.ac.jp

Abstract: Let M be a complex manifold. In the previous paper [1], Abe, Furushima and the author proved that for any family $\mathcal{F} \subset \mathcal{O}(M)$ the equicontinuity domain $E(\mathcal{F})$ and the Montel domain $B(\mathcal{F})$ of $\mathcal{F}$ satisfy the strong disk property in M. In the present paper, we discuss that the relation of these domains and Runge domains relative to M, when M is a Stein manifold.
Key Words and Phrases: *equicontinuity domain, Montel domain, Runge pair*
1991 Mathematics Subject Classification. 32E10, 32A17, 32F15.

1. Introduction

Throughout the paper all complex manifolds are supposed to be second countable. Let M be a complex manifold of dimension m and let $\mathcal{F}$ be a family of holomorphic functions on M.

The family $\mathcal{F}$ is said to be *normal* if for every connected component C of M and for every sequence $\{f_\nu\}_{\nu=1}^{\infty} \subset \mathcal{F}$ there exists a subsequence $\{f_{k_\nu}\}_{\nu=1}^{\infty}$ which converges in $\mathcal{O}(C)$ or diverges compactly on C (cf. Barth [2], Grauert-Reckziegel [3], Wu [4]). We denote by $D(\mathcal{F})$ the union of the open sets U of M such that $\mathcal{F}|_U \subset \mathcal{O}(U)$ is normal. Then $\mathcal{F}|_{D(\mathcal{F})}$ is also normal since M is second countable.

The open set $D(\mathcal{F})$ is said to be the *normality domain* of $\mathcal{F}$.

We denote by $B(\mathcal{F})$ the union of the open sets U of M such that $\mathcal{F}$ is uniformly bounded on U. The open set $B(\mathcal{F})$ is said to be the *Montel domain* of $\mathcal{F}$ (cf. Sato [5]). By Montel's theorem $B(\mathcal{F})$ coincides with the union of the open sets U such that $\mathcal{F}|_U$ is relatively compact in $\mathcal{O}(U)$.

We denote by $E(\mathcal{F})$ the set of all points $x \in M$ such that $\mathcal{F}$ is equicontinuous at x. Then the set $E(\mathcal{F})$ is open and is said to be the *equicontinuity domain* of $\mathcal{F}$.

In 1971, Barth [2] discussed higher dimensional cases, that is, he proved that the normality domain $D(\mathcal{F})$ is $(m-n)$-pseudoconvex in M where M is a complex manifold of dimension m, N is a complex manifold of dimension n which has a Stein holomorphic covering space and $\mathcal{F}$ is a family of holomorphic maps from M into N.

Especially if $N = \mathbf{C}$, then $D(\mathcal{F})$ is locally Stein in M. It is a generalization of Satz 12 of Cartan-Thullen [6] on the Julia problem.

On the other hand Sato [5] proved that both $B(\mathcal{F})$ and $E(\mathcal{F})$ are Stein if M is a Stein open set of $\mathbf{C}^m$ and M is the existence domain of $\mathcal{F}$.

In the previous paper [1], we had the stricter results than those in [5].

Before stating them, we prepare some definitions and the theorem by Abe [7], [8].

Definition 1.1 (Abe [7]) *Let $\Delta := \{t \in \mathbf{C} \mid |t| < 1\}$. Let M be a complex manifold and D an open set of M. We say that D satisfies the disk property in M if it satisfies the condition that if $\varphi : \overline{\Delta} \to M$ is a continuous map holomorphic in Δ such that $\varphi(\overline{\Delta}) \subset \overline{D}$ and $\varphi(\partial\Delta) \subset D$, then $\varphi(\overline{\Delta}) \subset D$.*

We say that D satisfies the strong disk property in M if it satisfies the condition that if $\varphi : \overline{\Delta} \to M$ is a continuous map holomorphic in Δ such that $\varphi(\partial\Delta) \subset D$, then $\varphi(\overline{\Delta}) \subset D$.

There exist the open sets which satisfy the disk property and do not satisfy the strong disk property (cf. [8])

Theorem 1.1 (Abe [8].) *If the open set D satisfies the disk property in M, then D is locally Stein in M.*

For the above theorem, the converse is not true. For example the Hartogs triangle $\{(z_1, z_2) \in \mathbf{C}^2 \mid |z_1| < |z_2| < 1\}$ is a Stein open set in $\mathbf{C}^2$ which does not have the disk property in $\mathbf{C}^2$.

The results [1] are as follows.

Theorem 1.2 (Abe - Furushima - Tsuji [1]) *Let M be a complex manifold and $\mathcal{F} \subset \mathcal{O}(M)$. Then both $B(\mathcal{F})$ and $E(\mathcal{F})$ satisfy the strong disk property in M.*

In the paper [1], we also proved that for any family $\mathcal{F}$ of holomorphic maps from a complex manifold M into a commutative complex Lie group L the equicontinuity domain $E(\mathcal{F})$ of $\mathcal{F}$ satisfies the strong disk property in M.

In the present paper, we prove that for a Stein manifold M, (M, Ω) is a Runge pair if and only if there exists a family of holomorphic functions $\mathcal{F}$ (resp. $\mathcal{F}'$) such that $\Omega = B(\mathcal{F})$ (resp. $\Omega = E(\mathcal{F}')$).

If (M, Ω) is Runge pair, then Ω has the strong disk property in M.

Therefore these are stricter results than those in the previous paper [1] when M is a Stein manifold M.

2. Main Results

Theorem 2.1 (Abe [9]) *Let M be a Stein manifold and $\mathcal{F} \subset \mathcal{O}(M)$. Then both $(M, B(\mathcal{F}))$ and $(M, E(\mathcal{F}))$ are Runge pairs.*

The outline of the proof is as follows.

First he proved that $(M, B(\mathcal{F}))$ is a Runge pair when M is a Stein open set in $\mathbf{C}^m$, by showing that for any compact set K of $B(\mathcal{F})$, $\hat{K}_M \cap B(\mathcal{F})$ is relatively compact in $B(\mathcal{F})$ where $\hat{K}_M$ is the convex hull with respect to holomorphic functions on M.

Let $z_1, \ldots, z_m$ be the coordinates of $\mathbf{C}^m$. Sato [5] proved that $E(\mathcal{F}) = \bigcap_{i=1}^m B(\partial\mathcal{F}/\partial z_i)$, where $\partial\mathcal{F}/\partial z_i := \{\partial f/\partial z_i \mid f \in \mathcal{F}\}$ for $i = 1, \ldots, m$.

Hence we have also $(M, E(\mathcal{F}))$ is a Runge pair.

Finally he proved the theorem, using the Remmert embedding theorem (cf. [10, p. 257]).

Theorem 2.2 *Let M be a Stein manifold and Ω be a Stein subset. If (M, Ω) is a Runge pair, then there exists $\mathcal{F} \subset \mathcal{O}(M)$ such that $\Omega = B(\mathcal{F})$.*

Proof. Let $\{K_n\}_{n\geq 1}$ be a sequence of compact domains such that $\cup_{n=1}^{\infty} K_n = \Omega$, $K_n \subset K_{n+1}$ and $\hat{K}_{n\Omega} = K_n$.

Let f be a holomorphic function on Ω such that for any boundary point a, there is a sequence of points $\{a_m\}_{m\geq 1} \subset \Omega$ which satisfies $|f(a_m)| \to \infty$ as $a_m \to a$.

Since (M, Ω) is a Runge pair, we have a sequence of holomorphic functions $\{f_n\}_{n\geq 1}$ on M such that $|f_n - f| < 1/n$ on K_n.

Let $\mathcal{H} = \{f_n : n \geq 1\}$. Then using the inequality $|f_n(x)| \leq |f_n(x) - f(x)| + |f(x)|$ on Ω, we have $\Omega \subset B(\mathcal{H})$ easily.

Also using the inequality $|f_n(a_m)| \geq |f(a_m)| - |f_n(a_m) - f(a_m)|$, we have $B(\mathcal{H}) \cap \partial\Omega = \phi$.

Let Ω_1 be the union of all the connected components of $B(\mathcal{H})$ except Ω and let $\{L_n\}_{n\geq 1}$ be a sequence of compact domains where $\cup_{n=1}^{\infty} L_n = \Omega_1 \cup \Omega$, $L_n \subset L_{n+1}$ and $\hat{L}_{n\Omega_1\cup\Omega} = L_n$.

For positive integer k, let h_k be the holomorphic function on $B(\mathcal{H})$ which takes the value 1 on Ω and the value k on Ω_1.

Since $(M, B(\mathcal{H}))$ is a Runge pair by Theorem 2.1, we have a sequence of holomorphic functions $\{h_{n,k}\}_{n\geq 1}$ on M such that $|h_{n,k} - h_k| < 1/n$ on L_n.

Thus by putting $\mathcal{F} = \{f_n, h_{m,k} : n, m, k \geq 1\}$ one has $\Omega = B(\mathcal{F})$.

Q.E.D

Theorem 2.3 *Let M be a Stein manifold and Ω be a Stein subset. If (M, Ω) is a Runge pair, then there exists $\mathcal{F} \subset \mathcal{O}(M)$ such that $\Omega = E(\mathcal{F})$.*

Proof. Let $\{K_n\}_{n\geq 1}$, f, $\{a_m\}_{m\geq 1}$, $\{f_n\}_{n\geq 1}$ and $\mathcal{H}$ be the same as those in Theorem 2.2.

Since for any positive number ϵ and any point x_0 of Ω there exists a neighborhood $U \subset\subset \Omega$ of x_0 such that $|f(x) - f(x_0)| < \epsilon$ for $x \in U$ and since the sequence $\{f_n\}$ uniformly approximate to f on $\bar{U}$, we have $|f_n(x) - f_n(x_0)| \leq |f_n(x) - f(x)| + |f(x) - f(x_0)| + |f(x_0) - f_n(x_0)| < 3\epsilon$ where $x \in U$ and n is sufficiently large.

Hence $\Omega \subset E(\mathcal{H})$.

Next we prove $E(\mathcal{H}) \cap \partial\Omega = \phi$.

Suppose the contrary. Then there exists a point $a \in \partial\Omega$ such that $a \in E(\mathcal{H})$. It follows that there exists a number m_0 such that for $m \geq m_0$ and $n \geq 1$, we have $|f_n(a) - f_n(a_m)| < 1$. Since $|f_n(a)| \leq |f_n(a) - f_n(a_{m_0})| + |f_n(a_{m_0}) - f(a_{m_0})| + |f(a_{m_0})|$, there exits n_0 such that $|f_n(a)| < 2 + |f(a_{m_0})|$ for $n \geq n_0$ and thus $|f_n(a_m)| \leq |f_n(a_m) - f_n(a)| + |f_n(a)| \leq 3 + |f(a_{m_0})|$ for $n \geq n_0$, $m \geq m_0$. This is a contradiction to the fact that $\{|f_n(a_m)|\}_{n,m\geq 1}$ is not uniformly bounded.

Let Ω_2 be the union of all the connected components of $E(\mathcal{H})$ except Ω and let $\{L'_n\}_{n\geq 1}$ be a sequence of compact domains where $\cup_{n=1}^{\infty} L'_n = \Omega_2 \cup \Omega$, $L'_n \subset L'_{n+1}$ and $\hat{L'}_{n\Omega_2\cup\Omega} = L'_n$.

By the Remmert embedding theorem we may assume that M is a closed complex submanifold of some $\mathbf{C}^N$ with the coordinates $z_1, \ldots, z_N$.

Let $\{h'_k\}_{k\geq 1}$ be the sequence of holomorphic functions on $B(\mathcal{H})$ which takes the value 1 on Ω and the value nz_1 on Ω_2.

Since $(M, B(\mathcal{H}))$ is a Runge pair by Theorem 2.1, we have a holomorphic function $\{h'_{n,k}\}_{n\geq 1}$ on M such that $|h'_{n,k} - h'_k| < 1/n$ on L'_n.

Thus by putting $\mathcal{F}' = \{f_n, h'_{n,k} : n \geq 1, k \geq 1\}$ we have $\Omega = E(\mathcal{F}')$.

Q.E.D

Acknowledgement

Research was supported by Grant-in-Aid for JSPS Research Fellow no. 10 0506 from the Ministry of Education, Science, Culture and Sports of Japan, 1998.

The author would like to express her hearty gratitude to Professor M. Abe for his helpful suggestions and his warm advices.

References

1. M. Abe, M. Furushima and M. Tsuji, in: Complex Variables Theory and Application, Gordon and Breach Science Publishers, (Netherlands, Amsterdam) (in press) (1998).
2. T.J. Barth, Math. Ann., **190**, 293–297 (1971).
3. H. Grauert and H. Reckziegel, Math. Z., **89**, 108–125 (1965).
4. H. Wu, Acta Math., **119**, 193–233 (1967).
5. S. Sato, Kumamoto J. Sci. Math., **9**, 75–86 (1973).
6. H. Cartan and P. Thullen, Math. Ann., **106**, 617–647 (1932).
7. M. Abe, Bull. Oshima Nat. College of Maritime Tech., **24**, 125–129 (1991).
8. M. Abe, Bull. Oshima Nat. College of Maritime Tech., **26**, 53–56 (1993).
9. M. Abe, personal communication (1998).
10. R.C. Gunning and H. Rossi, Analytic Functions of Several Complex Variables, Prentice-Hall, Englewood Cliffs (1965).
11. S.G. Krantz, Function Theory of Several Complex Variables, John Wiley, New York-Chichester-Brisbane-Toronto-Singapore (1982).
12. H.L. Royden, Proc. Amer. Math. Soc., **43**, 306–310 (1974).

9th Int. Coll. on Differential Equations, pp. 443–449
D. Bainov (Ed)

Riemann Conditions for Quasi-Abelian Varieties

TAKASHI UMENO
Department of Mathematics, Kyushu Sangyo University, 2-3-1, Matsukadai, Higashi-ku, Fukuoka, 813-8503, Japan
E-mail: umeno@ip.kyusan-u.ac.jp

Abstract: We shall show the conditions for a toroidal group to be a Quasi-Abelian variety in terms of the period matrix. These conditions include the Riemann conditions for an Abelian variety.
Key Words and Phrases: *Quasi-Abelian varieties, Toroidal groups, Riemann conditions.*

1. Introduction

Let Γ be a discrete subgroup of $\boldsymbol{C}^n$ generated by $\boldsymbol{R}$-lineary independent vectors $\{\lambda_1, \cdots, \lambda_{n+q}\}$ over $\boldsymbol{Z}$ $(1 \le q \le n)$. We say $\boldsymbol{C}^n/\Gamma$ is a toroidal group if $H^0(\boldsymbol{C}^n/\Gamma, \mathcal{O}) = \boldsymbol{C}$. A matrix $\Omega = [\lambda_1, \cdots, \lambda_{n+q}]$ is called a period matrix of $\boldsymbol{C}^n/\Gamma$. Put $\boldsymbol{R}_\Gamma :=$ the real vector subspace of $\boldsymbol{C}^n$ generated by Γ, and $\boldsymbol{C}_\Gamma := \boldsymbol{R}_\Gamma \cap \sqrt{-1}\boldsymbol{R}_\Gamma$, the maximal complex subspace of $\boldsymbol{R}_\Gamma$.

Definition. A toroidal group $\boldsymbol{C}^n/\Gamma$ is a *Quasi-Abelian variety* if there exists a Hermitian form H on $\boldsymbol{C}^n \times \boldsymbol{C}^n$ satisfying

(1) $H > 0$ on $\boldsymbol{C}_\Gamma \times \boldsymbol{C}_\Gamma$ and

(2) $\mathrm{Im}H|\Gamma \times \Gamma$ is a $\boldsymbol{Z}$-valued skew symmetric form.

On a Quasi-Abelian variety $\boldsymbol{C}^n/\Gamma$, we have a complex line bundle L such that $c_1(L) = E = \mathrm{Im}H|\Gamma \times \Gamma \in H^2(\boldsymbol{C}^n/\Gamma, \boldsymbol{Z})$ ([1]). Takayama [2] proved that a toroidal group $\boldsymbol{C}^n/\Gamma$ is a Quasi-Abelian variety if and only if there exists a positive line bundle on $\boldsymbol{C}^n/\Gamma$. Moreover, he proved the embedding theorem of a Quasi-Abelian variety $\boldsymbol{C}^n/\Gamma$ into a complex projective space.

In [3], we showed the conditions for $\boldsymbol{C}^n/\Gamma$ to be a Quasi-Abelian variety in terms of the period matrix. These conditions include the Riemann conditions I ([4]) in case of a complex torus. In this paper we shall prove the conditions for $\boldsymbol{C}^n/\Gamma$ to be a Quasi-Abelian variety, which generalize the Riemann conditions III ([4]), and describe the specified positive line bundles on $\boldsymbol{C}^n/\Gamma$.

2. Notations and Theorems

After a linear change of coordinates of $\boldsymbol{C}^n$, we see $\boldsymbol{C}^n/\Gamma$ has a period matrix of the form $\Omega = [I_n, V]$, where $I_n = [e_1, \cdots, e_n]$ is the (n, n) unit matrix and $V = [v_{ij}; 1 \le i \le n, 1 \le j \le q] = [v_1, \cdots, v_q]$ is a (n, q) matrix. Put $V_1 = [v_{ij}; 1 \le i, j \le q]$, and

$V_2 = [v_{ij}; q+1 \leq i \leq n, 1 \leq j \leq q]$. We may assume $\det V_1 \neq 0$. We put $v_i = \sqrt{-1}e_i$ for $q+1 \leq i \leq n$, $\beta_i = \operatorname{Im} v_i$ for $1 \leq i \leq n$ and $\beta = [\beta_{ij}] := [\beta_1, \cdots \beta_n]$. Then $\beta_1, \cdots \beta_n$ are lineary independent over $\boldsymbol{C}$. For any $z \in \boldsymbol{C}^n$, we define two coordinates $z_1, \cdots, z_n$ and $t_1, \cdots, t_{2n}$ by

$$(2.1) \qquad \begin{aligned} z &= z_1\beta_1 + \cdots + z_n\beta_n \\ &= t_1e_1 + \cdots + t_ne_n + t_{n+1}v_1 + \cdots + t_{2n}v_n . \end{aligned}$$

Then we have for $i = 1, \cdots, n$,

$$(2.2) \qquad t_i = \frac{1}{2\sqrt{-1}}\left(-\sum_{j=1}^{n} \overline{v}_{ij}z_j + \sum_{j=1}^{n} v_{ij}\overline{z}_j\right) \quad \text{and}$$

$$t_{n+i} = \frac{1}{2\sqrt{-1}}(z_i - \overline{z}_i).$$

For any $C^\infty d$-closed p-forms φ_1 and φ_2 on $\boldsymbol{C}^n/\Gamma$ we write $\varphi_1 \sim \varphi_2$ when φ_1 and φ_2 are d-cohomologous, namely, there is a $C^\infty(p-1)$-form ψ on $\boldsymbol{C}^n/\Gamma$ such that $\varphi_1 - \varphi_2 = d\psi$. Since t_{n+i} are global functions for $i = q+1, \cdots, n$ and from (2.2), we have

$$(2.3) \quad dt_i = \frac{1}{2\sqrt{-1}}\left(-\sum_{j=1}^{q} \overline{v}_{ij}dz_j + \sum_{j=1}^{q} v_{ij}d\overline{z}_j\right) \quad \text{for } i = 1, \cdots, q ,$$

$$dt_i \sim \frac{1}{2\sqrt{-1}}\left(-\sum_{j=1}^{q} \overline{v}_{ij}dz_j + 2\sqrt{-1}dz_i + \sum_{j=1}^{q} v_{ij}d\overline{z}_j\right) \text{ for } i = q+1, \cdots, n ,$$

$$dt_{n+i} = \frac{1}{2\sqrt{-1}}(dz_i - d\overline{z}_i) \quad \text{for } i = 1, \cdots, n, \text{ and in particular}$$

$$(2.4) \quad dz_i \sim d\overline{z}_i \quad \text{for } i = q+1, \cdots, n .$$

Let $T' := \boldsymbol{C}\left\{\frac{\partial}{\partial z_i}; i = 1, \cdots, n\right\}$ be the holomorphic tangent space of $\boldsymbol{C}^n/\Gamma$ at 0, $T_R := \boldsymbol{R}\left\{\frac{\partial}{\partial t_i}; i = 1, \cdots, 2n\right\}$ the real tangent space of $\boldsymbol{C}^n/\Gamma$ at 0, and $T_\Gamma := \boldsymbol{R}\left\{\frac{\partial}{\partial t_i}; i = 1, \cdots, n+q\right\}$.

For $\sigma = \sum_{i=1}^{n} \sigma_i \frac{\partial}{\partial z_i} \in T'$, we put $\hat{\sigma} := \sigma + \overline{\sigma} = \sum_{i=1}^{2n} s_i \frac{\partial}{\partial t_i} \in T_R$. Then the mapping $T' \ni \sigma \longmapsto \hat{\sigma} \in T_R$ is an $\boldsymbol{R}$-isomorhism. From (2.1) and (2.2), we have

$$\sigma_1\beta_1 + \cdots + \sigma_n\beta_n = s_1e_1 + \cdots + s_ne_n + s_{n+1}v_1 + \cdots + s_{2n}v_n.$$

Therefore, by the mappings

$$(2.5) \qquad \begin{aligned} &T' \ni \sigma \longmapsto {}^t(\sigma_1, \cdots, \sigma_n) \in \boldsymbol{C}^n \quad \text{and} \\ &T_R \ni \hat{\sigma} \longmapsto {}^t(s_1, \cdots, s_{2n}) \in \boldsymbol{R}^{2n} , \end{aligned}$$

we can identify T' with $\boldsymbol{C}^n$, T_R with $\boldsymbol{R}^{2n}$ and T_Γ with $\boldsymbol{R}_\Gamma$, respectively.

Then we have the following theorem I ([3]), which is a generalization of Riemann conditions I for a complex torus (cf. [4], p.303).

Theorem I. *Let $\boldsymbol{C}^n/\Gamma$ be a toroidal group, which has a period matrix of the form $\Omega = [I_n, V]$. Then $\boldsymbol{C}^n/\Gamma$ is a Quasi-Abelian variety if and only if there exists a $\boldsymbol{Z}$-valued skew-symmetric matrix $[E_{ij}] \in \boldsymbol{Z}^{(n+q)\times(n+q)}$ satisfying the following conditions:*

(1) ${}^tVE_1V + {}^tE_2V - {}^tVE_2 + E_3 = 0$ *and*

(2) $\frac{\sqrt{-1}}{2}({}^t\overline{V}E_1V + {}^tE_2V - {}^t\overline{V}E_2 + E_3) > 0,$

where $[E_{ij}] = \begin{bmatrix} E_1 & E_2 \\ -{}^tE_2 & E_3 \end{bmatrix}$, $E_1 \in Z^{n\times n}$, *and* $E_3 \in Z^{q\times q}$.

In the next section we shall prove the following Theorem II, which generalizes the Riemann conditions III (cf. [4], p.306).

Theorem II. *A toroidal group C^n/Γ with a period matrix Ω is a Quasi-Abelian variety if and only if there exist an integral basis $\lambda_1, \cdots, \lambda_{n+q}$ for Γ, a complex basis $e_1, \cdots, e_n$ for C^n and a Z-valued skew-symmetric matrix $F \in Z^{(n-q)\times(n-q)}$ satisfying the following conditions:*

(1) $\Omega = [\delta_1 e_1, \cdots, \delta_q e_q, e_{q+1}, \cdots, e_n, \lambda_{n+1}, \cdots, \lambda_{n+q}]$, *for positive integers δ_i* $(1 \le i \le q)$,

(2) $W_1 - {}^tW_1 = {}^tW_2FW_2$ *and*

(3) $\frac{\sqrt{-1}}{2}({}^t\overline{W}_1 - W_1 + {}^t\overline{W}_2FW_2) > 0,$

where $\begin{bmatrix} W_1 \\ W_2 \end{bmatrix} = [\lambda_{n+1}, \cdots, \lambda_{n+q}]$, *and $W_1 \in C^{q\times q}$ is a nonsingular matrix.*

3. Proof of Theorem II

Let C^n/Γ be a toroidal group with a period matrix of the form $\Omega = [I_n, V]$. We have the following results ([1], [3]), which we need for a proof of theorem I.

Proposition 3.1. *For any d-closed $C^\infty p$-form φ on C^n/Γ, there exist constant p-forms $\chi \in \bigwedge^p C\{dt_1, \cdots, dt_{n+q}\}$ and $\chi_C \in \bigwedge^p C\{dz_1, \cdots, dz_n, d\overline{z}_1, \cdots, d\overline{z}_q\}$ which are d-cohomologous to φ. Further these forms χ and χ_C are uniquely determined by φ.*

Proposition 3.2. *Let $E = \frac{1}{2}\sum_{1\le i,j\le n+q} E_{ij}dt_i \wedge dt_j \in H^2(C^n/\Gamma, Z)$ and $[E_{ij}] = \begin{bmatrix} E_1 & E_2 \\ -{}^tE_2 & E_3 \end{bmatrix}$, where $E_1 \in Z^{n\times n}$, and $E_3 \in Z^{q\times q}$. Then the following statements are equivalent.*

(1) ${}^tVE_1V + {}^tE_2V - {}^tVE_2 + E_3 = 0$

(2) *There exists a real $(1,1)$-form E_R such that E is d-cohomologous to E_R.*

Proposition 3.3. *For any constant 2-forms χ_1 and χ_2 on C^n/Γ, χ_1 is d-cohomologous to χ_2, if and only if $\chi_1 = \chi_2$ on $T_\Gamma \times T_\Gamma$.*

By (2.5), we identify any Z-valued skew-symmetric form E on $\Gamma\times\Gamma$ with a Z-valued 2-form $\frac{1}{2}\sum_{1\le i,j\le n+q} E_{ij}dt_i \wedge dt_j$, and these are elements of $H^2(C^n/\Gamma, Z)$.

To prove Theorem II, we need the following

Proposition 3.4. *Let C^n/Γ be a Quasi-Abelian variety. Then we have a Hermitian form H on $C^n \times C^n$ satisfying*

(1) $H > 0$ *on* $C^n \times C^n$ *and*

(2) $\mathrm{Im}H|\Gamma\times\Gamma$ *is a Z-valued skew-symmetric form.*

Proof Since C^n/Γ is a Quasi-Abelian variety, we have a Hermitian form H' on $C^n\times C^n$ satisfying $H' > 0$ on $C_\Gamma \times C_\Gamma$ and $E := \mathrm{Im}H'|\Gamma\times\Gamma$ is a Z-valued skew-symmetric form. Let $[G'_{ij}] := \frac{1}{2\sqrt{-1}}[H'(\beta_i, \beta_j)]$, $1 \le i, j \le n$, then $G' := \sum_{1\le i,j\le n} G'_{ij}dz_i \wedge dz_j$ is a real (1,1)-form and $G'|\Gamma \times \Gamma = \mathrm{Im}H|\Gamma \times \Gamma$, identifying T_Γ with $\boldsymbol{R}_\Gamma$ by (2.5). For any real symmetric matrix $[S_{ij}; q+1 \leq i,j \leq n]$, from (2.3) and (2.4),

$$\sqrt{-1} \sum_{q+1 \leq i,j \leq n} S_{ij} dz_i \wedge d\bar{z}_j = -2d\Big(\sum_{q+1 \leq i,j \leq n} S_{ij} t_{n+j} dz_i\Big).$$

Hence G' is d-cohomologous to

$$G := \sum_{1 \leq i,j \leq n} G_{ij} dz_i \wedge d\bar{z}_j = G' - \sqrt{-1} \sum_{q+1 \leq i,j \leq n} S_{ij} dz_i \wedge d\bar{z}_j.$$

If we choose $[S_{ij}]$ having sufficiently large positive eigenvalues, a Hermitian form H defined by $2\sqrt{-1}[G_{ij}]$ is positive definite on $\boldsymbol{C}^n \times \boldsymbol{C}^n$, and

$$\mathrm{Im}H|\Gamma \times \Gamma = G|\Gamma \times \Gamma = G'|\Gamma \times \Gamma$$

by Proposition 3.3. This proves the proposition.

Further we have the following

Proposition 3.5. *Let $\boldsymbol{C}^n/\Gamma$ be a Quasi-Abelian variety, which has a period matrix of the form $\Omega = [I_n, V]$, H a Hermitian form on $\boldsymbol{C}^n \times \boldsymbol{C}^n$ satisfying (1) and (2) in Proposition 3.4, and $E = \mathrm{Im}H|\Gamma \times \Gamma$. Then we have an integral base $\lambda_1, \cdots, \lambda_{n+q}$ for Γ such that*

(1) $E(\lambda_i, \lambda_{n+i}) = -\delta_i \ (1 \leq i \leq q)$,
$E(\lambda_i, \lambda_j) = 0 \ (1 \leq i \leq q,\ 1 \leq j \leq n+q,$ *and* $j \neq n+i)$, *and*
$E(\lambda_i, \lambda_{n+j}) = 0 \ (q+1 \leq i \leq n+q,\ 1 \leq j \leq q)$,

where δ_i are positive integers $(1 \leq i \leq q)$,

(2) $\lambda_1, \cdots, \lambda_n$ *are $\boldsymbol{C}$-linearly independent, and*

(3) $\lambda_{q+1}, \cdots, \lambda_{n+q}$ *are $\boldsymbol{C}$-linearly independent.*

Proof By Proposition 3.4, we have a real (1,1)-form $G = \sum_{1 \leq i,j \leq n} G_{ij} dz_i \wedge d\bar{z}_j$ such that $G|\Gamma \times \Gamma = E$ and $H = 2\sqrt{-1}[G_{ij}] > 0$ on $\boldsymbol{C}^n \times \boldsymbol{C}^n$. For any $z \neq 0 \in \boldsymbol{C}^n$, we have $H(z,z) = G(\sqrt{-1}\hat{z}, \hat{z}) > 0$. Hence for any $z \neq 0 \in \boldsymbol{C}_\Gamma \cap \Gamma$, we have $E(\sqrt{-1}\hat{z}, \hat{z}) = G(\sqrt{-1}\hat{z}, \hat{z}) > 0$, where $\hat{z} = z + \bar{z}$. We choose $\lambda_1,\ \lambda_{n+1} \in \Gamma$ satisfying

$$\delta_1 := E(\lambda_{n+1}, \lambda_1) = \min\{E(\lambda, \mu) > 0 \ ; \ \lambda, \mu \in \Gamma\}.$$

Then we have the following facts (cf. [4], p.305):

(3.1) λ_1, λ_{n+1} are $\boldsymbol{R}$-linearly independent,

(3.2) $\delta_1 | E(\lambda_1, \lambda)$ and $\delta_1 | E(\lambda_{n+1}, \lambda)$, for any $\lambda \in \Gamma$, and

(3.3) $\Gamma = \boldsymbol{Z}\{\lambda_1, \lambda_{n+1}\} \oplus \boldsymbol{Z}\{\lambda_1, \lambda_{n+1}\}^\perp \cap \Gamma$.

We can repeat this process to obtain $\lambda_1, \lambda_{n+1}, \cdots, \lambda_q, \lambda_{n+q} \in \Gamma$ with $E(\lambda_{n+i}, \lambda_i) = \delta_i$ for $1 \leq i \leq q$, and $E(\lambda_i, \lambda_j) = 0$ for $i \leq j$, and $j \neq n+i$,$(1 \leq i \leq q, n+1 \leq i \leq n+q, 1 \leq j \leq q, n+1 \leq j \leq n+q)$, where δ_i are positive integers. Then we can choose $\lambda_{q+1}, \cdots, \lambda_n \in \Gamma$ so that $\Gamma = \boldsymbol{Z}\{\lambda_1, \cdots, \lambda_{n+q}\}$ and $\lambda_{q+1}, \cdots, \lambda_n \in \boldsymbol{Z}\{\lambda_1, \cdots, \lambda_q, \lambda_{n+1}, \cdots, \lambda_{n+q}\}^\perp$. Since $H > 0$ on $\boldsymbol{C}^n \times \boldsymbol{C}^n$, we see $\lambda_1, \cdots, \lambda_q$ and $\lambda_{n+1}, \cdots, \lambda_{n+q}$ are $\boldsymbol{C}$-linearly independent, respectively. Further, we have

$$\lambda_{n+i} = \frac{G(\lambda_{n+i}, \sqrt{-1}\lambda_i) + \sqrt{-1}G(\lambda_i, \lambda_{n+i})}{G(\lambda_i, \sqrt{-1}\lambda_i)} \lambda_i,$$

for $1 \leq i \leq q$. Hence we have

$$C\{\lambda_1, \cdot\cdot, \lambda_q, \lambda_{n+1}, \cdot\cdot, \lambda_{n+q}\} \cong C^q.$$

On the other hand, since C^n/Γ is a toroidal group, we have

$$C\{\lambda_1, \cdots, \lambda_{n+q}\} \cong C^n.$$

Hence $\lambda_1, \cdots, \lambda_n$ and $\lambda_{q+1}, \cdots, \lambda_{n+q}$ are C-linearly independent, respectively, and the proposition is proved.

Now we shall prove Theorem II.

Proof of Theorem II Suppose C^n/Γ is a Quasi-Abelian variety. By Proposition 3.4, we have a positive definite Hermitian form H on $C^n \times C^n$, a real (1,1) form G and a Z-valued skew-symmetric form E such that

$$E = G|\Gamma \times \Gamma = \mathrm{Im}H|\Gamma \times \Gamma.$$

Then we have an integral basis $\lambda_1, \cdots, \lambda_{n+q}$ for Γ satisfying (1), (2), and (3) in Proposition 3.5. Put

$$\begin{aligned} &e_i := \frac{1}{\delta_i}\lambda_i \ (1 \leq i \leq q), \\ &e_i := \lambda_i \ (q+1 \leq i \leq q), \\ &\lambda_{n+i} := \sqrt{-1}e_i \ (q+1 \leq i \leq n), \text{ and} \\ &W := \begin{bmatrix} W_1 \\ W_2 \end{bmatrix} = [\lambda_{n+1}, \cdots, \lambda_{n+q}], \end{aligned}$$

where $W_1 \in C^{q\times q}$. Since $\lambda_1, \cdots, \lambda_n$ and $\lambda_{q+1}, \cdots, \lambda_{n+q}$ are C-linearly independent respectively, $e_1, \cdots, e_n$ form a complex basis for C^n and $\det W_1 \neq 0$. We put $\nu_i := \mathrm{Im}\, \lambda_{n+i}$, for $1 \leq i \leq n$. Set for any $w \in C^n$,

$$\begin{aligned} w &= w_1\nu_1 + \cdots + w_n\nu_n \\ &= s_1\lambda_1 + \cdots + s_{2n}\lambda_{2n}. \end{aligned} \tag{3.4}$$

We put

$$\begin{aligned} &W' := [\lambda_{n+1}, \cdots, \lambda_{2n}], \\ &\Delta_{\delta^{-1}} := [\delta_1^{-1}e_1, \cdot\cdot, \delta_q^{-1}e_q, e_{q+1}, \cdot\cdot, e_n], \quad \text{and} \\ &[w'_{ij}] := \Delta_{\delta^{-1}}W'. \end{aligned}$$

Then we get the following formulas similarly to (2.3) and (2.4)

$$ds_i = \frac{1}{2\sqrt{-1}}\Big(-\sum_{j=1}^{q}\overline{w'}_{ij}dz_j + \sum_{j=1}^{q}w'_{ij}d\overline{w}_j\Big) \quad \text{for } i = 1, \cdots, q\ , \tag{3.5}$$

$$ds_i \sim \frac{1}{2\sqrt{-1}}\Big(-\sum_{j=1}^{q}\overline{w'}_{ij}dw_j + 2\sqrt{-1}dw_i + \sum_{j=1}^{q}w'_{ij}d\overline{w}_j\Big), \text{ for } i = q+1, \cdots, n\ ,$$

$$ds_{n+i} = \frac{1}{2\sqrt{-1}}(dw_i - d\overline{w}_i) \quad \text{for } i = 1, \cdots, n, \text{ and}$$

$$dz_i \sim d\overline{z}_i \quad \text{for } i = q+1, \cdots, n\ . \tag{3.6}$$

Using coordinates (3.4) and by Proposition 3.5, we can write

$$E = -\sum_{i=1}^{q}\delta_i ds_i \wedge ds_{n+i} + \frac{1}{2}\sum_{q+1 \leq i,j \leq n} E_{ij}ds_i \wedge ds_j, \tag{3.7}$$

where $E_{ij} = E(\lambda_i, \lambda_j)$. We put $F = [E_{ij}, q+1 \leq i,j \leq n]$. By Proposition 3.3, E is d-cohomologous to the real (1,1)-form G. Substituting (3.5) and (3.6) to (3.7) , similarly to the proof of Theorem I([4]), we get (2) and (3) of Theorem II.

Conversely, suppose we have an integral basis $\lambda_1, \cdots, \lambda_{n+q}$ for Γ, a complex basis $e_1, \cdots, e_n$ for $\boldsymbol{C}^n$ and a $\boldsymbol{Z}$-valued skew-symmetric matrix $F \in \boldsymbol{Z}^{(n-q)\times(n-q)}$ satisfying (1), (2) and (3) in Theorem II. Put

$$(3.8) \qquad [E_{ij}] := \begin{bmatrix} E_1 & E_2 \\ -{}^tE_2 & E_3 \end{bmatrix}, \quad \text{where}$$

$$E_1 = \begin{bmatrix} O & O \\ O & F \end{bmatrix} \in \boldsymbol{Z}^{n\times n}, E_2 = [-\delta_1 e_1, \cdots, -\delta_q e_q] \in \boldsymbol{Z}^{n\times q}, \text{ and } E_3 = O \in \boldsymbol{Z}^{q\times q}.$$

By the condition (1) of the theorem, we have coordinates (3.4) and the formulas (3.5) and (3.6) hold. Let E be the $\boldsymbol{Z}$-valued skew-symmetric form defined by $[E_{ij}]$, then E is written as (3.7). Substituting (3.5) and (3.6) to (3.7), by the condition (2), we get a real (1,1)-form $G = \sum_{1\leq i,j\leq n} G_{ij} dw_i \wedge d\overline{w}_j$ which is d-cohomologous to E. Let H be the Hermitian form on $\boldsymbol{C}^n \times \boldsymbol{C}^n$ defined by G. Then by Proposition 3.3, $\mathrm{Im}H|\Gamma \times \Gamma = E$. Further by the condition (3), $H > 0$ on $\boldsymbol{C}_\Gamma \times \boldsymbol{C}_\Gamma$. Thus $\boldsymbol{C}^n/\Gamma$ is a Quasi-Abelian variety, and Theorm II is proved.

4. Positive Line Bundles

Let $\boldsymbol{C}^n/\Gamma$ be a Quasi-Abelian variety with an integral basis $\lambda_1, \cdots, \lambda_{n+q}$ for Γ and a complex basis $e_1, \cdots, e_n$ for $\boldsymbol{C}^n$ satisfying (1), (2) and (3) in Theorem II. Let E be a $\boldsymbol{Z}$-valued skew-symmetric form on $\Gamma \times \Gamma$ defined by $[E_{ij}]$ in (3.8). We extend E as a $\boldsymbol{R}$-valued skew-symmetric form on $\boldsymbol{R}_\Gamma \times \boldsymbol{R}_\Gamma$. In this section we use coordinates defined by $\lambda_1, \cdots, \lambda_{n+q}$ and $e_1, \cdots, e_n$ and write

$$(4.1) \qquad \begin{aligned} z &= z_1 e_1 + \cdots + z_n e_n \\ &= t_1 \lambda_1 + \cdots + t_{2n} \lambda_{2n}, \end{aligned}$$

where $\lambda_{n+i} = \sqrt{-1} e_i$ for $q+1 \leq i \leq n$.
Then we can write E as

$$(4.2) \quad E = -\sum_{i=1}^{q} \delta_i dt_i \wedge dt_{n+i} + \frac{1}{2} \sum_{q+1\leq i,j\leq n} F_{ij} dt_i \wedge dt_j.$$

Since E is d-cohomologous to a real (1,1)-form we have a complex line bundle L on $\boldsymbol{C}^n/\Gamma$ such that $c_1(L) = E$ ([1]). From the result of Takayama ([2]), we see L is positive. According to C.Vogt ([5]), we shall construct a theta bundle L_0 which is topologically equivalent to L as an application of Theorem II. Put

$$(4.3) \quad L_{\lambda_i}(z) = \frac{1}{2} \sum_{k=1}^{n} E(e_k, \lambda_i) z_k = 0, \quad \text{for} \quad 1 \leq i \leq q,$$

$$(4.4) \quad L_{\lambda_i}(z) = \frac{1}{2} \sum_{k=1}^{n} E(e_k, \lambda_i) z_k = \frac{1}{2} \sum_{k=q+1}^{n} f_{ki} z_k \quad \text{for } q+1 \leq i \leq n,$$

$$(4.5) \quad \begin{aligned} L_{\lambda_i}(z) &= \sum_{k=1}^{n} \frac{1}{\delta_k} (L_{\lambda_k}(\lambda_i) + E(\lambda_k, \lambda_i) z_k \\ &= -z_i + \frac{1}{2} \sum_{k,l=q+1}^{n} f_{lk} w_{li} z_k \quad \text{for} \quad n+1 \leq i \leq n+q, \end{aligned}$$

where $[w_{ij}; 1 \leq i \leq n, 1 \leq j \leq q] = [\lambda_{n+1}, \cdots, \lambda_{n+q}]$. For any $\lambda \in \{\lambda_1, \cdots, \lambda_{n+q}\}$, we put $e_\lambda(z) = \exp(2\pi\sqrt{-1}L_\lambda(z))$. Then we have the following

Proposition 4.1. *$\{e_\lambda(z)\}$ defines a complex line bundle L_0 on a Quasi-Abelian variety C^n/Γ, and $c_1(L_0) = E$.*

Proof To prove the above statement, we show that for any $\lambda, \mu \in \{\lambda_1, \cdots, \lambda_{n+q}\}$,

$$L_\lambda(z+\mu) + L_\mu(z) = L_\mu(z+\lambda) + L_\lambda(z) + E(\mu, \lambda).$$

This is equivalent to

$$L_\lambda(\mu) - L_\mu(\lambda) = E(\mu, \lambda).$$

To prove this, it suffices to show

$$L_{\lambda_{n+i}}(\lambda_{n+j}) - L_{\lambda_{n+j}}(\lambda_{n+i}) = E(\lambda_{n+j}, \lambda_{n+i}) \quad \text{for } 1 \leq i, j \leq q.$$

By the definition

$$\begin{aligned} L_{\lambda_{n+i}}(\lambda_{n+j}) - L_{\lambda_{n+j}}(\lambda_{n+i}) &= -w_{ij} + w_{ij} + \frac{1}{2} \sum_{k,l=q+1}^{n} f_{lk}(w_{li}w_{kj} - w_{lj}w_{ki}) \\ &= E(\lambda_{n+j}, \lambda_{n+i}), \end{aligned}$$

by the condition (2) of Theorem II. Thus the proposition is proved.

Acknowledgements

The author is supported in part by Grant-in-Aid for Scientific Research (c)(2) no.09640236 from the Ministry of Education, Science and Culture of Japan, 1998.

References

1. T.Umeno, Pusan Kyŏngnam Math. J., **9**, 295–311 (1993).

2. S.Takayama, (preprint).

3. T.Umeno, in: Proc. Fifth International Colloquium on Finite or Infinite Dimensional Complex Analysis, The Institute of Mathematics of Peking University (1997) pp. 323–326.

4. P.Griffiths and J.Harris, Principles of Algebraic Geometry, John Wiley & Sons (1978).

5. C.Vogt, J. Reine Angew. Math., **335**, 197–215 (1982).

9th Int. Coll. on Differential Equations, pp. 451–458
D. Bainov (Ed)

An Approximation of the Functional of a Free Boundary Problem

YOSHIHIKO YAMAURA
Department of Mathematics, College of Humanities and Sciences, Nihon University, Sakurajosui Tokyo 156-8550, Japan

Abstract: We construct a sequence of functionals which converges in the sense of Γ-convergence to the functional of the free boundary problem formulated by Alt and Caffarelli.
Key Words and Phrases: *Variational problem, Γ-convergence, free boundary problem, approximation.*

1. Introduction

Let Ω be a bounded Lipschitz domain in $\mathbf{R}^n$, where $n \geq 2$. For a non-negative and bounded function $\varphi \in W^{1,2}(\Omega)$, let $J : L^2(\Omega) \to [0, +\infty]$ be the functional defined by

$$J(v) = \begin{cases} \int_\Omega (|Dv|^2 + \chi(v))d\mathcal{L}^n & \text{if } v \in W^{1,2}(\Omega) \text{ and } v = \varphi \text{ on } \partial\Omega; \\ +\infty & \text{otherwise,} \end{cases}$$

where $\chi(t) = 1$ if $t > 0$; $= 0$ if $t \leq 0$. For each $\varepsilon \in (0, \frac{1}{10})$ we introduce the functional $J_\varepsilon : L^2(\Omega) \to [0, +\infty]$ as follows:

$$J_\varepsilon(v) = \begin{cases} \int_\Omega (|Dv|^2 + \chi_\varepsilon(v))d\mathcal{L}^n & \text{if } v \in C^\infty(\Omega) \cap W^{1,2}(\Omega) \text{ and } v = \varphi \text{ on } \partial\Omega; \\ +\infty & \text{otherwise,} \end{cases}$$

where

$$\chi_\varepsilon(t) = (F_\varepsilon)_{\frac{\varepsilon^3}{4}}(t) \equiv \frac{4}{\varepsilon^3} \int_R \eta \left(\frac{4(t-\tau)}{\varepsilon^3} \right) F_\varepsilon(\tau) d\tau \quad (t \in \mathbf{R}),$$

η is the 1-dimensional positive symmetric mollifier ([1, Page 147]), and

$$F_\varepsilon(t) = \begin{cases} 0 & \text{if } t < \frac{\varepsilon^3}{4}; \\ \frac{4t-\varepsilon^3}{4\varepsilon - 2\varepsilon^3} & \text{if } \frac{\varepsilon^3}{4} \leq t < \varepsilon - \frac{\varepsilon^3}{4}; \\ 1 & \text{if } t \geq \varepsilon - \frac{\varepsilon^3}{4}. \end{cases}$$

We state our conclusion of this paper.

Theorem 1 Let w_ε, $\varepsilon \in (0, \frac{1}{10})$, be functions in $L^2(\Omega)$ such that

$$\lim_{\varepsilon \to 0} [J_\varepsilon(w_\varepsilon) - \inf_{L^2(\Omega)} J_\varepsilon] = 0,$$

and let $\{w_{\varepsilon_j}\}_{j=1}^{\infty} \subset \{w_\varepsilon\}_{\varepsilon\in(0,\frac{1}{10})}$ be a subsequence satisfying $w_{\varepsilon_j} \to u$ in $L^2(\Omega)$ as $j \to \infty$. Then u is a minimizer of J in $L^2(\Omega)$, and $\lim_{j\to\infty} J_{\varepsilon_j}(w_{\varepsilon_j}) = J(u)$ holds.

The existence of such a subsequence as in Theorem 1 is guaranteed by Proposition 1 of Section 2. The variational problem "*Minimize the functional J in the class* $L^2(\Omega)$" is known as a free boundary problem formulated by Alt and Caffarelli ([2]). If we are able to determine numerically u_ε in Theorem 1, we obtain a numerical approximation of a minimizer of the free boundary problem. For this reason we consider that Theorem 1 gives us a motivation for the numerical experiment of the free boundary problem.

There exists a minimizer u_ε of J_ε in $L^2(\Omega)$ (see Remark 1 of Section 2), and so in particular it is possible to apply Theorem 1 to the minimizers u_ε, $\varepsilon \in (0, \frac{1}{10})$. This approximation is closely related to that in the paper [3] due to Berestycki, Caffarelli and Nirenberg. Because the minimizer u_ε satisfies the equation $\Delta u_\varepsilon = \frac{1}{2}\chi'_\varepsilon(u_\varepsilon)$ in Ω, which equation is of the type treated in their paper. Their results are applied to the investigation of a solution of a semilinear elliptic equation, while our goal is to approximate a minimizer of the free boundary problem. We stress here that it is impossible to find Euler-Lagrange equation which the minimizer should satisfy on the whole of Ω because of the singularity of χ.

The general theory of Γ-convergence of functionals has been introduced by E.De Giorgi ([4]) and applied to the functionals connected with physical problems ([5], [6], [7]), so that it has brought good results on the numerical analysis ([8], [9]). The idea of the function χ_ε in this paper was inspired by investigations due to S.Omata on the numerical analysis of the free boundary problem ([10], [11]).

2. On the Compactness in Theorem 1

In the sequel of this paper ε is a positive number less than $\frac{1}{10}$.

Proposition 1 Let w_ε be a functions satisfying $\lim_{\varepsilon\to 0}[J_\varepsilon(w_\varepsilon) - \inf_{L^2(\Omega)} J_\varepsilon] = 0$. Then there exist a subsequence $\{w_{\varepsilon_j}\}_{j=1}^{\infty} \subset \{w_\varepsilon\}$ and $u \in L^2(\Omega)$ such that

(i) $w_{\varepsilon_j} \to u$ in $L^2(\Omega)$ as $j \to \infty$;

(ii) $Dw_{\varepsilon_j} \to Du$ weakly in $L^2(\Omega)$ as $j \to \infty$.

In order to prove this we need the following which is proved by a variation of the proof of [12, Theorem 1.17]:

Lemma 1 Let $v \in W^{1,2}(\Omega)$. Then there exists a sequence $\{v_j\}_{j=1}^{\infty} \subset C^\infty(\Omega) \cap W^{1,2}(\Omega)$ such that $v_j \to v$ in $W^{1,2}(\Omega)$ as $j \to \infty$ and $v_j = v$ in $L^2(\partial\Omega)$ for $j = 1, 2, \cdots$.

The proof of Proposition 1 Set $\delta_\varepsilon = J_\varepsilon(w_\varepsilon) - \inf_{L^2(\Omega)} J_\varepsilon$. Then by the assumption $\lim_{\varepsilon\to 0}\delta_\varepsilon = 0$ holds. In particular from Lemma 1 it turns out that there exists $\tilde{\varphi} \in C^\infty(\Omega) \cap W^{1,2}(\Omega)$ such that $\tilde{\varphi} = \varphi$ on $\partial\Omega$, and so u_ε

$$J_\varepsilon(w_\varepsilon) \le J_\varepsilon(\tilde{\varphi}) + \delta_\varepsilon \le \int_\Omega |D\tilde{\varphi}|^2 d\mathcal{L}^n + \mathcal{L}^n(\Omega) + \delta_\varepsilon.$$

Hence we have $\sup_{\varepsilon\in(0,\frac{1}{10})} ||Dw_\varepsilon||_{L^2(\Omega)} < \infty$. By adopting Poincaré inequality to $w_\varepsilon - \varphi \in W_0^{1,2}(\Omega)$ we also have $\sup_{\varepsilon\in(0,\frac{1}{10})} ||w_\varepsilon||_{L^2(\Omega)} < \infty$. Thus by taking Rellich compactness theorem into accout there exists a subsequence $\{w_{\varepsilon_j}\}_{j=1}^{\infty}$ satisfying (i) and (ii).

Remark 1 We are able to show the existence of a minimizer of J_ε in $L^2(\Omega)$. In fact, a minimizer $v_\varepsilon \in W^{1,2}(\Omega)$ of the variational problem: "*Minimize* $\int_\Omega(|Dv|^2 + \chi_\varepsilon(v))d\mathcal{L}^n$ *in* $\{v \in W^{1,2}(\Omega) \,|\, v = \varphi$ *on* $\partial\Omega\}$" is the desired function. For verifying this, we have only to prove the fact that such v_ε belongs to $C^\infty(\Omega)$. By the minimality of v_ε it is shown that $0 \leq v_\varepsilon \leq \sup_{\partial\Omega}\varphi$ in Ω. Henceforth, since v_ε is a bounded weak solution of the partial differential equation $\Delta v_\varepsilon = \frac{1}{2}\chi'_\varepsilon(v_\varepsilon)$ in Ω, we are able to adopt the regularity theorem [13, Page 284, Theorem 6.4], so that we obtain the fact. In the following of this Remark we let u_ε be a minimizer of J_ε in $L^2(\Omega)$. The function β_ε defined by $\beta_\varepsilon(\tau) = \frac{1}{2}\chi'_\varepsilon(\tau)$, $\tau \in \mathbf{R}$, has its support in $[0, \varepsilon]$ and $0 \leq \beta_\varepsilon \leq \frac{1}{\varepsilon}$ in $\mathbf{R}$ (see Lemma 2 of Section 3), and u_ε satisfies $\Delta u_\varepsilon = \beta_\varepsilon(u_\varepsilon)$ in Ω. Such a circumstance enables us to demonstrate in the same way as the proofs in [3, Section 2,3] the following: For any subdomain $\tilde{\Omega} \subset\subset \Omega$ there exists a positive constant C, depending on $n, \mathrm{dist}(\tilde{\Omega}, \partial\Omega)$ and $\sup_{\partial\Omega}\varphi$, such that $\sup_{\tilde{\Omega}}|Du_\varepsilon| < C$ for any $\varepsilon < \min(\frac{1}{10}, \frac{1}{8}\mathrm{dist}(\tilde{\Omega}, \partial\Omega))$. We thus obtain a subsequence $\{u_{\varepsilon_j}\}_{j=1}^\infty \subset \{u_\varepsilon\}_{\varepsilon\in(0,\frac{1}{10})}$ and $u \in L^2(\Omega)$ satisfying the following convergency in addition to (i) and (ii) of Proposition 1 with w_{ε_j} replaced by u_{ε_j}, if necessary, by taking a subsequence furthermore:

(iii) $u_{\varepsilon_j} \to u$ locally uniformly in Ω as $j \to \infty$;

3. The Proof of Theorem 1

Theorem 1 is a direct conclusion of the following:

Theorem 2 $\lim\limits_{\varepsilon\to 0} J_\varepsilon = J$ in the sense of $\Gamma(L^2(\Omega))$-convergence, i.e. both the following (I) and (II) hold:

(I) For any $v \in L^2(\Omega)$ and $\{v_\varepsilon\}_{\varepsilon\in(0,\frac{1}{10})} \subset L^2(\Omega)$ which satisfies $\lim\limits_{\varepsilon\to 0} v_\varepsilon = v$ in $L^2(\Omega)$, $J(v) \leq \underline{\lim}_{\varepsilon\to 0} J_\varepsilon(v_\varepsilon)$;

(II) For any $v \in L^2(\Omega)$, there exists $\{v_\varepsilon\}_{\varepsilon\in(0,\frac{1}{10})} \subset L^2(\Omega)$ such that $\lim\limits_{\varepsilon\to 0} v_\varepsilon = v$ in $L^2(\Omega)$ and $\overline{\lim}_{\varepsilon\to 0} J_\varepsilon(v_\varepsilon) \leq J(v)$.

The proof of Theorem 1 Set $\delta_j = J_\varepsilon(w_{\varepsilon_j}) - \inf\limits_{L^2(\Omega)} J_\varepsilon$. Then by the assumption $\lim\limits_{j\to\infty}\delta_j = 0$. Let us take $v \in L^2(\Omega)$ arbitrarily. Owing to (II) of Theorem 2 there exists $\{v_j\}_{j=1}^\infty \subset L^2(\Omega)$ such that $v_j \to v$ in $L^2(\Omega)$ as $j \to \infty$ and

$$\overline{\lim_{j\to\infty}}\, J_{\varepsilon_j}(v_j) \leq J(v). \tag{1}$$

On the other hand, by (I) of Theorem 2

$$J(u) \leq \underline{\lim}_{j\to\infty}\, J_{\varepsilon_j}(w_{\varepsilon_j}). \tag{2}$$

Letting $j \to \infty$ in the inequality $J_{\varepsilon_j}(w_{\varepsilon_j}) \leq J_{\varepsilon_j}(v_j) + \delta_j$, we conclude with the help of (1) and (2) that $J(u) \leq J(v)$. This proves of the first assertion, and the second one is also obtained by repeating the above argument with v replaced by u. □

4. The Proof of Theorem 2

In the following, for a function v on Ω, $t \in \mathbf{R}$ and a set $A \subset \Omega$ we denote by $A(v > t)$ the set $\{x \in A \mid v(x) > t\}$, and other notations such as $A(v \geq t)$, $A(v < t)$ are defined analogously. The following Lemma 2 and 3 are needed in the proof of Theorem 2:

Lemma 2 χ_ε is infinitely differentiable on $\mathbf{R}$; $\chi_\varepsilon = 0$ in $(-\infty, 0], = 1$ in $[\varepsilon, \infty)$; $0 \leq \chi_\varepsilon \leq 1$ in $\mathbf{R}$; $\chi_\varepsilon > 0$ in $(0, \infty)$; $\chi_\varepsilon = F_\varepsilon$ in $[\frac{\varepsilon^3}{2}, \varepsilon - \frac{\varepsilon^3}{2}]$; spt $\chi'_\varepsilon \subset [0, \varepsilon]$; $\chi'_\varepsilon \geq 0$ in $\mathbf{R}$; $|\chi'_\varepsilon| \leq \frac{2}{\varepsilon}$ in $\mathbf{R}$.

Proof The properties of the mollifier enable us to show the assertions. □

Lemma 3 Let $\{\varepsilon_j\}_{j=1}^\infty \subset (0, \frac{1}{10})$ be a sequence such that $\varepsilon_j \to 0$ as $j \to \infty$ and $\{v_j\}_{j=1}^\infty \subset L^2(\Omega)$ satisfy $v_j \to v$ in $L^2(\Omega)$ as $j \to \infty$. Then for any measurable subset E of Ω

$$\int_E \chi(v) d\mathcal{L}^n \leq \varliminf_{j\to\infty} \int_E \chi_{\varepsilon_j}(v_j) d\mathcal{L}^n.$$

Proof Since for $j \in \mathbf{N}$

$$\int_E \chi_{\varepsilon_j}(v_j) d\mathcal{L}^n \geq \int_{E(v>0)} \chi_{\varepsilon_j}(v_j) d\mathcal{L}^n \geq \int_E \chi(v) d\mathcal{L}^n - \int_{E(v>0)} \left|\chi_{\varepsilon_j}(v_j) - \chi(v)\right| d\mathcal{L}^n,$$

it suffices to show

$$\lim_{j\to\infty} \int_{E(v>0)} \left|\chi_{\varepsilon_j}(v_j) - \chi(v)\right| d\mathcal{L}^n = 0. \tag{3}$$

It holds that

$$\begin{aligned} &\int_{E(v>0)} \left|\chi_{\varepsilon_j}(v_j) - \chi(v)\right| d\mathcal{L}^n \\ &= \int_{E(v>0)\cap E(v_j<\varepsilon_j)} \left|\chi_{\varepsilon_j}(v_j) - \chi(v)\right| d\mathcal{L}^n + \int_{E(v>0)\cap E(v_j\geq\varepsilon_j)} \left|\chi_{\varepsilon_j}(v_j) - \chi(v)\right| d\mathcal{L}^n \\ &\leq \mathcal{L}^n(E(v > 0) \cap E(v_j < \varepsilon_j)), \end{aligned} \tag{4}$$

where we use the fact that $\chi_{\varepsilon_j}(v_j) - \chi(v) = 1 - 1 = 0$ on $E(v > 0) \cap E(v_j \geq \varepsilon_j)$ (see Lemma 2). Let δ be an arbitrary positive number. By the continuity of the Lebesgue measure there exists a positive number β such that

$$\mathcal{L}^n(E(0 < v < 2\beta)) < \frac{\delta}{2}. \tag{5}$$

Since $v_j \to v$ in $L^2(\Omega)$ as $j \to \infty$, for β as in (5) there exists $j_0 \in \mathbf{N}$ such that

$$\mathcal{L}^n(E(|v_j - v| \geq \beta)) < \frac{\delta}{2} \quad \text{for } j > j_0. \tag{6}$$

Choosing here $j_1 \in \mathbf{N}$ such that both $j_0 < j_1$ and $\varepsilon_j < \beta$ for $j > j_1$ are valid, we deduce from (5) and (6) that $j > j_1$ implies

$$\mathcal{L}^n(E(v > 0) \cap E(v_j < \varepsilon_j)) \leq \mathcal{L}^n(E(0 < v < 2\beta)) + \mathcal{L}^n(E(|v_j - v| \geq \beta)) < \delta.$$

Combining this with (4), we achieve (3). □

The proof of Theorem 2 (I) We may assume $\varliminf_{\varepsilon\to 0} J_\varepsilon(v_\varepsilon) < \infty$. Choose a subsequence $\{v_{\varepsilon_j}\}_{j=1}^\infty \subset \{v_\varepsilon\}_{\varepsilon\in(0,\frac{1}{10})}$ such that $\lim_{j\to\infty} J_{\varepsilon_j}(v_{\varepsilon_j}) = \varliminf_{\varepsilon\to 0} J_\varepsilon(v_\varepsilon)$. In particular $v_{\varepsilon_j} \in W^{1,2}(\Omega)$

and $v_{\varepsilon_j} = \varphi$ on $\partial\Omega$ for each $j \in \mathbf{N}$. Then $\sup_{j\in N} \int_\Omega |Dv_{\varepsilon_j}|^2 d\mathcal{L}^n < \infty$, and so $v \in W^{1,2}(\Omega)$ and

$$v_{\varepsilon_j} \to v \quad \text{in } L^2(\Omega) \quad \text{as } j \to \infty; \tag{7}$$

$$Dv_{\varepsilon_j} \to Dv \quad \text{weakly in } L^2(\Omega) \quad \text{as } j \to \infty, \tag{8}$$

and moreover,

$$\int_\Omega |Dv|^2 d\mathcal{L}^n \leq \varliminf_{j\to\infty} \int_\Omega |Dv_{\varepsilon_j}|^2 d\mathcal{L}^n. \tag{9}$$

By applying Lemma 3 with E replaced by Ω

$$\int_\Omega \chi(v) d\mathcal{L}^n \leq \varliminf_{j\to\infty} \int_\Omega \chi_{\varepsilon_j}(v_{\varepsilon_j}) d\mathcal{L}^n. \tag{10}$$

Since from (7) and (8) it holds that $v = \varphi$ on $\partial\Omega$, we are able to verify from (9) and (10) that $J(v) \leq \lim_{j\to\infty} J_{\varepsilon_j}(v_{\varepsilon_j}) = \varliminf_{\varepsilon\to 0} J_\varepsilon(v_\varepsilon)$.

(II) We may assume $J(v) < \infty$. Then in particular $v \in W^{1,2}(\Omega)$ and $v = \varphi$ on $\partial\Omega$. Take a sequence $\{v_j\}_{j=1}^\infty \subset C^\infty(\Omega) \cap W^{1,2}(\Omega)$ as in Lemma 1. Set $\delta_1 = \frac{1}{10}$ and take $\{\delta_k\}_{k=1}^\infty \subset \mathbf{R}^+$ such that $\lim_{k\to\infty} \delta_k = 0$ and $\delta_1 > \delta_2 > \delta_3 > \cdots$. Since $v_j \to v$ in $L^2(\Omega)$ as $j \to \infty$, $\lim_{j\to\infty} \mathcal{L}^n(\Omega(|v_j - v| \geq \delta_1^3)) = 0$. Therefore there exists $j_1 \in \mathbf{N}$ such that $\mathcal{L}^n(\Omega(|v_{j_1} - v| \geq \delta_1^3)) \leq \delta_1$. By induction we define $\{v_{j_k}\}_{k=1}^\infty \subset \{v_j\}_{j=1}^\infty$ such that $j_k > j_{k-1}$ and $\mathcal{L}^n(\Omega(|v_{j_k} - v| \geq \delta_k^3)) \leq \delta_k$ holds for $k = 2, 3, \cdots$, and, for simplicity, we denote the subsequence by $\{v_k\}_{k=1}^\infty$. We define v_ε, $\varepsilon \in (0, \frac{1}{10}]$, as follows:

$$v_\varepsilon = v_k \quad \text{when } \varepsilon \in (\delta_k, \delta_{k-1}] \quad k = 2, 3, \cdots.$$

Then, since $\lim_{\varepsilon\downarrow 0} \int_\Omega |Dv_\varepsilon|^2 d\mathcal{L}^n = \int_\Omega |Dv|^2 d\mathcal{L}^n$ holds, we have only to demonstrate

$$\varlimsup_{\varepsilon\downarrow 0} \int_\Omega \chi_\varepsilon(v_\varepsilon) d\mathcal{L}^n \leq \int_\Omega \chi(v) d\mathcal{L}^n. \tag{11}$$

If we denote $A_k = \Omega(|v_k - v| \geq \delta_k^3)$, it holds that $\mathcal{L}^n(A_k) \leq \delta_k$ for $k = 1, 2, \cdots$. When $\varepsilon \in (\delta_k, \delta_{k-1}]$,

$$\begin{aligned}\int_\Omega \chi_\varepsilon(v_\varepsilon) d\mathcal{L}^n &= \int_{\Omega(v>0)} \chi_\varepsilon(v_\varepsilon) d\mathcal{L}^n + \int_{\Omega(v\leq 0)} \chi_\varepsilon(v_\varepsilon) d\mathcal{L}^n \\ &\leq \int_\Omega \chi(v) d\mathcal{L}^n + \int_{\Omega(v\leq 0)\cap A_k} \chi_\varepsilon(v_\varepsilon) d\mathcal{L}^n + \int_{\Omega(v\leq 0)\setminus A_k} \chi_\varepsilon(v_\varepsilon) d\mathcal{L}^n \\ &\leq \int_\Omega \chi(v) d\mathcal{L}^n + \delta_k + \int_{\Omega(v\leq 0)\setminus A_k} \chi_\varepsilon(v_k) d\mathcal{L}^n.\end{aligned} \tag{12}$$

If $x \in \Omega(v \leq 0) \setminus A_k$, then $v(x) \leq 0$ and $|v_k(x) - v(x)| < \delta_k^3$, and so $v_k(x) < \delta_k^3$. Therefore Since $\chi_\varepsilon(v_k(x)) \leq \chi_\varepsilon(\delta_k^3)$. Since $\delta_k^3 < \delta_k - \frac{\delta_k^3}{2} < \varepsilon - \frac{\varepsilon^3}{2}$, by recalling that, as stated in Lemma 2, χ_ε is non-decreasing and $\chi_\varepsilon = F_\varepsilon$ in $[\frac{\varepsilon^3}{2}, \varepsilon - \frac{\varepsilon^3}{2}]$,

$$\begin{aligned}\chi_\varepsilon(v_k(x)) \leq{}& \max(F_\varepsilon(\tfrac{\varepsilon^3}{2}), F_\varepsilon(\delta_k^3)) = \\ ={}& \max\left(\frac{\varepsilon^2}{4 - 2\varepsilon^2}, \frac{4\delta_k^3 - \varepsilon^3}{4\varepsilon - 2\varepsilon^3}\right) \leq \max\left(\frac{\delta_{k-1}^2}{4 - 2\delta_{k-1}^2}, \frac{3\delta_k^2}{4 - 2\delta_k^2}\right).\end{aligned} \tag{13}$$

If we denote by γ_k the last side of (13), we conclude from (12) that

$$\int_\Omega \chi_\varepsilon(v_\varepsilon) d\mathcal{L}^n \leq \int_\Omega \chi(v) d\mathcal{L}^n + \delta_k + \gamma_k \mathcal{L}^n(\Omega) \quad \text{for } \varepsilon \in (\delta_k, \delta_{k-1}], \quad k = 2, 3, \cdots. \tag{14}$$

Since $\lim_{k\to\infty} \gamma_k = 0$, we accomplish (11) by letting $\varepsilon \downarrow 0$ in (14). □

5. An Explicit Estimate of Convergency of Minimizers

Let $\{u_{\varepsilon_j}\}_{j=1}^{\infty}$ be a sequence constructed by applying Theorem 1 to u_ε, $\varepsilon \in (0, \frac{1}{10})$, a minimizer of J_ε in $L^2(\Omega)$ (see Remark 1). Then we may assume (i) and (ii) of Proposition 1 and (iii) of Remark 1 hold for u_{ε_j}. However, we give in this section an estimation of the value $|u_{\varepsilon_j}|$ on a subset of $\Omega(u = 0)$, independently of the convergency (iii) of Remark 1. For this purpose we remark the following lemma which is a direct conclusion of Theorem 1.

Lemma 4 Let $\{u_{\varepsilon_j}\}_{j=1}^{\infty}$ be as above and u be its limit in $L^2(\Omega)$. Then for any measurable set $E \subset \Omega$

$$\lim_{j\to\infty} \int_E \chi_{\varepsilon_j}(u_{\varepsilon_j}) d\mathcal{L}^n = \int_E \chi(u) d\mathcal{L}^n.$$

Proof As noted above (ii) of Proposition 1 holds, and so

$$\int_\Omega |Du|^2 dx \le \varliminf_{j\to\infty} \int_\Omega |Du_{\varepsilon_j}|^2 dx.$$

Therefore we infer from Theorem 1 that

$$\int_\Omega (|Du|^2 + \chi(u)) d\mathcal{L}^n \ge \int_\Omega |Du|^2 d\mathcal{L}^n + \varlimsup_{j\to\infty} \int_\Omega \chi_{\varepsilon_j}(u_{\varepsilon_j}) d\mathcal{L}^n.$$

We thus have

$$\varlimsup_{j\to\infty} \int_\Omega \chi_{\varepsilon_j}(u_{\varepsilon_j}) d\mathcal{L}^n \le \int_\Omega \chi(u) d\mathcal{L}^n.$$

Recalling the inequality of Lemma 3 with E replaced by Ω, we conclude

$$\lim_{j\to\infty} \int_\Omega \chi_{\varepsilon_j}(u_{\varepsilon_j}) d\mathcal{L}^n = \int_\Omega \chi(u) d\mathcal{L}^n. \tag{15}$$

Let E be a measurable set. For proving the assertion of Lemma we have only to show

$$\varlimsup_{j\to\infty} \int_E \chi_{\varepsilon_j}(u_{\varepsilon_j}) d\mathcal{L}^n \le \int_E \chi(u) d\mathcal{L}^n \tag{16}$$

owing to Lemma 3. However, (16) is able to be derived from (15) and the inequality given by adopting Lemma 3 with E replaced by $\Omega \setminus E$. □

Applying Lemma 4 to the set $\Omega(u = 0)$, we in particular have

$$\lim_{j\to\infty} \int_{\Omega(u=0)} \chi_{\varepsilon_j}(u_{\varepsilon_j}) d\mathcal{L}^n = 0. \tag{17}$$

By applying the following Theorem 3 to the subsequence $\{u_{\varepsilon_j}\}_{j=1}^{\infty}$ as in (17), we are able to show that the value of $|u_{\varepsilon_j}|$ on a subset of $\Omega(u = 0)$ is estimated by a constant which depends on the convergency (17) and ε_j.

Theorem 3 Let $u_\varepsilon \in L^2(\Omega)$ be a minimizer of the functional J_ε. Then if we set

$$\delta(\varepsilon) = \int_{\Omega(u=0)} \chi_\varepsilon(u_\varepsilon) d\mathcal{L}^n,$$

there exists a positive constant C, depending only on n, such that the estimate

$$0 \le u_\varepsilon(x) \le C(n)(\varepsilon + \delta(\varepsilon)^{\frac{1}{n}})$$

holds for $x \in N_\rho \equiv \{x \in \Omega \,|\, \overline{B}_\rho(x) \subset \Omega(u = 0)\}$ provided N_ρ is non-empty, where $\rho = (\frac{2}{\omega_n}\delta(\varepsilon))^{\frac{1}{n}}$, and ω_n is the volume of the n-dimensional unit ball.

Proof As stated in Remark 1 of Section 2, the first inequality holds. Let us show the second one. Let $B_r(x_0) \subset\subset \Omega$, and, for notational simplicity, we put $x_0 = 0$. Let $v_\varepsilon \in W^{1,2}(B_r(0))$ be a solution of the generalized Dirichlet problem

$$\begin{cases} \int_{B_r(0)} Dv_\varepsilon D\zeta d\mathcal{L}^n = 0 \quad \text{for any } \zeta \in C_0^1(B_r(0)); \\ v_\varepsilon - u_\varepsilon \in W_0^{1,2}(B_r(0)). \end{cases}$$

We here remark that since $u_\varepsilon \in C^\infty(\overline{B}_r(0))$, it holds that $v_\varepsilon \in C^\infty(\overline{B}_r(0))$ owing to [1, Theorem 8.14]. For each $\xi \in \partial B_r(0)$, we set $\omega_\xi = \frac{\xi}{r}$, and $R_\xi = \{\tau \in [\frac{r}{8}, r] \,|\, u_\varepsilon(\tau\omega_\xi) \le \varepsilon\}$. Define

$$r_\xi = \begin{cases} \min R_\xi & \text{if } R_\xi \ne \phi; \\ r & \text{if } R_\xi = \phi. \end{cases}$$

Then for each $\xi \in \partial B_r(0)$ with $R_\xi \ne \phi$

$$\int_{r_\xi}^{r} \frac{d}{d\tau}[(u_\varepsilon - v_\varepsilon)(\tau\omega_\xi)]d\tau = -u_\varepsilon(r_\xi\omega_\xi) + v_\varepsilon(r_\xi\omega_\xi) \ge -\varepsilon + v_\varepsilon(r_\xi\omega_\xi). \tag{18}$$

Recalling Poisson integral formula [1, (2.26)] we have the estimate

$$v_\varepsilon(x) \ge \left(\frac{1}{r} \fint_{\partial B_r(0)} u_\varepsilon d\mathcal{H}^{n-1}\right)(r - |x|).$$

Hence from (18)

$$\left(\frac{1}{r} \fint_{\partial B_r(0)} u_\varepsilon d\mathcal{H}^{n-1}\right)(r - r_\xi) - \varepsilon \le \sqrt{r}\left(\int_{r_\xi}^{r} |D(u_\varepsilon - v_\varepsilon)(\tau\omega_\xi)|^2 d\tau\right)^{\frac{1}{2}}. \tag{19}$$

The estimate (19) remains valid for $\xi \in \partial B_r(0)$ with $R_\xi = \phi$, and therefore we are able to integrate with respect to ξ on $\partial B_r(0)$, so that

$$\left(\frac{1}{r} \fint_{\partial B_r(0)} u_\varepsilon d\mathcal{H}^{n-1}\right)\left|(B_r(0) \setminus B_{\frac{r}{8}}(0))(u_\varepsilon \le \varepsilon)\right| - (n\omega_n r^{n-1})\varepsilon \le$$

$$\le \sqrt{r}(n\omega_n r^{n-1})^{\frac{1}{2}} \times \left(\int_{B_r(0)} |D(u_\varepsilon - v_\varepsilon)|^2 d\mathcal{L}^n\right)^{\frac{1}{2}}.$$

The technical calculation used in the proof of [2, Lemma 3.2] enables us to lead

$$\left(\frac{1}{r} \fint_{\partial B_r(0)} u_\varepsilon d\mathcal{H}^{n-1}\right)\left|B_r(0)(u_\varepsilon \le \varepsilon)\right| \le$$

$$\le C_1 r^{n-1}\varepsilon + C_2 r^{\frac{n}{2}}\left(\int_{B_r(0)} |D(u_\varepsilon - v_\varepsilon)|^2 d\mathcal{L}^n\right)^{\frac{1}{2}},$$

where and in the following C_j, $j = 1, 2, 3$, is a positive constant depending only on n. Since u_ε is subharmonic (see Remark 1 and [1, Theorem 2.1]),

$$\frac{1}{r}u_\varepsilon(0) \times \mathcal{L}^n(B_r(0)(u_\varepsilon \le \varepsilon)) \le C_1 r^{n-1}\varepsilon + C_2 r^{\frac{n}{2}}\left(\int_{B_r(0)} |D(u_\varepsilon - v_\varepsilon)|^2 d\mathcal{L}^n\right)^{\frac{1}{2}}. \tag{20}$$

From the fact that v_ε is harmonic and by the minimality of u_ε we deduce that

$$\int_{B_r(0)} |D(u_\varepsilon - v_\varepsilon)|^2 d\mathcal{L}^n \le \int_{B_r(0)} (\chi_\varepsilon(v_\varepsilon) - \chi_\varepsilon(u_\varepsilon)) d\mathcal{L}^n \le \mathcal{L}^n(B_r(0)).$$

Combining this with (20), we finally obtain

$$u_\varepsilon(x_0) \times \mathcal{L}^n(B_r(x_0)(u_\varepsilon \le \varepsilon)) \le C_1 r^n \varepsilon + C_3 r^{n+1}. \tag{21}$$

Set ρ as in the statement, and suppose that $x \in N_\rho$. Then by the assumption

$$\mathcal{L}^n(B_\rho(x)(u_\varepsilon > \varepsilon)) \le \int_{B_\rho(x)} \chi_\varepsilon(u_\varepsilon) d\mathcal{L}^n \le \delta(\varepsilon) = \frac{1}{2}\mathcal{L}^n(B_\rho(x))$$

and hence by applying (21) with $B_r(x_0)$ replaced by $B_\rho(x)$ we accomplish the desired estimate. □

References

1. D. Gilbarg and N.S.Trudinger, Elliptic Partial Differential Equations of Second Order, 2nd edition, Grund. Math. Wiss., **224**, Springer-Verlag, Berlin-Heidelberg-New York-Tokyo (1983).

2. H.W. Alt and L.A. Caffarelli, J. Reine Angew. Math., **325**, 105-144 (1981).

3. H. Belstycki, L.A. Caffarelli and L.Nirenberg, Publications du Laboratorie D'Analyse Numerique, R89022, 1-53 (1989).

4. G. Dal Maso, An Introduction to Γ-Convergence, Progress in Nonlinear Differential Equations and Their Applications, PNLDE 8, Birkhäuser, Boston-Basel-Berlin (1993).

5. L. Ambrosio and V.M. Tortorelli, Bollettino U.M.I.(7), **6-B**, 105-123 (1992).

6. L. Ambrosio and V.M. Tortorelli, Comm. Pure Appl. Math., **XLIII**, 999-1036 (1990).

7. L. Modica, Arch. Rat. Mech. Anal., **98**, 123-142 (1987).

8. S. Baldo and S.-G. Bellettini, Richerche Math., **XL**, 33-64 (1991).

9. R. March, Image and Vision Computing, **10**, 30-38 (1992).

10. S. Omata, Theoretical Appl. Mech., **45**, 189-194 (1996).

11. S. Omata, Nonlinear Analysis, **30**, 2181-2187 (1997).

12. E. Giusti, Minimal Surfaces and Functions of Bounded Variation, Monographs Math., **80**, Birkhäuser, Boston-Basel-Stuttgart (1984).

13. O.A. Ladyzhenskaya and N.N. Ural'tseva, Linear and Quasilinear Elliptic Equations, Academic Press, New York-London-Toronto-Sydney-San Francisco (1968).

9th Int. Coll. on Differential Equations, pp. 459-466
D. Bainov (Ed)

A Mathematical Expression of Vertical Distribution of Annual Ring Widths of Trees Based on Height Growth and Inflow of Solute Carbohydrate from Roots

TAKAHISA YOSHIDA
University Forest in Aichi, The University of Tokyo, Goizuka 11-44, Seto, Aichi 489-0031, Japan

Abstract: A height growth is incorporated into the linear growth process, and the resultant equation is transformed into the random walk. The lower part of a crown is fixed at a constant height. The upper part of the crown is assumed to grow simultaneously with the height growth so that both the apex of the stem and the uppermost point of the crown remain the same throughout the growth period. Also, the solute carbohydrate is assumed to flow into the phloem of the stem only from the roots. A series of these procedures are applied to the formation of the vertical distribution of the annular areas.
Key Words and Phrases: *Annual ring, Linear growth process, Photosynthetic substance, Random walk, Solute carbohydrate, Translocation.*

1. Introduction

In forestry, an annual ring width is generally defined as the mean of thickening increments along the four cardinal directions in the annular area. Therefore, at first, we get the value of the annular area. As the distance from the stem axis at the starting time of annual growth is already known, if we get the value of the annular area, the annual ring width is evaluated immediately.

Figure 1 shows the distribution of the specimen in this report and the curves fitted to this specimen. In this specimen, the curve of $j = 0$ to $j = 5$ was estimated as the exponential distribution and the curve of $j = 28$ to $j = 55$ was estimated as the gamma distribution. Therefore, on the premise that this specimen can be expressed by the synthesis of the gamma and the exponential distributions, I began this study.

The vertical distribution of annular areas is formed on the assumption that for a unit quantity of the fluid substance composed of the photosynthetic substance and the solute carbohydrate at each point of the phloem and at each point of time, the substances are fixed as annular areas at a constant rate which takes its particular value near the butt.

In this report, the specimens, the descriptions, and the assumptions for the formation of annular areas are generally the same as in the preceding report [1]. Therefore, the concrete values of coefficients ρ, κ and μ in the transition probabilities concerning the gamma distribution, the concrete value of the coefficient r_2 in the exponential distribution and r_0 as the fixation ratio of the fluid substance are the same as in the preceding report.

These are

$$\rho = 0.90000 \times 10^{-2} \quad (1)$$
$$\kappa = 0.53184 \times 10^{-3} \quad (2)$$
$$\mu = 0.26931 \times 10^{-1} \quad (3)$$
$$r_2 = 0.65585 \times 10^{-1} \quad (4)$$
$$r_0 = 0.50000 \times 10^{-5} \quad (5)$$

Any differences from the preceding report are described in this current report when they take place.

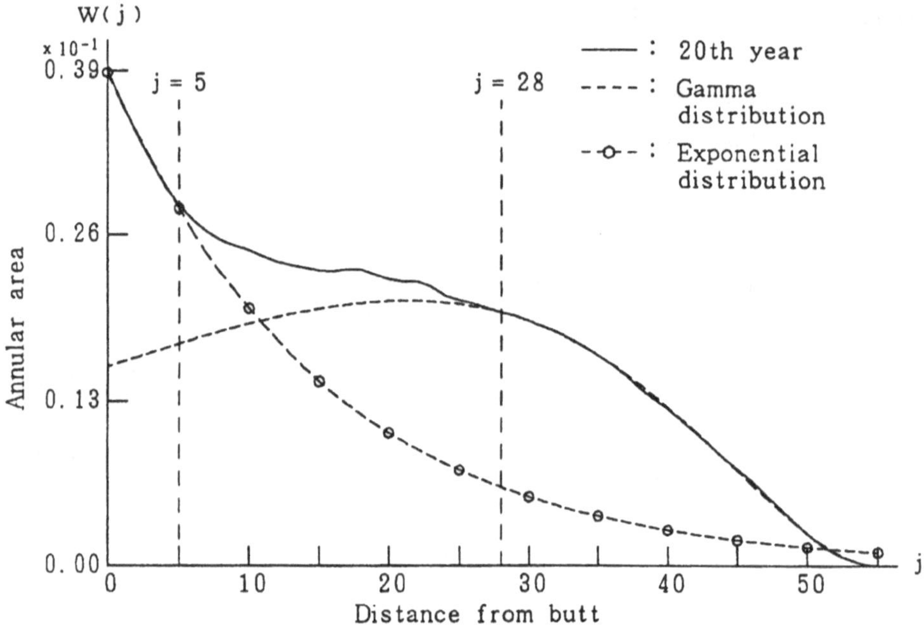

Figure 1. Distribution of specimen, the gamma and the exponential distributions.

2. Random Walk

Nagae et al. investigate the seasonal height growth for 3 years using 5-year-old trees of sugi (*Cpiptomeria japonica D. Don*) [2]. From their resultant figures, it is understood that the curve of the height growth for one year i.e. $\Delta h(t)$ is expressed approximately by the modified exponential curve i.e. Mitscherlich's curve. It is

$$\Delta h(t) = h_1\{1 - \exp(-h_2 t)\} \quad (6)$$

where h_1 and h_2 are positive constants and t is time and $t \geq 0$. In this report, it is assumed that this expression can be adapted also to mature trees.

The height growth for one year is assumed to begin at $j = 50$ and finish at $j = 55$. Therefore, setting

$$h_0 = 50 \quad (7)$$
$$h_1 = 5 \quad (8)$$

expression (6) is modified to (9) which expresses the tree height in this report.

$$h(t) = h_0 + h_1\{1 - \exp(-h_2 t)\}. \tag{9}$$

The coefficient h_2 is evaluated simply from

$$4.9 = 5\{1 - \exp(-h_2 i_p)\} \tag{10}$$

where i_p denotes the period of the inflow of the photosynthetic substance from the crown to the phloem of the stem and means the finishing time of the formation of annular areas as in the preceding report. If the left member in expression (10) is 5, then i_p has only one value i.e. ∞. Therefore, 4.9 is adopted as the most simple value which can be rounded off to 5 such that i_p has various finite values to allow the simulation.

The linear growth process is defined by

$$\frac{\partial u}{\partial t} = \frac{1}{2}\frac{\partial^2}{\partial x^2}(2\rho x u) - \frac{\partial}{\partial x}\{(\mu - \kappa x)u\},\ 0 < x < \infty \tag{11}$$

where ρ and κ are positive constants. When $\mu > \rho$, the steady distribution is the gamma distribution. Its starting point is $x = 0$, its peak is at some point on the x axis and the value trails off as x goes toward infinity. Therefore, to incorporate expression (9) to equation (11), let $y(x,t)$ be

$$y(x,t) = -\{x - h(t)\} \tag{12}$$

then the equation (11) is transformed to an equation with respect to y. Rewriting the symbol y as x, we get the following equation. It is

$$\begin{aligned}\frac{\partial u}{\partial t} &= \frac{1}{2}\frac{\partial^2}{\partial x^2}\Big[2\rho\{h(t) - x\}u\Big] \\ &+\frac{\partial}{\partial x}\left[\left\{\mu - \kappa\{h(t) - x\} - h_1 h_2 \exp(-h_2 t)\right\}u\right].\end{aligned} \tag{13}$$

The random walk with a step-size of 1 which is equivalent to equation (13) is

$$\begin{aligned}u(j, i+1) =& p(j-1, i)u(j-1, i) \\ &+ r(j,i)u(j,i) + q(j+1,i)u(j+1,i)\end{aligned} \tag{14}$$

where $p(j,i)$, $q(j,i)$ and $r(j,i)$ are the probabilities for one unit of the fluid substance to move by one unit of distance above or below, or to stay at the same point at intervals of a unit of time respectively. These are

$$p(j,i) = (\rho + \frac{1}{2}\kappa)\{h(i) - j\} - \frac{1}{2}\{\mu - h_1 h_2 \exp(-h_2 i)\} \tag{15}$$

$$r(j,i) = 1 - 2\rho\{h(i) - j\} \tag{16}$$

$$q(j,i) = (\rho - \frac{1}{2}\kappa)\{h(i) - j\} + \frac{1}{2}\{\mu - h_1 h_2 \exp(-h_2 i)\}. \tag{17}$$

As the step-size of the random walk is 1, $h(t)$ is allowed to take only integers from 50 to 55. Corresponding to this procedure, let all the values of $h(i)$ of expression (9) and in expressions (15) to (17) be rounded off to the nearest integers from 50 to 55.

The boundary condition at the apex of the stem is as follows: As the annular area at the apex is taken to be 0, let the substance at a point one unit of distance below the apex stay at the same point or move by one unit of distance below the point, without letting it move to the apex.

3. Crown Growth and Source

$f_c(j)$ denotes the crown form at the beginning of height growth for one year, letting the leaves be distributed from $j = 30$ to $j = 50$ on the stem. It is

$$f_c(j) = 0.60150 \left(\frac{j-30}{20}\right)\left(\frac{50-j}{20}\right)^2 \tag{18}$$

where the constant 0.60150 is the value that allows the sum of $f_c(j)$, j equals 30 to 50, to be 1. As decribed in the preceding report, this expression is the discrete form to which the beta distribution is changed and has its maximum value at 1/3 of the height from the base of the crown to the uppermost point i.e. at $j = 110/3$. In this report, this crown form is divided into two parts at $j = 110/3$. The lower part of the crown is fixed. And let the upper part of the crown be displaced in parallel to the j axis toward the apex of the stem in accordance with the height growth denoted by expression (9) and let both the maximum values of the lower and the upper parts of the crown be linked by a straight line. This crown form is generally observed in planted forests. In this manner, the crown is allowed to grow. As described above, expression (9) here is rounded off to the nearest integer. This crown form is denoted by $f_c(j,i)$.

The temporal change of the photosynthetic substance which flows into the phloem of the stem from each point within the above mentioned crown form is assumed to be expressed by (19)

$$f_r(i) = a_0 \left(\frac{i}{i_p}\right)^{\frac{2}{3}}\left(\frac{i_p - i}{i_p}\right)^2 \tag{19}$$

where the constant a_0 of $f_r(i)$ is the value for the adjustment that allows the sum of $f_r(i)$, i equals 0 to i_p, to be 1. Consequently, the source of the random walk $f(j,i)$ is expressed by (20)

$$f(j,i) = f_c(j,i) f_r(i). \tag{20}$$

4. Inflow of Solute Carbohydrate from Roots

The following phenomenon has been observed by some researchers [3, 4]: When a living stump bearing stool shoots and an adjacent tree are connected by root grafts, the living stump receives carbohydrates and other growth requirements from the adjacent tree through these root grafts. The stump continues to grow and forms xylem rings, which allows the stool shoots to grow vigorously.

If we understand the movement of carbohydrate as in the above description, there is a possibility that the solute carbohydrate originated in roots is translocated from the roots to the stem and contributes to the diameter growth of the stem. It is assumed that this process can be adapted also to mature trees. In this report, the term "solute carbohydrate" means only the reserve substance in the roots. And part of the vertical distribution of annular areas is formed based on the following two assumptions: 1) The *solute carbohydrate* flows into the phloem of the stem only from the roots. 2) The *solute carbohydrate* functions just like the *photosynthetic substance* in the formation of annular areas.

Because the productivity of photosynthesis is very low in early spring, the reserve carbohydrate is greatly consumed to form organs instead of the photosynthetic substance in this period [5]. To express this phenomenon simply, the solute carbohydrate is assumed to flow into the butt from roots at an exponentially decreasing rate. As the point of roots and the butt are $j = -1$ and $j = 0$ respectively, we get the following expression,

$$-\{u(0,i) - u(-1,i)\} = c_0 e^{-c_1 i} \tag{21}$$

where c_0 and c_1 are positive constants and c_1 is determined by the following expression,

$$e^{-c_1 i_p} = c_2. \tag{22}$$

The concrete values of c_0 and c_2 are determined in a later section i.e. in V, Simulation.

Substituting $u(-1,i)$ of expression (21) in the first term of the right members of expression (14), we get the random walk at $j = 0$. It is

$$\begin{aligned} u(j,i+1) &= p(j-1,i)\{u(j,i) + c_0 e^{-c_1 i}\} \\ &\quad + r(j,i)u(j,i) + q(j+1,\ i)u(j+1,i). \end{aligned} \tag{23}$$

5. Butt Swelling

The butt swelling of the mature tree is generally clear. The butt swelling is recognized even in younger trees. In this model, the butt swelling is formed mainly by both the inflow and the fixation of the solute carbohydrate from the roots.

In the preceding report, the fixation ratio was a constant value i.e. r_0 from the apex of the stem to the butt. In this report, to express butt swelling, the fixation ratio is allowed to take its particular value near the butt. Taking Figure 1 into account, the value of the fixation ratio is allowed to change from $j = 0$ to $j = 5$ in accordance with the exponential distribution. I adopted expression (24) for this part

$$f_{ix}(j) = r_1 e^{-r_2 j} \tag{24}$$

the constant r_1 is determined by

$$r_1 e^{-r_2 5} = r_0 \tag{25}$$

consequently, the value of r_1 is

$$r_1 = 0.69404 \times 10^{-5}. \tag{26}$$

For $j \geq 5$, the constant value r_0 is set in the same way as in the preceding report. Namely, it is

$$f_{ix}(j) = r_0. \tag{27}$$

6. Simulation

$S(j,i)$ denotes the annular area at point j and at time i. It is

$$\boldsymbol{S}(j,i) = \sum_{k=1}^{i} u(j,k) f_{ix}(j). \tag{28}$$

The initial value of i_p as the period of the inflow of the photosynthetic substance was set to be 1,500. The step-size which was added to the initial value for each trial in modifying i_p was set to be 500. The initial value of coefficients c_0 and c_2 in the exponential function for the inflow of the solute carbohydrate was set to be 0.1×10^{-3}. The step-size by which the initial value was multiplied for each trial in modifying c_0 and c_2 was set to be 0.1. According to this procedure, I simulated the vertical distribution of the annular areas by giving concrete values to i_p, c_0 and c_2. After the simulated curve approached the

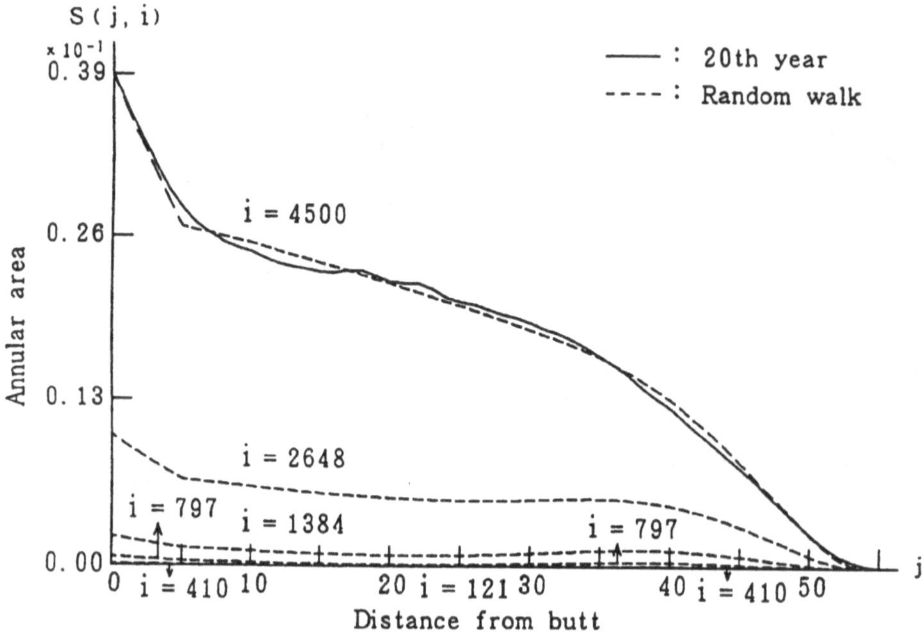

Figure 2. Process forming a series of annular areas.

curve of the specimen, it was precisely adjusted further by modifying the step-sizes of i_p, c_0 and c_2 little by little.

By trial and error, I obtained the values of $i_p = 4,500$, $c_0 = 0.25 \times 10^{-5}$ and $c_2 = 0.1 \times 10^{-7}$ which were suitable to depict the vertical distribution of the annular areas of this specimen. The simulated curves are shown by the broken lines in Figure 2. Each time shown in Figure 2 is the last time when the tree height stays between 50 and 55. The goodness of fit seems to show that this simulation succeeds in reflecting the phenomenon of actual annular formation.

This figure shows that this model depicts how the vertical distribution of the annular areas is formed as time passes.

Namely, in the first stage when $i = 121$ and $i = 410$, the two distribution curves are almost superimposed on the j axis. At the time when $i = 797$, all the annular areas from the apex to the butt grow a little. And the distribution curve is flat. But the annular areas have a trend such that they increase gradually to a maximum, then decrease gradually and increase gradually again. At the time when $i = 1,384$, all the annular areas grow bigger and this trend becomes clearer. In the middle stage when $i = 2,648$, all the annular areas continue to enlarge. And near the apex they increase, forming an upwards concave curve. From the end of this concave curve to the lower point of the crown, they have an upwards convex curve whose peak is within the crown and is at some point below the middle of the crown, although this convex curve and its peak are not so clear. From the lower point of the crown to the point where $j = 5$, they increase gradually and have a concave curve. From this point to the butt where $j = 0$, they increase steeply. In the last stage when $i = i_p$ (i.e. 4,500), all the annular areas enlarge greatly. And from the apex to the point where $j = 5$, they increase, forming a concave curve near the apex and next a convex curve, tracing and fitting this specimen. From this point to the butt where $j = 0$, they increase steeply, tracing this specimen.

The distributions of the annular areas of the random walk at each time shown in Figure 2 are converted into the distributions of the annual ring widths at each corresponding time, which are shown in Figure 3. $r_w(j,i)$ shows the annual ring width at point j and at time i.

This figure shows that this model depicts how the vertical distribution of the annual ring widths is formed as time passes. In this report, I only explain the last stage when $i = i_p$ i.e. 4,500. The annual rings increase steeply to some point near the apex and, forming an arc, decrease to the lower point of the crown, then form a concave curve with a gentle slope which is at first smooth and next fluctuate to the point where $j = 5$, and increase steeply to the butt where $j = 0$. This simulated curve fits this specimen well.

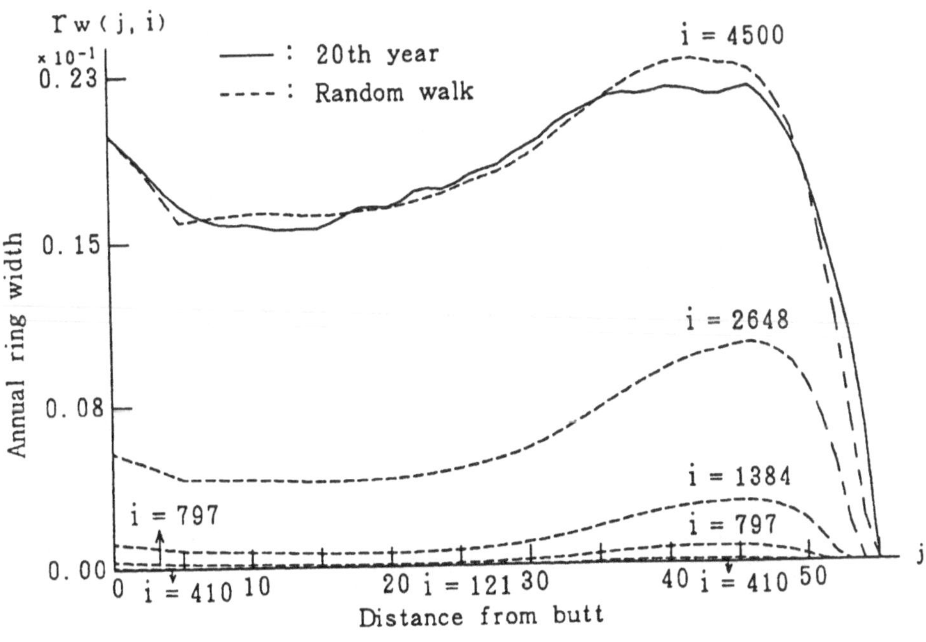

Figure 3. Process forming a series of annual ring widths.

7. Discussion

In the preceding model [1], annular areas around the crown are formed in the first stage and the annular area at the butt is formed in the last stage.

In the present model, because the inflow of the solute carbohydrate from the roots and the inflow of the photosynthetic substance from the crown occur simultaneously, all the annular areas from the apex to the butt begin to grow at almost the same time. Taking it into account that the butt swelling at younger tree is well simulated, this process of the present model reflects the actual growth better than that of the preceding model.

References

1. T. Yoshida, in: Proc. 8-th Int. Coll. Diff. Eq., D. Bainov (Ed.), (in print).

2. S. Nagae, S. Watanabe, Y. Arakida and K. Moroto, Chubu Branch Jpn. For. Soc., **20**, 169–175 (1971).

3. T. Satoo, Misc. Inform. Tokyo Univ. For., **15**, 54–60, (1964).

4. M. L. Wold and R. M. Lanner, Ecology, **46**, 755–756 (1965).

5. K. Negishi, in: Tree-Morphology and functions, T. Satoo and T. Tsutsumi (Eds.), Eibundo, Tokyo (1980) pp. 113–145.

Kolibri 360 GmbH
Ferdinand-Jühlke-Straße 7
99095 Erfurt, Deutschland
produktsicherheit@kolibri360.de